Operator Theory: Advances and Applications

Vol. 185

Commutative Algebras of Toeplitz Operators on the Bergman Space

Nikolai L. Vasilevski

Birkhäuser
Basel · Boston · Berlin

Author:

Nikolai L. Vasilevski
Departamento de Matemáticas
CINVESTAV del I.P.N.
Apartado Postal 14-740
07360 Mexico, D.F.
Mexico
e-mail: nvasilev@math.cinvestav.mx

2000 Mathematical Subject Classification: 30C40, 46E22, 47A25, 47B10, 47B35, 47C15, 47L15, 81S10

Library of Congress Control Number: 2008930643

Bibliographic information published by Die Deutsche Bibliothek.
Die Deutsche Bibliothek lists this publication in the Deutsche Nationalbibliografie;
detailed bibliographic data is available in the Internet at http://dnb.ddb.de

ISBN 978-3-7643-8725-9 Birkhäuser Verlag AG, Basel - Boston - Berlin

© 2008 Birkhäuser Verlag AG
Basel · Boston · Berlin
P.O. Box 133, CH-4010 Basel, Switzerland
Part of Springer Science+Business Media
Printed on acid-free paper produced from chlorine-free pulp. TCF∞
Printed in Germany

ISBN 978-3-7643-8725-9   e-ISBN 978-3-7643-8726-6
9 8 7 6 5 4 3 2 1   www.birkhauser.ch

Contents

Appendices

Preface

The book is devoted to the spectral theory of commutative C^*-algebras of Toeplitz operators on Bergman spaces, and its applications. For each such commutative algebra we construct a unitary operator which reduces each Toeplitz operator from this algebra to a certain multiplication operator, thus also providing its spectral type representation. This gives us a powerful research tool allowing direct access to the majority of the important properties of the Toeplitz operators studied herein. The presence and exploitation of these spectral type representations forms the basis for an essential part of the results presented in this book.

We give a criterion of when the algebras are commutative on each commonly considered weighted Bergman space. For Toeplitz operators generating such commutative algebras we describe their boundedness, compactness, and spectral properties. Furthermore, the above commutative algebras serve as model or local cases for a number of problems treated in the book, thus making their solutions possible.

We note that in the Bergman space case considered in the book the underlying manifold (the unit disk equipped with the hyperbolic metric) possesses a richer geometric structure in comparison with the Hardy space case (the unit circle). This fact has an important reflection in the presented theory.

We mention as well that from the general operator point of view the Toeplitz operators, both on Hardy space and on Bergman space, are compressions of multiplication operators onto certain subspaces, and thus they represent two interesting different models of operators having similar structure. At the same time the presented results show clearly essential differences between the theories for these two species of operators.

The book is addressed to a wide audience of mathematicians, from graduate students to researchers, whose primary interests lie in complex analysis and operator theory. The prerequisites for reading this book include a basic knowledge in one-dimensional complex analysis, functional analysis, and operator theory. An acquaintance with some facts of the theory of Banach and C^*-algebras will be useful as well. Among various excellent sources which may serve for the preliminary reading we mention, for example, the books by I. Gohberg, S. Goldberg, and M. Kaashoek [81], and R. Douglas [58].

The author is greatly indebted to his colleagues Serguei Grudsky, his coauthor in many papers, and Michael Porter who read the manuscript and made many important suggestions essentially improving the book. The author would like to address special words of gratitude to Olga Grudskaia who tragically died in a car accident in February 2004. She generously assisted in the preparation of this book, beautifully elaborating all the figures of the last three chapters.

The author is sincerely grateful to Israel Gohberg, the editor of the "Operator Theory: Advances and Applications" book series, for his invitation to publish the book in this series and whose friendly remarks and advice greatly assisted in the final stage of its preparation.

The author thankfully acknowledges CONACYT (Consejo Nacional de Ciencia y Tecnologia), Mexico, which has supported him in various projects.

Introduction

The notion of a Toeplitz operator goes back to the work of Otto Toeplitz at the beginning of the 20th century. The Toeplitz operator is normally defined in terms of the so-called Toeplitz matrix

$$
A = \begin{pmatrix}
a_0 & a_{-1} & a_{-2} & \cdots \\
a_1 & a_0 & a_{-1} & \cdots \\
a_2 & a_1 & a_0 & \cdots \\
\cdots & \cdots & \cdots & \cdots
\end{pmatrix},
$$

where $a_n \in \mathbb{C}$, $n \in \mathbb{Z}$. The result by O. Toeplitz of 1911 says that the matrix A defines a bounded operator on $l_2 = l_2(\mathbb{Z}_+)$, where $\mathbb{Z}_+ = \{0\} \cup \mathbb{N}$, if and only if the numbers $\{a_n\}$ are the Fourier coefficients of a function $a \in L_\infty(S^1)$, where S^1 is the unit circle. The (discrete) inverse Fourier transform is a unitary operator which maps $l_2(\mathbb{Z})$ onto $L_2(S^1)$ and $l_2(\mathbb{Z}_+)$ onto the so-called Hardy space $H^2_+(S^1)$. The last one consists of all $L_2(S^1)$-functions admitting an analytic continuation on the unit disk and whose discrete Fourier transform vanishes for all negative indices. That is, the operator defined by the matrix A is unitary equivalent to the operator T_a which acts on the Hardy space $H^2_+(S^1)$ by the rule

$$
T_a : f(t) \in H^2_+(S^1) \longmapsto (P_+ a f)(t) \in H^2_+(S^1),
$$

where P_+ is the so-called Szegö orthogonal projection of $L_2(S^1)$ onto $H^2_+(S^1)$, and the Fourier coefficients of the function a are given by the sequence $\{a_n\}$. The operator T_a is also called the Toeplitz operator, and the function $a \in L_\infty(S^1)$ is called the defining symbol[1] of T_a. As the Szegö projection P_+ is given in terms of the one-dimensional singular integral operator with the Cauchy kernel, the theories of the Toeplitz and singular integral operators are very related, especially in questions connected with the algebras generated by such operators. The first fundamental results on this direction for continuous symbols go back to the paper by I. Gohberg of 1952, and to papers by L. Coburn, R. Douglas, I. Gohberg and N. Krupnik of late 1960s – early 1970s for more general classes of defining symbols.

Another concept of a subspace of analytic functions of the L_2-space and of the corresponding orthogonal projection was introduced by S. Bergman in 1950. For a bounded simply connected domain D in $\mathbb{C}$, the Bergman space $\mathcal{A}^2(D)$ consists of all $L_2(D)$-functions which are analytic in D. We denote by B_D the orthogonal (Bergman) projection of $L_2(D)$ onto the Bergman space $\mathcal{A}^2(D)$. Then the Toeplitz operator T_a with defining symbol $a \in L_\infty(D)$ is given on the Bergman space in quite the same way,

$$
T_a : \varphi(z) \in \mathcal{A}^2(D) \longmapsto (B_D a \varphi)(z) \in \mathcal{A}^2(D).
$$

We note that in the Bergman space case the underlying manifold possesses a richer geometric structure in comparison with the Hardy space case. The unit disk

---

[1]The term "symbol" and its usage in the book is discussed in Section 2.1

equipped with the hyperbolic metric is both symplectic and Kähler. Therefore the Toeplitz operators on the Bergman space on the unit disk, and more generally on the bounded symmetric domains in $\mathbb{C}^n$, appear naturally, for example, under the Berezin quantization procedure.

At the same time, contrary to the Hardy space case, such Toeplitz operators do not admit the above matrix form with respect to the standard monomial basis even for the case of the unit disk.

We note as well that from the general point of view the Toeplitz operator is nothing but the compression of a bounded operator (in our case a multiplication operator) onto a certain subspace of a Hilbert space, representing thus an important model case in operator theory.

Although the basic properties and some results in the theories of Toeplitz operators acting on the Hardy space and on the Bergman space look similar, these two species of operators turn out to be quite different, and many deep results in both theories reflect and make clear their differences.

A quite unexpected and recently discovered phenomenon in the theory of Toeplitz operators on the Bergman space is the existence of a rich family of commutative C^*-algebras generated by Toeplitz operators with non-trivial defining symbols. The main goal of the book is a systematic exploration of this phenomenon.

A reader familiar with the theory of Toeplitz operators on the Hardy space may ask a legitimate question: *How can it be possible? What is this book about?* Indeed, the only possible commutative C^*-algebra of Toeplitz operators on the Hardy space over the unit circle is the algebra generated by Toeplitz operators with *constant* defining symbols. In this case the Toeplitz operators are just scalar multiples of the identity. Another possible option is the C^*-algebra with identity (Toeplitz operator with defining symbol 1) generated by a single self-adjoint element (a Toeplitz operator with real-valued defining symbol). But this is just a general operator theory statement which does not use any specific feature of Toeplitz operators. Furthermore, it is unknown whether such an algebra contains any other Toeplitz operator apart from its initial generator, nor how to single out such operators from the algebra in case of their presence.

In 1995 B. Korenblum and K. Zhu proved that the Toeplitz operators with radial defining symbols acting on the Bergman space over the unit disk can be diagonalized with respect to the standard monomial basis in this Bergman space. The C^*-algebra generated by such Toeplitz operators is therefore obviously commutative. Four years later the author of this book showed that the C^*-algebra generated by Toeplitz operators acting on the Bergman space over the upper half-plane and with defining symbols depending only on $\operatorname{Im} z$ is commutative as well. Further it was understood that there exists a rich family of commutative C^*-algebras of Toeplitz operators. Moreover it turned out that the smoothness properties of the symbols do not play any role in the commutativity: the symbols can be merely measurable. Surprisingly everything is governed by the geometry of the underlying manifold, the unit disk equipped with the hyperbolic metric. The precise description is as follows. *Each pencil of hyperbolic geodesics determines* *the set of symbols which are constant on the corresponding cycles, the orthogonal trajectories to geodesics forming a pencil. The C^*-algebra generated by Toeplitz operators with such defining symbols is commutative.*

An important feature of such algebras is that they remain commutative for Toeplitz operators acting on each of the commonly considered weighted Bergman spaces. Moreover, assuming some natural conditions on "richness" of symbol classes, we have the following complete characterization. *A C^*-algebra generated by Toeplitz operators is commutative on each weighted Bergman space if and only if the corresponding defining symbols are constant on cycles of some pencil of hyperbolic geodesics.* Apart from its inherent beauty this result reveals an extremely deep influence of the geometry of the underlying manifold on the properties of Toeplitz operators over this manifold.

In each case of the above commutative algebra we construct a unitary operator which reduces the corresponding Toeplitz operators to certain multiplication operators, thus also providing their spectral type representations. This gives us a powerful tool in the subject, in particular, yielding direct access to the majority of the important properties (such as boundedness, compactness, spectral properties, invariant subspaces) of the Toeplitz operators under study.

Furthermore, the presence and exploitation of these spectral type representations forms the core for many results presented in the book. Some of these results are a systematic study of Toeplitz operators with unbounded defining symbols; a clarification of the difference between compactness of commutators and semi-commutators of Toeplitz operators; the theory of Toeplitz and related operators with defining symbols having more than two limit values at boundary points; and a kind of semi-classical analysis of spectral properties of Toeplitz operators.

A detailed description of the main results of the book is given in the next section "Highlights of the chapters".

Highlights of the chapters

Chapter 1. Preliminaries

The chapter contains some algebraic material which will be used substantially in the book. Two themes are treated in the chapter: the local principle for C^*-algebras and the structure of the C^*-algebras generated by orthogonal projections.

Let $\mathcal{A}$ be a C^*-algebra with identity e, and let $\mathcal{Z}$ be its central commutative C^*-subalgebra, containing e. Denote by T the compact space of maximal ideals of the algebra $\mathcal{Z}$. For each point $t \in T$ denote by J_t the maximal ideal of the algebra $\mathcal{Z}$, which corresponds to the point t, and denote by $J(t) = \mathcal{A} \cdot J_t$ the closed two-sided ideal in the algebra $\mathcal{A}$, generated by J_t. Finally, introduce the system of ideals $J_T = \{J(t) : t \in T\}$.

The local algebra $\mathcal{A}(t)$, which corresponds to a point $t \in T$, is defined as $\mathcal{A}(t) = \mathcal{A}/J(t)$. We construct a canonical C^*-bundle defined by the C^*-algebra $\mathcal{A}$ and the system of its closed two-sided ideals J_T. Its base coincides with T and the fiber over each $t \in T$ coincides with the corresponding algebra $\mathcal{A}(t)$.

A very practical tool to describe various operator C^*-algebras considered in this book is the so-called Douglas-Varela local principle. It reads as follows. Let $\mathcal{A}$ be a C^*-algebra with identity, let $\mathcal{Z}$ be a central commutative C^*-subalgebra with the same identity, and let T be the compact set of maximal ideals of the algebra $\mathcal{Z}$. Then the algebra $\mathcal{A}$ is *-isomorphic and isometric to the algebra of all (global) continuous sections of the C^*-bundle, defined by the algebra $\mathcal{A}$ and the system of ideals $J_T = \{J(t) : t \in T\}$.

The local descriptions of many operator C^*-algebras considered in the book lead to algebras generated by a finite number of orthogonal projections. It is well known that the C^*-algebras generated by a finite number of orthogonal projections are wild in general. The lucky exception here is the C^*-algebra generated by two (arbitrary) orthogonal projections. This algebra is tame and has the following (after an inessential simplification) description. Let H be an infinite dimensional Hilbert space and let P and Q be two orthogonal projections on H. Denote by Δ the spectrum of $(P - Q)^2$. It is known that $\Delta \in [0, 1]$. Then the C^*-algebra with identity generated by P and Q is isomorphic and isometric to the algebra of all 2×2 matrix-functions continuous on Δ and diagonal at the points of $\Delta \cap \{0, 1\}$. This isomorphism is generated by the mapping

$$P \longmapsto \begin{pmatrix} 1 & 0 \\ 0 & 0 \end{pmatrix},$$

$$Q \longmapsto \begin{pmatrix} 1-t & \sqrt{t(1-t)} \\ \sqrt{t(1-t)} & t \end{pmatrix}, \qquad t \in \Delta = \operatorname{sp}(P - Q)^2.$$

Two and more orthogonal projections naturally appear in the study of various algebras generated by the Bergman projection and piece-wise continuous functions having two or more different limit values at a point. Although such projections, say $P, Q_1, \ldots, Q_n$, obey an extra condition $Q_1 + \ldots + Q_n = I$, they still generate a wild C^*-algebra in general. In all cases considered in the book and directly connected with commutative algebras of Toeplitz operators, these projections have another specific property: the operators PQ_1P, ..., PQ_nP mutually commute. This property makes the algebra tame, and it has a nice and simple description as the algebra of all $n \times n$ matrix-functions continuous on the joint spectrum Δ of the operators PQ_1P, ..., PQ_nP and having certain degeneration on the boundary of Δ.

Chapter 2. Prologue

This is an introductory chapter in which we introduce the main objects treated in the book, as well as present some preliminary material. Let D be a simply connected domain in the complex plane $\mathbb{C}$, being the unit disk $\mathbb{D}$ or the upper half-plane Π in the majority of cases. We introduce the space $L_2(D)$ with respect to the standard Lebesgue plane measure $dv(z) = dxdy$, where $z = x + iy$, together with its subspace, the Bergman space $\mathcal{A}^2(D)$, which consists of all functions analytic in D. The orthogonal Bergman projection B_D of $L_2(D)$ onto the Bergman space $\mathcal{A}^2(D)$ has the form

$$(B_D f)(z) = \int_D K_D(z, \zeta) f(\zeta) dv(\zeta),$$

where the function $K_D(z, \zeta)$ is called the Bergman kernel function of the domain D. We mention the following formulas for $K_D(z, \zeta)$,

$$K_D(z, \zeta) = \sum_{k=1}^{\infty} \varphi(z)\overline{\varphi(\zeta)} = -\frac{2}{\pi} \frac{\partial^2 g(z, \zeta)}{\partial z \partial \overline{\zeta}},$$

where $\{\varphi(z)\}_{k \in \mathbb{N}}$ is an orthonormal basis in $\mathcal{A}^2(D)$ and $g(z, \zeta)$ is the Green function of the domain D, and the following integral representation of the Bergman projection

$$B_D = I - S_D S_D^* + L, \tag{0.1}$$

where the two-dimensional singular integral operators S_D and S_D^* have the form

$$(S_D\varphi)(z) = -\frac{1}{\pi} \int_D \frac{\varphi(\zeta)}{(\zeta - z)^2} dv(\zeta), \qquad (S_D^*\varphi)(z) = -\frac{1}{\pi} \int_D \frac{\varphi(\zeta)}{(\overline{\zeta} - \overline{z})^2} dv(\zeta), \tag{0.2}$$

and L is a compact operator.

The main object of the book, the Toeplitz operator T_a with defining symbol $a \in L_\infty(D)$, acts on the Bergman space $\mathcal{A}^2(D)$ via

$$T_a : \varphi \in \mathcal{A}^2(D) \longmapsto B_D(a\varphi) \in \mathcal{A}^2(D).$$

We describe a number of operator algebras and their Fredholm symbol algebras. Among them there are the C^*-algebra generated by the Bergman projection B_D and the multiplication operators aI acting on $L_2(D)$, and the C^*-algebra generated by Toeplitz operators T_a acting on the Bergman space $\mathcal{A}^2(D)$. We study them for continuous and piece-wise continuous functions a.

Chapter 3. Bergman and Poly-Bergman Spaces

In this chapter we elaborate the Bergman space version of the following classical result. The Fourier transform F gives an isometric isomorphism of $L_2(\mathbb{R})$ under which

$$F: \; H^2_\pm \longrightarrow L^2(\mathbb{R}_\pm),$$

where $H^2_\pm$ are the Hardy spaces on the upper and lower half-planes respectively, $\mathbb{R}_\pm$ are the positive and negative half-axis, and furthermore

$$L_2(\mathbb{R}) = H^2_+ \oplus H^2_- .$$

We construct the unitary operator U acting on $L_2(\Pi) = L_2(\mathbb{R}) \otimes L_2(\mathbb{R}_+)$, such that

$$U: \; \mathcal{A}^2(\Pi) \longrightarrow L_2(\mathbb{R}_+) \otimes L_0,$$

where L_0 is the one-dimensional subspace of $L_2(\mathbb{R}_+)$ generated by $\ell_0(y) = e^{-y/2}$.

This shows us how much room the Bergman space occupies inside $L_2(\mathbb{R})$, and how big its orthogonal complement is. The following classes of functions turn out to be useful under the description of this orthogonal complement. We define the spaces $\mathcal{A}^2_n(\Pi)$ and $\widetilde{\mathcal{A}}^2_n(\Pi)$, $n \in \mathbb{N}$, of n-analytic and n-anti-analytic functions as the subspaces of $L_2(\Pi)$ which consist of all functions which satisfy the equations

$$\left(\frac{\partial}{\partial \bar{z}}\right)^n \varphi = \frac{1}{2^n}\left(\frac{\partial}{\partial x} + i\frac{\partial}{\partial y}\right)^n \varphi = 0, \quad \left(\frac{\partial}{\partial z}\right)^n \varphi = \frac{1}{2^n}\left(\frac{\partial}{\partial x} - i\frac{\partial}{\partial y}\right)^n \varphi = 0.$$

We introduce as well the spaces $\mathcal{A}^2_{(n)}(\Pi)$ and $\widetilde{\mathcal{A}}^2_{(n)}(\Pi)$ of true-n-analytic and true-n-anti-analytic functions,

$$\mathcal{A}^2_{(n)}(\Pi) = \mathcal{A}^2_n(\Pi) \ominus \mathcal{A}^2_{n-1}(\Pi), \quad \widetilde{\mathcal{A}}^2_{(n)}(\Pi) = \widetilde{\mathcal{A}}^2_n(\Pi) \ominus \widetilde{\mathcal{A}}^2_{n-1}(\Pi).$$

Then we have

$$U: \; \mathcal{A}^2_{(n)}(\Pi) \longrightarrow L_2(\mathbb{R}_+) \otimes L_{n-1}, \quad U: \; \widetilde{\mathcal{A}}^2_{(n)}(\Pi) \longrightarrow L_2(\mathbb{R}_-) \otimes L_{n-1},$$

where L_{n-1} is the one-dimensional subspace of $L_2(\mathbb{R}_+)$ generated by the function $\ell_{n-1}(y)$. We note that the system of functions $\ell_n(y) = e^{-y/2}L_n(y)$, where $L_n(y)$ is the Laguerre polynomial of degree n, $n = 0, 1, 2, \ldots$, forms the standard orthonormal basis in $L_2(\mathbb{R}_+)$.

This immediately yields the following direct sum decomposition:

$$L_2(\Pi) = \bigoplus_{k=1}^{\infty} \mathcal{A}^2_{(k)}(\Pi) \oplus \bigoplus_{k=1}^{\infty} \widetilde{\mathcal{A}}^2_{(k)}(\Pi).$$

We show that the action of the operators (0.2) is extremely transparent with respect to this direct sum decomposition, and that each of these operators admits a simple functional model: it is unitary equivalent to a direct sum of two unilateral shifts, backward and forward, both taken with infinite multiplicity. This permits us to obtain simple integral representations of projections onto the poly-Bergman spaces generalizing the representation (0.1).

In the chapter we also construct an operator $R : L_2(\Pi) \longrightarrow L_2(\mathbb{R}_+)$ whose restriction onto the Bergman space $\mathcal{A}^2(\Pi)$ is an isometric isomorphism between $\mathcal{A}^2(\Pi)$ and $L_2(\mathbb{R}_+)$, and

$$
\begin{aligned}
R R^* &= I &:\quad & L_2(\mathbb{R}_+) \longrightarrow L_2(\mathbb{R}_+), \\
R^* R &= B_\Pi &:\quad & L_2(\Pi) \longrightarrow \mathcal{A}^2(\Pi).
\end{aligned}
$$

Chapter 4. Bergman Type Spaces on the Unit Disk

This chapter is devoted to a unit disk version of the results for the upper half-plane of Chapter 3. The major difference compared with the previous case is that now

$$
\mathcal{A}^2_{(n)} = \ker\left(\overline{z}\,\frac{\partial}{\partial \overline{z}}\right)^n \ominus \ker\left(\overline{z}\,\frac{\partial}{\partial \overline{z}}\right)^{n-1}, \quad
\widetilde{\mathcal{A}}^2_{(n)} = \ker\left(z\,\frac{\partial}{\partial z}\right)^n \ominus \ker\left(z\,\frac{\partial}{\partial z}\right)^{n-1}.
$$

We again construct an operator R acting in this case from $L_2(\Pi)$ onto l_2^+ whose restriction onto the Bergman space $\mathcal{A}^2(\mathbb{D})$ is an isometric isomorphism between the spaces $\mathcal{A}^2(\mathbb{D})$ and l_2^+, and

$$
\begin{aligned}
R R^* &= I &:\quad & l_2^+ \longrightarrow l_2^+, \\
R^* R &= B_\mathbb{D} &:\quad & L_2(\mathbb{D}) \longrightarrow \mathcal{A}^2(\mathbb{D});
\end{aligned}
$$

here l_2^+ is the subspace of two-sided l_2, consisting of sequences supported on $\mathbb{Z}_+ = \mathbb{N} \cup \{0\}$.

Chapter 5. Toeplitz Operators with Commutative Symbol Algebras

Using the operator R of Chapter 3 we prove that each Toeplitz operator T_a with bounded measurable defining symbol $a(y)$, depending only on $y = \operatorname{Im} z$ and acting on $\mathcal{A}^2(\Pi)$, is unitary equivalent to the multiplication operator $\gamma_a I$ acting on $L_2(\mathbb{R}_+)$. The function $\gamma_a = \gamma_a(x)$ is given by

$$
\gamma_a(x) = \int_{\mathbb{R}_+} a\left(\frac{\eta}{2x}\right) e^{-\eta}\, d\eta, \quad x \in \mathbb{R}_+.
$$

This implies, in particular, that the C^*-algebra generated by such operators is commutative.

Based on this fact, we study the difference between compactness of the commutator $[T_a, T_b] = T_a T_b - T_b T_a$ and semi-commutator $[T_a, T_b) = T_a T_b - T_{ab}$ of two Toeplitz operators. Given a C^*-subalgebra $\mathcal{A}(\mathbb{D})$ of $L_\infty(\mathbb{D})$, consider the following statements (which can be false or true depending on $\mathcal{A}(\mathbb{D})$):

1) For each $a \in \mathcal{A}(\mathbb{D})$ the commutator $[B_\mathbb{D}, aI] = B_\mathbb{D} aI - a B_\mathbb{D}$ is compact.

2) For each pair $a, b \in \mathcal{A}(\mathbb{D})$ the semi-commutator $[T_a, T_b)$ is compact.

3) For each pair $a, b \in \mathcal{A}(\mathbb{D})$ the commutator $[T_a, T_b]$ is compact.

The first two statements turn out to be equivalent, that is, they can be true only simultaneously, and each of them implies the third statement. At the same time, as examples show, the third statement does not imply the first two statements. It is interesting and very important to understand the gap between the third and the first two properties, which actually turns out to be quite substantial.

Another related concept of the same problem is as follows.

Given a C^*-subalgebra $\mathcal{A}(\mathbb{D})$ of $L_\infty(\mathbb{D})$, introduce two operator C^*-algebras: the algebra $\mathcal{T}(\mathcal{A}(\mathbb{D}))$ which is generated by all Toeplitz operators

$$T_a : \varphi \in \mathcal{A}^2(\mathbb{D}) \longmapsto B_{\mathbb{D}} a\varphi \in \mathcal{A}^2(\mathbb{D})$$

with symbols $a \in \mathcal{A}(\mathbb{D})$, and the algebra $\mathcal{R}(\mathcal{A}(\mathbb{D}), B_{\mathbb{D}})$, generated by all operators of the form

$$A = aI + bB_{\mathbb{D}},$$

where $a, b \in \mathcal{A}(\mathbb{D})$, acting on $L_2(\mathbb{D})$.

One of the main features of those algebras $\mathcal{A}(\mathbb{D})$ having the first two properties is that both the operator algebras $\mathcal{T}(\mathcal{A}(\mathbb{D}))$ and $\mathcal{R}(\mathcal{A}(\mathbb{D}), B_{\mathbb{D}})$ admit a commutative symbolic calculus, i.e., both Fredholm symbol algebras $\operatorname{Sym}\mathcal{T}(\mathcal{A}(\mathbb{D})) = \mathcal{T}(\mathcal{A}(\mathbb{D}))/\mathcal{K}$ and $\operatorname{Sym}\mathcal{R}(\mathcal{A}(\mathbb{D}), B_{\mathbb{D}}) = \mathcal{R}(\mathcal{A}(\mathbb{D}), B_{\mathbb{D}})/\mathcal{K}$ are commutative, where $\mathcal{K}$ is the ideal of compact operators. At the same time, under (only) condition 3) the Fredholm symbol algebra $\operatorname{Sym}\mathcal{R}(\mathcal{A}(\mathbb{D}), B_{\mathbb{D}})$ of the algebra $\mathcal{R}(\mathcal{A}(\mathbb{D}), B_{\mathbb{D}})$ is *non-commutative*, while the Fredholm symbol algebra $\operatorname{Sym}\mathcal{T}(\mathcal{A}(\mathbb{D}))$ still remains *commutative*.

Another way to understand the difference between properties 2) and 3) is by comparison of the representations of these algebras. In other words it is important to understand how complicated the Fredholm symbol algebra of the algebra $\mathcal{R}(\mathcal{A}(\mathbb{D}), B_{\mathbb{D}})$ can be while the Toeplitz operator algebra $\mathcal{T}(\mathcal{A}(\mathbb{D}))$ still admits a commutative symbolic calculus.

To answer this question we show that for each finite set of integers $\Lambda = \langle n_0, n_1, \ldots, n_m \rangle$, where $1 = n_0 < n_1 < \ldots < n_m \leq \infty$, and $n_k \in \mathbb{N} \cup \{\infty\}$, there are algebras $\mathcal{A}_\Lambda$ (with the only property 3)), such that the Fredholm symbol algebras $\operatorname{Sym}\mathcal{T}(\mathcal{A}_\Lambda)$ of the algebras $\mathcal{T}(\mathcal{A}_\Lambda)$ are *commutative*, while the Fredholm symbol algebras $\operatorname{Sym}\mathcal{R}(\mathcal{A}_\Lambda, B_{\mathbb{D}})$ of the algebras $\mathcal{R}(\mathcal{A}_\Lambda, B_{\mathbb{D}})$ have *irreducible representations exactly of predefined dimensions* $n_0, n_1, \ldots, n_m$.

Unitary equivalence of Toeplitz operators T_a, with defining symbols depending on $y = \operatorname{Im} z$, with multiplication operators $\gamma_a I$ gives us, in particular, direct access to their spectral properties. We prove among others the following rather unexpected result.

Given a measurable set $M \subset \mathbb{R}_+$, the spectrum of the Toeplitz operator T_{χ_M} is the segment $[\alpha, \beta]$, where $\alpha = \inf_{x \in \mathbb{R}_+} \gamma_{\chi_M}(x) \geq 0$, $\beta = \sup_{x \in \mathbb{R}_+} \gamma_{\chi_M}(x) \leq 1$.

Moreover, for every α_0, β_0 with $0 \leq \alpha_0 < \beta_0 \leq 1$, there is a set M_0 such that $\operatorname{sp} T_{\chi_{M_0}} = [\alpha_0, \beta_0]$.

This result is drastically different from its Hardy space analog, in which $\operatorname{sp} T_\chi = [0, 1]$ for each characteristic function χ.

Chapter 6. Toeplitz Operators on the Unit Disk with Radial Symbols

Using the operator R of Chapter 4 we prove that each Toeplitz operator T_a with measurable radial defining symbol $a(r)$ acting on $\mathcal{A}^2(\mathbb{D})$ is unitary equivalent to the multiplication operator $\gamma_a I$ acting on l_2^+. The sequence $\gamma_a = \{\gamma_a(n)\}_{n \in \mathbb{Z}_+}$ is given by

$$\gamma_a(n) = (n+1) \int_0^1 a(\sqrt{r})\, r^n\, dr, \qquad n \in \mathbb{Z}_+.$$

This implies, in particular, that the C^*-algebra generated by such operators is commutative, and provides us with a simple description of boundedness, compactness, and spectral properties of the corresponding Toeplitz operators.

We note that the Toeplitz operators with radial defining symbols $a(r)$, continuous at the boundary point 1, are quite trivial, nothing but the compact perturbations of scalar operators, $T_{a(r)} = a(1)I + K$. At the same time Toeplitz operators with (generally unbounded) radial defining symbols have very interesting and rich structure.

We show that the behaviour near the boundary of a certain average of symbols, rather than the behaviour of the symbols themselves, is responsable for the boundedness and compactness properties of the corresponding Toeplitz operators. Moreover in order to generate a bounded or compact Toeplitz operator, its unbounded defining symbol must necessarily have sufficiently sophisticated oscillating behaviour near the unit circle. Surprisingly, there exist compact Toeplitz operators whose defining symbols are unbounded near each point of the unit circle $\partial\mathbb{D}$. We show that the essential spectra of Toeplitz operators with radial defining symbols are always connected, and even may be massive (i.e., have positive plane measure). For bounded operators T_a and T_b, whose defining symbols are unbounded, the operator $T_{a \cdot b}$ may not be bounded at all. That is, contrary to commonly considered cases, the set of defining symbols for which corresponding Toeplitz operators are bounded neither forms an algebra (under the point-wise multiplication) nor admits any natural norm. The unique natural structure on the set of such symbols is just a linear space.

Chapter 7. Toeplitz Operators with Homogeneous Symbols

In this chapter we study the third model case of Toeplitz operators generating a commutative C^*-algebra. We show that the Toeplitz operator T_a, with a bounded measurable defining symbol $a = a(\theta)$ which depends only on the polar angle θ, acting on the Bergman space on the upper half-plane Π is unitary equivalent to the multiplication operator $\gamma_a I$ acting on $L_2(\mathbb{R})$. The function $\gamma_a = \gamma_a(\lambda)$ is given by

$$\gamma_a(\lambda) = \frac{2\lambda}{1 - e^{-2\pi\lambda}} \int_0^\pi a(\theta)\, e^{-2\lambda\theta}\, d\theta, \qquad \lambda \in \mathbb{R}. \tag{0.3}$$

That is, the C^*-algebra generated by Toeplitz operators with defining symbols homogeneous of zero order ($\equiv$ symbols depending only on θ) is commutative. We show as well that, for many different classes of homogeneous of order zero defining symbols, the C^*-algebras generated by the corresponding Toeplitz operators are the same, and even this common C^*-algebra with identity can be generated by just one Toeplitz operator. The last fact will play a key role later on in the description of Toeplitz operators which belong to the algebra generated by Toeplitz operators with piece-wise continuous defining symbols. As an illustration we present here the results of just one of the examples of Section 7.7.

For each $\alpha \in [0, \pi]$ we introduce the characteristic function $\chi_\alpha(\theta)$ of the angle $\pi\alpha$ on the upper half-plane measured from the positive semi-axes $\{(x, 0) \in \Pi : x \geq 0\}$. We note that $\chi_{1/2}(\theta) = \chi_+$ is nothing but the characteristic function of the right quarter-plane. Consider as well the family of continuous functions f_α of the form

$$f_\alpha(x) = x^{2(1-\alpha)}\, \frac{(1 - x)^{2\alpha} - x^{2\alpha}}{(1 - x) - x}, \qquad x \in [0, 1].$$

Then

$$f_\alpha(T_{\chi_+}) = T_{\chi_\alpha},$$

and thus each Toeplitz operator T_{χ_α} belongs to the C^*-algebra with identity generated by T_{χ_+}.

We note that homogeneous functions of order zero are very convenient for the local description of functions having a finite number of limit values at a boundary point of discontinuity. The easy access to the properties of Toeplitz operators with such defining symbols via formula (0.3) permit us to achieve an essential extension of the results of Chapter 2. We describe the Fredholm symbol algebra of the C^*-algebra generated by the Bergman projection and the multiplication operators by piece-wise continuous functions having a finite number of limit values at some fixed boundary points of discontinuity. This Fredholm symbol algebra turns out to have finite dimensional irreducible representations only, the dimensions of which depend on the number of limit values achieved by piece-wise continuous functions at the boundary points of discontinuity.

We describe as well the Fredholm symbol algebra of the C^*-algebra generated by Toeplitz operators whose defining symbols admit a finite number of limit values at some fixed boundary points of discontinuity. This algebra appears to be commutative, and thus we have here another example of algebras clarifying the difference between compactness properties of commutators and semi-commutators of Toeplitz operators, as considered in Chapter 5.

The following surprising property is worth mentioning. As it turns out, in spite of the very different sets of initial generators, the C^*-algebras of Toeplitz operators considered in Chapter 2 and in this chapter are the same. But there was no way to see this fact before. This leads in turn to the following interesting corollary. Considering Toeplitz operators with piece-wise continuous defining symbols, we have that both the curves supporting symbol discontinuities and the number of such curves meeting at a boundary point of discontinuity do not play actually any essential role for the Toeplitz operator algebra studied. We can start from very different sets of defining symbols and obtain exactly the same operator algebra as a result.

Chapter 8. Anatomy of the Algebra Generated by Toeplitz Operators with Piece-wise Continuous Symbols

By construction, the C^*-algebra generated by Toeplitz operators with piece-wise continuous defining symbols consists of

- its initial generators, the Toeplitz operators T_a with piece-wise continuous defining symbols,

- all elements of the form

$$\sum_{k=1}^{p} \prod_{j=1}^{q_k} T_{a_{j,k}}, \tag{0.4}$$

- uniform limits of sequences of the above elements.

Although the description of the Fredholm symbol algebra of this algebra was already given in Chapter 2, many deep questions on its structure were left unanswered. For example, it is important to understand an explicit form of an arbitrary element of this Toeplitz operator algebra, how to recover an operator by its image in the Fredholm symbol algebra, the diversity of Toeplitz operators belonging to this algebra, etc.

As it turns out this algebra is indeed very rich in Toeplitz operators. It contains Toeplitz operators with bounded and even unbounded defining symbol which are drastically different from the symbols of the initial generators. To see this we consider, for example, the defining symbol which, after a Möbius transformation and as a function on the upper half-plane, has in polar coordinates the form

$$\widehat{a}_0(re^{i\theta}) = (\sin\theta)^{-\beta}\sin(\sin\theta)^{-\alpha},$$

where $0 < \beta < 1$ and $\alpha > 0$. This symbol does not have a limit value at any point of the boundary and moreover is unbounded at every boundary point. At the same time we show that the Toeplitz operator with this defining symbol does belong to the algebra generated by Toeplitz operators whose piece-wise continuous defining symbols have discontinuities at just two opposite boundary points.

We show that each operator A of the C^*-algebra generated by Toeplitz operators with piece-wise continuous defining symbols admits the following transparent canonical representation,

$$A = T_{s_A} + \sum_{k=1}^{m} T_{v_k} f_{A,k}(T_{\chi_k}) T_{v_k} + K,$$

where the functions s_A and $f_{A,k}$ are defined by the image of A in the Calkin algebra (see for details Section 8.3), and that A is a compact perturbation of a Toeplitz operator with *bounded* defining symbol if and only if all the operators $f_{A,k}(T_{\chi_k})$ are Toeplitz. This permits us to uncover as well many Toeplitz operators with *unbounded* defining symbols which belong to this algebra. As it turns out, all these Toeplitz operators, whose bounded and unbounded defining symbols are quite different from the initial piece-wise continuous functions, belong to the closure of the elements of the form (0.4), while none of the operators (0.4) is (a compact perturbation of) a Toeplitz operator.

Chapter 9. Commuting Toeplitz Operators and Hyperbolic Geometry

Three different commutative C^*-algebras generated by Toeplitz operators were studied in Chapters 5, 6, and 7. The special features of defining symbols, which make these algebras commutative, were *symbols depending only on the imaginary part of a variable* for Toeplitz operators on the upper half-plane, *radial symbols* for Toeplitz operators on the unit disk, and *symbols depending only on the angular part of a variable* for Toeplitz operators on the upper half-plane, respectively. A natural question appears: *whether there exist other classes of defining symbols which generate commutative Toeplitz operator C^*-algebras, and how to classify them.*

The invariance of the Bergman kernel under biholomorphic automorphisms of a domain suggests immediately other classes of defining symbols, which are obtained by means of biholomorphic automorphisms from those already studied. Surprisingly, the key to understanding the nature of such classes of symbols and of classifying them lies in the hyperbolic geometry of the unit disk endowed with the Poincaré metric, the hyperbolic plane.

Recall that any two geodesics in the hyperbolic plane can be either parallel, or intersecting, or disjoint. In each case the pair of geodesics defines the so-called *pencil of hyperbolic geodesics*, which consists of all geodesics having the same behaviour with respect to the initial ones defining a pencil.

The bold lines on Figure 1 are called *cycles* and are the orthogonal trajectories to geodesics forming a pencil.

Now the sets of defining symbols which generate commutative C^*-algebras of Toeplitz operators can be classified as follows. Each pencil of hyperbolic geodesics defines the set of symbols consisting of functions which are constant on corresponding cycles. The C^*-algebra generated by Toeplitz operators with such defining symbols is commutative.

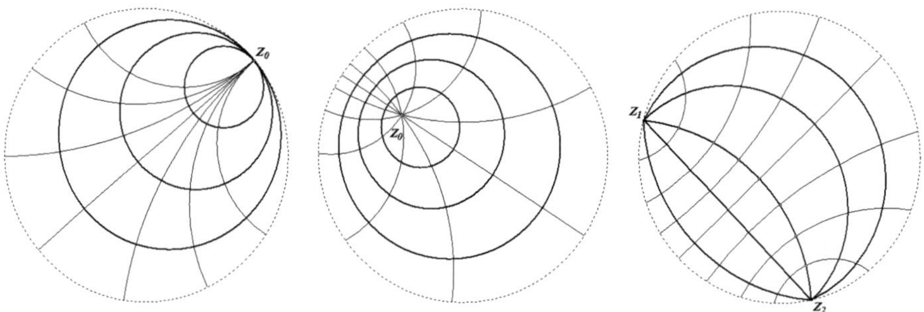

Figure 1: Three types of pencils of geodesics.

We note that this commutativity feature does not depend at all on smoothness properties of symbols; everything is defined here by the geometric configuration of the level sets, the defining symbols themselves can be merely measurable.

Yet another approach to the above classification result is based on the movements of the hyperbolic plane, the Möbius transformations. There is one-to-one correspondence between the pencils of hyperbolic geodesics and the maximal Abelian subgroups of the Möbius transformations ($\equiv$ one-parametric groups generated by non-identical Möbius transformation). For each such group the cycles of the corresponding pencils are exactly the lines which remain invariant under the action of the group.

That is, each maximal Abelian subgroup of the Möbius transformations generates a commutative Toeplitz operator algebra. The algebra is generated by Toeplitz operators whose (measurable) defining symbols are invariant under the action of this group.

Chapter 10. Weighted Bergman Spaces

This chapter is rather technical. We extend here some results of the previous chapters to the case of weighted Bergman spaces. The family of the weighted Bergman spaces appears naturally both in operator theory and in quantization. The definition of the weighted Bergman space follows the same lines as in Chapter 2, the only difference being that we substitute the standard Lebesgue measure $dv(z) = dxdy$ for another one depending on a weight parameter. In operator theory this parameter is normally $\lambda \in (-1, +\infty)$, and the corresponding measure for the case of the unit disk has the form

$$d\mu_\lambda(z) = \frac{\lambda+1}{\pi}\,(1 - |z|^2)^\lambda dv(z), \tag{0.5}$$

while in quantization one uses $h \in (0, 1)$ and

$$d\mu_h(z) = \left(\frac{1}{h} - 1\right)(1 - |z|^2)^{1/h}\frac{dv(z)}{\pi(1 - |z|^2)^2}. \tag{0.6}$$

For $\lambda + 2 = \frac{1}{h}$ we have the same spaces, and that for $\lambda = 0$ or $h = \frac{1}{2}$ we return to the classical weightless case (with the normalized measure).

As in Chapters 5, 6, and 7 we consider the Toeplitz operators with defining symbols depending only on the imaginary part of a variable on the upper half-plane, radial defining symbols on the unit disk, and defining symbols depending only on the angular part of a variable on the upper half-plane. We show that such Toeplitz operators in each case continue to generate commutative C^*-algebras on *each weighted Bergman space*, and that the corresponding Topelitz operators in all cases are unitary equivalent to the multiplication operators $\gamma_{a,\lambda} I$.

For the Toeplitz operators on the unit disk with radial symbols, $\gamma_{a,\lambda} = \{\gamma_{a,\lambda}(n)\}$ is the sequence with the following entries,

$$\gamma_{a,\lambda}(n) = \frac{1}{B(n+1, \lambda+1)} \int_0^1 a(\sqrt{r}) r^n (1-r)^\lambda dr, \qquad n \in \mathbb{Z}_+, \qquad (0.7)$$

while for Toeplitz operators on the upper half-plane with symbols depending on $y = \operatorname{Im} z$ we have

$$\gamma_{a,\lambda}(x) = \frac{x^{\lambda+1}}{\Gamma(\lambda+1)} \int_0^\infty a(t/2) t^\lambda e^{-xt} dt, \qquad x \in \mathbb{R}_+, \qquad (0.8)$$

and with symbols depending on the polar angle θ,

$$\gamma_{a,\lambda}(\xi) = \left(\int_0^\pi e^{-2\xi\theta} \sin^\lambda \theta \, d\theta \right)^{-1} \int_0^\pi a(\theta) \, e^{-2\xi\theta} \sin^\lambda \theta \, d\theta, \qquad \xi \in \mathbb{R}. \qquad (0.9)$$

We give as well the formulas for the Wick symbols of the above Toeplitz operators and the formulas for the star-product of the Wick symbols.

Formulas (0.7), (0.8), and (0.9) gives us an easy access to the boundedness, compactness, spectral properties, and to the description of the invariant subspaces of the above Toeplitz operators.

Chapter 11. Commutative Algebras of Toeplitz Operators

The commutative C^*-algebras of Toeplitz operators on the classical (weightless) Bergman space were classified in Chapter 9 by pencils of geodesics on the unit disk, considered as the hyperbolic plane. In Chapter 10 we extended this result to the case of weighted Bergman space. The main question, *whether the above cases are the only possible sets of defining symbols which might generate the commutative C^*-algebras of Toeplitz operators on each weighted Bergman space*, has been left open. In the chapter we give an affirmative answer to this question.

At the same time there is a trivial case having in fact no connection with specific properties of Toeplitz operators. Each C^*-algebra with identity (Toeplitz operators with the defining symbol $e(z) \equiv 1$) generated by a self-adjoint element (Toeplitz operator with a real-valued defining symbol $a = a(z)$) is obviously commutative. We exclude this obvious case from our considerations.

We note that the commutativity of the Toeplitz operator algebras on *each* weighted Bergman space is of great importance and permits us to make use of the Berezin quantization procedure. As it turns out, to obtain the necessary information about potential symbols, we need to calculate the tree term asymptotic expansion of the commutator of the Wick symbols of Toeplitz operators. The result is as follows:

$$
\widetilde{a}_h \star \widetilde{b}_h - \widetilde{b}_h \star \widetilde{a}_h = \frac{ih}{2\pi}\,\{a,b\}
$$
$$
+ \frac{h^2}{2}\left[\frac{i}{8\pi^2}\left(\Delta\{a,b\} + \{a,\Delta b\} + \{\Delta a,b\}\right) + \frac{i}{\pi}\{a,b\}\right]
$$
$$
+ h^3\left[\frac{i}{192\pi^3}\left(\{\Delta a,\Delta b\} + \{a,\Delta^2 b\} + \{\Delta^2 a,b\}\right.\right.
$$
$$
\left. + \Delta^2\{a,b\} + \Delta\{a,\Delta b\} + \Delta\{\Delta a,b\}\right)
$$
$$
\left. + \frac{7i}{48\pi^2}\left(\Delta\{a,b\} + \{a,\Delta b\} + \{\Delta a,b\}\right) + \frac{i}{2\pi}\{a,b\}\right] + o(h^3),
$$

where the Poisson bracket and the Laplace-Beltrami operator have the form

$$
\{a,b\} = 2\pi i(1 - z\bar{z})^2\left(\frac{\partial a}{\partial z}\frac{\partial b}{\partial \bar{z}} - \frac{\partial a}{\partial \bar{z}}\frac{\partial b}{\partial z}\right)
$$

and

$$
\Delta = 4\pi(1 - z\bar{z})^2\frac{\partial^2}{\partial z\partial \bar{z}}.
$$

The first, second, and third terms of the asymptotic expansion together provide us with the exact geometric information: *in order to generate a commutative C^*-algebra of Toeplitz operators on each weighted Bergman space, their defining symbols must be constant on the orthogonal trajectories to the geodesics of a certain pencil.*

At the same time we show that there exist non-typical, in a sense, C^*-algebras of Toeplitz operators which are commutative *only on a single* Bergman space.

Chapters 12, 13, and 14. Dynamics of Properties of Toeplitz Operators

Given a smooth defining symbol $a = a(z)$, the family of Toeplitz operators $T_a = \{T_a^{(h)}\}$, where $h \in (0,1)$, was considered in the previous chapter under the Berezin quantization procedure. For a fixed h the Toeplitz operator $T_a^{(h)}$ acts on the weighted Bergman space $\mathcal{A}_h^2(\mathbb{D})$, where the parameter h characterizes the weight (0.6) on $\mathcal{A}_h^2(\mathbb{D})$. In these chapters we will use another parameterization (0.5) of the weighted Bergman spaces i.e., we will consider the spaces $\mathcal{A}_\lambda^2(\mathbb{D})$, where the weight parameter $\lambda \in [0,+\infty)$ is connected with $h \in (0,1)$ by the rule $\lambda + 2 = \frac{1}{h}$.

In these three chapters we study the behavior of different properties (boundedness, compactness, spectral properties, etc.) of $T_a^{(\lambda)}$ in dependence on λ, and compare their limit behavior under $\lambda \to \infty$ ($h \to 0$) with corresponding properties of the initial symbol a.

The word "dynamics" is used to emphasize our main question: *what happens to the properties of Toeplitz operators acting on weighted Bergman spaces when the weight parameter varies.*

It seems to be quite impossible to get a reasonably complete answer to the above problem for general (smooth) defining symbols. At the same time the classes of commutative C^*-algebras of Toeplitz operators described suggest the classes of symbols for which a satisfactory complete answer can be given. Indeed, the key feature of defining symbols constant on cycles, which allows us to get much more complete information than studying general symbols, is that in each such case the corresponding Toeplitz operators admit a spectral type representation i.e., they are unitary equivalent to certain multiplication operators. Depending on the case considered, formula (0.7) or (0.8) or (0.9) gives us direct access to the properties we are interested in.

Given a symbol a constant on cycles, we denote by $B(a)$ the set of all $\lambda \in [0, +\infty)$ for which the Toeplitz operator $T_a^{(\lambda)}$ is bounded on $\mathcal{A}_\lambda^2(\mathbb{D})$. We carry out a detailed study of boundedness of Toeplitz operators with unbounded symbols, and show, in particular, that the set $B(a)$, for given a, necessarily has one of the forms

$$[0, +\infty), \qquad [0, \lambda_0], \qquad [0, \lambda_0).$$

For Toeplitz operators with radial symbols (where compact operators may appear) we show that the set $K(a)$ (consisting of all $\lambda \in [0, +\infty)$ for which the Toeplitz operator $T_a^{(\lambda)}$ is compact on $\mathcal{A}_\lambda^2(\mathbb{D})$) can have the same three forms. We give as well a number of conditions of the Schatten class membership for Toeplitz operators with radial symbols.

Then, as a kind of semiclassical analysis, we study the limit behaviour of the spectra $\operatorname{sp} T_a^{(\lambda)}$ when $\lambda \to +\infty$. We denote by $M_\infty(a)$ the (partial) limit set of the family $\{\operatorname{sp} T_a^{(\lambda)}\}_{\lambda \in [0, +\infty)}$ and study its properties for different classes of symbols (constant on cycles).

We show that for continuous symbols

$$\lim_{\lambda \to +\infty} \operatorname{sp} T_a^{(\lambda)} = M_\infty(a) = \operatorname{Range} a,$$

while for the piece-wise continuous a, the limit set $M_\infty(a)$ is the union of $\operatorname{Range} a$ and the straight line segments connecting the one-sided limit values of the symbol at the points of its discontinuities. We give as well a number of examples which illustrate the results.

In particular for the continuous radial symbol (hypocycloid)

$$a(r) = \frac{3}{4}(r + i\sqrt{1 - r^2})^8 + (r - i\sqrt{1 - r^2})^4,$$

we have Figure 2 below.

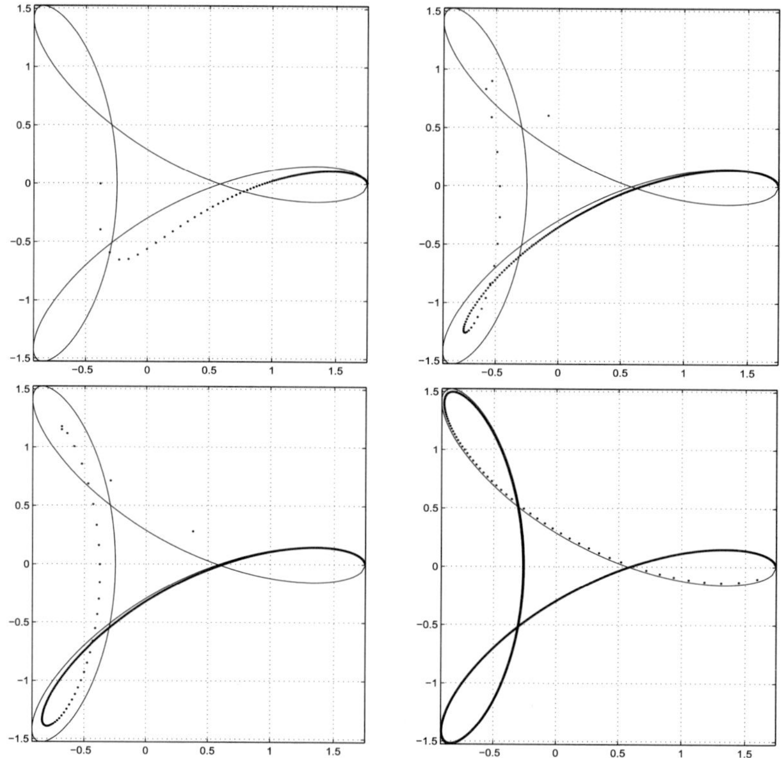

Figure 2: The hypocycloid and $\operatorname{sp} T_a^{(\lambda)}$ for $\lambda = 0$, $\lambda = 5$, $\lambda = 12$, and $\lambda = 200$.

The situation for oscillating symbols is more complicated but more interesting. Consider, for example, two oscillating symbols

$$a_1(y) = (1 + 2y)^i = e^{i \ln(1+2y)} \quad \text{and} \quad a_2(y) = e^{i2y},$$

where $y = \operatorname{Im} z \in [0, \infty)$.

Both symbols are continuous at the point $y = 0$ and have oscillation type discontinuity at infinity. Both of them are of the form

$$a_k(y) = e^{i\varphi_k(y)}, \qquad k = 1, 2,$$

where the functions $\varphi_k(y)$ are continuous and increasing on $[0, +\infty]$ with $\varphi_k(0) = 0$ and $\varphi_k(+\infty) = +\infty$. The only difference between them is the speed of growth at infinity. This difference leads to a drastic difference between the spectral behaviour of the corresponding Toeplitz operators.

All spectra $\operatorname{sp} T_{a_1}^{(\lambda)}$ and $\operatorname{sp} T_{a_2}^{(\lambda)}$, for $\lambda \in [0, +\infty)$, are spirals (Figure 3), but

$$M_\infty(a_1) = \partial\mathbb{D}, \qquad \text{while} \qquad M_\infty(a_2) = \overline{\mathbb{D}}.$$

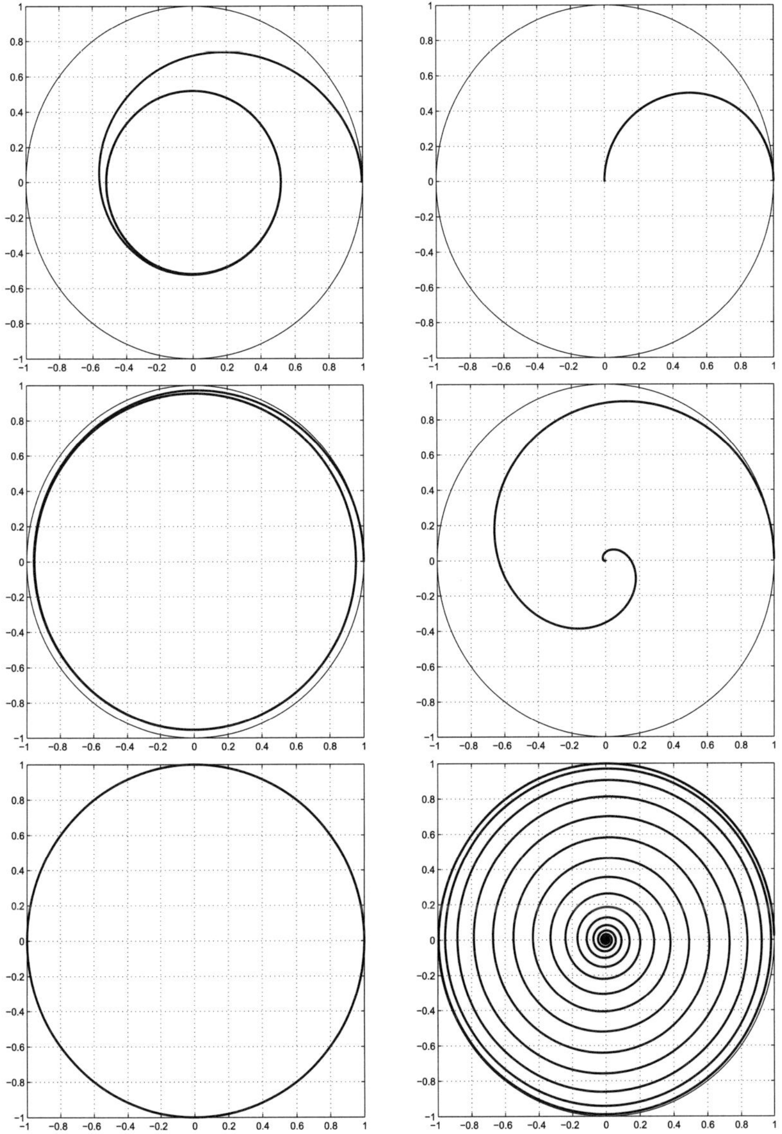

Figure 3: The spectra $\mathrm{sp}\,T_{a_1}^{(\lambda)}$ and $\mathrm{sp}\,T_{a_2}^{(\lambda)}$ for $\lambda = 0$, $\lambda = 10$, and $\lambda = 1000$.

Chapter 1

Preliminaries

The chapter contains some algebraic material which will be used substantially in the book and which can not be easily found in the standard C^*-algebraic sources. Those immediately interested in the main content of the book may skip this chapter at the first reading.

1.1 General local principle for C^*-algebras

1.1.1 C^*-bundles

We start with some necessary definitions. The triple $\xi = (p, E, T)$, where E and T are topological spaces, and $p : E \longrightarrow T$ is a surjective map, is called a *bundle*. The set T is called the *base* of the bundle, the set $\xi(t) = p^{-1}(t)$ is called the *fiber over the point $t \in T$*.

Let V be an open set in T. A function $\sigma : V \longrightarrow E$ is called a *local section* of the bundle ξ if $p(\sigma(t)) = t$ for all $t \in V$ (or $\sigma(t) \in \xi(t)$ for all $t \in V$). If $V = T$ the section is called *global*, or just a section. Denote by $\Gamma(\xi)$ the set of all *continuous* sections of the bundle ξ. If not specifically mentioned, in what follows all sections are supposed to be continuous.

Let
$$E \vee E = \{(x, y) \in E \times E : \ p(x) = p(y)\}.$$

The bundle $\xi = (p, E, T)$ is called a *C^*-bundle* if each fiber $\xi(t)$ has a structure of a C^*-algebra, and

1. the functions
$$(x, y) \mapsto x + y : \ E \vee E \longrightarrow E,$$
$$(x, y) \mapsto x \cdot y : \ E \vee E \longrightarrow E,$$
$$(\alpha, x) \mapsto \alpha x : \ \mathbb{C} \times E \longrightarrow E,$$
$$x \mapsto x^* : \ E \longrightarrow E$$

   are continuous;

2. the subsets

$$U_V(\sigma, \varepsilon) = \{x \in E : \ p(x) \in V \text{ and } \|x - \sigma(p(x))\| < \varepsilon\},$$

where V is an open subset in T, σ is a continuous section of ξ over V, $\varepsilon > 0$, form a basis of open sets in the space E.

For $V = T$ the set

$$U(\sigma, \varepsilon) = U_T(\sigma, \varepsilon) = \{x \in E : \|x - \sigma(p(x))\| < \varepsilon\},$$

is called a *tubular neighborhood of the section σ of radius ε.*

Let s be a continuous section of the bundle ξ, then the set

$$\begin{aligned} V = s^{-1}(U(\sigma, \varepsilon)) &= \{t \in T : \ \|s(t) - \sigma(t)\| < \varepsilon\} \\ &= \{t \in T : \ \|(s - \sigma)(t)\| < \varepsilon\} \end{aligned}$$

is open in T. Thus for each $a \in \Gamma(\xi)$ the set

$$V(a, \varepsilon) = \{t \in T : \ \|a(t)\| < \varepsilon\}$$

is open in T.

If each C^*-algebra $\xi(t)$ has an identity $e(t)$ and the section $e : \ t \longrightarrow e(t)$ is continuous, then the C^*-bundle ξ is a bundle with identity.

1.1.2 C^*-algebra defined by a bundle

Given a C^*-bundle $\xi = (p, E, T)$, there is the canonically defined C^*-algebra associated with ξ. Namely, the set of all bounded continuous sections σ of $\xi = (p, E, T)$ with componentwise operations and the norm

$$\|\sigma\| = \sup_{t \in T} \|\sigma(t)\|$$

is obviously a C^*-algebra. We will call this algebra the *C^*-algebra defined by a C^*-bundle $\xi = (p, E, T)$,* and will denote it by $\Gamma^b(\xi)$.

Lemma 1.1.1. *For each section $\sigma \in \Gamma(\xi)$ the function $\|\cdot\| : \ t \longmapsto \|\sigma(t)\|$ is upper semi-continuous.*

Proof. For each point $t_0 \in T$ and each positive ε it is sufficient to find a neighborhood V of the point t_0 such that for each $t \in V$

$$\|\sigma(t)\| < \|\sigma(t_0)\| + \varepsilon.$$

Take

$$V = V(\sigma, \|\sigma(t_0)\| + \varepsilon) = \{t \in T : \ \|\sigma(t)\| < \|\sigma(t_0)\| + \varepsilon\}. \qquad \square$$

We note that the sections are not necessarily norm continuous.

Denote by $C^b(T)$ the C^*-algebra of all bounded continuous functions on T.

Lemma 1.1.2. *The algebra $\Gamma^b(\xi)$ is a $C^b(T)$-module.*

Proof. We will prove only that $a\sigma \in \Gamma^b(\xi)$, for each $\sigma \in \Gamma^b(\xi)$ and each $a \in C^b(T)$. Here $(a\sigma)(t) = a(t)\sigma(t)$, for all $t \in T$.

Given σ and a, we check the continuity of the section $a\sigma$ at a point $t_0 \in T$. Let $M = \|\sigma\| = \sup_t \|\sigma(t)\|$. We show that for each $U_V(s, \varepsilon)$ (an open set from the base of topology on E), such that $(a\sigma)(t_0) \in U_V(s, \varepsilon)$, there is an open set $V_0(\subset T)$, which contains the point t_0, such that $(a\sigma)(t) \in U_V(s, \varepsilon)$ for each $t \in V_0$.

From $(a\sigma)(t_0) \in U_V(s, \varepsilon)$ it follows that $t_0 \in V$ and $\varepsilon_0 = \|a(t_0)\sigma(t_0) - s(t_0)\| < \varepsilon$. Let $\delta = \varepsilon - \varepsilon_0$. Introduce

$$V_1 = \{t \in T : |a(t) - a(t_0)| < \frac{\delta}{3M}\},$$

$$V_2 = V(a(t_0)\sigma - s, \varepsilon_0 + \frac{\delta}{3})$$

$$= \{t \in T : \|a(t_0)\sigma(t) - s(t)\| < \varepsilon_0 + \frac{\delta}{3}\}.$$

Then the set $V_0 = V \cap V_1 \cap V_2$ is open in T and contains t_0. For each $t \in V_0$ we have

$$\|a(t)\sigma(t) - s(t)\| \leq \|a(t)\sigma(t) - a(t_0)\sigma(t)\| + \|a(t_0)\sigma(t) - s(t)\|$$

$$\leq |a(t) - a(t_0)| \cdot \|\sigma(t)\| + \|a(t_0)\sigma(t) - s(t)\|$$

$$< \frac{\delta}{3M} \cdot M + \varepsilon_0 + \frac{\delta}{3} < \varepsilon.$$

That is, $a(t)\sigma(t) \in U_V(s, \varepsilon)$ for each $T \in V_0$. $\qquad\square$

Recall that a topological space is called quasi-compact if each of its open covers has a finite subcover, and is called compact if it is Hausdorff and quasi-compact.

Lemma 1.1.3. *If the space T is quasi-compact, then $\Gamma^b(\xi) = \Gamma(\xi)$ and $C^b(T) = C(T)$.*

Proof. We will prove only the first statement. Let $\sigma \in \Gamma(\xi)$. The set $V_n = V(\sigma, n) = \{t \in T : \|\sigma(t)\| < n\}$ is open in T, and the system $\{V_n\}_{n \in \mathbb{N}}$ covers T. Then by quasi-compactness we have that $\|\sigma(t)\|$ is bounded, and thus $\sigma \in \Gamma^b(\xi)$. $\qquad\square$

Recall that a space T is called *quasi-completely regular* if for each $t_0 \in T$ and each closed set $Y(\subset T)$ which does not contain t_0, there exists a continuous function $f : T \longrightarrow [0, 1]$ such that

$$f(t_0) = 0, \quad f|_Y \equiv 1.$$

Theorem 1.1.4 (Stone-Weierstrass). *Let $\xi = (p, E, T)$ be a C^*-bundle over a quasi-compact, quasi-completely regular space T. Let $\mathcal{A}$ be a closed $C(T)$-submodule of $\Gamma(\xi)$ $(= \Gamma^b(\xi))$, and let for each $t \in T$ the set $\mathcal{A}(t) = \{a(t) : a \in \mathcal{A}\}$ be dense in the fiber $\xi(t) = p^{-1}(t)$. Then $\mathcal{A} = \Gamma(\xi)$.*

Proof. It is sufficient to prove that for each section $\sigma \in \Gamma(\xi)$ and each $\varepsilon > 0$ there is an element $a \in \mathcal{A}$ such that

$$\|\sigma - a\| = \sup_{t \in T} \|\sigma(t) - a(t)\| < \varepsilon.$$

Fix σ and ε. For each t_0 we select an element $a_{t_0} \in \mathcal{A}$ such that $\|\sigma(t_0) - a_{t_0}(t_0)\| < \varepsilon$. Then the set

$$V'_{t_0} = V(\sigma - a_{t_0}, \varepsilon) = \{t \in T : \|\sigma(t) - a_{t_0}(t)\| < \varepsilon\}$$

is open in T and contains the point t_0. For the point t_0 and the closed set $Y_{t_0} = T \setminus V'_{t_0}$ denote by f_{t_0} a continuous on T function $f_{t_0} : T \longrightarrow [0, 1]$, such that

$$f_{t_0}(t_0) = 1, \qquad f_{t_0}|_{Y_{t_0}} \equiv 0.$$

The set

$$V_{t_0} = \{t \in T : f_{t_0}(t) > 0\}$$

is open and contains t_0. The system $\{V_t\}_{t \in T}$ covers T; let $\{V_{t_k}\}_{k=1}^n$ be a finite subcover. Introduce the functions

$$\psi_k = \frac{f_{t_k}}{\sum_{j=1}^n f_{t_j}}, \qquad k = \overline{1, n}.$$

All functions $\psi_k : T \to [0, 1]$ are continuous, and $\sum_{k=1}^n \psi_k \equiv 1$. Furthermore,

$$\|\psi_k(t)\sigma(t) - \psi_k(t)a_{t_k}(t)\| = \psi_k(t)\|\sigma(t) - a_{t_k}(t)\| < \psi_k(t)\varepsilon,$$

for each $t \in T$. Finally, the element $a = \sum_{k=1}^n \psi_k a_{t_k}$ belongs to $\mathcal{A}$, and

$$\|\sigma - a\| = \sup_{t \in T} \|\sum_{k=1}^n \psi_k(t)(\sigma(t) - a_{t_k}(t))\|$$

$$\leq \sup_{t \in T} \sum_{k=1}^n \|\psi_k(t)(\sigma(t) - a_{t_k}(t))\|$$

$$= \sup_{t \in T} \sum_{k=1}^n \psi_k(t)\|\sigma(t) - a_{t_k}(t)\|$$

$$< \sup_{t \in T} \sum_{k=1}^n \psi_k(t) \cdot \varepsilon = \varepsilon.$$

$\square$

1.1.3 C^*-bundle defined by a C^*-algebra and a system of its ideals

Let $\mathcal{A}$ be a C^*-algebra and let $J_T = \{J(t) : t \in T\}$ be a system of its closed two-sided ideals, parameterized by points of a set T. We describe now a canonical procedure for constructing the C^*-bundle defined by the above data.

For each $t \in T$ introduce the quotient algebra $\mathcal{A}(t) = \mathcal{A}/J(t)$. We will denote by $a(t)$ the image of an element $a \in \mathcal{A}$ in the quotient algebra $\mathcal{A}(t)$. Let

$$E = \bigsqcup_{t \in T} \mathcal{A}(t)$$

be the disjoint union of the C^*-algebras $\mathcal{A}(t)$. We define the action of the (additive) group $\mathcal{A}$ on the set E: each element $a \in \mathcal{A}$ generates the mapping

$$g_a : E \longrightarrow E$$

by the rule

$$g_a : x(t) \longmapsto (x + a)(t).$$

It is easy to see that the orbit of each point $x = x(t)$ $(\in \mathcal{A}(t))$ under the action of the group $\mathcal{A}$ coincides with the whole algebra $\mathcal{A}(t)$, and the collection of orbits is parameterized by points of T. The partition of E into disjoint orbits generates the projection

$$p : x(t) \in E \longmapsto t \in T,$$

with $p^{-1}(t) = \mathcal{A}(t)$.

We endow the sets E and T with appropriate topologies in order for the triple $\xi = (p, E, T)$ to be a C^*-bundle. Each element $a \in \mathcal{A}$ generates the section $\tilde{a} : T \to E$ by the rule

$$\tilde{a} : t \longmapsto a(t).$$

Denote by $\widetilde{\mathcal{A}}$ the set of all the above sections $\tilde{a}$. For each $\varepsilon > 0$ and each $\tilde{a} \in \widetilde{\mathcal{A}}$ introduce the set

$$U(\tilde{a}, \varepsilon) = \{x \in E : \ \|x - \tilde{a}(p(x))\| < \varepsilon\},$$

and endow the set E with the topology whose prebase consists of all sets $U(\tilde{a}, \varepsilon)$.

Remark 1.1.5. The topology of each fiber $\xi(t) = \mathcal{A}(t)$, generated by the quotient norm $\|x(t)\| = \inf_{z \in J(t)} \|x + z\|$, coincides with the topology of the fiber $\xi(t)$, generated by the restriction onto $\xi(t)$ of the topology on E.

Endow the base T of the bundle $\xi = (p, E, T)$ with the orbit space topology (or the quotient topology): the *strongest* topology under which the projection $p : E \to T$ is continuous.

Lemma 1.1.6. 1. *The mapping* $p : E \to T$ *is open.*

2. *The topology on T coincides with the weakest topology under which all the mappings $\tilde{a} \in \widetilde{\mathcal{A}}$ are continuous.*

3. *A prebase of the topology on T is given by the system of sets*

$$V(\tilde{a}, \varepsilon) = \{t \in T : \|\tilde{a}(t)\| < \varepsilon\}.$$

Proof. For each $a \in \mathcal{A}$ the mapping

$$g_a : \; x(t) \longmapsto (x + a)(t)$$

is one-to-one, and $g_a^{-1} = g_{-a}$. Furthermore

$$g_a^{-1}(U(\tilde{b}, \varepsilon)) = g_{-a}(U(\tilde{b}, \varepsilon))$$
$$= \{x \in E : \|x + a(p(x)) - \tilde{b}(p(x))\| < \varepsilon\}$$
$$= U(\widetilde{(b - a)}, \varepsilon).$$

Thus each mapping g_a is a homeomorphism of the space E. We show now that the mapping $p : \; E \to T$ is open. Let U be an open set in E. We need to check that $V = p(U)$ is open in T. By the property $V = p(p^{-1}(V))$ and the definition of the quotient topology it is sufficient to prove that the set $p^{-1}(V)$ is open. We have

$$p^{-1}(V) = \bigcup_{a \in \mathcal{A}} g_a(U).$$

Thus the set $p^{-1}(V)$ is open, and the first statement is proved.

Now, since the mapping $p : \; E \to T$ is open, the prebase of the topology on T can be described as the image of the prebase in E under the mapping p. Consider the prebase in E consisting of all sets of the form

$$U = U(\tilde{a}, \varepsilon_1) \cap U(\tilde{b}, \varepsilon_2)$$
$$= \{x \in E : \|x - \tilde{a}(p(x))\| < \varepsilon_1 \;\; \text{and} \;\; \|x - \tilde{b}(p(x))\| < \varepsilon_2\},$$

where a and b are elements of $\mathcal{A}$, and ε_1 and ε_2 are positive.

Then the prebase of the topology on T consists of the sets

$$V = p(U) = \{t \in T : \|\tilde{a}(t) - \tilde{b}(t)\| < \varepsilon_1 + \varepsilon_2\}$$
$$= \{t \in T : \|(a - b)(t)\| < \varepsilon\}$$
$$= V(\widetilde{a - b}, \varepsilon);$$

here $\varepsilon = \varepsilon_1 + \varepsilon_2$.

The prebase of the weakest topology on T such that all the sections $\tilde{a} \in \widetilde{\mathcal{A}}$ are continuous can be described as follows: for all the sections $\tilde{a} \in \widetilde{\mathcal{A}}$ and each open set U from a prebase in E the set $\tilde{a}^{-1}(U)$ must be open. Let $U = U(\tilde{b}, \varepsilon)$, then

$$\tilde{a}^{-1}(U) = p(U \cap \tilde{a}(T))$$
$$= \{t \in T : \|\tilde{a}(t) - \tilde{b}(t)\| < \varepsilon\}$$
$$= \{t \in T : \|(a - b)(t)\| < \varepsilon\}$$
$$= V(\widetilde{a - b}, \varepsilon).$$

Thus we have proved the second statement of the lemma. To prove the third statement it is sufficient to use $a - b \in \mathcal{A}$ in place of $a \in \mathcal{A}$ in the above description of the prebase of the topology on T. $\qquad\square$

The introduced topology on T is called the *-bundle topology*.

Lemma 1.1.7. *The constructed triple $\xi = (p, E, T)$ is a C^*-bundle.*

Proof. We will check only the second property in the definition of a C^*-bundle. Moreover it is sufficient to prove that each point of the set

$$U_V(s, \varepsilon) = \{x \in E : \ p(x) \in V \ \text{ and } \ \|x - s(p(x))\| < \varepsilon\}$$

is interior. Here V is open in T, $\varepsilon > 0$, s is a continuous section over V.

Given $x_0 \in U_V(s, \varepsilon)$, we prove that x_0 is interior. Introduce $t_0 = p(x_0)$, $s_0 = s(t_0)$, $\sigma_0 = \frac{1}{2}(s_0 + x_0)$, $\delta = \|x_0 - s_0\| < \varepsilon$, and $\varepsilon_0 = \frac{1}{2}(\delta + \frac{1}{2}(\varepsilon - \delta)) < \frac{1}{2} \cdot \varepsilon$. Denote by σ an arbitrary element of the algebra $\mathcal{A}$ such that $\sigma(t_0) = \sigma_0$ ($\in \mathcal{A}(t_0)$), and let $U = U(\tilde{\sigma}, \varepsilon_0)$. It is easy to see that $x_0 \in U$ and $s_0 \in U$. The section s is continuous, thus the set

$$V_0 = s^{-1}(U) = p(U \cap s(V))$$

is open, and $V_0 \subset V$. The projection $p : \ E \to T$ is continuous, thus the set $p^{-1}(V_0)$ is open in E, and thus the set

$$U_0 = U(\tilde{\sigma}, \varepsilon_0) \cap p^{-1}(V_0)$$

is open as well. It is easy to see that

$$x_0 \in U_0 \subset U_V(s, \varepsilon),$$

and thus the point x_0 is interior. $\qquad\square$

Given a C^*-algebra $\mathcal{A}$ and a system of its closed two-sided ideals $J_T = \{J(t) : t \in T\}$, the C^*-bundle $\xi = (p, E, T)$ described above is called the *canonical C^*-bundle defined by the C^*-algebra $\mathcal{A}$ and the system of ideals J_T.*

1.1.4 Main theorem

The next theorem can be treated as a non-commutative generalization of the Gelfand representation of a commutative Banach algebra.

Theorem 1.1.8. *Let $\mathcal{A}$ be a C^*-algebra, $J_T = \{J(t) : \ t \in T\}$ be a system of its closed two-sided ideals, $\xi = (p, E, T)$ be the canonical C^*-bundle defined by $\mathcal{A}$ and J_T, $\Gamma^b(\xi)$ be the C^*-algebra defined by the bundle $\xi = (p, E, T)$. Then the mapping*

$$\tilde{\pi} : \ a \in \mathcal{A} \longmapsto \tilde{a} \in \Gamma^b(\xi)$$

is a morphism of the C^*-algebras $\mathcal{A}$ and $\Gamma^b(\xi)$, such that

1. $\ker \tilde{\pi} = \bigcap_{t \in T} J(t)$,

2. $\operatorname{Im} \tilde{\pi} = \widetilde{\mathcal{A}}$.

In particular, the mapping $\tilde{\pi} : \mathcal{A} \to \widetilde{\mathcal{A}}$ is an isometric $*$-isomorphism if and only if

$$\bigcap_{t \in T} J(t) = \{0\}.$$

Proof. Indeed, $\tilde{\pi}(a) = 0$ if and only if $a(t) = 0$ for each $t \in T$, or $a \in J(t)$ for each $t \in T$, or $a \in \bigcap_{t \in T} J(t)$. To finish the proof we apply the result on the canonical decomposition of a C^*-algebra morphism (see, for example, [56], Corollary 1.8.3). $\qquad\qquad\qquad\qquad\qquad\qquad\qquad\qquad\qquad\qquad\qquad\qquad\quad\Box$

1.1.5   General Local Principle

The construction of the C^*-bundle defined by a C^*-algebra and a system J_T of its ideals, together with Theorem 1.1.8, gives in fact a general conception of the local principles in the theory of C^*-algebras: an isomorphic description of a C^*-algebra $\mathcal{A}$ (or $\mathcal{A}/\bigcap_{t \in T} J(t)$) collecting the information about "simpler objects", the so-called local algebras.

The case when $\bigcap_{t \in T} J(t) = \{0\}$, is most important. We obtain here an isomorphic description of the initial algebra $\mathcal{A}$. Studying the representations of an operator algebra $\mathcal{A}$ which contains the ideal $\mathcal{K}$ of all compact operators, or studying the Fredholm theory for operators from $\mathcal{A}$, the case $\bigcap_{t \in T} J(t) = \mathcal{K}$ is important as well. It yields an isomorphic description of the Calkin (or Fredholm symbol) algebra $\operatorname{Sym} \mathcal{A} = \mathcal{A}/\mathcal{K}$ of the algebra $\mathcal{A}$.

Remark 1.1.9. It is well known that the identical representation of an operator C^*-algebra $\mathcal{A}$ which contains the ideal $\mathcal{K}$ of all compact operators is irreducible. At the same time the Calkin algebra $\operatorname{Sym} \mathcal{A} = \mathcal{A}/\mathcal{K}$ carries essential information on the structure of the algebra $\mathcal{A}$ since all other irreducible representations of $\mathcal{A}$ are described in therms of the irreducible representations of $\operatorname{Sym} \mathcal{A}$.

The algebra $\operatorname{Sym} \mathcal{A}$ is often also called the Fredholm symbol algebra of the algebra $\mathcal{A}$, as an operator $A \in \mathcal{A}$ is Fredholm if and only if its image in $\operatorname{Sym} \mathcal{A}$ is invertible.

In what follows we will restrict ourselves to the above two most important for us cases.

Having a C^*-algebra $\mathcal{A}$ and a system J_T of its closed two-sided ideals (with or without the property $\bigcap_{t \in T} J(t) = \{0\}$) we say that we will *localize by points of the set T*. Elements a_1 and a_2 of the algebra $\mathcal{A}$ are called *locally equivalent at the point $t \in T$* $(a_1 \overset{t}{\sim} a_2)$ if and only if $a_1 - a_2 \in J(t)$. The natural projections $\pi_t : \mathcal{A} \to \mathcal{A}(t)$ identify the elements of the algebra $\mathcal{A}$ locally equivalent at the point t, and the algebra $\mathcal{A}(t)$ is called the *local algebra at the point $t \in T$*. Having

$a \in \mathcal{A}$, the element $\pi_t(a) = a(t)$ sometimes will be called the *local representative of a in the local algebra* $\mathcal{A}(t)$.

We mention that

- the algebra we are interested in (either $\mathcal{A}$, if $\bigcap_{t \in T} J(t) = \{0\}$, or $\mathrm{Sym}\,\mathcal{A}$, if $\bigcap_{t \in T} J(t) = \mathcal{K}$) is described via its local algebras using the *canonical construction*: as an algebra of continuous sections of the canonically defined C^*-bundle over T;

- the norm of elements in the described algebra ($\mathcal{A}$ or $\mathrm{Sym}\,\mathcal{A}$) is equal to supremum of the norms of all its local representatives $\sup_{t \in T} \|a(t)\|$.

Given a C^*-algebra $\mathcal{A}$, each concrete choice of the system of ideals J_T defines then a concrete version of the local principle. There are several known "canonical" cases, when one can state *a priori* that $\bigcap_{t \in T} J(t) = \{0\}$. One of these cases, the so-called Douglas-Varela local principle, will be treated in detail in the next section; for another one see the example below.

The following observation describes an advantage of the approach proposed. There are situations when one can not use a "canonical" principle. For example, the Douglas-Varela principle is not applicable for an algebra with trivial (scalar) center. But this does not exclude, in general, the possibility of a localization. In such a case, using the special features of the C^*-algebra under consideration, one can try to find a system J_T of its ideals permitting one to carry out the localization procedure.

Example 1.1.10. One of the canonical cases with the desired property is when $J_T = \mathrm{Prim}\,\mathcal{A}$, the set of all primitive ideals $J(t) = P_t$ of the algebra $\mathcal{A}$. As well known, independently of the algebra $\mathcal{A}$ we have

$$\bigcap_{\mathrm{Prim}\,\mathcal{A}} P_t = \{0\}.$$

In this case all the local algebras are irreducible, and we have, in some sense, the most "detailed" localization. The algebra $\mathcal{A}$ is described as an algebra of continuous sections over the space $T = \mathrm{Prim}\,\mathcal{A}$, endowed with the *-bundle topology.

This case presents the following peculiarities:

- the *-bundle topology on T, in general, does not coincide with the Jacobson topology on $T = \mathrm{Prim}\,\mathcal{A}$ (the last one is weaker);

- the algebra $\widetilde{\mathcal{A}}$ (which is isomorphic and isometric to $\mathcal{A}$) is, in general, a proper subalgebra of $\Gamma^b(\xi)$;

- if the space $T = \mathrm{Prim}\,\mathcal{A}$ is endowed with the Jacobson topology, then the mappings

$$\tilde{a} : \ t \longmapsto a(t)$$

from $\widetilde{\mathcal{A}}$ are not necessarily continuous,

and thus may not be optimal in a variety of cases.

1.1.6   Douglas-Varela local principle

This section is devoted to a very important and frequently used particular case of
the general situation described by Theorem 1.1.8. We combine here the Douglas
approach to a local principle [58, 59] with the Varela results [197]. The major
advantages obtained are as follows. First, we put a topology on the Douglas direct
sum of local algebras and on the parameter space, which permits us to describe
the algebra in terms of *continuous* sections; and second, the Stone-Weierstrass
theorem guarantees that in this specific case the algebra under study is always
isomorphic to the algebra of *all* continuous sections of the corresponding C^*-
bundle. The local principle obtained after such a unification can be naturally
called the Douglas-Varela local principle.

Let $\mathcal{A}$ be a C^*-algebra with identity e, and let $\mathcal{Z}$ be its *central* commutative
C^*-subalgebra, containing e. Denote by T the compact set of maximal ideals of
the algebra $\mathcal{Z}$; then, of course, $\mathcal{Z} \cong C(T)$. For each point $t \in T$ denote by J_t the
maximal ideal of the algebra $\mathcal{Z}$ which corresponds to the point t, and denote by
$J(t) = \mathcal{A} \cdot J_t$ the closed two-sided ideal generated by J_t in the algebra $\mathcal{A}$. Finally,
we introduce the system of ideals $J_T = \{ J(t) : \ t \in T \}$.

Lemma 1.1.11. *We have*

$$\bigcap_{t \in T} J(t) = \{0\}.$$

Proof. Let $P \in \operatorname{Prim} \mathcal{A}$ be a primitive ideal of the algebra $\mathcal{A}$. Show first, that
there exists a point $t \in T$ such that $P \cap \mathcal{Z} = J_t$. Indeed, let π be the irreducible
representation of the C^*-algebra $\mathcal{A}$ with $\ker \pi = P$; then

$$\mathbb{C} \cdot I \subset \pi(\mathcal{Z}) \subset \pi(\mathcal{A})' = \mathbb{C} \cdot I.$$

Thus

$$\mathbb{C} \cong \pi(\mathcal{Z}) = (\mathcal{Z} + P)/P \cong \mathcal{Z}/(\mathcal{Z} \cap P),$$

and therefore $\mathcal{Z} \cap P$ is a maximal ideal of the algebra $\mathcal{Z}$, thus equal to J_t for some
point $t \in T$. It is easy to see that the conditions $P \cap \mathcal{Z} = J_t$ and $P \supset J(t)$ are
equivalent for each primitive ideal P. Furthermore, by [56], 2.9.7,

$$J(t) = \bigcap_{P \supset J(t)} P.$$

Thus

$$\{0\} \subset \bigcap_{t \in T} J(t) \subset \bigcap_{\operatorname{Prim} \mathcal{A}} P = \{0\},$$

or

$$\bigcap_{t \in T} J(t) = \{0\}. \qquad \square$$

Introduce now the C^*-bundle $\xi = (p, E, T)$, defined by the algebra $\mathcal{A}$ and the system of ideals $J_T = \{J_t : t \in T\}$ under consideration. The space T has two natural topologies:

- $*$-bundle topology of the bundle ξ,

- the topology of the compact space of maximal ideals of the algebra $\mathcal{Z}$, or, which is the same, the hull kernel topology of the C^*-algebra $\mathcal{Z}$.

Let us compare these topologies. Given $t \in T$, let u_0 and u_1 be two neighborhoods of the point t in the hull kernel topology on T, such that $t \in u_1 \subset \bar{u}_1 \subset u_0$. By the Tietze theorem there are functions $f : T \to [0, 1]$, such that

$$f|_{u_1} \equiv 1, \quad f|_{T \setminus u_0} \equiv 0.$$

Denote by F_t the set of all such functions f for all possible pairs (u_0, u_1) with the above properties.

Lemma 1.1.12. *For each element $a \in \mathcal{A}$ and each point $t \in T$, identifying $\mathcal{Z}$ with $C(T)$, we have*

$$\|a(t)\| = \|\tilde{a}(t)\| = \inf_{f \in F_t} \|f \cdot a\|.$$

Proof. Fix $\varepsilon > 0$, and let $x \in J(t)$ satisfy the property

$$\|a + x\| < \|a(t)\| + \varepsilon.$$

Select now elements $y_k \in J_t$ and $b_k \in \mathcal{A}$, $k = \overline{1, n}$ with

$$\|x - (y_1 b_1 + \ldots + y_n b_n)\| < \varepsilon,$$

and a function $f \in F_t$ with

$$\|f(y_1 b_1 + \ldots + y_n b_n)\| < \varepsilon.$$

Then

$$
\begin{aligned}
\|a(t)\| + \varepsilon > \|a + x\| &\geq \|fa + fx\| \\
&\geq \|fa + f(y_1 b_1 + \ldots + y_n b_n)\| - \|f(y_1 b_1 + \ldots + y_n b_n) - fx\| \\
&\geq \|fa\| - \|f(y_1 b_1 + \ldots + y_n b_n)\| - \|(y_1 b_1 + \ldots + y_n b_n) - x\| \\
&\geq \|fa\| - 2\varepsilon.
\end{aligned}
$$

Thus

$$\|fa\| \leq \|a(t)\| + 3\varepsilon,$$

which implies

$$\inf_{f \in F_t} \|fa\| \leq \|a(t)\|.$$

Conversely, we have $fa = a + (f - 1)a$, where $(f - 1)a \in J(t)$, thus

$$\|a(t)\| \leq \inf_{f \in F_t} \|fa\|. \qquad \square$$

Corollary 1.1.13. *Elements a_1 and a_2 of the algebra $\mathcal{A}$ are locally equivalent at the point $t \in T$: $a_1 \overset{t}{\sim} a_2$ if and only if*

$$\inf_{f \in F_t} \|f(a_1 - a_2)\| = 0.$$

Remark 1.1.14. In other approaches to the local principles, the statement of Corollary 1.1.13 is usually used as the definition of locally equivalent elements. This is exactly the case, for example, for the Gohberg-Krupnik local principle [85], which is formulated in terms of a system of so-called localizing classes.

Lemma 1.1.15. *The *-bundle topology and the hull kernel topology coincide on the space T.*

Proof. We show first that the set

$$V = V(\tilde{a}, \varepsilon) = \{t \in T : \ \|a(t)\| < \varepsilon\}$$

is open in the hull kernel topology. To do this we show that each point $t_0 \in V$ is interior. By Lemma 1.1.12 there exist a function $f \in F_{t_0}$ and neighborhoods $u_1 \subset u_0$ ($\overline{u}_1 \subset u_0$), such that $f|_{u_1} \equiv 1$ and $\|fa\| < \varepsilon$. Then for each point $t \in \text{Int}\, u_1$ we have
$$\|a(t)\| \leq \|fa\| < \varepsilon,$$

thus $t_0 \in \text{Int}\, u_1 \subset V$, and the point t_0 is interior. Thus the *-bundle topology is not stronger then the hull kernel topology.

Now, the *- bundle topology is the weakest one such that the sections $\tilde{a}$: $t \mapsto a(t)$ are continuous for all $a \in \mathcal{A}$, and hull kernel topology is the weakest one such that the sections of the form $\tilde{b}$: $t \mapsto b(t)e(t)$ are continuous for all $b \in C(T)$. Thus the hull kernel topology is no stronger then the *-bundle topology. $\qquad\square$

Theorem 1.1.16 (Douglas-Varela local principle). *Let $\mathcal{A}$ be a C^*-algebra with identity, $\mathcal{Z}$ be its central commutative C^*-subalgebra with the same identity, T be the compact set of maximal ideals of the algebra $\mathcal{Z}$. Further, let J_t be the maximal ideal of $\mathcal{Z}$ corresponding to a point $t \in T$, and $J(t) = J_t \cdot \mathcal{A}$ be the two-sided closed ideal generated by J_t in the algebra $\mathcal{A}$.*

Then the algebra $\mathcal{A}$ is *-isomorphic and isometric to the algebra of all (global) continuous sections of the C^*-bundle, defined by the algebra $\mathcal{A}$ and the system of ideals $J_T = \{J(t) : \ t \in T\}$; moreover the *-bundle topology on T coincides with the hull kernel topology of the compact T.

Proof. By Lemma 1.1.11 we have

$$\bigcap_{t \in T} J(t) = \{0\},$$

thus by Theorem 1.1.8 the algebra $\mathcal{A}$ is *-isomorphic and isometric to the C^*-algebra $\widetilde{\mathcal{A}}$, and, in particular, $\widetilde{\mathcal{A}}$ is closed.

Now for each $b(t) \in C(T)$ and the identity $\tilde{e} = e(t)$ of the algebra $\tilde{\mathcal{A}}$ there exists an element $b \in \mathcal{Z}$ such that $\tilde{b} = b\tilde{e}$, thus for each $\tilde{a} \in \tilde{\mathcal{A}}$ we have

$$b\tilde{a} = b\tilde{e}\tilde{a} = \tilde{b}\tilde{a} = \tilde{b}a \in \tilde{\mathcal{A}}.$$

Thus the algebra $\tilde{\mathcal{A}}$ is a $C(T)$-module. To finish the proof we apply Theorem 1.1.4. $\qquad\square$

The next theorem describes all irreducible representations of the algebra $\mathcal{A}$ in terms of irreducible representations of its local algebras.

Theorem 1.1.17. *Let $\mathcal{A}$ be a C^*-algebra with identity, $\mathcal{Z}$ be its central commutative subalgebra with the same identity, $T = \mathrm{sp}\,\mathcal{Z}$ be the compact set of maximal ideals of the algebra $\mathcal{Z}$. Introduce the C^*-bundle $\xi = (p, E, T)$ defined by the algebra $\mathcal{A}$ and the system of ideals J_T (we have $\mathcal{A} \cong \Gamma(\xi)$). For each point $t \in T$ and each irreducible representation $\pi \in \widehat{\mathcal{A}}(t)$ introduce the (irreducible) representation*

$$\rho_\pi : \quad a \longmapsto a(t) \longmapsto \pi(a(t))$$

of the C^*-algebra $\mathcal{A}$.
Then the mapping $\pi \longmapsto \rho_\pi$ identifies $\bigcup_{t \in T} \widehat{\mathcal{A}}(t)$ with the spectrum $\widehat{\mathcal{A}}$ of the algebra $\mathcal{A}$.

Proof. Let t_1 and t_2 be two different points of the set T, and let π_1 and π_2 be irreducible representations of the corresponding algebras $\mathcal{A}(t_1)$ and $\mathcal{A}(t_2)$. We show that the representations ρ_{π_1} and ρ_{π_2} of the C^*-algebra $\mathcal{A}$ cannot be equivalent. This will prove that the mapping $\pi \longmapsto \rho_\pi$ is injective.

Select an element $a \in \mathcal{A}$ with $\pi_1(a(t_1)) \neq 0$, and a function $b(t) \in C(T)$ with $b(t_1) = 1$ and $b(t_2) = 0$. Then $ba \in \mathcal{A}$, and for representations ρ_{π_1} and ρ_{π_2} we have

$$\rho_{\pi_1}(ba) = \pi_1(b(t_1)a(t_1)) \neq 0,$$
$$\rho_{\pi_2}(ba) = \pi_2(b(t_2)a(t_2)) = 0.$$

Thus the representations ρ_{π_1} and ρ_{π_2} are not equivalent, and the mapping $\pi \longmapsto \rho_\pi$ is injective.

We show now that this mapping is surjective, i.e., that for each representation $\rho \in \widehat{\mathcal{A}}$ there exist a point $t \in T$ and a representation π of the algebra $\mathcal{A}(t)$, such that $\rho = \rho_\pi$. It is easy to see that it is sufficient to find a point $t \in T$ with $J(t) \subset \ker\rho$. As follows from the proof of Lemma 1.1.11, for each primitive ideal (and thus for $P = \ker\rho$) there exists a point $t \in T$ with $P \cap \mathcal{Z} = J_t$. Thus for this t we have $J(t) \subset P = \ker\rho$. $\qquad\square$

1.2 C^*-Algebras generated by orthogonal projections

1.2.1 C^*-algebra generated by two self-adjoint idempotents

The aim of this section is to describe the C^*-algebra $\mathcal{R} = C^*(p, q)$ with identity e generated by two elements p and q having only the relations

$$p = p^* = p^2, \quad q = q^* = q^2. \tag{1.2.1}$$

To describe this algebra we will use the Douglas-Varela local principle. Introduce

$$s = p - q, \quad n = e - p - q,$$

then

$$p = \frac{1}{2}(e + s + n), \quad q = \frac{1}{2}(e - s - n),$$

and, of course, $C^*(p, q) = C^*(s, n)$. The elements s and n satisfy the conditions

$$s^2 + n^2 = e, \quad sn + ns = 0, \quad s = s^*, \quad n = n^*, \tag{1.2.2}$$

which are equivalent to (1.2.1).

From (1.2.2) it follows that the set of all elements of the form

$$a = P(s) + Q(s)n,$$

where $P(\cdot)$ and $Q(\cdot)$ are (complex) polynomials, is dense in the algebra $\mathcal{R}$.

Lemma 1.2.1. *The spectra of the elements s and n are contained in the segment* $[-1, 1]$.

Proof. Indeed, the elements s and n are self-adjoint and

$$(\lambda e - s)(\lambda e + s) = (n - \sqrt{1 - \lambda^2}e)(n + \sqrt{1 - \lambda^2}e). \qquad \square$$

It is easy to see that the algebra $\mathcal{Z} = C^*(s^2)$ is a central commutative subalgebra of the algebra $\mathcal{R}$, and its compact space of maximal ideals coincides with $T = \mathrm{sp}\, s^2$ $(\subset [0, 1])$.

We will apply the Douglas-Varela local principle, localizing by points $t \in T$, and will denote by $\mathcal{R}(t)$ the local algebra corresponding to the point t.

Let $\nabla = \mathrm{sp}\, s^2 \cap (0, 1)$ and $\Delta = \overline{\nabla}$.

Lemma 1.2.2. *Given $t \in \nabla$, the algebra $\mathcal{R}(t)$ is isomorphic to* $\mathrm{Mat}_2(\mathbb{C})$, *and under the mapping*

$$\mu_t : \ \mathcal{R} \longrightarrow \mathcal{R}(t) \cong \mathrm{Mat}_2(\mathbb{C})$$

the image of the element $a = P(s) + Q(s)n$ is given by

$$\begin{pmatrix} P(\tau) & Q(\tau)\sqrt{1 - \tau^2} \\ Q(-\tau)\sqrt{1 - \tau^2} & P(-\tau) \end{pmatrix},$$

where $\tau = \sqrt{t}$.

Proof. Denote by $s(t)$, $n(t)$ and $e(t)$ the images of elements s, n and e in the local algebra $\mathcal{R}(t)$. Then we have

$$s(t)^2 = te(t), \quad n(t)^2 = (1-t)e(t), \quad s(t)n(t) + n(t)s(t) = 0.$$

Introduce the elements

$$s_0 = \frac{1}{\tau} s(t) \quad \text{and} \quad n_0 = \frac{1}{\sqrt{1-\tau^2}} n(t),$$

where $\tau = \sqrt{t}$. Then

$$s_0^2 = e(t), \quad n_0^2 = e(t), \quad s_0 n_0 + n_0 s_0 = 0,$$

and $\mathcal{R}(t) \cong C^*(s_0, n_0)$. Now the algebra $\mathcal{R}(t)$ is naturally isomorphic to the algebra

$$\widetilde{\mathcal{R}}(t) = \frac{1}{\sqrt{2}} \begin{pmatrix} e(t) & e(t) \\ n_0 & -n_0 \end{pmatrix} \begin{pmatrix} \mathcal{R}(t) & 0 \\ 0 & s_0 \mathcal{R}(t) s_0 \end{pmatrix} \frac{1}{\sqrt{2}} \begin{pmatrix} e(t) & n_0 \\ e(t) & -n_0 \end{pmatrix}.$$

Under the isomorphism $\mathcal{R}(t) \longrightarrow \widetilde{\mathcal{R}}(t)$ the elements of the form $a = P(s) + Q(s)n$ are mapped to

$$\begin{pmatrix} P(\tau) & Q(\tau)\sqrt{1-\tau^2} \\ Q(-\tau)\sqrt{1-\tau^2} & P(-\tau) \end{pmatrix} p_+ + \begin{pmatrix} P(-\tau) & Q(-\tau)\sqrt{1-\tau^2} \\ Q(\tau)\sqrt{1-\tau^2} & P(\tau) \end{pmatrix} p_-,$$

where $p_\pm = \frac{1}{2}(e(t) \pm s_0)$. From the properties

$$p_\pm^2 = p_\pm \quad \text{and} \quad p_+ p_- = p_- p_+ = 0$$

it follows that the algebra $\widetilde{\mathcal{R}}(t)$ is isomorphic to the algebra of all pairs of complex matrices of the form

$$\left\{ \begin{pmatrix} a_{11} & a_{12} \\ a_{21} & a_{22} \end{pmatrix}, \begin{pmatrix} a_{22} & a_{21} \\ a_{12} & a_{11} \end{pmatrix} \right\}.$$

But

$$\begin{pmatrix} a_{22} & a_{21} \\ a_{12} & a_{11} \end{pmatrix} = \begin{pmatrix} 0 & -i \\ -i & 0 \end{pmatrix} \begin{pmatrix} a_{11} & a_{12} \\ a_{21} & a_{22} \end{pmatrix} \begin{pmatrix} 0 & i \\ i & 0 \end{pmatrix},$$

and thus $\widetilde{\mathcal{R}}(t) \cong \mathrm{Mat}_2(\mathbb{C})$. To finish the proof one only needs to follow all the above isomorphisms. $\qquad\square$

Corollary 1.2.3. *The sets* $\mathrm{sp}\, s \cap (-1, 1)$ *and* $\mathrm{sp}\, n \cap (-1, 1)$ *are symmetric with respect to zero, and moreover*

$$\mathrm{sp}\, s \cap (-1, 1) = \{\pm\tau : \ \tau^2 \in \nabla\},$$

$$\mathrm{sp}\, n \cap (-1, 1) = \{\pm\sqrt{1-\tau^2} : \ \tau^2 \in \nabla\}.$$

Lemma 1.2.4. *If* $1 \in \Delta = \overline{\nabla}$, *then* $\mathcal{R}(1) \cong \mathbb{C}^2$, *and under the mapping*

$$\mu_1 : \mathcal{R} \longrightarrow \mathcal{R}(1) \cong \mathbb{C}^2$$

we have

$$\mu_1 : \ a = P(s) + Q(s)n \longmapsto (P(1), P(-1)).$$

If $0 \in \Delta = \overline{\nabla}$, *then* $\mathcal{R}(0) \cong \mathbb{C}^2$, *and under the mapping*

$$\mu_0 : \mathcal{R} \longrightarrow \mathcal{R}(0) \cong \mathbb{C}^2$$

we have

$$\mu_0 : \ a = P(s) + Q(s)n \longmapsto (P(0) + Q(0), P(0) - Q(0)).$$

Proof. The proofs of both statements of the lemma are similar, we will prove only the first one. In the algebra $\mathcal{R}(1)$ we have $s(1)^2 = e(1)$ and $n(1)^2 = 0$. By Corollary 1.2.3 $s(1) \neq \pm 1$, and from $n(1) = n(1)^*$ it follows that $n(1) = 0$. Thus, it is easy to see that the element $a = P(s) + Q(s)n$ has the following image in $\mathcal{R}(1)$,

$$P(1)p_+ + P(-1)P_-,$$

where $P_\pm = \frac{1}{2}(e(1) + s(1))$. To finish the proof, apply the properties $p_\pm^2 = p_\pm$ and $p_+p_- = p_-p_+ = 0$. $\qquad\qquad\qquad\qquad\qquad\qquad\qquad\qquad\qquad\qquad\qquad\qquad\quad\square$

Remark 1.2.5. If the value 1 is an isolated point of $\operatorname{sp} s^2$, i.e., $1 \in T \setminus \Delta$, then besides the situation of Lemma 1.2.4, which occurs when $\{-1, 1\} \subset \operatorname{sp} s$, we may have the following possibilities:

$$\mathcal{R}(1) \cong \mathbb{C} \quad \text{and} \quad \mu_1(a) = P(1),$$
$$\mathcal{R}(1) \cong \mathbb{C} \quad \text{and} \quad \mu_1(a) = P(-1),$$

for the cases when $-1 \notin \operatorname{sp} s$, or $1 \notin \operatorname{sp} s$, respectively.

Analogously, if the value 0 is an isolated point of $\operatorname{sp} s^2$, i.e., $0 \in T \setminus \Delta$, then besides the situation of Lemma 1.2.4, which occurs when $\{-1, 1\} \subset \operatorname{sp} n$, we may have the following possibilities:

$$\mathcal{R}(0) \cong \mathbb{C} \quad \text{and} \quad \mu_0(a) = P(0) + Q(0),$$
$$\mathcal{R}(0) \cong \mathbb{C} \quad \text{and} \quad \mu_0(a) = P(0) - Q(0),$$

for the cases when $-1 \notin \operatorname{sp} n$, or $1 \notin \operatorname{sp} n$, respectively.

To formulate the final result introduce the following notation. Denote by Y the subset of $\{0, 1\} \times \{0, 1\}$ such that

$$
\begin{aligned}
(0,0) \in Y, \quad &\text{if} \quad 0 \in T \setminus \Delta \ \text{and} \ -1 \in \operatorname{sp} n, \\
(1,0) \in Y, \quad &\text{if} \quad 0 \in T \setminus \Delta \ \text{and} \ 1 \in \operatorname{sp} n, \\
(0,1) \in Y, \quad &\text{if} \quad 1 \in T \setminus \Delta \ \text{and} \ 1 \in \operatorname{sp} s, \\
(1,1) \in Y, \quad &\text{if} \quad 1 \in T \setminus \Delta \ \text{and} \ -1 \in \operatorname{sp} s.
\end{aligned}
$$

We denote by $\mathfrak{S}$ the set of all pairs $\sigma = (\sigma_1, \sigma_2)$, where $\sigma_1 \in C(Y)$ and $\sigma_2 \in C(\Delta, \mathrm{Mat}_2(\mathbb{C}))$, such that σ_2 is diagonal at the points $\{0,1\} \cap \Delta$ (if any). The set $\mathfrak{S}$ is a C^*-algebra with respect to componentwise operations and the norm

$$\|\sigma\| = \max\{\max_Y |\sigma_1(\cdot)|, \ \max_\Delta \|\sigma_2(\cdot)\|\}.$$

Theorem 1.2.6. *The algebra* $\mathcal{R} = C^*(p, g)$ *is* *-*isomorphic and isometric to the algebra* $\mathfrak{S}$. *The isomorphism*

$$\nu : \mathcal{R} \longrightarrow \mathfrak{S}$$

is generated by the following mapping of the generators of the algebra $\mathcal{R}$:

$$\nu(p) = \begin{cases} 1 - i, & (i,j) \in Y, \\ \begin{pmatrix} 1 & 0 \\ 0 & 0 \end{pmatrix}, & t \in \Delta, \end{cases}$$

$$\nu(q) = \begin{cases} 1 - |i - j|, & (i,j) \in Y, \\ \begin{pmatrix} 1-t & \sqrt{t(1-t)} \\ \sqrt{t(1-t)} & t \end{pmatrix}, & t \in \Delta. \end{cases}$$

Proof. For each $t \in \nabla$ introduce the isomorphism

$$\nu_t : \ \mathcal{R}(t) \xrightarrow{\ \mu_t\ } \mathrm{Mat}_2(\mathbb{C}) \xrightarrow{\ m_t\ } \mathrm{Mat}_2(\mathbb{C}),$$

where μ_t is the isomorphism of Lemma 1.2.2, and m_t is the following automorphism of the algebra $\mathrm{Mat}_2(\mathbb{C})$:

$$m_t : \ h \longmapsto m(\tau) h m(\tau),$$

where

$$m(\tau) = m(\tau)^{-1} = m(\tau)^* = \frac{1}{\sqrt{2}} \begin{pmatrix} \sqrt{1+\tau} & -\sqrt{1-\tau} \\ -\sqrt{1-\tau} & -\sqrt{1+\tau} \end{pmatrix}, \quad \tau = \sqrt{t}.$$

Now change τ for $\sqrt{t}$. Then under the homomorphism

$$\mathcal{R} \longrightarrow \mathcal{R}(t) \xrightarrow{\ \nu_t\ } \mathrm{Mat}_2(\mathbb{C})$$

we have

$$p \ \longmapsto \ \begin{pmatrix} 1 & 0 \\ 0 & 0 \end{pmatrix}, \quad t \in \nabla,$$

$$q \ \longmapsto \ \begin{pmatrix} 1-t & \sqrt{t(1-t)} \\ \sqrt{t(1-t)} & t \end{pmatrix}, \quad t \in \nabla.$$

To finish the proof apply Theorem 1.1.16, extending the above mapping to $\Delta = \overline{\nabla}$ and to Y, if $Y \neq \emptyset$. $\qquad\square$

Remark 1.2.7. The most common situation in applications is when either $T = \Delta$, or if $j \in \{0,1\}$ is an isolated of the spectrum $\operatorname{sp}(p-q)^2$. Then both points $(0,j)$ and $(1,j)$ belong to the set Y. In this case Theorem 1.2.6 is equivalent to the following one.

Theorem 1.2.8. *The algebra $\mathcal{R} = C^*(p,q)$ is isomorphic and isometric to the algebra of all 2×2 matrix-functions continuous on $T = \operatorname{sp}(p-q)^2 \,(\subset [0,1])$, and diagonal at the points of $\{0,1\} \cap T$. This isomorphism is generated by the following mapping of the generators of the algebra $\mathcal{R}$:*

$$p \;\longmapsto\; \begin{pmatrix} 1 & 0 \\ 0 & 0 \end{pmatrix},$$

$$q \;\longmapsto\; \begin{pmatrix} 1-t & \sqrt{t(1-t)} \\ \sqrt{t(1-t)} & t \end{pmatrix},$$

where $t \in T$.

1.2.2   Two orthogonal projections

Let a Hilbert space H be given in two different ways as a direct sum of subspaces

$$H = L \oplus L^\perp = M \oplus M^\perp.$$

Denote by P the orthogonal projection on H with image L, and by Q the orthogonal projection on H with image M. The subspaces

$$\begin{aligned}
H_{0,0} &= L \cap M, & H_{0,1} &= L \cap M^\perp, \\
H_{1,0} &= L^\perp \cap M^\perp, & H_{0,1} &= L^\perp \cap M,
\end{aligned}$$

are invariant with respect to the projections, and the restrictions of the projections onto $H_{i,j}$, $i,j = 0,1$, have the obvious structure. Thus it is natural to exclude these subspaces from the analysis.

Due to P. Halmos [98], the projections P and Q are in *generic position* if all the spaces $H_{i,j}$, $i,j = 0,1$, are trivial, or, equivalently, they satisfy the conditions

$$\begin{aligned}
\operatorname{Im} P \cap \operatorname{Im} Q &= \{0\}, & \ker P \cap \operatorname{Im} Q &= \{0\}, \\
\operatorname{Im} P \cap \ker Q &= \{0\}, & \ker P \cap \ker Q &= \{0\}.
\end{aligned}$$

Introduce

$$H_0 = H_{0,0} \oplus H_{0,1} \oplus H_{1,0} \oplus H_{1,1} \quad \text{and} \quad H_1 = H \ominus H_0.$$

Then the restrictions $P_1 = P|_{H_1}$ and $Q_1 = Q|_{H_1}$ are orthogonal projections on H_1 with images $L_1 = L \cap H_1$ and $M_1 = M \cap H_1$ respectively, and which are in generic position.

Lemma 1.2.9. *All the subspaces L_1, $L_1^{\perp} = H_1 \ominus L_1$, M_1 and $M_1^{\perp} = H_1 \ominus M_1$ have the same dimensions.*

Proof. Let $F_1 = P_1|_{M_1} : M_1 \to L_1$ be the restriction of the projection P_1 onto the space M_1. From $M_1 \cap L_1^{\perp} = \{0\}$ it follows that $\ker F_1 = \{0\}$. We prove now that $\operatorname{Im} F_1$ is dense in L_1. Indeed, let $x \in L_1 \cap (\operatorname{Im} F_1)^{\perp}$, then for all $y \in M_1$ we have $0 = (x, F_1 y) = (x, P_1 y) = (P_1 x, y) = (x, y)$, thus $x \in L_1 \cap M_1^{\perp} = \{0\}$. Thus the operator F_1 maps M_1 one-to-one onto a linear manifold dense in L_1, and by ([99], problem 56) $\dim M_1 = \dim L_1$. To finish the proof we apply the above arguments to other combinations of subspaces and projections. $\qquad\square$

In addition to the operator $F_1 = P_1|_{M_1} : M_1 \to L_1$, introduce the operator $F_2 = (I - P_1)|_{M_1} : M_1 \to L_1^{\perp}$, and their adjoints $F_1^* = Q_1|_{L_1} : L_1 \to M_1$ and $F_2^* = Q_1|_{L_1^{\perp}} : L_1^{\perp} \to M_1$. We use the polar decomposition $F_j = U_j \cdot D_j$, with a unitary operator U_j and a positive operator $D_j : M_1 \to M_1$, $j = 1, 2$. Introduce the unitary operator

$$U = \operatorname{diag}(U_1, U_2) : \ M_1^2 \longrightarrow H_1 = L_1 \oplus L_1^{\perp}.$$

Thus the projections $P_1' = U^* P_1 U$ and $Q_1' = U^* Q_1 U$, acting on the space M_1^2 admit the matrix form

$$P_1' = \begin{pmatrix} 1 & 0 \\ 0 & 0 \end{pmatrix},$$

$$Q_1' = \begin{pmatrix} D_1^2 & D_1 D_2 \\ D_2 D_1 & D_2^2 \end{pmatrix}.$$

Note that $D_1^2 + D_2^2 = I$. In fact,

$$D_1^2 + D_2^2 = D_1 U_1^* \cdot U_1 D_1 + D_2 U_2^* \cdot U_2 D_2 = F_1^* F_1 + F_2^* F_2$$
$$= Q_1(P_1^2 + (I - P_1)^2)Q_1|_{M_1} = Q_1|_{M_1} = I.$$

Introduce finally $C = D_2^2$, then $D_1^2 = I - C$.

The following theorem summarizes the results of this section.

Theorem 1.2.10. *Each pair of orthogonal projections P and Q on a Hilbert space H admit the following canonical representation:*

The Hilbert space H splits into direct sum $H = H_0 \oplus H_1$, where both summands are invariant with respect to projections.

The restrictions $P_0 = P|_{H_0}$ and $Q_0 = Q|_{H_0}$ are the "commutative part" of the projections and have the form

$$P_0 = I \oplus I \oplus 0 \oplus 0, \tag{1.2.3}$$
$$Q_0 = I \oplus 0 \oplus 0 \oplus I, \tag{1.2.4}$$

corresponding to the splitting $H_0 = H_{0,0} \oplus H_{0,1} \oplus H_{1,0} \oplus H_{1,1}$, *where*

$$H_{0,0} = \operatorname{Im} P \cap \operatorname{Im} Q, \qquad H_{0,1} = \operatorname{Im} P \cap \ker Q,$$
$$H_{1,0} = \ker P \cap \ker Q, \qquad H_{1,1} = \ker P \cap \operatorname{Im} Q.$$

The restrictions $P_1 = P|_{H_1}$ *and* $Q_1 = Q|_{H_1}$ *are the "generic position part" of the projections and have the form*

$$P_1 = U^* \begin{pmatrix} 1 & 0 \\ 0 & 0 \end{pmatrix} U, \tag{1.2.5}$$

$$Q_1 = U^* \begin{pmatrix} I - C & \sqrt{C(I-C)} \\ \sqrt{C(I-C)} & C \end{pmatrix} U, \tag{1.2.6}$$

where $C = Q_1(I - P_1)Q_1|_{\operatorname{Im} Q_1}$, $U = \operatorname{diag}(U_1, U_2)$, *where* $U_1 : \operatorname{Im} Q_1 \to \operatorname{Im} P_1$ *is the unitary multiple of the polar decomposition of the operator* $P_1|_{\operatorname{Im} Q_1}$, $U_2 :$ $\operatorname{Im} Q_1 \to (\operatorname{Im} P_1)^\perp$ *is the unitary multiple of the polar decomposition of the operator* $(I - P_1)|_{\operatorname{Im} Q_1}$.

Remark 1.2.11. If some spaces $H_{i,j}$ are trivial, i.e., $H_{i,j} = \{0\}$, for the given projections, then the corresponding summand in the representation (1.2.3), (1.2.4) should be omitted.

Remark 1.2.12. The operator C is positive, and $0 \le C \le I$. Using the spectral theorem in (1.2.5), (1.2.6) we immediately obtain the representations of Theorem 1.2.8 for generic position orthogonal projections P_1 and Q_1,

$$P_1 \longmapsto \begin{pmatrix} 1 & 0 \\ 0 & 0 \end{pmatrix}, \tag{1.2.7}$$

$$Q_1 \longmapsto \begin{pmatrix} 1 - t & \sqrt{t(1-t)} \\ \sqrt{t(1-t)} & t \end{pmatrix}, \tag{1.2.8}$$

where $t \in \operatorname{sp} C = \operatorname{sp}(P_1 - Q_1)^2$.

The description of the algebra $\mathcal{R} = C^*(P, Q)$ with identity I, generated by the orthogonal projections P and Q, is given by Theorem 1.2.6. We describe now the inverse isomorphism

$$\nu^{-1} : \mathfrak{S} \longrightarrow \mathcal{R}.$$

Let

$$\sigma = \begin{cases} \sigma_1(i,j), & (i,j) \in Y \\ \begin{pmatrix} \sigma_2^{1,1}(t) & \sigma_2^{1,2}(t) \\ \sigma_2^{2,1}(t) & \sigma_2^{2,2}(t) \end{pmatrix}, & t \in \Delta \end{cases}$$

be an element of the algebra $\mathfrak{S}$.

For each pair (i,j) with $H_{i,j} \ne \{0\}$, we have either $(i,j) \in Y$, or $j \in \Delta$. Introduce

$$\sigma_0(i,j) = \begin{cases} \sigma_1(i,j), & \text{if } (i,j) \in Y \\ \sigma_2^{i+1,i+1}(j), & \text{if } j \in \Delta \end{cases}.$$

Then the operator $\nu^{-1}(\sigma)$ of the algebra $\mathcal{R} = C^*(P, Q)$ has the form

$$\nu^{-1}(\sigma) = \left(\bigoplus_{(i,j)} \sigma_0(i,j) I_{H_{i,j}} \right) \bigoplus U^* \begin{pmatrix} \sigma_2^{1,1}(C) & \sigma_2^{1,2}(C) \\ \sigma_2^{2,1}(C) & \sigma_2^{2,2}(C) \end{pmatrix} U,$$

corresponding to the splitting $H = H_0 \oplus H_1$. The operator C, as before, is given by $C = Q_1(I - P_1)Q_1|_{\mathrm{Im}\, Q_1}$.

1.2.3   More than two projections

The situation becomes unpredictable in the general case of a C^*-algebra generated by a finite set of orthogonal projections (and even in the case of just three projections!). First of all, this algebra can be irreducible.

Theorem 1.2.13 (Ch. Davis [55], 1955). *Let H be a separable Hilbert space. There exist three orthogonal projections P_1, P_2 and P_3 on it such that the von Neumann algebra $W^*(P_1, P_2, P_3)$ generated by them coincides with the algebra $\mathcal{L}(H)$ of all bounded linear operators acting on the space H.*

Theorem 1.2.14 (V. Sunder [194], 1988). *Let H be a separable Hilbert space. Then for each n there exist n orthogonal projections P_1, ..., P_n on it such that*

(i) $W^*(P_1, \ldots, P_n) = \mathcal{L}(H)$,

(ii) $W^*(S) \subset \mathcal{L}(H)$, *whenever* $S \subset \{P_1, \ldots, P_n\}$.

Consider now a special case, important for applications, when all the projections but one mutually commute, or (which is equivalent but with a bit more simple setting) when all but one of these projections are mutually orthogonal. Start with three (independent) projections.

Let a Hilbert space H be given in two different ways as a direct sum of subspaces

$$H = L \oplus L^\perp = M_1 \oplus M_2 \oplus M_3,$$

P is the orthogonal projection on H with image L, and Q_k are the orthogonal projection on H with images M_k, $k = \overline{1,3}$. Even in this specific case the algebra generated by the projections can be not only irreducible, but can be even a simple algebra.

Example 1.2.15 (The Choi algebra). (M.-D. Choi [40], 1979)
In a Hilbert space $H = H_0 \oplus H_1$ with $H_1 = H_\alpha \oplus H_\beta$ and $\dim H_0 = \dim H_1 = \dim H_\alpha = \dim H_\beta$ consider two *unitary* operators U and V such that

$$U : \begin{cases} H_0 \to H_1 \\ H_1 \to H_0 \end{cases}, \quad V : \begin{cases} H_0 \to H_\alpha \\ H_\alpha \to H_\beta \\ H_\beta \to H_0 \end{cases},$$

and $U^2 = I$, $V^3 = I$.

Introduce the orthogonal projections

$$P = 1/2(I + U)$$

and

$$Q_k = 1/3(I + e^{2\pi i(k-1)/3}V + e^{4\pi i(k-1)/3}V^2), \quad k = \overline{1,3}.$$

We have

$$Q_j \cdot Q_k = \delta_{j,k}Q_j, \quad Q_1 + Q_2 + Q_3 = I,$$

and the C^*-algebra $C^*(U, V)$ generated by U and V coincides with the C^*-algebra with identity $C^*(P, Q_1, Q_2)$ generated by the projections P, Q_1 and Q_2.

Theorem 1.2.16. *The algebra $C^*(U, V)$ (and thus $C^*(P, Q_1, Q_2)$) is a simple and algebraically unique C^*-algebra.*

Our aim is to shed light on the general situation in the above-mentioned specific case. In the next few subsections we will study the following problem:

Consider a Hilbert space H and a finite set of orthogonal projections P, Q_1, ..., Q_n on it, with the conditions

$$Q_j \cdot Q_k = \delta_{j,k}Q_k, \quad j,k = \overline{1,n}, \quad Q_1 + \ldots + Q_n = I,$$
$$\operatorname{Im} P \cap \ker Q_k = \{0\}, \quad \operatorname{Im} Q_k \cap \ker P = \{0\}, \quad k = \overline{1,n}.$$

Describe the C^*-algebra $\mathcal{R} = C^*(P, Q_1, \ldots, Q_n)$ generated by these projections.

It turns out that the structure of the algebra $\mathcal{R}$ is determined by the mutual properties of some n positive injective contractions C_k with $\sum_{k=1}^{n} C_k = I$, and thus by the structure of the C^*-algebra generated by them. The principal difference between the case of two projections and the general case of a finite set of projections is now completely understandable: for $n = 2$ we have only *one* contraction and the spectral theorem leads directly to the representation (1.2.7), (1.2.8); for $n \geq 2$ we have to study the C^*-algebra generated by a finite set of arbitrary non-commuting positive injective contractions.

The study of the algebra $\mathcal{R} = C^*(P, Q_1, \ldots, Q_n)$ is closely connected with the description of the C^* free product $\mathbb{C}^2 * \mathbb{C}^n$, and, after the change of the generators P, Q_1, ..., Q_n for $u = 2P - 1$, $v = \sum_{k=1}^{n} e^{2\pi(k-1)/n}Q_k$, with the description of the (full) C^*-algebra of the free product $\mathbb{Z}_2 * \mathbb{Z}_n$ of two cyclic groups of order 2 and n respectively. Note that the Choi algebra [40] is isomorphic to the reduced C^*-algebra of this group for $n = 3$ (see [179] for the reduced algebra for arbitrary finite n). For the case $n = 3$ one can think of the algebra $\mathcal{R}$ as the C^*-algebra with identity generated by two self-adjoint elements P and $Q = Q_1 - Q_2$ with the conditions

$$P^2 = P, \quad Q^3 = Q.$$

1.2.4 Canonical form of all-but-one orthogonal projections

Let a Hilbert space H be given in two different ways as a direct sum of subspaces

$$H = L \oplus L^\perp = M_1 \oplus \ldots \oplus M_n.$$

Denote by P the orthogonal projection on H with image L, and by Q_k the orthogonal projection on H with image M_k, $k = \overline{1, n}$. We have obviously

$$Q_j \cdot Q_k = \delta_{jk} Q_k, \quad j, k = \overline{1, n}, \tag{1.2.9}$$

$$I = Q_1 + \ldots + Q_n. \tag{1.2.10}$$

We will call the system P, Q_1, ..., Q_n of projections (self-adjoint idempotents in a C^*-algebra) *all-but-one* if they satisfy the above conditions (1.2.9), (1.2.10).

The subspace

$$\bigoplus_{k=1}^{n} (L \cap M_k) \bigoplus_{k=1}^{n} (L^\perp \cap M_k) \tag{1.2.11}$$

is invariant with respect to all projections, and both the restrictions of the projections onto (1.2.11) and the (commutative) C^*-algebra generated by these restrictions have the obvious structure. Thus we assume that the subspace (1.2.11) is trivial, and that

$$\dim L = \dim L^\perp = \dim M_1 = \ldots = \dim M_n = \infty.$$

Corresponding to the decomposition

$$H = \bigoplus_{k=1}^{n} M_k.$$

the projections P and Q_k, $k = \overline{1, n}$ have the matrix form

$$P = (Q_j P Q_k)_{j,k=1}^{n}, \tag{1.2.12}$$

$$Q_k = \operatorname{diag}(0, \ldots, 0, \underset{k\text{-place}}{I}, 0, \ldots, 0), \quad k = \overline{1, n}.$$

For each $k = \overline{1, n}$ introduce an isometry $U_k : L \to M_k$, and the operator $D_k = P Q_k U_k : L \to L$. Then $D_k^* = U_k^* Q_k P|_L : L \to L$. For the unitary operator

$$U = \operatorname{diag}(U_1, \ldots, U_n) : \ L^n \longrightarrow H = \bigoplus_{k=1}^{n} M_k,$$

the projections $U^* P U$ and $U^* Q_k U$, $k = \overline{1, n}$, acting on the space L^n have the matrix form

$$U^* P U = (D_j^* \cdot D_k)_{j,k=1}^{n},$$

$$U^* Q_k U = \operatorname{diag}(0, \ldots, 0, \underset{k\text{-place}}{I}, 0, \ldots, 0), \quad k = \overline{1, n},$$

Lemma 1.2.17. *The operators D_k, $k = \overline{1,n}$, are injective and*

$$\sum_{k=1}^{n} D_k D_k^* = I.$$

Proof. The first statement follows from $L^\perp \cap M_k = \{0\}$, $k = \overline{1,n}$. Further,

$$\sum_{k=1}^{n} D_k D_k^* = \sum_{k=1}^{n} P Q_k U_k \cdot U_k^* Q_k P|_L = P \left(\sum_{k=1}^{n} Q_k \right) P|_L = P|_L = I. \qquad \square$$

Theorem 1.2.18. *For an appropriate choice of the isometry U_k, the operator D_k can be*

(i) *normal,*

(ii) *self-adjoint,*

(iii) *positive,*

if and only if $\operatorname{Im} P \cap \ker Q_k = \{0\}$.

Proof. Let $\operatorname{Im} P \cap \ker Q_k \neq \{0\}$, it is sufficient to prove that D_k cannot be normal. We have $\ker D_k^* = \operatorname{Im} P \cap \ker Q_k \neq \{0\}$ and thus $\ker D_k^* \neq \ker D_k = \{0\}$, so the operator D_k can not be normal.

Conversely, let $\operatorname{Im} P \cap \ker Q_k = \{0\}$. We will prove that the operator D_k can be positive. It is easy to see that the operator $F_k = Q_k|_L : L \to M_k$ is injective and with the dense range. Thus it has the polar decomposition $F_k = V_k \widetilde{D}_k$ with a unitary operator $V_k : L \to M_k$ and a positive operator $\widetilde{D}_k : L \to L$. Put $U_k = V_k$, then $D_k = P Q_k V_k = F_k^* V_k = \widetilde{D}_k$. $\qquad \square$

The theorem motivates the following definition. We say that all-but-one projections P, Q_1, ..., Q_n are in *generic position* if they satisfy the conditions

$$\operatorname{Im} P \cap \ker Q_k = \{0\}, \tag{1.2.13}$$

$$\operatorname{Im} Q_k \cap \ker P = \{0\}, \quad k = \overline{1,n}. \tag{1.2.14}$$

If $n = 2$, i.e., we have only two (independent) projections P and $Q = Q_1$, the notion of generic position is equivalent to one of Halmos [98] and means exactly that the space (1.2.11) is trivial. Note that this is the only case when the triviality of the space (1.2.11) permits us to have all the D_k positive, for $n > 2$ our notion of generic position is stronger and implies in particular the triviality of the space (1.2.11).

Assuming that all-but-one projections P, Q_1, ..., Q_n are in generic position, select all the isometries U_k in such a form that all D_k are positive and introduce $C_k = D_k^2$, $k = \overline{1,n}$. Then all C_k are positive and injective,

$$\sum_{k=1}^{n} C_k = I,$$

and for the projections U^*PU, U^*Q_kU, $k = \overline{1,n}$, acting on the space L^n we have

$$U^*PU = (C_j^{1/2} \cdot C_k^{1/2})_{j,k=1}^n,$$

$$U^*Q_kU = \operatorname{diag}(0,\ldots,0, \underset{k\text{-place}}{I}, 0,\ldots,0), \quad k = \overline{1,n}.$$

The following theorem summarizes the above results.

Theorem 1.2.19. *Each set of generic position orthogonal projections P, Q_1, ..., Q_n on a Hilbert space H with the properties*

$$Q_j \cdot Q_k = \delta_{jk} \cdot Q_k, \quad j,k = \overline{1,n}, \tag{1.2.15}$$
$$I = Q_1 + \ldots + Q_n$$

admits the following canonical representation:

There exists a Hilbert space L ($= \operatorname{Im} P$), a unitary operator U from the Hilbert space $L^n = L \times \ldots \times L$ onto H, and n positive injective operators C_1, ..., C_n on L with $C_1 + \ldots + C_n = I$, such that the operators U^*PU and U^*Q_kU, $k = \overline{1,n}$, acting on L^n are, in matrix form, given as

$$U^*PU = (C_j^{1/2} \cdot C_k^{1/2})_{j,k=1}^n, \tag{1.2.16}$$

$$U^*Q_kU = \operatorname{diag}(0,\ldots,0, \underset{k\text{-place}}{I}, 0,\ldots,0), \quad k = \overline{1,n}.$$

Conversely, each set of n positive injective operators C_1, ..., C_n acting on a Hilbert space L, with $C_1 + \ldots + C_n = I$ and each unitary operator U from L^n onto a Hilbert space H define, by formulas (1.2.16), a set of generic position orthogonal projections P, Q_1, ..., Q_n on a Hilbert space H with the properties (1.2.15).

Theorem 1.2.20. *Let $\{P', Q_1', \ldots, Q_n'\}$ and $\{P'', Q_1'', \ldots, Q_n''\}$ be two systems of all-but-one generic position orthogonal projections on the Hilbert spaces H' and H'' respectively, and let $\{C_1', \ldots, C_n'\}$ and $\{C_1'', \ldots, C_n''\}$ be the corresponding systems of positive contractions. Then the systems of projections are unitary equivalent, i.e., there exists a unitary operator $W : H' \to H''$ such that $P' = W^*P''W$, $Q_k' = W^*Q_k''W$, $k = \overline{1,n}$, if and only if the systems of contractions are unitary equivalent, i.e., there exists a unitary operator $V : \operatorname{Im} P' \to \operatorname{Im} P''$ such that $C_k' = V^*C_k''V$, $k = \overline{1,n}$.*

Proof. The "if" part is obvious. Now assume that the system of projections are given in their canonical forms (1.2.16). Then the conditions $Q_k' = W^*Q_k''W$, $k = \overline{1,n}$, imply that $W = \operatorname{diag}(W_1, W_2, \ldots, W_n)$, where $W_k : \operatorname{Im} P' \to \operatorname{Im} P''$, $k = \overline{1,n}$, are unitary operators. The condition $P' = W^*P''W$ implies that

$$(C_j')^{1/2}(C_k')^{1/2} = W_j^*(C_j'')^{1/2}(C_k'')^{1/2}W_k, \quad j,k = \overline{1,n},$$

and, in particular, that $C_k' = W_k^*C_k''W_k$ and $(C_k')^{1/2} = W_k^*(C_k'')^{1/2}W_k$, $k = \overline{1,n}$. Thus

$$(C_j')^{1/2}(C_k')^{1/2} = W_j^*(C_j'')^{1/2}W_j\, W_j^*(C_k'')^{1/2}W_k = (C_j')^{1/2}\, W_j^*(C_k'')^{1/2}W_k,$$

and the injectivity of C_j' gives

$$(C_k')^{1/2} = W_j^*(C_k'')^{1/2}W_k.$$

In the same manner

$$(C_k')^{1/2} = ((C_k')^{1/2})^* = W_k^*(C_k'')^{1/2}W_j = W_k^*(C_k'')^{1/2}W_kW_k^*W_j = (C_k')^{1/2}W_k^*W_j$$

gives $W_k^*W_j = I$, or $W_1 = \ldots = W_n \ (= V)$. $\qquad\qquad\qquad\square$

1.2.5 C^*-algebra generated by all-but-one projections

The C^*-algebra $\mathcal{R} = C^*(P, Q_1, \ldots, Q_n)$ generated by the generic position orthogonal projections P and Q_k, $k = \overline{1,n}$, having the property (1.2.15) is naturally isomorphic to the algebra $U^*\mathcal{R}U$. We will identify these algebras and thus will assume that the C^*-algebra $\mathcal{R}$ is generated by the projections

$$P = (C_j^{1/2} \cdot C_k^{1/2})_{j,k=1}^n,$$

$$Q_k = \mathrm{diag}\,(0, \ldots, 0, \underset{k\text{-place}}{I}, 0, \ldots, 0), \quad k = \overline{1,n},$$

where C_k, $k = \overline{1,n}$ are positive injective contractions with the property $\sum_{k=1}^n C_k = I$.

The difficulties begin when $n \geq 3$. In the case of two orthogonal projections we have in fact only one independent contraction, $C = C_2$, then $C_1 = I - C$, and thus using the spectral theorem we come to the known representation of two projections in the generic position,

$$P = \begin{pmatrix} 1 - x & \sqrt{x(1-x)} \\ \sqrt{x(1-x)} & x \end{pmatrix},$$

$$Q = Q_1 = \begin{pmatrix} 1 & 0 \\ 0 & 0 \end{pmatrix},$$

where $x \in \mathrm{sp}\, P(I - Q)|_{\mathrm{Im}\,P} = \mathrm{sp}\,(P - Q)^2 \subset [0,1]$.

In the case $n \geq 3$ the description of the C^*-algebra $\mathcal{R}$ depends essentially on the mutual properties of the contractions C_k, and thus on the structure of the C^*-algebra $\mathcal{R}_C = C^*(C_1, \ldots, C_n)$ generated by them.

Lemma 1.2.21. *The set (algebra) of all elements of the form*

$$A = \mathrm{diag}\,(\alpha_1, \ldots, \alpha_n)I + (C_j^{1/2}A_{j,k}C_k^{1/2})_{j,k=1}^n, \qquad (1.2.17)$$

where $\alpha_k \in \mathbb{C}$, $A_{j,k} \in \mathcal{R}_C$, is dense in the algebra $\mathcal{R}$.

Proof. It is sufficient to prove that all elements of the form (1.2.17) where $A_{j,k}$ are non-commutative polynomials of the variables C_1, $\ldots$, C_{n-1} belong to $\mathcal{R}$. Denote for a moment by $[A]_{j,k}$ the element of $\mathcal{R}$ with the only non-zero entry j, k equal to A. Now the proof follows directly from the equalities

$$[\, C_j^{1/2} C_k^{1/2}\,]_{j,k} = Q_j P Q_k,$$

$$[\, C_j^{1/2} C_m C_k^{1/2}\,]_{j,k} = [\, C_j^{1/2} C_m^{1/2}\,]_{j,m} [\, C_m^{1/2} C_k^{1/2}\,]_{m,k},$$

$$[\, C_j^{1/2} A_1 A_2 C_k^{1/2}\,]_{j,k} = \sum_{m=1}^{n} [\, C_j^{1/2} A_1 C_m^{1/2}\,]_{j,m} [\, C_m^{1/2} A_2 C_k^{1/2}\,]_{m,k},$$

$$(C_j^{1/2} A_{j,k} C_k^{1/2})_{j,k=1}^{n} = \sum_{j,k=1}^{n} [\, C_j^{1/2} A_{j,k} C_k^{1/2}\,]_{j,k}.$$

$\square$

Corollary 1.2.22. *If all the operators C_k are invertible, then the algebra $\mathcal{R}$ is *-isomorphic to $\mathcal{R}_C \otimes \mathrm{Mat}_n(\mathbb{C})$.*

Corollary 1.2.23. *Let $\mathcal{A}$ be a finitely generated C^*-algebra with identity, and let n_0 be a minimal number of self-adjoint elements generating $\mathcal{A}$. Then for each $n > n_0$ there exists a continuous family of non-UNITARILY equivalent systems of all-but-one self-adjoint idempotents (generic position orthogonal projections) p, q_1, ..., q_n, such that for each of these systems we have*

$$\mathcal{A} \otimes \mathrm{Mat}_n(\mathbb{C}) \cong C^*(p, q_1, \ldots, q_n).$$

Proof. Denote by s_1, ..., s_{n_0} the self-adjoint elements generating $\mathcal{A}$, and fix any $n > n_0$. Now we need a family of systems of positive invertible elements c_1, ..., c_n of the algebra $\mathcal{A}$, such that

$$\sum_{k=1}^{n} c_k = e \quad \text{and} \quad C^*(c_1, \ldots, c_n) = \mathcal{A}.$$

Of course, there are many possibilities to construct such a family. For example one may take, for $\alpha \in [0, 1]$,

$$c_k = c_k(\alpha) = \frac{1}{4n_0}\left(e + \frac{1+\alpha}{\|s_k\|} s_k\right), \quad k = \overline{1, n_0},$$

$$c_k = c_k(\alpha) = \frac{1}{n - n_0}\left(e - \sum_{j=1}^{n_0} c_j\right), \quad k = \overline{n_0 + 1, n}.$$

Finally, the family of the following systems of self-adjoint idempotents (projections) satisfies all the necessary properties,

$$p = p(\alpha) = (c_j^{1/2}(\alpha) \cdot c_k^{1/2}(\alpha))_{j,k=1}^{n},$$

$$q_k = q_k(\alpha) = \mathrm{diag}\,(0, \ldots, 0, \underset{k\text{-place}}{e}, 0, \ldots, 0), \quad k = \overline{1, n},$$

where $\alpha \in [0, 1]$.

$\square$

Example 1.2.24 (Cuntz algebras are generated by all-but-one self-adjoint idempotents). Recall that the Cuntz algebra [51] $\mathcal{O}_n$ is a simple (and algebraically unique) C^*-algebra generated by n isometries $S_1, \ldots, S_n$ whose ranges are orthogonal complements:

$$S_1 S_1^* + \ldots + S_n S_n^* = I.$$

We have $\mathcal{O}_n \otimes \mathrm{Mat}_n(\mathbb{C}) \cong \mathcal{O}_n$ (see [40] for $n = 2$), and thus $\mathcal{O}_n \cong \mathcal{O}_n \otimes \mathrm{Mat}_{n^k}(\mathbb{C})$, $k \in \mathbb{N}$.

Example 1.2.25 (Family of all-but-one generic position orthogonal projections, generating non-isomorphic simple C^*-algebras). Apply Corollary 1.2.23 to the irrational rotation algebra [168] $\mathcal{A}_\alpha$, α is an irrational number. This algebra is simple and is generated by two unitary operators

$$(U\varphi)(t) = t\varphi(t) \quad \text{and} \quad (V\varphi)(t) = \varphi(e^{i2\pi\alpha}t),$$

acting on $L_2(\mathbb{T})$. By [168], for irrational $\alpha, \beta \in [0, 1/2]$ the algebras $\mathcal{A}_\alpha \otimes \mathrm{Mat}_n(\mathbb{C})$ and $\mathcal{A}_\beta \otimes \mathrm{Mat}_m(\mathbb{C})$ are isomorphic if and only if $\alpha = \beta$ and $m = n$.

Example 1.2.26 (Non type I algebras generated by all-but-one generic position orthogonal projections). Apply Corollary 1.2.23 to the Toeplitz operator algebras [52] $C^*(\Omega_{\delta_1, \delta_2})$, with irrational $\ln(\delta_2)/\ln(\delta_1)$. These algebras are non type I and are generated by multiplication operators $z_1 I$, $z_2 I$, acting on the Bergman space $\mathcal{A}^2(\Omega_{\delta_1, \delta_2}, \mu)$, where

$$\Omega_{\delta_1, \delta_2} = \{(z_1, z_2) \in \mathbb{C}^2 : |z_1| < \delta_1 \text{ and } |z_2| < 1, \text{ or } |z_1| < 1 \text{ and } |z_2| < \delta_2\}.$$

Theorem 1.2.27. *Each irreducible representation $\rho : \mathcal{R} \to \mathcal{L}(Y)$ of the algebra $\mathcal{R}$ has the following structure:*

There exist an irreducible representation $\pi : \mathcal{R}_C \to \mathcal{L}(X)$ of the algebra $\mathcal{R}_C$ and a set of n orthogonal projections $P^{(j)} : X \to Y_j$, $j = \overline{1, n}$, with the property

$$P^{(j)}\pi(C_j^{1/2} C_k^{1/2}) = \pi(C_j^{1/2} C_k^{1/2})P^{(k)}, \quad j, k = \overline{1, n}, \tag{1.2.18}$$

such that

$$Y = \bigoplus_{j=1}^{n} \mathrm{Im}\, P^{(j)} \subset X^n$$

and

$$\rho : (A_{j,k})_{j,k=1}^n \in \mathcal{R} \longmapsto (\pi(A_{j,k}))_{j,k=1}^n |_Y. \tag{1.2.19}$$

Proof. The algebra $\mathcal{R}$ is a C^*-subalgebra of $\mathcal{R}_C \otimes \mathrm{Mat}_n(\mathbb{C})$, each irreducible representation of which is of the form

$$\pi^\# = \pi \otimes id : \mathcal{R}_C \otimes \mathrm{Mat}_n(\mathbb{C}) \to \mathcal{L}(X^n)$$

where $\pi : \mathcal{R}_C \to \mathcal{L}(X)$ is an irreducible representation of $\mathcal{R}_C$ and id is the identical representation of $\mathrm{Mat}_n(\mathbb{C})$.

Thus [56] for each irreducible representation $\rho : \mathcal{R} \to \mathcal{L}(Y)$ of the algebra $\mathcal{R}$ there exist an irreducible representation $\pi : \mathcal{R}_C \to \mathcal{L}(X)$ and a subspace Y of the Hilbert space X^n such that

$$\rho : (A_{j,k})^n_{j,k=1} \in \mathcal{R} \longmapsto \pi^{\#}\big((A_{j,k})^n_{j,k=1}\big)$$
$$= (\pi(A_{j,k})^n)_{j,k=1}|_Y \in \mathcal{L}(Y).$$

The orthogonal projection $P_Y : X^n \to Y$ should commute with all elements of $\pi^{\#}(\mathcal{R})$. Commutation with $\pi^{\#}(Q_j)$, $j = \overline{1,n}$ leads to

$$P_Y = \bigoplus_{j=1}^{n} P^{(j)},$$

where $P^{(j)}$ is the orthogonal projections of X onto its subspace Y_j, $j = \overline{1,n}$. Thus

$$Y = \bigoplus_{j=1}^{n} \operatorname{Im} P^{(j)} = \bigoplus_{j=1}^{n} Y_j.$$

Finally, the commuting of P_Y with all the elements of the form $\pi^{\#}(Q_j P Q_k)$, $j, k = \overline{1,n}$ leads to

$$P^{(j)} \pi(C_j^{1/2} C_k^{1/2}) = \pi(C_j^{1/2} C_k^{1/2}) P^{(k)}. \qquad \square$$

Remark 1.2.28. The condition (1.2.18) is necessary and sufficient for the mapping (1.2.19) to be a representation.

1.2.6 Pairwise commuting contractions C_k

In this section we consider the special case when all contractions C_k, $k = \overline{1,n}$, are pairwise commuting, and thus the C^*-algebra $\mathcal{R}_C$ generated by them is commutative. This is the simplest and more direct generalization of the two-projection situation. This case is itself interesting and its understanding is important for the general case when the commutator ideal of the algebra $\mathcal{R}_C$ is proper.

Denote by $\Delta = \Delta(C_1, \ldots, C_n) \subset \mathbb{R}^n$ the joint spectrum of the operators $C_1, \ldots, C_n$. Then the C^*-algebra $\mathcal{R}_C$ is naturally isomorphic to $C(\Delta)$, and the coordinate function t_k, $k = \overline{1,n}$, $(t = (t_1, \ldots, t_n) \in \Delta \subset \mathbb{R}^n)$ is the image of C_k under this isomorphism.

Denote by $\mathfrak{S}(\Delta)$ the C^*-algebra generated by the following $n \times n$ matrix-valued functions continuous on Δ:

$$p(t) = (\sqrt{t_j}\sqrt{t_k})^n_{j,k=1}, \tag{1.2.20}$$

$$q_k(t) = \operatorname{diag}(0, \ldots, 0, \underset{k\text{-place}}{1}, 0, \ldots, 0), \tag{1.2.21}$$

where $t = (t_1, \ldots, t_n) \in \Delta$.

To describe the algebra $\mathfrak{S}(\Delta)$ we introduce first a sort of "maximal" algebra. Let

$$\Delta_{n-1} = \{t = (t_1, \ldots, t_n) \in \mathbb{R}^n : t_k \geq 0, \quad k = \overline{1, n}, \quad \sum_{k=1}^{n} t_k = 1\}$$

be the standard $(n-1)$ - dimensional simplex. Thus $\mathfrak{S}(\Delta_{n-1})$ is the C^*-algebra generated by the elements (1.2.20), (1.2.21) with $t = (t_1, \ldots, t_n) \in \Delta_{n-1}$.

Denote by $\mathfrak{S}_{max}$ the C^*-algebra of all $n \times n$ matrix-valued functions continuous on Δ_{n-1} and having the following properties:

(i) they are diagonal at all vertices $(1, 0, \ldots, 0)$, $(0, 1, \ldots, 0)$, $\ldots$, $(0, \ldots, 0, 1)$ of Δ_{n-1};

(ii) fix a k-tuple $[k] = (i_1, \ldots, i_k)$ of indices $1 \leq i_1 < i_2 < \ldots < i_k \leq n$, and let $[n]\backslash[k]$ be the complementary $(n-k)$-tuple of indices; then for each k-tuple, $1 \leq k \leq n-1$,

$$\sigma_{ij}(t) = \sigma_{ji}(t) \equiv 0 \quad \text{for} \quad i \in [k], \quad j \neq i$$

and $\sigma_{ii}(t) = \sigma_i = const$ on $(n-k-1)$-face $t_{i_1} = t_{i_2} = \ldots = t_{i_k} = 0$ of Δ_{n-1}.

Standard arguments lead directly to the following lemma.

Lemma 1.2.29. *All irreducible representations of the algebra $\mathfrak{S}_{max}$ are up to unitary equivalence as follows:*

(i) *the family of n-dimensional representations parameterized by the points $t \in \text{Int}\,\Delta_{n-1}$ and defined by the rule*

$$\pi_t : \sigma \in \mathfrak{S}(\Delta_{n-1}) \longmapsto \sigma(t) \in \text{Mat}_n(\mathbb{C});$$

(ii) *C_n^k families $\{\pi_t^{[k]}\}$ of $(n-k)$-dimensional representations, $k = \overline{1, n-1}$, parameterized by the interior parts of the $(n-k-1)$-face $t_{i_1} = t_{i_2} = \ldots = t_{i_k} = 0$, $i_j \in [k]$ of Δ_{n-1}; then representations (for the corresponding points t) are defined by the rule*

$$\pi_t^{[k]} : \sigma \in \mathfrak{S}(\Delta_{n-1}) \mapsto (\sigma_{i,j}(t))_{i,j \in [n]\backslash[k]} \in \text{Mat}_{n-k}(\mathbb{C});$$

(iii) *$2n$ one-dimensional representations defined as follows:*

$$\pi'_{(j)} : \sigma \in \mathfrak{S}(\Delta_{n-1}) \quad \longmapsto \quad \sigma_{jj}(0, \ldots, 0, \underset{j\text{-place}}{1}, 0, \ldots, 0) \in \mathbb{C}$$

$$\pi''_{(j)} : \sigma \in \mathfrak{S}(\Delta_{n-1}) \quad \longmapsto \quad \sigma_{jj}(\tau_1, \ldots, \tau_{j-1}, 0, \tau_{j+1}, \ldots, \tau_n) \in \mathbb{C},$$

where $t = (\tau_1, \ldots, \tau_{j-1}, 0, \tau_{j+1}, \ldots, \tau_n) \in \partial\Delta_{n-1}$.

Theorem 1.2.30. *The algebra $\mathfrak{S}(\Delta_{n-1})$ coincides with the C^*-algebra $\mathfrak{S}_{max}$.*

Proof. The algebra $\mathfrak{S}(\Delta_{n-1})$ is obviously a C^*-subalgebra of $\mathfrak{S}_{max}$. We will prove that $\mathfrak{S}(\Delta_{n-1})$ is a massive [56] subalgebra of $\mathfrak{S}_{max}$, and then by the Stone-Weierstrass theorem for GCR-algebras [56] we will have $\mathfrak{S}(\Delta_{n-1}) = \mathfrak{S}_{max}$. To prove that, we need to check two properties:

(i) for each irreducible representation π of $\mathfrak{S}_{max}$ the representation $\pi|_{\mathfrak{S}(\Delta_{n-1})}$ is irreducible;

(ii) for each two non-equivalent irreducible representations π_1 and π_2 of $\mathfrak{S}_{max}$ their restrictions $\pi_1|_{\mathfrak{S}(\Delta_{n-1})}$ and $\pi_2|_{\mathfrak{S}(\Delta_{n-1})}$ are non-equivalent as well.

We check these two statements for n-dimensional representations only. Fix a point $t_0 \in \Delta_{n-1}$ and consider the corresponding representation π_{t_0}. To prove the first statement it is sufficient to find for each matrix $m = (m_{kj})_{k,j=1}^n \in \mathrm{Mat}_n(\mathbb{C})$ an element $a \in \mathfrak{S}(\Delta_{n-1})$ such that

$$\pi_{t_0}(a) = m.$$

It is easy to see that we can take a to be of the form

$$a = \left(m_{kj} \frac{1}{(p(t_0))_{kj}} (q_k p(t_0) q_j)_{kj} \right)_{k,j=1}^n.$$

To prove the second statement it is sufficient to find for every two fixed distinct points $t' = (t'_1, \ldots, t'_n)$ and $t'' = (t''_1, \ldots, t''_n)$ of $\mathrm{Int}\,\Delta_{n-1}$ an element $a \in \mathfrak{S}(\Delta_{n-1})$ such that

$$\pi_{t'}(a) = 0, \quad \pi_{t''}(a) \neq 0.$$

The corresponding element, for example, is

$$a = q_1(p - t'_1 e)q_1 + \ldots + q_n(p - t'_n e)q_n. \qquad \square$$

Corollary 1.2.31. *The algebra $\mathfrak{S}(\Delta)$, where $\Delta = \Delta(C, \ldots, C_n)$, is the restriction of $\mathfrak{S}(\Delta_{n-1})$ onto the set Δ, i.e.,*

$$\mathfrak{S}(\Delta) = \{\sigma|_\Delta : \sigma \in \mathfrak{S}(\Delta_{n-1})\}.$$

Remark 1.2.32. The set $\Delta = \Delta(C_1, \ldots, C_n)$ can be an arbitrary compact subset of the simplex Δ_{n-1}. Some examples of such a set are given in Section 7.5.

In the sequel we will need the following variant of the above situation. Consider the all-but-one projections $P, Q_1, \ldots, Q_n$, which satisfy the only property (1.2.5) (the property (1.2.6) does not hold). Thus for some indices k we have

$$\mathrm{Im}\,Q_k \cap \ker P \neq \{0\}.$$

Denote by m ($0 < m \leq n$) the number of these indices, and reorder the projections Q_k in such a form that

$$\mathrm{Im}\,Q_k \cap \ker P \neq \{0\}, \quad k = \overline{1, m},$$
$$\mathrm{Im}\,Q_k \cap \ker P = \{0\}, \quad k = \overline{m+1, n}.$$

We still assume that the operators $C_k = PQ_kP : \operatorname{Im} P \to \operatorname{Im} P$, $k = \overline{1,n}$, are pairwise commuting, and still denote by $\Delta = \Delta(C_1, \ldots, C_n)$ their joint spectrum.

Theorem 1.2.33. *Under the above conditions the C^*-algebra $C^*(P, Q_1, \ldots, Q_n)$, generated by the orthogonal projections P, Q_1, ..., Q_n is isomorphic and isometric to a subalgebra of the algebra $\mathfrak{S}(\Delta) \oplus \mathbb{C}^m$. The isomorphic imbedding*

$$\nu : C^*(P, Q_1, \ldots, Q_n) \longrightarrow \mathfrak{S}(\Delta) \oplus \mathbb{C}^m$$

is generated by the mapping of the projections

$$\begin{aligned}
\nu : P &\longmapsto (p(t), (0, 0, \ldots, 0)), \\
\nu : Q_k &\longmapsto (q_k(t), (0, \ldots, 0, \underset{k\text{-place}}{1}, 0, \ldots, 0)), \quad k = \overline{1, m}, \\
\nu : Q_k &\longmapsto (q_k(t), (0, 0, \ldots, 0)), \quad k = \overline{m+1, n},
\end{aligned}$$

where $p(t)$ and $q_k(t)$ are defined in (1.2.20) and (1.2.21), $t = (t_1, \ldots, t_n) \in \Delta$.

Proof. Split the space H into the direct sum

$$H = \bigoplus_{k=1}^{m} (\operatorname{Im} Q_k \cap \ker P) \oplus H', \tag{1.2.22}$$

and introduce the projections

$$P' = P|_{H'}, \qquad Q_k' = Q_k|_{H'}, \qquad k = \overline{1, n}.$$

Corresponding to the splitting (1.2.22) the initial projections have the form

$$\begin{aligned}
P &= (0 \oplus 0 \oplus \ldots \oplus 0) \oplus P', \\
Q_k &= (0 \oplus \ldots 0 \oplus \underset{k\text{-place}}{1} \oplus 0 \oplus \ldots \oplus 0) \oplus Q_k', \quad k = \overline{1, m}, \\
Q_k &= (0 \oplus 0 \oplus \ldots \oplus 0) \oplus Q_k', \quad k = \overline{m+1, n}.
\end{aligned}$$

The all-but-one projections P', Q_1', ..., Q_n' are in generic position, and the C^*-algebra generated by them is described in Corollary 1.2.31. Finally, note that $\operatorname{Im} P = \operatorname{Im} P'$, and $P'Q_k'P' = PQ_kP$ on this space. $\qquad\square$

Chapter 2

Prologue

2.1 On the term "symbol"

As it unfortunately happens in mathematics, some terms carry different meanings depending on the context in which they are used. The term "symbol" is not exceptional in this sense. It will be used systematically in the book and will be supplied with different adjectives clarifying its different meanings: Fredholm symbol, defining symbol, Wick symbol, anti-Wick symbol, etc. That is why we would like to comment first on its meanings and usage.

To the best of the author's knowledge, the term "symbol" was introduced by S. Mikhlin in the 1930s in the theory of multidimensional singular integral operators (SIO). Being a continuous function, it already carried two different semantic meanings: a function which is responsible for the Fredholm properties of a SIO, and a function which defines the SIO in terms of the Fourier transform. The situation was repeated in the 1950s–1960s in the study of Toeplitz operators defined by continuous functions and acting on the Hardy space. This function, called "symbol", on the one hand defines the operator itself and, on the other hand, carries all the information about its Fredholm properties. In that period an intensive study of the different operator algebras and the Fredholm properties of their elements started. The terms "symbol algebra" and "symbol of an operator" were used mainly for the quotient algebra of the operator algebra under study, modulo the ideal of the compact operators (the Calkin algebra) and for the image of an operator on the Calkin algebra, respectively. Consequently the meaning of the term "symbol" shifted to the objects responsible for the Fredholm properties. The results obtained showed that for more general classes of functions defining the operators ("symbols" in their second meaning) the objects which correspond to different meanings of the term "symbol" become quite different. We mention here, for example, the pioneering works of I. Gohberg and N. Krupnik on the algebra generated by one-dimensional singular integral operators with piece-wise continuous coefficients and on the algebra generated by Toeplitz operators on the Hardy space with piece-wise continuous defining functions of the late 1960s. We mention as well that, as we will see later in the book (and this is also an old result of the 1970s), even for Toeplitz operators with continuous defining functions and acting on the Bergman space, the symbol (as a defining function) is a continuous function on a closed domain, while the symbol (as an object responsible for the Fredholm properties) is the restriction of the defining function to the boundary of the domain.

At the same time in the theory of Toeplitz operators both on the Hardy and on the Bergman space the term "symbol" in its meaning of a function which defines the operator survived in the literature, and its usage in this sense became quite traditional in the last two decades.

Trying to preserve an existing reality and tradition, in this book we will follow the next rule. The quotient algebra of the operator algebra under study modulo the ideal of the compact operators will be called the *Fredholm symbol algebra*. The image of an operator on the Fredholm symbol algebra will be called the *Fredholm symbol* of an operator. A function which defines the Toeplitz operator will be called the *defining symbol* of a Toeplitz operator.

The terms *Wick symbol* and *anti-Wick symbol* were introduced by F. Berezin in the 1970s, and in this book we give their exact definitions.

Sometimes we will omit the corresponding adjectives when the usage of the term "symbol" is clear from the context.

2.2   Bergman space and Bergman projection

Let D be a bounded simply connected domain in the complex plane $\mathbb{C}$ with a smooth boundary γ. Consider the space $L_2(D)$ with respect to the standard Lebesgue plane measure $dv(z) = dxdy$, $z = x + iy$, and its subspace $\mathcal{A}^2(D)$ consisting of analytic functions in D, which is called the *Bergman space over the domain D*.

An important property of the Bergman space is contained in the following statement.

Statement 2.2.1. *Given a compact set $K \subset D$, there is a constant C_K, depending on K, such that*

$$\sup_{z \in K} |f(z)| \leq C_K \|f\|_{\mathcal{A}^2(D)}, \tag{2.2.1}$$

for all $f \in \mathcal{A}^2(D)$.

Proof. Denote by $D_{z,r}$ the open disk centered at z and having radius r. For any fixed $f \in \mathcal{A}^2(D)$ and any $D_{z,r} \subset D$ we consider the Taylor series

$$f(\zeta) = f(z) + \sum_{n=1}^{\infty} a_n(\zeta - z)^n, \qquad \zeta \in D_{z,r}.$$

This series converges uniformly on each closed disk $\overline{D}_0 = \overline{D}_{z,r_0}$, where $r_0 < r$. Thus

$$\int_{\overline{D}_0} f(\zeta)dv(\zeta) = f(z)\int_{\overline{D}_0} dv(\zeta) + \sum_{n=1}^{\infty} a_n \int_{\overline{D}_0} (\zeta - z)^n dv(\zeta) = \pi r_0^2 f(z).$$

Then

$$|f(z)| = \frac{1}{\pi r_0^2} \left| \int_D f(\zeta)\chi_{\overline{D}_0}(\zeta)dv(\zeta) \right|$$

$$\leq \frac{1}{\pi r_0^2} \left(\int_D |f(\zeta)|^2 dv(\zeta) \right)^{1/2} \left(\int_{\overline{D}_0} dv(\zeta) \right)^{1/2}$$

$$= \frac{1}{\sqrt{\pi} r_0} \|f\|_{\mathcal{A}^2(D)},$$

where $\chi_{\overline{D}_0}$ is the characteristic function of $\overline{D}_0$.

To finish the proof we cover the compact set K by a finite number of such disks. $\qquad\square$

Corollary 2.2.2. *The Bergman space $\mathcal{A}^2(D)$ is a closed subspace of $L_2(D)$.*

Proof. Let $\{f_n\}$ be a fundamental sequence of analytic functions from $\mathcal{A}^2(D)$ converging on $L_2(D)$ to a certain function $f \in L_2(D)$. For any compact $K \subset D$ by (2.2.1) we have

$$|f_n(z) - f_m(z)| \leq C_K \|f_n - f_m\|_{\mathcal{A}^2(D)}.$$

Thus the sequence $\{f_n\}$ converges uniformly on every compact subset of D to the function f. Hence f is analytic in D and belongs to $\mathcal{A}^2(D)$. $\qquad\square$

From the statement it follows as well that for any fixed point $z \in D$ the evaluation functional $\varphi_z : f \longmapsto f(z)$ is linear and bounded. Thus by the Riesz representation theorem there exists a unique element $k_z \in \mathcal{A}^2(D)$ such that $\varphi_z = \langle \cdot, k_z \rangle$; that is

$$f(z) = \int_D f(\zeta)\,\overline{k_z(\zeta)}\,dv(\zeta). \tag{2.2.2}$$

The function $K_D(z,\zeta) = \overline{k_z(\zeta)}$ is called the *Bergman kernel function* of the domain D, and it has the following reproducing property:

$$f(z) = \int_D K_D(z,\zeta)\,f(\zeta)\,dv(\zeta), \tag{2.2.3}$$

for all $f(z) \in \mathcal{A}^2(D)$.

We note that from the above definition it follows that the function $K_D(z,\zeta)$ is analytic in z and anti-analytic in ζ (analytic in $\overline{\zeta}$).

Lemma 2.2.3. *The Bergman kernel function is hermitian symmetric*

$$K_D(z,\zeta) = \overline{K_D(\zeta,z)}.$$

Let $\{\varphi_k\}_{k\in\mathbb{N}}$ be any orthonormal basis in $\mathcal{A}^2(D)$. Then

$$K_D(z,\zeta) = \sum_{k=1}^{\infty} \varphi_k(z)\overline{\varphi_k(\zeta)}. \tag{2.2.4}$$

Proof. The function $k_\zeta(z) = \overline{K_D(\zeta,z)}$ belongs to $\mathcal{A}^2(D)$, thus by (2.2.3) we have

$$\overline{K_D(\zeta,z)} = \int_D K_D(z,\tau)\,\overline{K_D(\zeta,\tau)}\,dv(\tau)$$

$$= \overline{\int_D K_D(\zeta,\tau)\,\overline{K_D(z,\tau)}\,dv(\tau)}$$

$$= \overline{\overline{K_D(z,\zeta)}} = K_D(z,\zeta).$$

Let now $\{\varphi_k\}$ be an orthonormal basis in $\mathcal{A}^2(D)$. Then the analytic function $k_\zeta(z) = \overline{K_D(\zeta,z)} = K_D(z,\zeta)$ admits the expansion

$$K_D(z,\zeta) = \sum_{k=1}^{\infty} c_k(\zeta)\,\varphi_k(z),$$

where the coefficients $c_k(\zeta)$ are given by

$$c_k(\zeta) = \langle k_\zeta(z), \varphi_k(z)\rangle = \int_D K_D(z,\zeta)\,\overline{\varphi_k(z)}\,dv(z)$$

$$= \overline{\int_D K_D(\zeta,z)\,\varphi_k(z)\,dv(z)} = \overline{\varphi_k(\zeta)}.$$

That is,

$$K_D(z,\zeta) = \sum_{k=1}^{\infty} \varphi_k(z)\overline{\varphi_k(\zeta)}. \qquad \square$$

Example 2.2.4 (Kernel function for the unit disk). It is well known (and easy to check) that the system of functions $\varphi_k(z) = \sqrt{\frac{k}{\pi}}\, z^{k-1}$, $k \in \mathbb{N}$, forms an orthonormal basis in $\mathcal{A}^2(\mathbb{D})$, where $\mathbb{D}$ is the unit disk. Therefore

$$K_{\mathbb{D}}(z,\zeta) = \sum_{k=1}^{\infty} \frac{k}{\pi}\,(z\overline{\zeta})^{k-1} = \frac{1}{\pi(1 - z\overline{\zeta})^2}.$$

The explicit formula for the Bergman kernel function is known for a very few domains. Apart from the formula (2.2.4) the next lemma gives additional information about the Bergman kernel function.

Lemma 2.2.5. *Let the function $w = \alpha(z)$ be a biholomorphic mapping of the domain D onto a domain G. Then*

$$K_D(z,\zeta) = K_G(w,\omega) \cdot \alpha'(z) \cdot \overline{\alpha'(\zeta)}. \tag{2.2.5}$$

Proof. Assume that the system $\{f_k(w)\}_{k\in\mathbb{N}}$ is an orthonormal basis in the space $\mathcal{A}^2(G)$. Then the equality

$$\delta_{j,k} = \int_G f_j(w)\,\overline{f_k(w)}\,dv(w) = \int_D f_j(\alpha(z))\,\overline{f_k(\alpha(z))}\,\alpha'(z)\,\overline{\alpha'(z)}\,dv(z)$$

shows that the system $\varphi_k(z) = f_k(\alpha(z))\,\alpha'(z)$, $k \in \mathbb{N}$, is an orthonormal basis in $\mathcal{A}^2(D)$. Thus

$$K_D(z,\zeta) = \sum_{k=1}^{\infty} \varphi_k(z)\,\overline{\varphi_k(\zeta)} = \sum_{k=1}^{\infty} f_k(\alpha(z))\alpha'(z)\,\overline{f_k(\alpha(\zeta))\alpha'(\zeta)}$$

$$= \sum_{k=1}^{\infty} f_k(\alpha(z))\,\overline{f_k(\alpha(\zeta))} \cdot \alpha'(z) \cdot \overline{\alpha'(\zeta)} = K_G(w,\omega) \cdot \alpha'(z) \cdot \overline{\alpha'(\zeta)}.$$

$\square$

Example 2.2.6 (Kernel function for the upper half-plane). The Möbius transformation

$$z = \frac{w - i}{1 - iw} \tag{2.2.6}$$

maps the upper half-plane Π onto the unit disk $\mathbb{D}$. Thus the Bergman kernel function for the upper half-plane has the form

$$K_\Pi(w,\omega) = K_\mathbb{D}(z,\zeta) \cdot \alpha'(w) \cdot \overline{\alpha'(\omega)}$$

$$= \frac{1}{\pi\left(1 - \frac{w-i}{1-iw}\,\frac{\overline{w}+i}{1+i\overline{w}}\right)^2} \cdot \frac{2}{(1-iw)^2} \cdot \frac{2}{(1+i\overline{w})^2}$$

$$= -\frac{1}{\pi\,(w-\overline{w})^2}.$$

The Bergman space $\mathcal{A}^2(D)$ is a closed subspace of the Hilbert space $L_2(D)$. Thus there exists the orthogonal projection B_D from $L_2(D)$ onto $\mathcal{A}^2(D)$. This projection is called the *Bergman projection*, and, as the following theorem shows, coincides with the integral operator with a Bergman kernel.

Theorem 2.2.7. *The Bergman projection B_D has the integral representation*

$$(B_D f)(z) = \int_D K_D(z,\zeta) f(\zeta) dv(\zeta). \tag{2.2.7}$$

Proof. For any $f \in L_2(D)$ the function $(B_D f)(z)$ belongs to $\mathcal{A}^2(D)$; thus, using (2.2.2), we have

$$(B_D f)(z) = \langle B_D f, k_z \rangle = \langle f, B_D k_z \rangle = \langle f, k_z \rangle$$
$$= \int_D f(\zeta)\overline{k_z(\zeta)}dv(\zeta) = \int_D K_D(z,\zeta)f(\zeta)dv(\zeta). \qquad \square$$

Let again the function $\omega = \alpha(z)$ be a biholomorphic mapping of the domain D onto a domain G, and let $z = \beta(\omega)$ be the inverse mapping. Introduce the unitary operator $W : L_2(G) \to L_2(D)$ by the rule

$$(W\varphi)(z) = \alpha'(z)\varphi[\alpha(z)];$$

its inverse (and adjoint) operator $W^{-1} = W^* : L_2(D) \to L_2(G)$ is given by

$$(W^{-1}\varphi)(\omega) = \beta'(\omega)\varphi[\beta(\omega)].$$

Lemma 2.2.8. *We have $B_D = W B_G W^{-1}$.*

Proof. Using the equality $\beta'(\alpha(\zeta))\,\alpha'(\zeta) = 1$ and the property (2.2.5) of the Bergman kernel function, we have

$$(W B_G W^{-1}\varphi)(z) = \alpha'(z) \int_G K_G(\alpha(z),\omega)\,\beta'(\omega)\varphi(\beta(\omega))\,dv(\omega)$$
$$= \int_D K_G(\alpha(z),\alpha(\zeta))\,\alpha'(z)\,\beta'(\alpha(\zeta))\,\varphi(\zeta)\,\alpha'(\zeta))\,\overline{\alpha'(\zeta))}\,dv(\zeta)$$
$$= \int_D K_G(\alpha(z),\alpha(\zeta))\,\alpha'(z)\,\overline{\alpha'(\zeta))}\,\varphi(\zeta)\,dv(\zeta)$$
$$= \int_D K_D(z,\zeta)\,\varphi(\zeta)\,dv(\zeta) = (B_D\varphi)(z). \qquad \square$$

2.3   Representations of the Bergman kernel function

We start with recalling the definition (see, for example, [100]) of the classical (harmonic) Green function. Let D a bounded connected domain with a smooth boundary. Denote by $h(z,\zeta)$ the (unique) harmonic function of z in D, for each $\zeta \in D$, and having on ∂D the boundary values

$$h(z,\zeta) = \ln|z - \zeta|, \qquad z \in \partial D,$$

for each $\zeta \in D$. Then the Green function $g(z,\zeta)$ of the domain D is defined in $D \times D \setminus diag$, where $diag = \{(z,z) : z \in D\}$, by

$$g(z,\zeta) = \ln \frac{1}{|z - \zeta|} + h(z,\zeta), \qquad (2.3.1)$$

and has the following properties:

(i) $g(z, \zeta)$ is harmonic in $D \times D \setminus diag$,

(ii) $g(z, \zeta)$ is continues in $\overline{D} \times \overline{D} \setminus diag$,

(iii) $g(z, \zeta) = 0$ for $(z, \zeta) \in \partial D \times D$,

(iv) $g(z, \zeta) = g(\zeta, z)$ for $(z, \zeta) \in D \times D \setminus diag$.

Example 2.3.1. The Green function for the unit disk $\mathbb{D}$ is given by (see, for example, [64, 100])

$$g(z, \zeta) = \ln \frac{1}{|z - \zeta|} + \ln |1 - z\overline{\zeta}| = \ln \left| \frac{1 - z\overline{\zeta}}{z - \zeta} \right|.$$

We mention as well the following two formulas:
the two-dimensional Stokes (Cauchy-Green) formula (see, for example, [64, 219])

$$\int_D \frac{\partial f}{\partial \overline{\zeta}} \, dv(\zeta) = \frac{1}{2i} \int_{\partial D} f(\zeta) d\zeta, \tag{2.3.2}$$

and the Pompeiu formula (see, for example, [64, 219])

$$f(z) = \frac{1}{2\pi i} \int_{\partial D} \frac{f(\zeta)}{\zeta - z} \, d\zeta - \frac{1}{\pi} \int_D \frac{\partial f}{\partial \overline{\zeta}} \frac{1}{\zeta - z} \, dv(\zeta), \tag{2.3.3}$$

where, say, $f \in C^1(D) \cap C(\overline{D})$ for both formulas.

Theorem 2.3.2. *Let $g(z, \zeta)$ be the Green function of a domain D, then*

$$K_D(z, \zeta) = -\frac{2}{\pi} \cdot \frac{\partial^2 g(z, \zeta)}{\partial z \, \partial \overline{\zeta}}. \tag{2.3.4}$$

Proof. Before starting the proof itself we make several remarks. Denote by $k_D(z, \zeta)$ the right-hand side of (2.3.4), i.e.,

$$k_D(z, \zeta) = -\frac{2}{\pi} \cdot \frac{\partial^2 g(z, \zeta)}{\partial z \, \partial \overline{\zeta}}.$$

From

$$\begin{aligned}
k_D(z, \zeta) &= -\frac{2}{\pi} \left(\frac{\partial^2}{\partial z \, \partial \overline{\zeta}} \ln \frac{1}{|z - \zeta|} + \frac{\partial^2 h(z, \zeta)}{\partial z \, \partial \overline{\zeta}} \right) \\
&= -\frac{2}{\pi} \cdot \frac{\partial^2 h(z, \zeta)}{\partial z \, \partial \overline{\zeta}}
\end{aligned}$$

it follows that the function $k_D(z, \zeta)$ is well defined in $D \times D$. Next, from $\frac{\partial}{\partial z} \frac{\partial}{\partial \overline{z}} = \frac{1}{4} \Delta$ it follows that the function $k_D(z, \zeta)$ is analytic in z and anti-analytic in ζ. Finally, from the symmetry of the Green function it follows that $k_D(z, \zeta)$ is hermitian symmetric,

$$k_D(\zeta, z) = \overline{k_D(z, \zeta)}.$$

Denote by $D_{z,\varepsilon}$ the domain D with the disk $\{\zeta \in D : \ |\zeta - z| \leq \varepsilon\}$ excluded. Then for any φ analytic in D and continuous on $\overline{D}$, by (2.3.2) applied to $f(\zeta) = \frac{\partial g(z,\zeta)}{\partial z}\varphi(\zeta)$, we have

$$\int_{D_{z,\varepsilon}} k_D(z,\zeta)\varphi(\zeta)dv(\zeta) = -\frac{1}{\pi i}\int_{\partial D}\frac{\partial g(z,\zeta)}{\partial z}\varphi(\zeta)d\zeta + \frac{1}{\pi i}\int_{|\zeta-z|=\varepsilon}\frac{\partial g(z,\zeta)}{\partial z}\varphi(\zeta)d\zeta.$$

The first integral on the right-hand side is equal to zero by properties (iii) and (iv) of the Green function. Calculating the second integral we have

$$\int_{D_{z,\varepsilon}} k_D(z,\zeta)\varphi(\zeta)dv(\zeta) = \frac{1}{2\pi i}\int_{|\zeta-z|=\varepsilon}\left(\frac{1}{\zeta-z} + 2\frac{\partial h}{\partial z}\right)\varphi(\zeta)d\zeta = \varphi(z) + O(\varepsilon).$$

Passing to the limit when $\varepsilon \to 0$ we have

$$\int_D k_D(z,\zeta)\varphi(\zeta)dv(\zeta) = \varphi(z),$$

for all φ analytic in D and continuous on $\overline{D}$. That is, like the Bergman kernel function the function $k_D(z,\zeta)$ has the reproducing property, and thus by the uniqueness of the reproducing property kernel coincides with the $K_D(z,\zeta)$. $\quad\Box$

We note that in spite of (2.3.4) the explicit formula for the Bergman kernel function is known only for a few domains. Our next aim is to get an integral representation of the Bergman kernel function for an arbitrary bounded domain with a smooth boundary.

Following [26] and [45], Appendix by M. Schiffer, we introduce the function

$$L_D(z,\zeta) = -\frac{2}{\pi}\frac{\partial^2 g(z,\zeta)}{\partial z\partial\zeta},$$

being a combination of derivatives of the Green function different from $K_D(z,\zeta)$, see (2.3.4).

From (2.3.1) it follows that

$$L_D(z,\zeta) = \frac{1}{\pi(z-\zeta)^2} - l_D(z,\zeta), \tag{2.3.5}$$

where the function

$$l_D(z,\zeta) = l_D(\zeta,z) = \frac{2}{\pi}\frac{\partial^2 h(z,\zeta)}{\partial z\partial\zeta}$$

is analytic on both z and ζ in D.

We note that for the case of the unit disk $\mathbb{D}$, by Example 2.3.1 we have

$$L_{\mathbb{D}}(z,\zeta) = \frac{1}{\pi(z-\zeta)^2}, \qquad l_{\mathbb{D}}(z,\zeta) \equiv 0. \tag{2.3.6}$$

Lemma 2.3.3. *For each function $\varphi \in \mathcal{A}^2(D)$, we have*

$$\int_D \varphi(\zeta)\overline{L_D(z,\zeta)}\, dv(\zeta) = 0$$

for all $z \in D$.

Proof. As in the proof of Theorem 2.3.2, denote by $D_{z,\varepsilon}$ the domain D with the disk $\{\zeta \in D : |\zeta - z| \leq \varepsilon\}$ excluded. Then by the Cauchy-Green formula (2.3.2), we have

$$\int_{D_{z,\varepsilon}} \varphi(\zeta)\overline{L_D(z,\zeta)}\, dv(\zeta) = -\frac{2}{\pi}\int_{D_{z,\varepsilon}} \varphi(\zeta)\frac{\partial^2 g(z,\zeta)}{\partial \overline{z}\partial\overline{\zeta}}\, dv(\zeta)$$

$$= -\frac{1}{\pi i}\int_{\partial D} \varphi(\zeta)\frac{\partial g}{\partial \overline{z}}\, d\zeta + \frac{1}{\pi i}\int_{|\zeta-z|=\varepsilon} \varphi(\zeta)\frac{\partial g}{\partial \overline{z}}\, d\zeta.$$

By properties (iv) and (iii) of the Green function $g(z,\zeta)$ the first integral in the last line is equal to zero, while

$$\frac{1}{\pi i}\int_{|\zeta-z|=\varepsilon} \varphi(\zeta)\frac{\partial g}{\partial \overline{z}}\, d\zeta = \frac{1}{2\pi i}\int_{|\zeta-z|=\varepsilon} \varphi(\zeta)\left(\frac{1}{\overline{\zeta}-\overline{z}} + 2\frac{\partial h}{\partial \overline{z}}\right) d\zeta$$

$$= \frac{1}{2\pi i}\int_{|\zeta-z|=\varepsilon} \varphi(\zeta)\left(\frac{\zeta - z}{\varepsilon^2} + 2\frac{\partial h}{\partial \overline{z}}\right) d\zeta = O(\varepsilon).$$

The result follows under the limit when $\varepsilon \to 0$. $\qquad\square$

By the above lemma and representation (2.3.5), we have

$$\frac{1}{\pi}\int_D \frac{\varphi(\zeta)}{(\overline{z} - \overline{\zeta})^2}\, dv(\zeta) = \int_D \varphi(\zeta)\overline{l_D(z,\zeta)}dv(\zeta),$$

for all $\varphi \in \mathcal{A}^2(D)$. In particular, for $\varphi(\zeta) = l_D(\zeta,\eta)$, we have

$$\frac{1}{\pi}\int_D \frac{l_D(\zeta,\eta)}{(\overline{z} - \overline{\zeta})^2}\, dv(\zeta) = \int_D l_D(\zeta,\eta)\overline{l_D(z,\zeta)}dv(\zeta). \tag{2.3.7}$$

Theorem 2.3.4. *The Bergman kernel function admits the integral representation*

$$K_D(z,\zeta) = \frac{1}{2\pi^2 i}\int_{\partial D} \frac{dt}{(t - z)^2(\overline{t} - \overline{\zeta})} + l_1(z,\zeta), \tag{2.3.8}$$

where the function $l_1(z,\zeta)$ is analytic on z and $\overline{\zeta}$ in $D \times D$, and is continuous on $\overline{D} \times \overline{D}$.

Proof. By properties (iv) and (iii) of the Green function we have that

$$\frac{\partial g(z,\zeta)}{\partial z} = 0$$

for all $\zeta \in \partial D$ and all $z \in D$. Differentiation of this equality by the length parameter s gives

$$\frac{\partial^2 g(z,\zeta)}{\partial z \partial \zeta} \zeta'(s) + \frac{\partial^2 g(z,\zeta)}{\partial z \partial \bar{\zeta}} \overline{\zeta'(s)} \equiv 0,$$

which is equivalent to

$$L_D(z,\zeta)\zeta'(s) + K_D(z,\zeta)\overline{\zeta'(s)} \equiv 0,$$

or to

$$L_D(z,\zeta)\zeta'(s) = -\overline{K_D(\zeta,z)\zeta'(s)}.$$

By the Cauchy integral formula we have

$$\frac{1}{2\pi i} \int_{\partial D} \frac{L_D(z,\zeta)}{\bar{\zeta} - \bar{\eta}} d\zeta = -\frac{1}{2\pi i} \int_{\partial D} \frac{\overline{K_D(\zeta,z)\zeta'(s)}}{\bar{\zeta} - \bar{\eta}} ds$$

$$= \frac{1}{2\pi i} \int_{\partial D} \frac{\overline{K_D(\zeta,z)}}{\zeta - \eta} d\zeta = \overline{K_D(\eta,z)} = K_D(z,\eta),$$

and thus, using (2.3.5) and changing letters for variables,

$$K_D(z,\zeta) = \frac{1}{2\pi i} \int_{\partial D} \frac{L_D(z,t)}{\bar{t} - \bar{\zeta}} dt$$

$$= \frac{1}{2\pi^2 i} \int_{\partial D} \frac{dt}{(t-z)^2(\bar{t}-\bar{\zeta})} - \frac{1}{2\pi i} \int_{\partial D} \frac{l_D(z,t)}{\bar{t} - \bar{\zeta}} dt.$$

Now, using the Cauchy-Green formula (2.3.2) and applying the above type arguments to $D_{z,\varepsilon}$, we get

$$\frac{1}{\pi} \int_{D_{\eta,\varepsilon}} \frac{l_D(t,z)}{(\bar{t} - \bar{\zeta})^2} dv(t) = -\frac{1}{2\pi i} \int_{\partial D} \frac{l_D(t,z)}{\bar{t} - \bar{\zeta}} dt + O(\varepsilon).$$

Finally, passing to the limit when $\varepsilon \to 0$ and using (2.3.7), we have

$$K_D(z,\zeta) = \frac{1}{2\pi^2 i} \int_{\partial D} \frac{dt}{(t-z)^2(\bar{t} - \bar{\zeta})} + l_1(z,\zeta),$$

where

$$l_1(z,\zeta) = \int_D l_D(t,z)\overline{l_D(t,\zeta)}dv(t). \qquad \square$$

2.4   Some integral operators and representation of the Bergman projection

Let D be a bounded connected domain with a smooth boundary. We introduce the following integral operators bounded on $L_2(D)$

$$(T_D\varphi)(z) = -\frac{1}{\pi} \int_D \frac{\varphi(\zeta)}{\zeta - z} dv(\zeta)$$

and

$$(S_D \varphi)(z) = -\frac{1}{\pi} \int_D \frac{\varphi(\zeta)}{(\zeta - z)^2} \, dv(\zeta), \tag{2.4.1}$$

where the last integral is understood in the principal value sense. Their adjoint operators are given respectively by

$$(T_D^* \varphi)(z) = -\frac{1}{\pi} \int_D \frac{\varphi(\zeta)}{\overline{\zeta} - \overline{z}} \, dv(\zeta)$$

and

$$(S_D^* \varphi)(z) = -\frac{1}{\pi} \int_D \frac{\varphi(\zeta)}{(\overline{\zeta} - \overline{z})^2} \, dv(\zeta). \tag{2.4.2}$$

Note that the operators S_D and S_D^* are the restrictions to the domain D of the following classical two-dimensional singular integral operators over $\mathbb{C} = \mathbb{R}^2$:

$$(S_{\mathbb{R}^2} \varphi)(z) = -\frac{1}{\pi} \int_{\mathbb{R}^2} \frac{\varphi(\zeta)}{(\zeta - z)^2} \, dv(\zeta) \quad \text{and} \quad (S_{\mathbb{R}^2}^* \varphi)(z) = -\frac{1}{\pi} \int_{\mathbb{R}^2} \frac{\varphi(\zeta)}{(\overline{\zeta} - \overline{z})^2} \, dv(\zeta),$$

which have the following representations in terms of the Fourier transform

$$S_{\mathbb{R}^2} = F^{-1} \frac{\overline{\xi}}{\xi} F \quad \text{and} \quad S_{\mathbb{R}^2}^* = S_{\mathbb{R}^2}^{-1} = F^{-1} \frac{\xi}{\overline{\xi}} F. \tag{2.4.3}$$

Here $\xi = \xi_1 + i\xi_2 = (\xi_1, \xi_2)$, and the Fourier transform F is given by

$$(F\varphi)(\xi) = \frac{1}{2\pi} \int_{\mathbb{R}^2} e^{-i\xi \cdot x} \, \varphi(x) \, dx.$$

Some important connections between the above operators appear in the next theorem, the proof of which can be found, for example, in [219].

Theorem 2.4.1. *For each $f \in L_1(D)$ we have*

$$\frac{\partial}{\partial \overline{z}} T_D f = f, \qquad \frac{\partial}{\partial z} T_D^* f = f,$$

$$\frac{\partial}{\partial z} T_D f = S_D f, \qquad \frac{\partial}{\partial \overline{z}} T_D^* f = S_D^* f.$$

The next fundamental result has been established by A. Dzhuraev (see, for example, [64]) for operators acting on $L_p(D)$, $p > 1$, but in what follows we will use it for the space $L_2(D)$ only.

Theorem 2.4.2. *The following equality holds,*

$$B_D = I - S_D S_D^* + L, \tag{2.4.4}$$

where the operators S_D and S_D^* are given by (2.4.1) and (2.4.2), respectively, and L is a compact operator.

In the case of the unit disk $\mathbb{D}$ the last (compact) summand in (2.4.4) is equal to zero.

Proof. We prove first the equality (2.4.4) on a dense set of functions $f \in C^1(D) \cap C(\overline{D})$ and then extend it by continuity to all of $L_2(D)$.

Using the integral representation (2.3.8) we have

$$
\begin{aligned}
(B_D\varphi)(z) &= \int_D K_D(z,\zeta)\varphi(\zeta)dv(\zeta) \\
&= \frac{1}{2\pi i}\int_{\partial D}\left(-\frac{1}{\pi}\int_D \frac{\varphi(\zeta)}{\overline{\zeta}-\overline{t}}\,dv(\zeta)\right)\frac{dt}{(t-z)^2} + \int_D l_1(z,\zeta)\varphi(\zeta)dv(\zeta) \\
&= \frac{1}{2\pi i}\int_{\partial D}(T_D^*\varphi)(t)\frac{dt}{(t-z)^2} + \int_D l_1(z,\zeta)\varphi(\zeta)dv(\zeta) \\
&= \frac{\partial}{\partial z}\left(\frac{1}{2\pi i}\int_{\partial D}(T_D^*\varphi)(t)\frac{dt}{t-z}\right) + \int_D l_1(z,\zeta)\varphi(\zeta)dv(\zeta).
\end{aligned}
$$

Making use of the Pompeiu formula (2.3.3) we get

$$
\begin{aligned}
(B_D\varphi)(z) &= \frac{\partial}{\partial z}\left((T_D^*\varphi)(z) + \frac{1}{\pi}\int_D \frac{1}{t-z}\frac{\partial}{\partial \overline{t}}(T_D^*\varphi)(t)dv(t)\right) \\
&\quad + \int_D l_1(z,\zeta)\varphi(\zeta)dv(\zeta) \\
&= \varphi(z) + \frac{1}{\pi}\int_D \frac{1}{(t-z)^2}\,(S_D^*\varphi)(t)dv(t) + \int_D l_1(z,\zeta)\varphi(\zeta)dv(\zeta) \\
&= (I - S_D S_D^* + L)\varphi,
\end{aligned}
$$

where the operator

$$
(L\varphi)(z) = \int_D l_1(z,\zeta)\varphi(\zeta)dv(\zeta)
$$

is obviously compact.

For the case of the unit disk $\mathbb{D}$, by (2.3.6), we have $l_1(z,\zeta) \equiv 0$, and thus $L = 0$. $\qquad\square$

Note that Theorem 3.5.7, in particular, states that for the case of the upper half-plane Π the last compact summand of (2.4.4) is equal to zero as well.

Remark 2.4.3. Formula (2.4.4) shows that the Bergman projection B_D belongs to the algebra generated by two-dimensional singular integral (zero-order pseudodifferential) operators on the domain D. This enables us to use corresponding results and tools studying the properties of the Bergman projection.

Lemma 2.4.4. *For each function $a(z) \in C(\overline{D})$ the commutator $[B_D, a(z)I] = B_D a(z)I - a(z)B_D$ is compact.*

Proof. Follows immediately from the representation (2.4.4) and the results of Mikhlin for two-dimensional singular integral operators (see, for example, [146]). $\qquad\square$

Theorem 2.4.5.   1. *Let $a(z) \in L_\infty(D)$ and be continuous at all points of $\gamma = \partial D$ with $a|_\gamma \equiv 0$, then the operators $a(z)B_D$ and $B_D a(z)I$ are compact.*

2. *Let $a(z) \in L_\infty(D)$ and be continuous at all points of $\gamma = \partial D$, then the commutator $[B_D, a(z)I] = B_D a(z)I - a(z)B_D$ is compact.*

Proof. Because of $B_D a(z)I = (\overline{a(z)}B_D)^*$ it is sufficient to prove the first statement for the operator $a(z)B_D$ only. Given $a(z)$ and $\varepsilon > 0$, introduce

$$a_\varepsilon(z) = \begin{cases} 0, & \text{if } \inf_{t \in \gamma} |z - t| < \varepsilon \\ a(z), & \text{otherwise} \end{cases}.$$

The integral operator $a_\varepsilon(z)B_D$ has a bounded kernel, and thus is compact on $L_2(D)$. Finally, the operator $a(z)B_D$ is compact as a uniform limit (under $\varepsilon \to 0$) of the compact operators $a_\varepsilon(z)B_D$.

To prove the second statement for given $a(z)$ introduce a *continuous* function $b(z)$ on $\overline{D}$ such that $a|_\gamma = b|_\gamma$. Then the operator

$$[B_D, a(z)I] = [B_D, b(z)I] - (a - b)B_D + B_D(a - b)$$

is compact as a sum of compact operators. $\qquad\qquad\square$

2.5   "Continuous" theory and local properties of the Bergman projection

Let $\mathcal{A}$ be a C^*-algebra of functions on $\overline{D}$ with the properties

1. $C(\overline{D}) \subset \mathcal{A} \subset L_\infty(D)$,

2. each function from $\mathcal{A}$ is continuous at all points of $\gamma = \partial D$.

We will study the C^*-algebra $\mathcal{R} = \mathcal{R}(\mathcal{A}, B_D)$ generated by all operators of the form

$$A = a(z)I + b(z)B_D,$$

where $a, b \in \mathcal{A}$.

Lemma 2.5.1. *The C^*-algebra $\mathcal{R} = \mathcal{R}(\mathcal{A}, B_D)$ is irreducible.*

Proof. To prove the lemma we show that the algebra $\mathcal{R}$ does not have any nontrivial invariant subspace.

Each invariant subspace of its subalgebra

$$\{aI : a \in \mathcal{A}\} \cong \mathcal{A}$$

obviously has the form

$$X_M = \{\psi_M(z) = \chi_M(z)f(z) : f \in L_2(D)\},$$

where χ_M is the characteristic function of a measurable set M having positive measure. Now, the space X_M is invariant for the whole algebra $\mathcal{R}$ if and only if $\chi_M B_D = B_D \chi_M I$. At the same time the function

$$\chi_M(z) = \chi_M \cdot 1 = (\chi_M B_D)1 = (B_D \chi_M I)1 = (B_D \chi_M)(z)$$

is analytic in D and

$$\chi_M(z) = \left\{ \begin{array}{ll} 1, & z \in M \\ 0, & z \in D \setminus M \end{array} \right. .$$

Thus, modulo a zero measure set, either $M = \emptyset$, or $M = D$. That is, each invariant subspace of the algebra $\mathcal{R}$ is trivial, being either $\{0\}$, or $L_2(D)$. $\hspace{1cm}$ $\square$

Corollary 2.5.2. *The C^*-algebra $\mathcal{R} = \mathcal{R}(\mathcal{A}, B_D)$ contains the ideal $\mathcal{K}$ of all compact operators on $L_2(D)$.*

Proof. The algebra $\mathcal{R} = \mathcal{R}(\mathcal{A}, B_D)$ is irreducible and, by Theorem 2.4.5, contains non-trivial compact operators (commutators). Thus (see, for example, [148] Theorem 2.4.9) it contains the whole ideal $\mathcal{K}$ of compact operators on $L_2(D)$. $\hspace{0.3cm}$ $\square$

A good approach to the final result gives the next assertion followed by Lemma 2.5.3.

The operators B_D and $I - B_D$ are orthogonal and complementary projections, and have compact commutators with multiplication operators aI, $a \in \mathcal{A}$, by Theorem 2.4.5. Thus, given two operators

$$A_n = a_n(z)I + b_n(z)B_D + K_n = a_n(I - B_D) + (a_n + b_n)B_D + K_n,$$

where $a_n, b_n \in \mathcal{A}$, K_n are compact operators, $n = 1, 2$, we have

$$\begin{array}{rl} A_1 + A_2 = & (a_1 + a_2)(I - B_D) + [(a_1 + b_1) + (a_2 + b_2)]B_D + K', \\ A_1 \cdot A_2 = & a_1 \cdot a_2(I - B_D) + (a_1 + b_1) \cdot (a_2 + b_2)B_D + K'', \end{array}$$

where K' and K'' are compact operators.

Lemma 2.5.3. *Let $a, b \in \mathcal{A}$. If the function a is invertible in $\mathcal{A}$ and the function $(a + b)|_\gamma$ is invertible in $C(\gamma)$, then the operator*

$$A = a(z)I + b(z)B_D + K,$$

where K is a compact operator, is Fredholm on the space $L_2(D)$.

Proof. Let a^{-1} be the inverse to a, and $(a + b)^{(-1)}$ be a function from $\mathcal{A}$ such that $(a + b)^{(-1)}|_\gamma$ is the inverse to $(a + b)|_\gamma$ in $C(\gamma)$. Then for the operator

$$R = a^{-1}(I - B_D) + (a + b)^{(-1)}B_D$$

we have

$$
\begin{aligned}
AR &= (I - B_D) + (a + b)(a + b)^{(-1)} B_D + K' \\
&= I + ((a + b)(a + b)^{(-1)} - 1) B_D + K', \\
RA &= (I - B_D) + (a + b)(a + b)^{(-1)} B_D + K'' \\
&= I + ((a + b)(a + b)^{(-1)} - 1) B_D + K'',
\end{aligned}
$$

where K' and K'' are compact.

By Theorem 2.4.5 the operator $((a + b)(a + b)^{(-1)} - 1) B_D$ is compact, and thus R is the two-sided regularizer for A. $\qquad\square$

Let $\operatorname{Sym} \mathcal{R} = \widehat{\mathcal{R}} = \mathcal{R}/\mathcal{K}$ be the Fredholm symbol algebra of the algebra $\mathcal{R} = \mathcal{R}(\mathcal{A}, B_D)$. The letter $\mathcal{K}$ will always denote the ideal of all compact operators on the space under consideration. Denote by π the natural projections

$$
\pi : \mathcal{R} \longrightarrow \widehat{\mathcal{R}} = \mathcal{R}/\mathcal{K}.
$$

To describe the algebra $\operatorname{Sym} \mathcal{R} = \widehat{\mathcal{R}}$ we use the Douglas-Varela local principle (see Subsection 1.1.6).

It is easy to see that the algebra $\pi(\mathcal{A}) \cong \mathcal{A}$ is a central commutative C^*-subalgebra of the algebra $\widehat{\mathcal{R}}$. Denote by $\widehat{D}$ the compact set of maximal ideals of the algebra $\mathcal{A}$, and denote by $\widetilde{a} \in C(\widehat{D})$ the Gelfand image of $a \in \mathcal{A}$. It is obvious that $\gamma \subset \widehat{D}$ and $\pi(\mathcal{A}) \cong C(\widehat{D})$; thus we will localize by the points of $\widehat{D}$. For each point $t_0 \in \widehat{D}$, introduce the corresponding maximal ideal J_{t_0} of the algebra $\pi(\mathcal{A}) \cong \mathcal{A}$ and the closed two-sided ideal $J(t_0)$ of the algebra $\widehat{\mathcal{R}}$ generated by J_{t_0}. Then $\widehat{\mathcal{R}}(t_0) = \widehat{\mathcal{R}}/J(t_0)$ is the local algebra which corresponds to the point $t_0 \in \widehat{D}$. We denote by $\pi(t_0)$ the natural projection

$$
\pi(t_0) : \widehat{\mathcal{R}} \longrightarrow \widehat{\mathcal{R}}(t_0).
$$

Let further

$$
\pi_{t_0} = \pi(t_0) \circ \pi : \mathcal{R} \to \widehat{\mathcal{R}}(t_0),
$$

and $\pi(B_D) = \widehat{B}_D$, $\pi_{t_0}(B_D) = \pi(t_0)(\widehat{B}_D) = \widehat{B}_D(t_0)$.

There are two different types of local algebras $\widehat{\mathcal{R}}(t_0)$, which correspond to the cases $t_0 \in \widehat{D} \setminus \gamma$ and $t_0 \in \gamma \subset \widehat{D}$, respectively.

We start with the first case. Let $t_0 \in \widehat{D} \setminus \gamma$. It is easy to see that $\pi(a_1 I) \overset{t_0}{\sim} \pi(a_2 I)$ if and only if $\widetilde{a}_1(t_0) = \widetilde{a}_2(t_0)$, and thus $\pi_{t_0}(aI) = \widetilde{a}(t_0)$. Denote by $e_{t_0}(z)$ a function from $\mathcal{A}$ such that $e_{t_0}|_\gamma \equiv 0$ and $\widetilde{e}_{t_0}(t_0) = 1$. Then the operator $e_{t_0}(z) B_D$ is compact, and we have

$$
\widehat{B}_D = 1 \cdot \widehat{B}_D \overset{t_0}{\sim} \pi(e_{t_0} I)\pi(B_D) = \pi(e_{t_0}(z) B_D) = 0.
$$

Thus we arrive at the following lemma.

Lemma 2.5.4. *Let $t_0 \in \widehat{D} \setminus \gamma$. Then the local algebra $\widehat{\mathcal{R}}(t_0)$ is isomorphic to $\mathbb{C}$. Identifying them we have*

$$\pi_{t_0} : A = a(z)I + b(z)B_D \longmapsto \widetilde{a}(t_0) \in \mathbb{C}.$$

Let now $t_0 \in \gamma \subset \widehat{D}$. First of all there are not too many possibilities for local algebras $\widehat{\mathcal{R}}(t_0)$ in this case. The operator B_D, being an orthogonal projection, has only three possibilities: $\widehat{B}_D(t_0)$ is equal either to zero, to the identity, or to a self-adjoint idempotent in $\widehat{\mathcal{R}}(t_0)$. Knowing that $\pi_{t_0}(aI) = \widetilde{a}(t_0) \in \mathbb{C}$, the algebra $\widehat{\mathcal{R}}(t_0)$ is isomorphic either to $\mathbb{C}$ (in the two first cases), or to $\mathbb{C}^2$ (in the third case).

Lemma 2.5.5. *For each $t_0 \in \gamma \subset \widehat{D}$ the local algebras $\widehat{\mathcal{R}}(t_0)$ are isomorphic among themselves.*

Proof. Keeping in mind the local algebras we may assume that $\mathcal{A} = C(\overline{D})$. Let $\alpha(z)$ be a conformal mapping from D onto unit disk $\mathbb{D}$ with $1 = \alpha(t_0)$. Then the operator $(W\varphi)(z) = \alpha'(z)\varphi[\alpha(z)]$ generates by Lemma 2.2.8 the isomorphism $A \mapsto W^*AW$ of the algebras $\mathcal{R}(C(\overline{D}); B_D)$ and $\mathcal{R}(C(\overline{\mathbb{D}}); B_{\mathbb{D}})$, which in turn induces the isomorphism of each local algebra $\widehat{\mathcal{R}}(t_0)$ onto the specific one $\widehat{\mathcal{R}}^*(C(\overline{\mathbb{D}}); B_{\mathbb{D}})(1)$.
$\qquad\square$

Lemma 2.5.6. *Let $t_0 \in \gamma$. Then the local algebra $\widehat{\mathcal{R}}(t_0)$ is isomorphic to $\mathbb{C}^2$. Identifying them we have*

$$\pi_{t_0} : A = a(z)I + b(z)B_D \longmapsto (a(t_0), a(t_0) + b(t_0)) \in \mathbb{C}^2. \qquad (2.5.1)$$

Proof. Both operators B_D and $I - B_D$ are non-compact, thus $\pi(B_D) = \widehat{B}_D$ is a self-adjoint idempotent in $\widehat{\mathcal{R}}$. The form of local algebras $\widehat{\mathcal{R}}(t_0)$ shows that the algebra $\widehat{\mathcal{R}}$ is commutative. Its maximal ideal space $\mathfrak{M}$ can be identified with the union of one-dimensional representations of all local algebras, and must be compact. By previous considerations it must be the disjoint union of $\widehat{D}$ and some (closed) subset of γ. By Lemma 2.5.5 this subset must be either $\emptyset$, or the whole γ. But the Gelfand transform of $\widehat{B}_D$ is a characteristic function of some non-empty (closed and open) subset of $\mathfrak{M}$, which is constant on $\widehat{D}$. Thus $\mathfrak{M} = \widehat{D} \sqcup \gamma$. Hence for each $t_0 \in \gamma$, $\widehat{B}_D(t_0)$ is an idempotent, and $\widehat{\mathcal{R}}(t_0) \cong \mathbb{C}^2$. To check (2.5.1) one needs to use the fact that $\widetilde{a}(t_0) = a(t_0)$ for the points $t_0 \in \gamma$.
$\qquad\square$

The descriptions of the local algebras $\widehat{\mathcal{R}}(t_0)$ together with standard arguments lead directly to the following theorem.

Theorem 2.5.7. *The algebra $\mathcal{R} = \mathcal{R}(\mathcal{A}, B_D)$ coincides with the set of all the operators of the form*

$$A = a(z)I + b(z)B_D + K,$$

where $a, b \in \mathcal{A}$, K is a compact operator.

The Fredholm symbol algebra $\operatorname{Sym}\mathcal{R}$ of the algebra $\mathcal{R}$ is isomorphic to the algebra $\mathcal{A}\oplus C(\gamma)$. Identifying them, the symbol homomorphism

$$\operatorname{sym} : \mathcal{R} \longrightarrow \operatorname{Sym}\mathcal{R} = \mathcal{A}\oplus C(\gamma)$$

is given by

$$\operatorname{sym} : A = a(z)I + b(z)B_D + K \longmapsto (a(z), a(t) + b(t)),$$

where $z \in \overline{D}$, $t \in \gamma$.

The operator $A = a(z)I + b(z)B_D + K$ is Fredholm if and only if its symbol is invertible, i.e., the function a is invertible in $\mathcal{A}$ and the function $(a + b)|_\gamma$ is invertible in $C(\gamma)$.

Passing to the index calculation we start with the following Dzhuraev result (see, for example, [62]). Let $a, b \in C(\overline{D})$, and $a(z) \neq 0$ on $\overline{D}$, $a(t) + b(t) \neq 0$ on γ. Then the index of the Fredholm operator

$$A = a(z)I + b(z)B_D + K,$$

is calculated by the formula

$$\operatorname{Ind} A = -\frac{1}{2\pi}\{\arg(a(t) + b(t))\}_\gamma.$$

A simple trick shows that this formula contains a hidden summand. Indeed,

$$A = a(z)I + b(z)B_D + K = a(z)\left(I + \frac{b(z)}{a(z)}B_D + K_1\right) = a(z)A_1,$$

then obviously

$$
\begin{aligned}
\operatorname{Ind} A \;&=\; \operatorname{Ind} A_1 = -\frac{1}{2\pi}\left\{\arg\frac{a(t) + b(t)}{a(t)}\right\}_\gamma \\
&=\; -\frac{1}{2\pi}\{\arg(a(t) + b(t))\}_\gamma + \frac{1}{2\pi}\{\arg a(t)\}_\gamma.
\end{aligned}
$$

Any continuous function on $\overline{D}$ is homotopic to a constant, thus for the case of continuous coefficients the last summand is always equal to zero. At the same time for more general coefficients it has to be present.

Theorem 2.5.8. *Let the operator*

$$A = a(z)I + b(z)B_D + K,$$

be Fredholm, where $a, b \in \mathcal{A}$, K is a compact operator. Then

$$\operatorname{Ind} A = \frac{1}{2\pi}\{\arg a(t)\}_\gamma - \frac{1}{2\pi}\{\arg(a(t) + b(t))\}_\gamma.$$

Proof. We start with

$$A = a(z)I + b(z)B_D + K = a(z)\left(I + \frac{b(z)}{a(z)}B_D + K_1\right) = a(z)A_1,$$

and consider the operator $A_2 = I + c(z)B_D$, where $c(z)$ is a continuous on $\overline{D}$ function whose restriction on γ coincides with $\frac{b}{a}\big|_\gamma$. Then obviously $\operatorname{Ind} A = \operatorname{Ind} A_1 = \operatorname{Ind} A_2$. For the last operator we apply the Dzhuraev formula. $\hfill\square$

2.6 Model discontinuous case

Let $\mathbb{D}$ be the unit disk. Then

$$(B_{\mathbb{D}}\varphi)(z) = (B\varphi)(z) = \frac{1}{\pi}\int_{\mathbb{D}}\frac{\varphi(\zeta)\,d\zeta}{(1 - \overline{\zeta}z)^2}.$$

Let ℓ be a simple piece-wise smooth curve in $\mathbb{D}$ connecting two points of the unit circle $\Gamma = \partial\mathbb{D}$. Denote by $\mathbb{D}_+$ and $\mathbb{D}_-$ the parts of the unit disk $\mathbb{D}$ which are separated by ℓ, and denote by P the orthogonal projection of $L_2(\mathbb{D})$ onto $L_2(\mathbb{D}_+)$.

Introduce the algebra $C(\mathbb{D}_\pm)$ of all piece-wise constant functions of the form

$$a(z) = \begin{cases} a_1, & \text{if } z \in \mathbb{D}_+ \\ a_2, & \text{if } z \in \mathbb{D}_- \end{cases}.$$

We study the algebra $\mathcal{R}_0 = \mathcal{R}(C(\mathbb{D}_\pm), B)$, generated by all operators of the form

$$A = a(z)I + b(z)B,$$

where $a, b \in C(\mathbb{D}_\pm)$. The algebra $\mathcal{R}_0$ is obviously generated by two orthogonal projections B and P. To describe it we prove first that $\operatorname{sp}(P - B)^2 = [0, 1]$, and then apply the two-projection theorem (Theorem 1.2.8).

To do this introduce the operator $B_1 = PB\big|_{L_2(\mathbb{D}_+)}$, acting on $L_2(\mathbb{D}_+)$, and the algebra $\mathcal{R}_1 = \mathcal{R}(C(\overline{\mathbb{D}}_+), B_1, \mathcal{K})$ generated by all operators of the form

$$A = a(z)I + b(z)B_1 + K,$$

where $a(z)$, $b(z) \in C(\overline{\mathbb{D}}_+)$, and K is a compact operator.

It is easy to see that the algebra $\pi(C(\overline{\mathbb{D}}_+)) \cong C(\overline{\mathbb{D}}_+)$ is a central commutative subalgebra of the algebra $\widehat{\mathcal{R}}_1 = \operatorname{Sym}\mathcal{R}_1$. We use the Douglas-Varela local principle, localizing by the points of $\overline{\mathbb{D}}_+$.

There are three different types of the local algebras $\widehat{\mathcal{R}}_1(t_0)$, depending on $t_0 \in \overline{\mathbb{D}}_+$.

Let $t_0 \in \overline{\mathbb{D}}_+ \setminus \Gamma$. Then as in Lemma 2.5.4, we have $\widehat{\mathcal{R}}_1(t_0) \cong \mathbb{C}$, and

$$\pi_{t_0} : A = a(z)I + b(z)B_1 + K \longmapsto a(t_0) \in \mathbb{C}.$$

Let $t_0 \in (\Gamma \cap \overline{\mathbb{D}}_+) \setminus \partial(\Gamma \cap \overline{\mathbb{D}}_+)$. Then as in Lemma 2.5.6, we have $\widehat{\mathcal{R}}_1(t_0) \cong \mathbb{C}^2$, and

$$\pi_{t_0} : A = a(z)I + b(z)B_1 + K \longmapsto (a(t_0), a(t_0) + b(t_0)) \in \mathbb{C}^2.$$

Let finally $t_0 \in \partial(\Gamma \cap \overline{\mathbb{D}}_+)$. Each coefficient is locally equivalent to its value at the point t_0, and thus the C^*-algebra $\widehat{\mathcal{R}}_1(t_0)$ is generated by one self-adjoint element $\widehat{B}_1(t_0)$. Hence $\widehat{\mathcal{R}}_1(t_0) \cong C(\mathrm{sp}\,\widehat{B}_1(t_0))$, and

$$\pi_{t_0} : A = a(z)I + b(z)B_1 + K \longmapsto a(t_0) + b(t_0)x_3,$$

where $x_3 \in \mathrm{sp}\,\widehat{B}_1(t_0)$.

Denote by Δ the set of points $(x_1, x_2, x_3) \in \mathbb{R}^3$ satisfying one of the following conditions:

1. $x_1 + ix_2 \in \overline{\mathbb{D}}_+$, $x_3 = 0$,

2. $x_1 + ix_2 \in \Gamma \cap \overline{\mathbb{D}}_+$, $x_3 = 1$,

3. $x_1 + ix_2 \in \partial(\Gamma \cap \overline{\mathbb{D}}_+)$, $x_3 \in \mathrm{sp}\,\widehat{B}_1(x_1 + ix_2)$.

Collecting now all the descriptions of the local algebras $\widehat{\mathcal{R}}_1(t_0)$, and using the fact that commutative C^*-algebra $\mathrm{Sym}\,\mathcal{R}_1$ is isomorphic and isometric to the algebra of all continuous functions on its maximal ideals space, we arrive at the following theorem.

Theorem 2.6.1. *The Fredholm symbol algebra* $\mathrm{Sym}\,\mathcal{R}_1$ *of the algebra* $\mathcal{R}_1$ *is isomorphic and isometric to the algebra* $C(\Delta)$. *The homomorphism*

$$\mathrm{sym} : \mathcal{R}_1 \longrightarrow \mathrm{Sym}\,\mathcal{R}_1 = C(\Delta)$$

is generated by the following mapping of generators of the algebra $\mathcal{R}_1$*: if* $A = a(z)I + b(z)B_1 + K$*, where* $a, b \in C(\overline{\mathbb{D}}_+)$ *and* K *is a compact operator, then*

$$\mathrm{sym}\, A = a(x_1 + ix_2) + b(x_1 + ix_2) \cdot x_3,$$

where $(x_1, x_2, x_3) \in \Delta$.

Corollary 2.6.2. *For the points* $t_0 \in \partial(\Gamma \cap \overline{\mathbb{D}}_+)$ *we have*

$$\mathrm{sp}\,\widehat{B}_1(t_0) = \mathrm{ess\text{-}sp}\,B_1 = [0, 1].$$

Proof. If for any $t_0 \in \partial(\Gamma \cap \overline{\mathbb{D}}_+)$ the spectrum $\mathrm{sp}\,\widehat{B}_1(t_0)$ is a proper subset of $[0, 1]$ then the compact of maximal ideals Δ of the algebra $\mathrm{Sym}\,\mathcal{R}_1$ splits into contractible components. Thus for the homotopy group of invertible elements in $\mathrm{Sym}\,\mathcal{R}_1$ we have $[\mathrm{Sym}^{-1}\mathcal{R}_1] \cong 0$, which implies that all Fredholm operators of $\mathcal{R}_1$ must have the same index. However, we give an example of two Fredholm operators in $\mathcal{R}_1$ with different indices. For $A_1 = I$ we have obviously $\mathrm{Ind}\,A_1 = 0$. To construct the second operator consider a function $c(z) \in C(\overline{\mathbb{D}}_+)$ such that

$c(z) = 1$ for all $z \in \overline{\partial \mathbb{D}_+} \setminus \Gamma$ and $c(z) \neq 0$ for all $z \in \Gamma \cap \partial \overline{\mathbb{D}}_+$. For such a function $c(z)$ the integer

$$\kappa(c) = \frac{1}{2\pi} \{\arg c(z)\}_{\Gamma \cap \partial \overline{\mathbb{D}}_+}$$

is well defined. In addition select a function $c(z)$ such that $\kappa(c) = 1$. Now the Fredholm operator $A_2 = I + [c(z) - 1]B_1$, acting on the space $L_2(\mathbb{D}_+)$ has the same index as the Fredholm operator $A_2' = I + b'(z)B$, where

$$b'(z) = \begin{cases} c(z) - 1, & \text{if } z \in \overline{\mathbb{D}}_+ \\ 0, & \text{if } z \in \mathbb{D}_- \end{cases},$$

acting on the space $L_2(\mathbb{D})$. The coefficients of the operator A_2' are continuous, $\kappa(c) = 1$, thus by Theorem 2.5.8 we have $\text{Ind } A_2' = -1$. Hence $\text{Ind } A_2 = \text{Ind } A_2' \neq \text{Ind } A_1$, and $\text{sp } \widehat{B}_1(t_0) = [0, 1]$. Finally,

$$\text{ess-sp } B_1 = \bigcup \text{sp } \widehat{B}_1(t_0) = [0, 1]. \qquad \square$$

Corollary 2.6.3. *We have*

$$\text{sp}\,(P - B)^2 = \text{ess-sp}\,(P - B)^2 = \text{sp}\,(\widehat{P} - \widehat{B})^2 = [0, 1].$$

Proof. By Theorem 1.2.8 and Corollary 2.6.2,

$$\text{sp}\,(\widehat{P} - \widehat{B})^2 = \text{sp}\,\widehat{P}\widehat{B}\widehat{P} = \text{ess-sp } B_1 = [0, 1].$$

Further, the algebra $\mathcal{R}_0 = \mathcal{R}(\mathbb{C}, P, B)$ does not contain any non-zero compact operator, and thus

$$\text{sp}\,(P - B)^2 = \text{ess-sp}\,(P - B)^2 = [0, 1]. \qquad \square$$

Denote by L the closed contour in Δ, which consists of two arcs $\Gamma \cap \overline{\mathbb{D}}_+$, lying on the planes $x_3 = 0$ and $x_3 = 1$, and of segments $x_3 \in [0, 1]$, connecting their endpoints. We orient L according to the positive direction of $\Gamma \cap \overline{\mathbb{D}}_+$, lying on the plane $x_3 = 0$.

Corollary 2.6.4. *An operator $A \in \mathcal{R}_1 = \mathcal{R}(C(\overline{\mathbb{D}}_+); B_1; \mathcal{K})$ is Fredholm if and only if*

$$\text{sym } A\,(x) \neq 0, \quad x = (x_1, x_2, x_3) \in \Delta.$$

The index of the Fredholm operator $A \in \mathcal{R}_1$ is given by

$$\text{Ind } A = \frac{1}{2\pi} \{\arg \text{sym } A\}_L.$$

Return now to the algebra $\mathcal{R} = \mathcal{R}(C(\mathbb{D}_\pm), B) = \mathcal{R}(\mathbb{C}, P, B)$.

Theorem 2.6.5. *The algebra* $\operatorname{Sym}\mathcal{R}$ *of the algebra* $\mathcal{R}_0$ *is isomorphic and isometric to the algebra* $\mathfrak{S}$ *of all* 2×2 *matrix-functions continuous on* $[0,1]$ *and diagonal at the points* 0 *and* 1*. The isomorphism*

$$\nu_0 : \mathcal{R} \longrightarrow \mathfrak{S}$$

is generated by the following mapping of the generators of the algebra $\mathcal{R}$*:*

$$A = a(z)I + b(z)B \longmapsto \begin{pmatrix} a_1 x + a_2(1-x) & (c_1 - c_2)\sqrt{x(1-x)} \\ (a_1 - a_2)\sqrt{x(1-x)} & c_1(1-x) + c_2 x \end{pmatrix}, \quad x \in [0,1],$$

where $a(z)$*,* $b(z) \in C(\mathbb{D}_\pm)$*, and* $c(z) = a(z) + b(z)$*.*

Proof. The first statement follows directly from Corollary 2.6.3 and Theorem 1.2.8. Define now the isomorphism ν_0 as

$$\nu_0 : \mathcal{R} \xrightarrow{\ \nu\ } \mathfrak{S} \xrightarrow{\ \nu_1\ } \mathfrak{S},$$

where ν is the isomorphism of Theorem 1.2.8, and ν_1 is given by

$$\sigma(x) \longmapsto \begin{pmatrix} \sqrt{x} & -\sqrt{1-x} \\ \sqrt{1-x} & \sqrt{x} \end{pmatrix} \sigma(x) \begin{pmatrix} \sqrt{x} & \sqrt{1-x} \\ -\sqrt{1-x} & \sqrt{x} \end{pmatrix},$$

where $x \in [0,1]$. Direct calculation proves the second statement of the theorem. $\square$

Corollary 2.6.6. *An operator* A *from the algebra* $\mathcal{R}_0 = \mathcal{R}(C(\mathbb{D}_\pm), B)$ *is invertible if and only if the element* $\nu_0(A)$ *is invertible in* $\mathfrak{S}$*, i.e.,*

$$(\det \nu_0(A))(x) \neq 0, \quad x \in [0,1].$$

2.7 Symbol algebra

Let D be a bounded simply connected domain in $\mathbb{C}$ with smooth boundary γ. Denote by ℓ a piece-wise smooth curve in $\overline{D}$ satisfying the following properties: there is a finite number of points (nodes), which divide ℓ into simple oriented smooth curves ℓ_j, $j = \overline{1,k}$. We assume that the endpoints of ℓ are among the nodes. Denote by U the set of all nodes of the curve ℓ. We will refer to nodes using symbols u_q^r, where r is the number of lines meeting at the node, and q is the index of the node. We assume as well that the boundary γ of the domain D intersects with ℓ only at (some) nodes of the type u_q^1. Let the intersection $\gamma \cap \ell$ consist of m points.

Denote by $PC(\overline{D}, \ell)$ the algebra of all functions $a(z)$, continuous in $\overline{D} \setminus \ell$, and having one-sided limit values at all points of ℓ_j: $a^+(z)$ and $a^-(z)$. Thus at the nodes of the type u_q^r (except $u_q^1 \in \gamma \cap \ell$) functions from $PC(\overline{D}, \ell)$ have r limit values. We will denote them by $a^{(1)}(u_q^r)$, ..., $a^{(r)}(u_q^r)$. Our aim is to study the algebra $\mathcal{R} = \mathcal{R}(PC(\overline{D}, \ell), B_D)$ generated by all operators of the form

$$A = a(z)I + b(z)B_D,$$

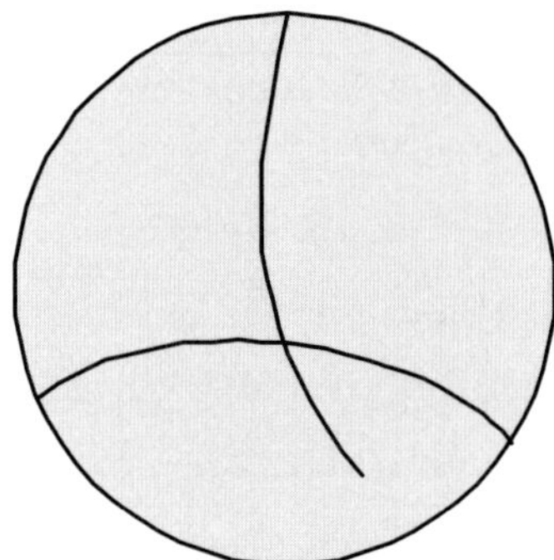

Figure 2.1: An example of a domain D with a curve ℓ.

where $a(z), b(z) \in PC(\overline{D}, \ell)$ and B_D is the Bergman projection of the domain D, acting on the space $L_2(D)$.

The algebra $\mathcal{R}$ contains the ideal $\mathcal{K}$ of all compact operators (its subalgebra $\mathcal{R}(C(\overline{D}), B_D)$ already does), and is irreducible.

Remark 2.7.1. Let $\omega = \alpha(z)$ be a biholomorphic mapping of the domain D onto a domain G, and let $\ell' = \alpha(\ell)$. Then by Lemma 2.2.8 the operator algebras $\mathcal{R}(PC(\overline{D}, \ell), B_D)$ and $\mathcal{R}(PC(\overline{G}, \ell'), B_G)$ are unitary equivalent, and their Fredholm symbol algebras are isomorphic and isometric.

To describe the Fredholm symbol algebra $\operatorname{Sym}\mathcal{R}$ of $\mathcal{R} = \mathcal{R}(PC(\overline{D}, \ell), B_D)$ we use the Douglas-Varela local principle (Subsection 1.1.6). By Lemma 2.4.4 the C^*-algebra $\pi(C(\overline{D})) \cong C(\overline{D})$ is a central commutative subalgebra of $\widehat{\mathcal{R}} = \operatorname{Sym}\mathcal{R}$; thus we will localize by the points of $\overline{D}$. There are five cases for local algebras $\widehat{\mathcal{R}}(t_0)$ depending on the point $t_0 \in \overline{D}$:

a) Let $t_0 \in \overline{D} \setminus (\gamma \cup \ell)$. Then as in Lemma 2.5.4, $\widehat{\mathcal{R}}(t_0) \cong \mathbb{C}$, and for the operator $A = a(z)I + b(z)B_D$ we have $\pi_{t_0}(A) = a(t_0)$.

b) Let $t_0 \in \gamma \setminus \ell$. Then as in Lemma 2.5.6, $\widehat{\mathcal{R}}(t_0) \cong \mathbb{C}^2$, and for the operator $A = a(z)I + b(z)B_D$ we have $\pi_{t_0}(A) = (a(t_0), a(t_0) + b(t_0))$.

c) Let $t_0 \in \ell \setminus U$. Then as in Lemma 2.5.4, $\widehat{B}_D(t_0) = 0$, but $a_1(z) \overset{t_0}{\sim} a_2(z)$ if and only if $a_1^+(t_0) = a_2^+(t_0)$ and $a_1^-(t_0) = a_2^-(t_0)$. Thus in this case $\widehat{\mathcal{R}}(t_0) \cong \mathbb{C}^2$, and for the operator $A = a(z)I + b(z)B_D$ we have $\pi_{t_0}(A) = (a^+(t_0), a^-(t_0))$.

d) Let $t_0 \in U \setminus \gamma$ and $t_0 = u_q^r$ for some q. Then again $\widehat{B}_D(t_0) = 0$, but a function $a(z) \in PC(\overline{D}, \ell)$ now has r limit values at the point t_0. Thus $\widehat{\mathcal{R}}(t_0) \cong \mathbb{C}^r$, and for the operator $A = a(z)I + b(z)B_D$ we have $\pi_{t_0}(A) = (a^{(1)}(t_0), \ldots, a^{(r)}(t_0))$.

e) Let finally $t_0 \in \gamma \cap \ell$. Combining the arguments of Lemmas 2.5.5 and 2.5.6 with the results of Corollary 2.6.3 and Theorem 2.6.5 we arrive at the following result.

Lemma 2.7.2. *Let $t_0 \in \gamma \cap \ell$. Then the local algebra $\widehat{\mathcal{R}}(t_0)$ is isomorphic and isometric to the algebra of all 2×2 matrix-functions continuous on $[0,1]$ and diagonal at the points 0 and 1. Identifying them, we have that if*

$$A = a(z)I + b(z)B_D \in \mathcal{R} = \mathcal{R}(PC(\overline{D}, \ell), B_D),$$

then

$$\pi_{t_0}(A) = \begin{pmatrix} a(t_0+0)x + a(t_0-0)(1-x) & (c(t_0+0) - c(t_0-0))\sqrt{x(1-x)} \\ (a(t_0+0) - a(t_0-0))\sqrt{x(1-x)} & c(t_0+0)(1-x) + c(t_0-0)x \end{pmatrix},$$

where $x \in [0,1]$, and $c(z) = a(z) + b(z)$.

The descriptions of the local algebras $\widehat{\mathcal{R}}(t_0)$ show that the Fredholm symbol algebra $\operatorname{Sym}\mathcal{R} = \widehat{\mathcal{R}}$ is a C^*-algebra which has one- and two-dimensional irreducible representations only. To formulate the final result we introduce some notation.

Denote by $\widehat{D}$ the compactification of the set $\overline{D}$, cut along the line ℓ. Under this each point from $(\ell \backslash U) \cup (\ell \cap \gamma)$ will correspond to a pair of points in $\widehat{D}$; similarly each node of the form u_q^r ($\not\in \ell \cap \gamma$) will correspond to r points of $\widehat{D}$. The pair of points, which correspond to a point $t_p \in \ell \cap \gamma$, $p = \overline{1,m}$, we will denote by $t_p' - 0$ and $t_p' + 0$, following the positive orientation on γ. The set $\widehat{D}$ coincides obviously with the compact set of maximal ideals of the algebra $PC(\overline{D}, \ell)$. Analogously, denote by $\widehat{\gamma}$ the compactification of the curve γ, cut by points $t_p \in \ell \cap \gamma$. The pair of points which correspond to a point $t_p \in \ell \cap \gamma$, $p = \overline{1,m}$, we will denote by $t_p'' - 0$ and $t_p'' + 0$, following the positive orientation on γ.

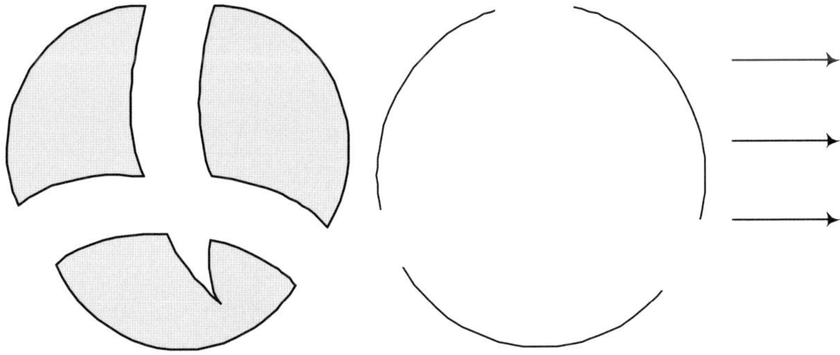

Figure 2.2: The sets $\widehat{D}$, $\widehat{\gamma}$, and $\overline{X} = \cup_{p=1}^m \Delta_p$.

Let $Y = \widehat{D} \cup \widehat{\gamma}$, and let $\overline{X} = \cup_{p=1}^{m} \Delta_p$ be a disjoint union of segments $\Delta_p = [0,1]$. Denote by μ the mapping which identifies the points of $\partial \overline{X}$ with the pair of points of Y by the following rule:

$$\mu(0_p) = (t'_p - 0, t''_p + 0), \quad \text{where} \quad 0_p \in \Delta_p, \; t'_p - 0 \in \widehat{D}, \; t''_p + 0 \in \widehat{\gamma},$$
$$\mu(1_p) = (t'_p + 0, t''_p - 0), \quad \text{where} \quad 1_p \in \Delta_p, \; t'_p + 0 \in \widehat{D}, \; t''_p - 0 \in \widehat{\gamma}.$$

From the descriptions of the local algebras $\widehat{\mathcal{R}}(t_0)$ it follows that the set $\mathfrak{M} = \overline{X} \cup_{\mu} Y$ is the spectrum of the C^*-algebra $\widehat{\mathcal{R}} = \operatorname{Sym} \mathcal{R}$.

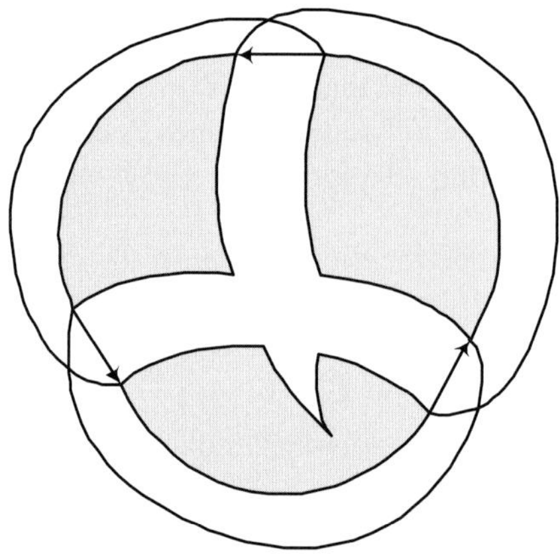

Figure 2.3: The set $\mathfrak{M} = \overline{X} \cup_{\mu} Y$.

Denote by $\mathfrak{S}$ the algebra of all pairs $\sigma = (\sigma_1, \sigma_2)$, where $\sigma_1 \in C(Y)$, $\sigma_2 \in C(\overline{X}, \operatorname{Mat}_2(\mathbb{C}))$, satisfying the following condition:
if $\mu(x_0) = (y_1, y_2)$, $x_0 \in \partial \overline{X}$, $y_1 \in \widehat{D}$, $y_2 \in \widehat{\gamma}$ then

$$\lim_{\substack{x \to x_0 \\ x \in X}} \sigma_2(x) = \begin{pmatrix} \sigma_1(y_1) & 0 \\ 0 & \sigma_1(y_2) \end{pmatrix}; \qquad (2.7.1)$$

the norm in the algebra $\mathfrak{S}$ is determined as follows:

$$\|\sigma\| = \max\{\sup_{Y} |\sigma_1(y)|, \sup_{\overline{X}} \|\sigma_2(x)\|\},$$

where $\|\sigma_2(x)\|^2$ is the largest eigenvalue of the matrix $\sigma_2(x)\sigma_2^*(x)$.

We note that each above pair $\sigma = (\sigma_1, \sigma_2)$ defines a continuous object on $\mathfrak{M} = \overline{X} \cup_{\mu} Y$, which is a continuous function on $Y = \widehat{D} \cup \widehat{\gamma}$, a continuous 2×2 matrix-function on $\overline{X} = \cup_{p=1}^{m} \Delta_p$, and a diagonal 2×2 matrix at the points of $\partial \overline{X}$, whose scalar diagonal values are glued with certain values of the function on $Y = \widehat{D} \cup \widehat{\gamma}$ according to (2.7.1).

The descriptions of the local algebras $\widehat{\mathcal{R}}(t_0)$ together with the Douglas-Varela local principle lead to the next theorem.

Theorem 2.7.3. *The Fredholm symbol algebra* $\mathrm{Sym}\,\mathcal{R}$ *of the* C^**-algebra* $\mathcal{R} = \mathcal{R}(PC(\overline{D}, \ell), B_D)$ *is isomorphic and isometric to the algebra* $\mathfrak{S}$*. Identifying them, the homomorphism*

$$\mathrm{sym} : \mathcal{R} \to \mathfrak{S}$$

is generated by the following mapping of generators of the algebra $\mathcal{R}$*:*

$$\mathrm{sym} : A = a(z)I + b(z)B_D + K \longmapsto$$

$$\left\{ \begin{array}{l} a(t), \quad t \in \widehat{D}, \\ c(t), \quad t \in \widehat{\gamma}, \\ \begin{pmatrix} a(t_p + 0)(1 - x) + a(t_p - 0)x & (c(t_p + 0) - c(t_p - 0))\sqrt{x(1 - x)} \\ (a(t_p + 0) - a(t_p - 0))\sqrt{x(1 - x)} & c(t_p + 0)(1 - x) + c(t_p - 0)x \end{pmatrix}, \end{array} \right.$$

$$x \in [0, 1], \quad t_p \in \gamma \cap \ell,$$

where $c(z) = a(z) + b(z)$.

Corollary 2.7.4. *An operator* A *from* $\mathcal{R} = \mathcal{R}(PC(\overline{D}, \ell), B_D)$ *is Fredholm if and only if its symbol is invertible, i.e.,*

$$\mathrm{sym}\,A \neq 0 \quad \text{on} \quad Y, \quad \det \mathrm{sym}\,A \neq 0 \quad \text{on} \quad \overline{X}.$$

2.8 Toeplitz operators

Let D be a bounded simply connected domain in $\mathbb{C}$ with smooth boundary $\gamma = \partial D$. We return to the Bergman space $\mathcal{A}^2(D)$, the closed subspace of $L_2(D)$ consisting of analytic functions, and to the orthogonal Bergman projection B_D (2.2.7) of $L_2(D)$ onto $\mathcal{A}^2(D)$.

Given a function $a = a(z) \in L_\infty(D)$, introduce the Toeplitz operator T_a with *defining symbol* $a = a(z)$ acting on the Bergman space $\mathcal{A}^2(D)$ as follows,

$$T_a = B_D\, a(z)|_{\mathcal{A}^2(D)} : \mathcal{A}^2(D) \longrightarrow \mathcal{A}^2(D),$$

or, in the integral form,

$$(T_a \varphi)(z) = \int_D K_D(z, \zeta) a(\zeta) \varphi(\zeta) d\nu(\zeta).$$

In the case of the unit disk $\mathbb{D}$ we have obviously

$$(T_a \varphi)(z) = \frac{1}{\pi} \int_{\mathbb{D}} \frac{a(\zeta) \varphi(\zeta)}{(1 - z\overline{\zeta})^2}\, d\nu(\zeta).$$

The next theorem summarizes the obvious properties of Toeplitz operators.

Theorem 2.8.1. *Let $\alpha,\ \beta \in \mathbb{C}$ and $a,\ b \in L_\infty(D)$, then*

(i) *the operator T_a is bounded on $\mathcal{A}^2(D)$ and $\|T_a\| \leq \|a\|_\infty$,*

(ii) $T_{\alpha a + \beta b} = \alpha T_a + \beta T_b,$

(iii) $T_a^* = T_{\overline{a}}.$

The next theorem shows that there is a one-to-one correspondence between the Toeplitz operators and their defining symbols.

Theorem 2.8.2. *For any $a \in L_\infty(D)$, $T_a = 0$ if and only if $a \equiv 0$ almost everywhere.*

Proof. If $a \equiv 0$ a.e., than obviously $T_a = 0$. Assume now that $T_a = 0$. Then for each $\varphi \in \mathcal{A}^2(\mathbb{D})$ we have $T_a \varphi = 0$, or $a\varphi \in (\mathcal{A}^2(\mathbb{D}))^\perp$. Thus, in particular, for each $n, m = 0, 1, 2, \ldots$, we have

$$
\begin{aligned}
0 &= \langle az^n, z^m \rangle = \int_D a(z) z^n \overline{z^m} dv(z) \\
  &= \int_D a(z) \overline{z^m \overline{z^n}} dv(z) = \langle a, z^m \overline{z^n} \rangle_{L_2(D)}.
\end{aligned}
$$

The linear span of all monomials $z^m \overline{z^n}$ is dense in $L_2(D)$; thus the function $a(z)$ must be 0 almost everywhere. $\qquad\square$

By Theorem 2.4.5 we have as well

Theorem 2.8.3. 1. *Let $a(z) \in L_\infty(D)$ and be continuous at all points of $\gamma = \partial D$ with $a|_\gamma \equiv 0$. Then the Toeplitz operators T_a is compact.*

2. *Let $a(z),\ b(z) \in L_\infty(D)$ and be continuous at all points of $\gamma = \partial D$. Then the so-called* semi-commutator *$[T_a, T_b) = T_a T_b - T_{ab}$ is compact.*

3. *Let $a(z),\ b(z) \in L_\infty(D)$ and be continuous at all points of $\gamma = \partial D$. Then the commutator $[T_a, T_b] = T_a T_b - T_b T_a$ is compact.*

Note that if two L_∞-functions $a(z)$ and $b(z)$ are continuous at all points of the boundary $\gamma = \partial D$ and coincide on γ, $a|_\gamma \equiv b|_\gamma$, then by the first statement of the previous theorem the corresponding Toeplitz operators differ on a compact summand,

$$
T_a = T_b + K,
$$

where K is compact.

That is, passing to the study of the C^*-algebra generated by Toeplitz operators with defining symbols continuous on the boundary ∂D, it is irrelevant to start either with symbols from any algebra $\mathcal{A}$, defined at the beginning of Section 2.5, or just with the algebra $C(\overline{D})$ of continuous symbols; the resulting algebras will be the same.

Thus it is sufficient to introduce the C^*-algebra $\mathcal{T} = \mathcal{T}(C(\overline{D}))$ generated by all Toeplitz operators with defining symbols $a(z) \in C(\overline{D})$.

The next lemma has been essentially established by Coburn [42].

Lemma 2.8.4. *The C^*-algebra $\mathcal{T} = \mathcal{T}(C(\overline{D}))$ is irreducible and contains the whole ideal $\mathcal{K}$ of compact operators acting on $\mathcal{A}^2(D)$.*

Proof. To prove that the algebra $\mathcal{T} = \mathcal{T}(C(\overline{D}))$ is irreducible we show that each orthogonal projection P commuting with all Toeplitz operators T_a, where $a \in C(\overline{D})$, has to be either 0, or I. Having such a projection, introduce the function $g = P1$, which is obviously in $\mathcal{A}^2(D)$.

Now for each function $\varphi \in C(\overline{D}) \cap \mathcal{A}^2(D)$ we have

$$P\varphi = PT_\varphi 1 = T_\varphi P1 = T_\varphi g = g \cdot \varphi.$$

That is, on the dense set $C(\overline{D}) \cap \mathcal{A}^2(D)$ of $\mathcal{A}^2(D)$ the projection P acts as the multiplication operator by the analytic function $g(z)$.

It is well known (see, for example, [102] p. 25, for the case of the unit ball), that for an analytic function g one has $g\,\mathcal{A}^2(D) \subset \mathcal{A}^2(D)$ if and only if g is bounded.

The projection P, being a bounded linear operator, can be obviously extended from $C(\overline{D}) \cap \mathcal{A}^2(D)$ onto the whole Bergman space $\mathcal{A}^2(D)$. Thus the function $g(z)$ has to be bounded, and $Pf = g \cdot f$, for all $f \in \mathcal{A}^2(D)$.

Now the bounded analytic function g satisfies the property

$$g = P1 = P^2 1 = P(P1) = Pg = g^2.$$

Thus either $g(z) \equiv 0$, or $g(z) \equiv 1$, or equivalently, either $P = 0$, or $P = I$. That is, the C^*-algebra $\mathcal{T} = \mathcal{T}(C(\overline{D}))$ is irreducible.

Being irreducible the algebra $\mathcal{T} = \mathcal{T}(C(\overline{D}))$ contains non-trivial compact operators. Thus (see, for example, [148] Theorem 2.4.9) it contains the whole ideal $\mathcal{K}$ of compact operators on $\mathcal{A}^2(D)$. $\qquad\square$

We note that the Toeplitz operator T_a, with $a \in C(\overline{D})$ and acting on the Bergman space $\mathcal{A}^2(D)$, coincides with the restriction onto the invariant subspace $\mathcal{A}^2(D)$ of the operator $B_D a B_D \in \mathcal{R}(C(\overline{D}), B_D)$, acting on $L_2(D)$. Thus as a corollary of Theorems 2.5.7 and 2.5.8 we have

Theorem 2.8.5. *The algebra $\mathcal{T}$ coincides with the set of all operators of the form*

$$T = T_a + K,$$

where $a(z) \in C(\overline{\mathbb{D}})$ and K is a compact operator. The Fredholm symbol algebra $\mathrm{Sym}\,\mathcal{T} = \mathcal{T}/\mathcal{K}$ of the algebra $\mathcal{T}$ is isomorphic to the algebra $C(\gamma)$, where $\gamma = \partial \mathbb{D}$, as previously. Identifying them, the symbol homomorphism

$$\mathrm{sym} : \mathcal{T} \longrightarrow \mathrm{Sym}\,\mathcal{T} = C(\gamma)$$

is given by

$$\mathrm{sym} : T = T_a + K \longmapsto a(t),$$

where $t \in \gamma$.

The operator $T = T_a + K$ is Fredholm if and only if its symbol is invertible,
i.e., $a(t) \neq 0$ on γ, and

$$\operatorname{Ind} T = -\frac{1}{2\pi}\{\arg a(t)\}_\gamma.$$

We continue to consider a simply connected domain D with smooth boundary $\gamma = \partial D$. As in Section 2.7 we introduce a piece-wise smooth curve ℓ and the algebra $PC(\overline{D}, \ell)$ of piece-wise continuous defining symbols.

We note first a number of qualitative differences between the properties of Toeplitz operators with continuous and piece-wise defining continuous symbols. Contrary to Lemma 2.4.4 and the second assertion of Theorem 2.8.3 we have now

Theorem 2.8.6.   1. *Let $a(z) \in PC(\overline{D}, \ell)$. Then commutator $[B_D, aI] = B_D aI - aB_D$ is not compact in general.*

2. *Let $a(z)$, $b(z) \in PC(\overline{D}, \ell)$. Then the semi-commutator $[T_a, T_b) = T_a T_b - T_{ab}$ is not compact in general.*

The proof of the above two statements just consists of calculating the Fredholm symbols of the operators involved using the recipe of Theorem 2.7.3 and the fact that the compactness of an operator is equivalent to vanishing of its Fredholm symbol.

At the same time we still have

Theorem 2.8.7. *Let $a(z)$, $b(z) \in PC(\overline{D}, \ell)$. Then the commutator $[T_a, T_b] = T_a T_b - T_b T_a$ is compact.*

Consider now the C^*-algebra $\mathcal{T}_{PC} = \mathcal{T}(PC(\overline{D}, \ell))$ generated by all Toeplitz operators T_a with defining symbols $a(z) \in PC(\overline{D}, \ell)$. As in Section 2.7 we assume that $\ell \cap \gamma = \{t_1, t_2, \ldots, t_m\}$, and let $\widehat{\gamma}$ be the set γ, cut by points $t_p \in \ell \cap \gamma$. The pair of points which correspond to a point $t_p \in \ell \cap \gamma$, $p = \overline{1, m}$, we denote by $t_p - 0$ and $t_p + 0$, following the positive orientation of γ. Let $\overline{X} = \bigsqcup_{p=1}^m \Delta_p$ be the disjoint union of segments $\Delta_p = [0, 1]$. Denote by Γ the union $\widehat{\gamma} \cup \overline{X}$ with the point identification

$$t_p - 0 \equiv 0_p, \qquad t_p + 0 \equiv 1_p,$$

where $t_p \pm 0 \in \widehat{\gamma}$, 0_p and 1_p are the boundary points of Δ_p, $p = 1, 2, \ldots, m$.

Then as a corollary of Theorem 2.7.3, Corollary 2.7.4, and the last part of Theorem 2.8.5 we have

Theorem 2.8.8. *The C^*-algebra $\mathcal{T}_{PC}$ is irreducible and contains the ideal $\mathcal{K}$ of compact operators. The Fredholm symbol algebra $\operatorname{Sym} \mathcal{T}_{PC} = \mathcal{T}_{PC}/\mathcal{K}$ is isomorphic to the algebra $C(\Gamma)$. Identifying them, the symbol homomorphism*

$$\operatorname{sym} : \mathcal{T}_{PC} \to \operatorname{Sym} \mathcal{T}_{PC} = C(\Gamma)$$

is generated by the following mapping of generators of $\mathcal{T}_{PC}$,

$$\operatorname{sym} : T_a \longmapsto \begin{cases} a(t), & t \in \widehat{\gamma} \\ a(t_p - 0)(1 - x) + a(t_p + 0)x, & x \in [0, 1] \end{cases},$$

where $t_p \in \ell \cap \gamma$, $p = 1, 2, \ldots, m$.

The operator $T \in \mathcal{T}_{PC}$ is Fredholm if and only if its symbol is invertible, i.e., $\operatorname{sym} T \neq 0$ on Γ, and

$$\operatorname{Ind} T = -\frac{1}{2\pi} \{\operatorname{sym} T\}_\Gamma.$$

The result of the above theorem repeats the corresponding result of [82] for the Hardy space Toeplitz operators. This is a direct consequence of the two facts: the same local description at each point of γ where the defining symbol is continuous, and

$$\operatorname{sp}(P - B)^2 = [0, 1] = \operatorname{sp}(\chi_{\gamma_0} I - P_+)^2,$$

where B is the Bergman projection, P is defined in Section 2.6 (see Corollary 2.6.3), P_+ is the Szegö projection, and χ_{γ_0} is the characteristic function of a connected subarc γ_0 of γ.

We note that for piece-wise continuous defining symbols the product of two Toeplitz operators is not anymore a compact perturbation of a Toeplitz operator, in general. The algebra $\mathcal{T}_{PC}$ does not coincide with the set of all operators of the form $T_a + K$, as in the case of continuous defining symbols. It has a much more complicated structure, coinciding with the uniform closure of the set of all elements of the form

$$\sum_{k=1}^{p} \prod_{j=1}^{q_k} T_{a_{j,k}}, \tag{2.8.1}$$

where $a_{j,k} \in PC(\overline{D}, \ell)$, $p, q_k \in \mathbb{N}$.

It is very interesting and important to understand the nature of the operators forming this algebra and, in particular, to know *whether this Toeplitz operator algebra contains any Toeplitz operator other then its initial generators*. We note that this question has remained unanswered since the earliest work on the subject.

Chapter 8 is devoted to a systematic study of this question. We give a simple and transparent representation of each operator from the above C^*-algebra. We show as well that none of the elements of the form (2.8.1) is (a compact perturbation of) a Toeplitz operator, but in spite of that the uniform closure of the set of all elements of the form (2.8.1) contains many Toeplitz operators, whose (unbounded in general) defining symbols belong to a much wider class of discontinuous functions, as compared with defining symbols of the initial generators.

2.9 Some further results on compactness

We list here more results on compactness properties of the Toeplitz operators. Recall that the first statement of Theorem 2.8.3 gives a simple and evident sufficient condition for compactness: *if a function $a(z)$ is in $L_\infty(D)$ and is continuous at all points of $\gamma = \partial D$ with $a|_\gamma \equiv 0$, then the Toeplitz operator T_a is compact.*

At first glance this condition seems to be "very near" to be necessary as well. At the same time we have an example (see Lemma 5.3.5) of a defining symbol a such that

(i) a is continuous in each point of the boundary, *except a single point $t_0 \in \partial D$*,

(ii) the function a is continuous *along the boundary*, and $a|_{\partial D} \equiv 0$,

(iii) for each $t \in \partial D \setminus \{t_0\}$,

$$\lim_{D \ni z \to t} a(z) = 0,$$

(iv) all non-tangential limits of a at the point t_0 are equal to zero,

but nevertheless the Toeplitz operator T_a is non-compact.

On the other hand there are defining symbols (see Example 6.1.7) which are oscillating bounded or even *unbounded* near each point of the boundary (no limit at each boundary point) for which nevertheless the corresponding Toeplitz operators are compact.

The above examples suggest that the behaviour near the boundary of a certain average of the defining symbol, rather then of the defining symbol itself, is responsible for compactness of the corresponding Toeplitz operator. The final result for bounded defining symbols was given by S. Axler and D. Zheng [12] in terms of the so-called *Berezin transform*, or the *Wick symbol*. In fact they gave a much more general result, the criterion of compactness of operators of the form (2.8.1).

The general definition and properties of the Wick symbol and the Berezin transform in terms of coherent states are given in Appendix A, Section A.1. In order to formulate the S. Axler and D. Zheng result we give here a definition for operators acting on the Bergman space $\mathcal{A}^2(D)$.

Let A be a bounded linear operator acting on $\mathcal{A}^2(D)$. Then its Wick symbol is the function $\widetilde{A}(z)$ defined in D as follows (see for details Appendix A):

$$\widetilde{A}(z) = \frac{\langle A k_z, k_z \rangle}{\langle k_z, k_z \rangle},$$

where $k_z(\zeta) = \overline{K_D(z, \zeta)} = K_D(\zeta, z)$.

If $A = T_a$ is the Toeplitz operator with defining symbol $a(z)$, then its Wick symbol, or the Berezin transform of the function a, is calculated by the formula

$$\widetilde{a}(z) \quad = \quad \widetilde{T_a}(z) = \frac{\langle B_D a k_z, k_z \rangle}{\langle k_z, k_z \rangle} = \frac{\langle a k_z, B_D k_z \rangle}{\langle k_z, k_z \rangle} = \frac{\langle a k_z, k_z \rangle}{\langle k_z, k_z \rangle}$$

$$= \quad \frac{1}{K_D(z, z)} \int_D a(\zeta) |K_D(z, \zeta)|^2 dv(\zeta). \tag{2.9.1}$$

For the case of the unit disk $\mathbb{D}$ we have

$$\widetilde{a}(z) = \widetilde{T_a}(z) = \frac{(1 - z\overline{z})^2}{\pi} \int_{\mathbb{D}} \frac{a(\zeta)\, dv(\zeta)}{(1 - \zeta\overline{z})^2 (1 - z\overline{\zeta})^2}.$$

The following result has been proved in [12].

Theorem 2.9.1. *Let*

$$A = \sum_{k=1}^{p} \prod_{j=1}^{q_k} T_{a_{j,k}},$$

where $a_{j,k} \in L_\infty(D)$, $p, q_k \in \mathbb{N}$. Then the operator A is compact if and only if

$$\lim_{z \to \partial D} \widetilde{A}(z) = 0.$$

In particular, the Toeplitz operator T_a is compact if and only if

$$\lim_{z \to \partial D} \frac{1}{K_D(z,z)} \int_D a(\zeta) \, |K_D(z,\zeta)|^2 dv(\zeta) = 0.$$

Chapter 3

Bergman and Poly-Bergman Spaces

We start by recalling an old and well-known result. Let $H_+^2(\mathbb{R})(\subset L_2(\mathbb{R}))$ be the Hardy space on the upper half-plane Π in $\mathbb{C}$, which by definition consists of all functions φ on $\mathbb{R}$ admitting analytic continuation in Π and satisfying the condition

$$\sup_{v>0} \int_{\mathbb{R}+iv} |\varphi(u+iv)|^2 du < \infty.$$

Let $P_{\mathbb{R}}^+$ be the (orthogonal) Szegö projection of $L_2(\mathbb{R})$ onto $H_+^2(\mathbb{R})$. Then: *the Fourier transform F gives an isometric isomorphism of the space $L_2(\mathbb{R})$, under which*

1. *the Hardy space $H_+^2(\mathbb{R})$ is mapped onto $L_2(\mathbb{R}_+)$,*

$$F : H_+^2(\mathbb{R}) \longrightarrow L_2(\mathbb{R}_+),$$

2. *the Szegö projection $P_{\mathbb{R}}^+ : L_2(\mathbb{R}) \longrightarrow H_+^2(\mathbb{R})$ is unitary equivalent to the projection*

$$F P_{\mathbb{R}}^+ F^{-1} = \chi_+ I.$$

Furthermore, if $H_-^2(\mathbb{R})$ is the Hardy space on the lower half-plane and $P_{\mathbb{R}}^-$ is the (orthogonal Szegö) projection of $L_2(\mathbb{R})$ onto $H_-^2(\mathbb{R})$, then

$$L_2(\mathbb{R}) = H_+^2(\mathbb{R}) \oplus H_-^2(\mathbb{R}),$$

and the *same* Fourier transform gives

1. $\quad F(H_-^2(\mathbb{R})) = L_2(\mathbb{R}_-),$

2. $\quad F P_{\mathbb{R}}^- F^{-1} = \chi_- I.$

Here we use the obvious notation: $\mathbb{R}_+$ ($\mathbb{R}_-$) is the positive (negative) half-line, and χ_+ (χ_-) is the characteristic function of $\mathbb{R}_+$ ($\mathbb{R}_-$).

Consider now the space $L_2(D)$ and its Bergman subspace $\mathcal{A}^2(D)$. The aim of this and the next chapter is to present a "Bergman space version" of the above Hardy space result. That is, we will answer the following natural questions:

  - how much room does the Bergman space $\mathcal{A}^2(D)$ occupy inside $L_2(D)$?

  - does there exist a complete decomposition of $L_2(D)$ into analytic and analytic like spaces?

  - does there exist a unitary operator on $L_2(D)$, which *simultaneously* reduces all the pieces of decomposition to some simple transparent form?

  - are there any connections between these pieces and the Hardy spaces?

In this chapter we present the results for the case when D is the upper half-plane Π, and in the next chapter – for the case of the unit disk $\mathbb{D}$.

3.1 Bergman space and Bergman projection

Let Π be the upper half-plane in $\mathbb{C}$. Consider the space $L_2(\Pi)$ with the usual Lebesgue plane measure $dv(w) = du\,dv$, $w = u + iv$, and its Bergman subspace $\mathcal{A}^2(\Pi)$. The Bergman projection B_Π of $L_2(\Pi)$ onto $\mathcal{A}^2(\Pi)$ has the form

$$(B_\Pi\varphi)(w) = \int_\Pi K_\Pi(w,\omega)\varphi(\omega)\,dv(\omega),$$

where the Bergman kernel function $K_\Pi(w,\omega)$ of the upper half-plane is given (see Example 2.2.6) by

$$K_\Pi(w,\omega) = -\frac{1}{\pi}\cdot\frac{1}{(w-\bar{\omega})^2}.$$

We note that the Bergman space $\mathcal{A}^2(\Pi)$ can be described alternatively as the (closed) subspace of $L_2(\Pi)$, which consists of all functions satisfying the equation

$$2\frac{\partial}{\partial\bar{w}}\varphi = \left(\frac{\partial}{\partial u} + i\frac{\partial}{\partial v}\right)\varphi = 0.$$

Introduce the unitary operator

$$U_1 = F \otimes I : L_2(\Pi) = L_2(\mathbb{R}) \otimes L_2(\mathbb{R}_+) \longrightarrow L_2(\mathbb{R}) \otimes L_2(\mathbb{R}_+), \qquad (3.1.1)$$

where the Fourier transform $F : L_2(\mathbb{R}) \to L_2(\mathbb{R})$ is given by

$$(Ff)(u) = \frac{1}{\sqrt{2\pi}}\int_\mathbb{R} e^{-iu\xi}f(\xi)\,d\xi.$$

Then the image of the Bergman space $\mathcal{A}_1^2 = U_1(\mathcal{A}^2(\Pi))$ can be described as the (closed) subspace of $L_2(\Pi)$, which consists of all functions satisfying the equation

$$(F \otimes I)2\frac{\partial}{\partial \overline{w}}(F^{-1} \otimes I)\,\varphi = i(u + \frac{\partial}{\partial v})\,\varphi = 0. \tag{3.1.2}$$

The equation (3.1.2) is easy to solve, and its general solution has the form

$$\varphi(u, v) = \psi(u) \cdot e^{-uv}.$$

The condition $\varphi \in L_2(\Pi)$ suggests (and we will check this right now) that the space $\mathcal{A}_1^2$ consists of all functions of the form

$$\varphi_0(u, v) = \chi_+(u)\,\sqrt{2u}\,f(u) \cdot e^{-uv}, \tag{3.1.3}$$

where $f(u) \in L_2(\mathbb{R})$, and $\chi_+(u)$ is the characteristic function of the positive half-line. Here, in addition

$$\|\varphi_0(u, v)\|_{\mathcal{A}_1^2} = \|f(u)\|_{L_2(\mathbb{R}_+)}.$$

The orthogonal projection $B_1 : L_2(\Pi) \to \mathcal{A}_1^2$ obviously has the form

$$B_1 = (F \otimes I)B_\Pi(F^{-1} \otimes I).$$

Calculate

$$\begin{aligned}
(F^{-1} \otimes I)B_1\varphi &= B_\Pi(F^{-1} \otimes I)\varphi = \langle (F^{-1} \otimes I)\varphi, \overline{K_\Pi(w, \omega)} \rangle \\
&= \langle \varphi, (F \otimes I)\overline{K_\Pi(w, \omega)} \rangle = \langle \varphi, (F \otimes I)K_\Pi(\omega, w) \rangle.
\end{aligned}$$

By [19] the Fourier transform (with respect to $\xi = \operatorname{Re}\omega$) of the function

$$K_\Pi(\omega, w) = -\frac{1}{\pi} \cdot \frac{1}{(\omega - \overline{w})^2} = \frac{1}{\pi} \cdot \frac{1}{(i\overline{w} + \eta - i\xi)^2}$$

is equal to

$$\sqrt{\frac{2}{\pi}}\,\chi_+(\xi)\,\xi \cdot e^{-(\eta + i\overline{w})\xi}.$$

Thus

$$\begin{aligned}
(F^{-1} \otimes I)B_1\varphi &= \langle \varphi, (F \otimes I)K_\Pi(\omega, \overline{w}) \rangle \\
&= \sqrt{\frac{2}{\pi}} \int_\Pi \varphi(\xi, \eta)\,\chi_+(\xi)\,\xi \cdot e^{-(\eta - iw)\xi}\,d\xi\,d\eta \\
&= \sqrt{\frac{2}{\pi}} \int_\Pi \varphi(\xi, \eta)\,\chi_+(\xi)\,\xi \cdot e^{-(\eta + v)\xi} \cdot e^{iu\xi}\,d\xi\,d\eta \\
&= \frac{1}{\sqrt{2\pi}} \int_\mathbb{R} \left(\int_{\mathbb{R}_+} 2\xi\,\chi_+(\xi)\,\varphi(\xi, \eta) \cdot e^{-(\eta + v)\xi}\,d\eta \right) e^{iu\xi}\,d\xi \\
&= (F^{-1} \otimes I)\left(2u\,\chi_+(u) \cdot e^{-uv} \int_{\mathbb{R}_+} \varphi(u, \eta) \cdot e^{-\eta u}\,d\eta \right).
\end{aligned}$$

Or

$$(B_1\varphi)(u,v) = 2u\,\chi_+(u)\cdot e^{-uv}\int_{\mathbb{R}_+}\varphi(u,\eta)\cdot e^{-\eta u}\,d\eta.$$

Thus for each function $\varphi \in L_2(\Pi)$ its image $(B_1\varphi)(u,v)$ has the form (3.1.3) with

$$f(u) = \sqrt{2u}\int_{\mathbb{R}_+}\varphi(u,\eta)\cdot e^{-\eta u}\,d\eta,$$

and furthermore, for each function φ_0 of the form (3.1.3) we have

$$(B_1\varphi_0)(u,v) = \varphi_0(u,v).$$

Indeed,

$$\begin{aligned}
(B_1\varphi_0)(u,v) &= 2u\,\chi_+(u)\cdot e^{-uv}\int_{\mathbb{R}_+}\varphi_0(u,\eta)\cdot e^{-\eta u}\,d\eta \\
&= 2u\,\chi_+(u)\cdot e^{-uv}\,\sqrt{2u}\,f(u)\int_{\mathbb{R}_+}e^{-2\eta u}\,d\eta \\
&= \chi_+(u)\sqrt{2u}\,f(u)\cdot e^{-uv}\int_{\mathbb{R}_+}2u\,e^{-2\eta u}\,d\eta \\
&= \chi_+(u)\sqrt{2u}\,f(u)\cdot e^{-uv} = \varphi_0(u,v).
\end{aligned}$$

Thus $\mathcal{A}_1^2$ coincides with the space of all functions of the form (3.1.3).

Introduce the unitary operator

$$U_2 : L_2(\Pi) = L_2(\mathbb{R})\otimes L_2(\mathbb{R}_+) \longrightarrow L_2(\mathbb{R})\otimes L_2(\mathbb{R}_+)$$

by the rule

$$U_2 : \varphi(u,v) \longmapsto \frac{1}{\sqrt{2|x|}}\,\varphi\!\left(x,\frac{y}{2|x|}\right). \tag{3.1.4}$$

Then the inverse operator $U_2^{-1} = U_2^* : L_2(\mathbb{R})\otimes L_2(\mathbb{R}_+) \longrightarrow L_2(\mathbb{R})\otimes L_2(\mathbb{R}_+)$ is given by

$$U_2^{-1} : \varphi(x,y) \longmapsto \sqrt{2|u|}\,\varphi(u,2|u|\cdot v).$$

Denote $\mathcal{A}_2^2 = U_2(\mathcal{A}_1^2)$. The operator $B_2 = U_2\,B_1\,U_2^{-1}$ is obviously the orthogonal projection of $L_2(\Pi)$ onto $\mathcal{A}_2^2$. Calculate

$$\begin{aligned}
(B_2\varphi)(x,y) &= U_2\!\left(2u\,\chi_+(u)\,e^{-uv}\int_{\mathbb{R}_+}e^{-u\eta}\sqrt{2u}\,\varphi(u,2u\eta)\,d\eta\right) \\
&= U_2\!\left(\sqrt{2u}\,\chi_+(u)\,e^{-uv}\int_{\mathbb{R}_+}\varphi(u,\nu)\,e^{-\nu/2}\,d\nu\right) \\
&= \chi_+(x)\,e^{-y/2}\int_{\mathbb{R}_+}\varphi(x,\nu)\,e^{-\nu/2}\,d\nu.
\end{aligned}$$

Introduce $\ell_0(y) = e^{-y/2}$. We have that $\ell_0(y) \in L_2(\mathbb{R}_+)$ and $\|\ell_0(y)\| = 1$. Denote by L_0 the one-dimensional subspace of $L_2(\mathbb{R}_+)$ generated by $\ell_0(y)$, then the one-dimensional projection P_0 of $L_2(\mathbb{R}_+)$ onto L_0 has the form

$$(P_0\psi)(y) = \langle \psi, \ell_0 \rangle \cdot \ell_0 = e^{-y/2} \int_{\mathbb{R}_+} \psi(\nu)\, e^{-\nu/2}\, d\nu. \qquad (3.1.5)$$

Thus

$$B_2 = \chi_+(x) I \otimes P_0.$$

This leads to the following theorem.

Theorem 3.1.1. *The unitary operator $U = U_2 U_1$ gives an isometric isomorphism of the space $L_2(\Pi) = L_2(\mathbb{R}) \otimes L_2(\mathbb{R}_+)$ under which*

1. *the Bergman space $\mathcal{A}^2(\Pi)$ is mapped onto $L_2(\mathbb{R}_+) \otimes L_0$,*

$$U : \mathcal{A}^2(\Pi) \longrightarrow L_2(\mathbb{R}_+) \otimes L_0,$$

   where L_0 is the one-dimensional subspace of $L_2(\mathbb{R}_+)$, generated by $\ell_0(y) = e^{-y/2}$,

2. *the Bergman projection B_Π is unitary equivalent to*

$$U\, B_\Pi\, U^{-1} = \chi_+ I \otimes P_0,$$

   where P_0 is the one-dimensional projection (3.1.5) of $L_2(\mathbb{R}_+)$ onto L_0.

Remark 3.1.2. The above result describes the structure of the Bergman space inside $L_2(\Pi)$, and is a "Bergman space version" of the following, mentioned in the introduction to this chapter, "Hardy space version" result: *the Fourier transform F gives an isometric isomorphism of the space $L_2(R)$, under which*

1. *the Hardy space $H_+^2(\mathbb{R})$ is mapped onto $L_2(\mathbb{R}_+)$,*

$$F : H_+^2(\mathbb{R}) \longrightarrow L_2(\mathbb{R}_+),$$

2. *the Szegö projection $P_{\mathbb{R}}^+ : L_2(\mathbb{R}) \longrightarrow H_+^2(\mathbb{R})$ is unitary equivalent to*

$$F\, P_{\mathbb{R}}^+\, F^{-1} = \chi_+ I.$$

Introduce the isometric imbedding

$$R_0 : L_2(\mathbb{R}_+) \longrightarrow L_2(\mathbb{R}) \otimes L_2(\mathbb{R}_+)$$

by the rule

$$(R_0 f)(x, y) = \chi_+(x)\, f(x)\, \ell_0(y);$$

here the function $f(x)$ is extended to an element of $L_2(\mathbb{R})$ by setting $f(x) \equiv 0$, for $x < 0$. The image of R_0 obviously coincides with the space $\mathcal{A}_2^2$. The adjoint operator

$$R_0^* : L_2(\Pi) \to L_2(\mathbb{R}_+)$$

is given by

$$(R_0^*\varphi)(x) = \chi_+(x) \int_{\mathbb{R}_+} \varphi(x, \eta)\, \ell_0(\eta)\, d\eta,$$

and

$$
\begin{aligned}
R_0^* R_0 &= I &&: & L_2(\mathbb{R}_+) &\longrightarrow L_2(\mathbb{R}_+), \\
R_0 R_0^* &= B_2 &&: & L_2(\Pi) &\longrightarrow \mathcal{A}_2^2 = L_2(\mathbb{R}_+) \otimes L_0.
\end{aligned}
$$

Now the operator $R = R_0^* U$ maps the space $L_2(\Pi)$ onto $L_2(\mathbb{R}_+)$, and the restriction

$$R|_{\mathcal{A}^2(\Pi)} : \mathcal{A}^2(\Pi) \longrightarrow L_2(\mathbb{R}_+)$$

is an isometric isomorphism. The adjoint operator

$$R^* = U^* R_0 : L_2(\mathbb{R}_+) \longrightarrow \mathcal{A}^2(\Pi) \subset L_2(\Pi)$$

is an isometric isomorphism of $L_2(\mathbb{R}_+)$ onto the subspace $\mathcal{A}^2(\Pi)$ of the space $L_2(\Pi)$.

Remark 3.1.3. We have

$$
\begin{aligned}
R R^* &= I &&: & L_2(\mathbb{R}_+) &\longrightarrow L_2(\mathbb{R}_+), \\
R^* R &= B_\Pi &&: & L_2(\Pi) &\longrightarrow \mathcal{A}^2(\Pi).
\end{aligned}
$$

Theorem 3.1.4. *The isometric isomorphism*

$$R^* = U^* R_0 : L_2(\mathbb{R}_+) \longrightarrow \mathcal{A}^2(\Pi)$$

is given by

$$(R^* f)(z) = \frac{1}{\sqrt{\pi}} \int_{\mathbb{R}_+} \sqrt{\xi}\, f(\xi)\, e^{iz \cdot \xi}\, d\xi. \qquad (3.1.6)$$

Proof. Calculate

$$
\begin{aligned}
(R^* f)(z) &= (U_1^* U_2^* R_0 f)(z) \\
&= (F^{-1} \otimes I)(\chi_+(\xi)\, f(\xi)\, \sqrt{2\xi}\, e^{-xy}) \\
&= \frac{1}{\sqrt{2\pi}} \int_{\mathbb{R}} \chi_+(\xi)\, f(\xi)\, \sqrt{2\xi}\, e^{-xy}\, e^{ix\xi}\, d\xi \\
&= \frac{1}{\sqrt{\pi}} \int_{\mathbb{R}_+} \sqrt{\xi}\, f(\xi)\, e^{i(x+iy) \cdot \xi}\, d\xi.
\end{aligned}
$$

$\square$

Corollary 3.1.5. *The inverse isomorphism*

$$R : \mathcal{A}^2(\Pi) \longrightarrow L_2(\mathbb{R}_+)$$

is given by

$$(R\varphi)(x) = \sqrt{x}\,\frac{1}{\sqrt{\pi}}\int_{\Pi}\varphi(w)\,e^{-i\,\overline{w}\cdot x}\,d\mu(w). \tag{3.1.7}$$

3.2 Connections between Bergman and Hardy spaces

By Theorem 3.1.1 and the statement inside Remark 3.1.2 we have immediately

Theorem 3.2.1. *The unitary operator $W = (F^{-1}\otimes I)U_2(F\otimes I)$ gives an isometric isomorphism of the space $L_2(\Pi) = L_2(\mathbb{R})\otimes L_2(\mathbb{R}_+)$ under which*

1. *the Bergman $\mathcal{A}^2(\Pi)$ and the Hardy $H_+^2(\mathbb{R})$ spaces are connected by the formula*

$$W(\mathcal{A}^2(\Pi)) = H_+^2(\mathbb{R})\otimes L_0,$$

2. *the Bergman B_Π and the Szegö $P_{\mathbb{R}}^+$ projections are connected by the formula*

$$W\,B_\Pi\,W^{-1} = P_{\mathbb{R}}^+\otimes P_0,$$

where P_0 is the one-dimensional projection (3.1.5) of $L_2(\mathbb{R}_+)$ onto the one-dimensional space L_0 generated by $\ell_0(y) = e^{-y/2} \in L_2(\mathbb{R}_+)$.

For a domain D (which will be Π in this chapter) besides the Bergman space $\mathcal{A}^2(D)$ introduce the space $\widetilde{\mathcal{A}}^2(D)$ being the (closed) subspace of $L_2(D)$ consisting of all functions anti-analytic in D. It is well known (see, for example, [64]) that if $K_D(z,\zeta)$ is the Bergman kernel-function of the domain D, then the orthogonal projection $\widetilde{B}_D$ of $L_2(D)$ onto $\widetilde{\mathcal{A}}^2(D)$ is given by

$$(\widetilde{B}_D\varphi)(z) = \int_D K_D(\zeta,z)\,\varphi(\zeta)\,d\mu(\zeta).$$

Theorem 3.2.2. *The unitary operator $U = U_2U_1$ gives an isometric isomorphism of the space $L_2(\Pi) = L_2(\mathbb{R})\otimes L_2(\mathbb{R}_+)$ under which*

1. *the space $\widetilde{\mathcal{A}}^2(\Pi)$ is mapped onto $L_2(\mathbb{R}_-)\otimes L_0$,*

$$U : \widetilde{\mathcal{A}}^2(\Pi) \longrightarrow L_2(\mathbb{R}_-)\otimes L_0,$$

where L_0 is the one-dimensional subspace of $L_2(\mathbb{R}_+)$ generated by $\ell_0(y) = e^{-y/2}$,

2. *the projection $\widetilde{B}_\Pi$ is unitary equivalent to*

$$U\,\widetilde{B}_\Pi\,U^{-1} = \chi_- I \otimes P_0,$$

where χ_- is the characteristic function of the negative half-line, and P_0 is the one-dimensional projection 3.1.5 of $L_2(\mathbb{R}_+)$ onto L_0.

Proof. Follows all the steps of the proof of Theorem 3.1.1. We note only that all $L_2(\Pi)$-solutions of the equation

$$(u - \frac{\partial}{\partial v})\varphi = 0$$

are of the form

$$\varphi(u,v) = \chi_-(u)\,\sqrt{2|u|}\,f(u)\,e^{uv},$$

where $f(u) \in L_2(\mathbb{R})$, and that the Fourier image (with respect to $\xi = \mathrm{Re}\,\omega$) of the function

$$\overline{K_\Pi(\omega,w)} = -\frac{1}{\pi}\,\frac{1}{(\overline{\omega} - w)^2}$$

is (see, for example, [19]) equal to the function

$$-\sqrt{\frac{2}{\pi}}\,\xi\,\chi_-(\xi)\,e^{(\eta - iw)\cdot\xi}.$$

The projection $\widetilde{B}_1 = U_1\,\widetilde{B}_\Pi\,U_1^{-1}$ onto the space $\widetilde{\mathcal{A}}_1^2 = U_1(\widetilde{\mathcal{A}}^2(\Pi))$ has the form

$$(\widetilde{B}_1\varphi)(u.v) = 2|u|\,\chi_-(u)\,e^{uv}\int_{\mathbb{R}_+}\varphi(u,\eta)\,e^{u\eta}\,d\eta. \qquad \square$$

Introduce now the Hardy space $H_-^2(\mathbb{R})$ as the subspace of $L_2(\mathbb{R})$ which consists of the functions φ having analytic extension to the lower half-plane and satisfying the condition

$$\sup_{v<0}\int_{\mathbb{R}+iv}|\varphi(u+iv)|^2 < \infty.$$

It is well known that the operator $P_{\mathbb{R}}^- = F^{-1}\chi_-(\xi)\,F$ is the orthogonal (Szegö) projection of $L_2(\mathbb{R})$ onto $H_-^2(\mathbb{R})$.

Corollary 3.2.3. *The unitary operator $W = (F^{-1} \otimes I)U_2(F \otimes I)$ gives an isometric isomorphism of the space $L_2(\Pi) = L_2(\mathbb{R}) \otimes L_2(\mathbb{R}_+)$ under which*

1. *the spaces $\widetilde{\mathcal{A}}^2(\Pi)$ and $H_-^2(\mathbb{R})$ are connected by the formula*

$$W(\widetilde{\mathcal{A}}^2(\Pi)) = H_-^2(\mathbb{R}) \otimes L_0,$$

2. *the projections $\widetilde{B}_\Pi$ and $P_{\mathbb{R}}^-$ are connected by the formula*

$$W\,\widetilde{B}_\Pi\,W^{-1} = P_{\mathbb{R}}^- \otimes P_0,$$

where P_0 is the one-dimensional projection (3.1.5) of $L_2(\mathbb{R}_+)$ onto one-dimensional space L_0 generated by $\ell_0(y) = e^{-y/2} \in L_2(\mathbb{R}_+)$.

Remark 3.2.4. The direct sum of both Hardy spaces coincides with the whole space

$$L_2(\mathbb{R}) = H_+^2(\mathbb{R}) \oplus H_-^2(\mathbb{R}),$$

while the direct sum $\mathcal{A}^2(\Pi) \oplus \widetilde{\mathcal{A}}^2(\Pi)$ has the (sufficiently large) orthogonal complement

$$\mathcal{A}^\perp(\Pi) = L_2(\Pi) \ominus (\mathcal{A}^2(\Pi) \oplus \widetilde{\mathcal{A}}^2(\Pi)),$$

whose image under the mapping W is equal to

$$W(\mathcal{A}^\perp(\Pi)) = L_2(\mathbb{R}) \otimes L_0^\perp.$$

3.3   Poly-Bergman spaces, decomposition of $L_2(\Pi)$

Analogously to the Bergman spaces $\mathcal{A}^2(\Pi)$ and $\widetilde{\mathcal{A}}^2(\Pi)$, of respectively analytic and anti-analytic functions from $L_2(\Pi)$, introduce the spaces of poly-analytic and poly-anti-analytic functions (see, for example, [17, 18, 63, 64]), the poly-Bergman spaces.

Define the space $\mathcal{A}_n^2(\Pi)$ of n-analytic functions as the (closed) subspace of $L_2(\Pi)$ of all functions $\varphi = \varphi(w, \overline{w}) = \varphi(u, v)$, which satisfy the equation

$$\left(\frac{\partial}{\partial \overline{w}}\right)^n \varphi = \frac{1}{2^n}\left(\frac{\partial}{\partial u} + i\frac{\partial}{\partial v}\right)^n \varphi = 0.$$

Similarly, define the space $\widetilde{\mathcal{A}}_n^2(\Pi)$ of n-anti-analytic functions as the (closed) subspace of $L_2(\Pi)$ of all functions $\varphi = \varphi(w, \overline{w}) = \varphi(u, v)$, which satisfy the equation

$$\left(\frac{\partial}{\partial w}\right)^n \varphi = \frac{1}{2^n}\left(\frac{\partial}{\partial u} - i\frac{\partial}{\partial v}\right)^n \varphi = 0.$$

Of course, we have $\mathcal{A}_1^2(\Pi) = \mathcal{A}^2(\Pi)$ and $\widetilde{\mathcal{A}}_1^2(\Pi) = \widetilde{\mathcal{A}}^2(\Pi)$, for $n = 1$, as well as $\mathcal{A}_n^2(\Pi) \subset \mathcal{A}_{n+1}^2(\Pi)$ and $\widetilde{\mathcal{A}}_n^2(\Pi) \subset \widetilde{\mathcal{A}}_{n+1}^2(\Pi)$, for each $n \in \mathbb{N}$.

Recall (see, for example, [20]), that the Laguerre polynomial $L_n(y)$ of degree n, where $n = 0, 1, 2, \ldots$, and type 0 is defined by

$$
\begin{aligned}
L_n(y) &= L_n^0(y) = \frac{e^y}{n!}\frac{d^n}{dy^n}(e^{-y}\,y^n) \\[2mm]
&= \sum_{k=0}^{n} \frac{n!}{k!(n-k)!}\frac{(-y)^k}{k!}, \qquad y \in \mathbb{R}_+,
\end{aligned}
\tag{3.3.1}
$$

and that the system of functions

$$\ell_n(y) = e^{-y/2} L_n(y), \qquad n = 0, 1, 2, \ldots \tag{3.3.2}$$

forms an orthonormal basis in the space $L_2(\mathbb{R}_+)$.

Denote by L_n, $n = 0, 1, 2, \ldots$, the one-dimensional subspace of $L_2(\mathbb{R}_+)$, generated by the function $\ell_n(y)$. Note, that for $n = 0$ this definition gives exactly the space L_0 already defined. And let

$$L_n^{\oplus} = \bigoplus_{k=0}^{n} L_k$$

be the direct sum of the first $(n + 1)$ spaces.

Theorem 3.3.1. *The unitary operator $U = U_2 U_1 : L_2(\mathbb{R}) \otimes L_2(\mathbb{R}_+) \to L_2(\mathbb{R}) \otimes L_2(\mathbb{R}_+)$ maps the space $\mathcal{A}_n^2(\Pi)$ of n-analytic functions onto the space $L_2(\mathbb{R}_+) \otimes L_{n-1}^{\oplus}$.*

Proof. The space $U(\mathcal{A}_n^2(\Pi))$ obviously coincides with the set of all functions from $L_2(\Pi) = L_2(\mathbb{R}) \otimes L_2(\mathbb{R}_+)$, which satisfy the equation

$$U \left(\frac{\partial}{\partial u} + i \frac{\partial}{\partial v} \right)^n U^{-1} \varphi = i^n U_2 \left(u + \frac{\partial}{\partial v} \right)^n U_2^{-1} \varphi$$

$$= i^n |x|^n \left(\operatorname{sign}(x) + 2 \frac{\partial}{\partial y} \right)^n \varphi = 0.$$

It is easy to see that the intersection of the general solution of this equation with the space $L_2(\mathbb{R}) \otimes L_2(\mathbb{R}_+)$ coincides with the set of all functions

$$\sum_{k=0}^{n-1} \chi_+(x)\, \psi_k(x)\, y^k e^{-y/2},$$

where $\psi_k(x) \in L_2(\mathbb{R})$, for all $k = \overline{0, n-1}$, or rearranging polynomials on y, it coincides with the set of all functions

$$\sum_{k=0}^{n-1} \chi_+(x)\, f_k(x)\, L_k(y) e^{-y/2} = \sum_{k=0}^{n-1} \chi_+(x)\, f_k(x)\, \ell_k(y),$$

where $f_k(x) \in L_2(\mathbb{R})$, for all $k = \overline{0, n-1}$. $\qquad \square$

Introduce the space $\mathcal{A}_{(n)}^2$ of true-n-analytic functions by

$$\mathcal{A}_{(n)}^2 = \mathcal{A}_n^2 \ominus \mathcal{A}_{n-1}^2,$$

for $n > 1$, and by $\mathcal{A}_{(1)}^2 = \mathcal{A}_1^2$, for $n = 1$, then, of course,

$$\mathcal{A}_n^2 = \bigoplus_{k=1}^{n} \mathcal{A}_{(k)}^2.$$

Corollary 3.3.2. *The unitary operator $U : L_2(\mathbb{R}) \otimes L_2(\mathbb{R}_+) \to L_2(\mathbb{R}) \otimes L_2(\mathbb{R}_+)$ maps the space $\mathcal{A}^2_{(n)}(\Pi)$ of true-n-analytic functions onto the space $L_2(\mathbb{R}_+) \otimes L_{n-1}$.*

Analogously in the anti-analytic situations we have the following assertion.

Theorem 3.3.3. *The unitary operator $U : L_2(\mathbb{R}) \otimes L_2(\mathbb{R}_+) \to L_2(\mathbb{R}) \otimes L_2(\mathbb{R}_+)$ maps the space $\widetilde{\mathcal{A}}^2_n(\Pi)$ of n-anti-analytic functions onto the space $L_2(\mathbb{R}_-) \otimes L^{\oplus}_{n-1}$.*

Proof. Follow all the steps of the proof of Theorem 3.3.1. The space $U(\widetilde{\mathcal{A}}^2_n(\Pi))$ coincides with the set of all functions from $L_2(\Pi) = L_2(\mathbb{R}) \otimes L_2(\mathbb{R}_+)$, which satisfy the equation

$$U \left(\frac{\partial}{\partial u} - i \frac{\partial}{\partial v} \right)^n U^{-1} \varphi \;=\; i^n\, U_2 \left(u - \frac{\partial}{\partial v} \right)^n U_2^{-1} \varphi$$

$$=\; i^n\, |x|^n \left(\text{sign}(x) - 2 \frac{\partial}{\partial y} \right)^n \varphi = 0.$$

Now the intersection of the general solution of this equation with the space $L_2(\mathbb{R}) \otimes L_2(\mathbb{R}_+)$ coincides with the set of all functions

$$\sum_{k=0}^{n-1} \chi_-(x)\, \psi_k(x)\, y^k e^{-y/2},$$

where $\psi_k(x) \in L_2(\mathbb{R})$, for all $k = \overline{0, n-1}$, or rearranging polynomials on y, it coincides with the set of all functions

$$\sum_{k=0}^{n-1} \chi_-(x)\, f_k(x)\, L_k(y) e^{-y/2} = \sum_{k=0}^{n-1} \chi_-(x)\, f_k(x)\, \ell_k(y),$$

where $f_k(x) \in L_2(\mathbb{R})$, for all $k = \overline{0, n-1}$. $\qquad\square$

Symmetrically, introduce the space $\widetilde{\mathcal{A}}^2_{(n)}$ of true-n-anti-analytic functions by

$$\widetilde{\mathcal{A}}^2_{(n)} = \widetilde{\mathcal{A}}^2_n \ominus \widetilde{\mathcal{A}}^2_{n-1},$$

for $n > 1$, and by $\widetilde{\mathcal{A}}^2_{(1)} = \widetilde{\mathcal{A}}^2_1$, for $n = 1$, analogously,

$$\widetilde{\mathcal{A}}^2_n = \bigoplus_{k=1}^{n} \widetilde{\mathcal{A}}^2_{(k)}.$$

Corollary 3.3.4. *The unitary operator $U : L_2(\mathbb{R}) \otimes L_2(\mathbb{R}_+) \to L_2(\mathbb{R}) \otimes L_2(\mathbb{R}_+)$ maps the space $\widetilde{\mathcal{A}}^2_{(n)}(\Pi)$ of true-n-anti-analytic functions onto the space $L_2(\mathbb{R}_-) \otimes L_{n-1}$.*

The above results lead up to the following theorem.

Theorem 3.3.5. *We have the following isometric isomorphisms and decompositions of spaces:*

1. *Isomorphic images of poly-analytic spaces*

$$W \quad : \quad \mathcal{A}^2_{(n)}(\Pi) \longrightarrow H^2_+(\mathbb{R}) \otimes L_{n-1},$$

$$W \quad : \quad \mathcal{A}^2_n(\Pi) \longrightarrow H^2_+(\mathbb{R}) \otimes \bigoplus_{k=0}^{n-1} L_k,$$

$$W \quad : \quad \bigoplus_{k=1}^{\infty} \mathcal{A}^2_{(k)}(\Pi) \longrightarrow H^2_+(\mathbb{R}) \otimes L_2(\mathbb{R}_+).$$

2. *Isomorphic images of poly-anti-analytic spaces*

$$W \quad : \quad \widetilde{\mathcal{A}}^2_{(n)}(\Pi) \longrightarrow H^2_-(\mathbb{R}) \otimes L_{n-1},$$

$$W \quad : \quad \widetilde{\mathcal{A}}^2_n(\Pi) \longrightarrow H^2_-(\mathbb{R}) \otimes \bigoplus_{k=0}^{n-1} L_k,$$

$$W \quad : \quad \bigoplus_{k=1}^{\infty} \widetilde{\mathcal{A}}^2_{(k)}(\Pi) \longrightarrow H^2_-(\mathbb{R}) \otimes L_2(\mathbb{R}_+).$$

3. *Decomposition of the space $L_2(\Pi)$*

$$L_2(\Pi) \quad = \quad \bigoplus_{k=1}^{\infty} (\mathcal{A}^2_{(k)}(\Pi) \oplus \widetilde{\mathcal{A}}^2_{(k)}(\Pi)) \qquad\qquad (3.3.3)$$

$$= \quad \bigoplus_{k=1}^{\infty} \mathcal{A}^2_{(k)}(\Pi) \oplus \bigoplus_{k=1}^{\infty} \widetilde{\mathcal{A}}^2_{(k)}(\Pi).$$

Here L_n is the one-dimensional subspace of $L_2(\mathbb{R}_+)$, generated by the function $\ell_n(y) = e^{-y/2} L_n(y)$, where the Laguerre polynomial $L_n(y)$ of degree n is given by (3.3.1).

3.4 Projections onto the poly-Bergman spaces

The reverse procedure allows us to give exact formulas for the orthogonal projections onto each summand in the decomposition (3.3.3). Of course it is sufficient to describe the projections onto the spaces of true-n-analytic and true-n-anti-analytic functions only. We start with the projection onto the space $\mathcal{A}^2_{(n)}(\Pi)$ of true-n-analytic functions.

Theorem 3.4.1. *The orthogonal projection $B_\Pi^{(n)}$ of $L_2(\Pi)$ onto the space $\mathcal{A}_{(n)}^2(\Pi)$ of true-n-analytic functions is given by*

$$(B_\Pi^{(n)}\varphi)(w) = \int_\Pi K_\Pi^{(n)}(w,\omega)\,\varphi(\omega)\,d\mu(\omega),$$

where

$$K_\Pi^{(n)}(w,\omega) = K_\Pi(w,\omega)\cdot\sum_{j=0}^{n-1}\sum_{k=0}^{n-1}\kappa_{j,k}^{n-1}\left(\frac{w-\overline{w}}{w-\overline{\omega}}\right)^j\left(\frac{\omega-\overline{\omega}}{w-\overline{\omega}}\right)^k,\qquad(3.4.1)$$

and

$$K_\Pi(w,\omega) = -\frac{1}{\pi}\cdot\frac{1}{(w-\overline{\omega})^2}$$

is the Bergman kernel function of the domain Π,

$$\kappa_{j,k}^{n-1} = (-1)^{j+k}\left(\frac{n!}{j!\,k!}\right)^2\frac{(j+k+1)!}{(n-j)!\,(n-k)!}.\qquad(3.4.2)$$

Proof. The orthogonal projection $U\,B_\Pi^{(n)}\,U^{-1} = B_2^{(n)} : L_2(\Pi) \to L_2(\mathbb{R}_+)\otimes L_{n-1}$ is obviously given by

$$B_2^{(n)} = \chi_+(x)\otimes P_{n-1},$$

where

$$(P_{n-1}\psi)(y) = \langle\psi,\ell_{n-1}\rangle\,\ell_{n-1} = \ell_{n-1}(y)\int_{\mathbb{R}_+}\psi(\nu)\,\ell_{n-1}(\nu)\,d\nu\qquad(3.4.3)$$

is the orthogonal projection of $L_2(\mathbb{R}_+)$ onto the one-dimensional space L_{n-1}, generated by the function $\ell_{n-1}(y)$ given by (3.3.2). Or,

$$(B_2^{(n)}\varphi)(x,y) = \chi_+(x)\ell_{n-1}(y)\int_{\mathbb{R}_+}\varphi(x,\nu)\,\ell_{n-1}(\nu)\,d\nu.$$

Let $w = u+iv$, $\omega = \xi+i\eta$. Calculate the projection $B_1^{(n)} = U_2^{-1}\,B_2^{(n)}\,U_2$,

$$\begin{aligned}
(B_1^{(n)}\varphi)(u,v) &= \sqrt{2|u|}\,\chi_+(u)\,\ell_{n-1}(2|u|v)\int_{\mathbb{R}_+}\frac{1}{\sqrt{2|u|}}\varphi\left(u,\frac{\nu}{2|u|}\right)\ell_{n-1}(\nu)\,d\nu\\
&= \chi_+(u)\,2u\,e^{-uv}\,L_{n-1}(2uv)\int_{\mathbb{R}_+}\varphi(u,\eta)\,e^{-u\eta}\,L_{n-1}(2u\eta)\,d\eta.
\end{aligned}$$

Calculate now

$$\begin{aligned}
(F^{-1}\otimes I)B_1^{(n)}\varphi &= \frac{1}{\sqrt{2\pi}}\int_\mathbb{R}\left(\int_{\mathbb{R}_+}\chi_+(\xi)2\xi e^{-v\xi}L_{n-1}(2\xi v)L_{n-1}(2\xi\eta)\right.\\
&\qquad\left.\cdot\,\varphi(\xi,\eta)e^{-\eta\xi}d\eta\right)e^{iu\xi}d\xi\\
&= \sqrt{\frac{2}{\pi}}\int_\Pi\varphi(\xi,\eta)\chi_+(\xi)\xi L_{n-1}(2\xi v)L_{n-1}(2\xi\eta)e^{-(\eta-iw)\xi}d\xi d\eta.
\end{aligned}$$

From $B_2^{(n)} = U_2(F \otimes I)B_\Pi^{(n)}(F^{-1} \otimes I)U_2^{-1}$ it follows that

$$B_1^{(n)} = U_2^{-1} B_2^{(n)} U_2 = (F \otimes I)B_\Pi^{(n)}(F^{-1} \otimes I),$$

and that

$$\begin{aligned}
(F^{-1} \otimes I)B_1^{(n)}\varphi &= B_\Pi^{(n)}(F^{-1} \otimes I)\varphi \\
&= \langle (F^{-1} \otimes I)\varphi, \overline{K_\Pi^{(n)}(w,\omega)} \rangle \\
&= \langle \varphi, (F \otimes I)\overline{K_\Pi^{(n)}(w,\omega)} \rangle.
\end{aligned}$$

Thus

$$(F \otimes I)\overline{K_\Pi^{(n)}(w,\overline{\omega})} = \sqrt{\frac{2}{\pi}}\,\chi_+(\xi)\,\xi\,L_{n-1}(2\xi v)\,L_{n-1}(2\xi \eta)\,e^{-(\eta+i\overline{w})\xi}.$$

By (3.3.1)

$$L_{n-1}(y) = \sum_{k=0}^{n-1} \frac{n!}{k!\,(n-k)!}\,\frac{(-y)^k}{k!} = \sum_{k=0}^{n-1} \lambda_k^{n-1}\,y^k,$$

where

$$\lambda_k^{n-1} = \frac{(-1)^k\,n!}{k!\,(n-k)!\,k!}. \tag{3.4.4}$$

And thus

$$(F \otimes I)\overline{K_\Pi^{(n)}(w,\omega)} = \sum_{j=0}^{n-1}\sum_{k=0}^{n-1} \lambda_j^{n-1}\,\lambda_k^{n-1}\,2^{j+k}\,v^j\,\eta^k\,\left(\sqrt{\frac{2}{\pi}}\,\chi_+(\xi)\,\xi^{j+k+1}\,e^{-(\eta+i\overline{w})\xi}\right).$$

By [19] the function

$$\sqrt{\frac{2}{\pi}}\,\chi_+(\xi)\,\xi^m\,e^{-(\eta+i\overline{w})\xi}$$

is the Fourier transform of the function

$$\frac{1}{\pi}\,\frac{m!}{(i\overline{w}+\eta-i\xi)^{m+1}}.$$

Thus

$$\overline{K_\Pi^{(n)}(w,\omega)} = \frac{1}{\pi}\sum_{j=0}^{n-1}\sum_{k=0}^{n-1} \lambda_j^{n-1}\,\lambda_k^{n-1}\,2^{j+k}\,v^j\,\eta^k\,\frac{(j+k+1)!}{(i\overline{w}+\eta-i\xi)^{j+k+2}},$$

or

$$K_\Pi^{(n)}(w,\omega) = -\frac{1}{\pi}\,\frac{1}{(w-\overline{\omega})^2}\sum_{j=0}^{n-1}\sum_{k=0}^{n-1} \lambda_j^{n-1}\,\lambda_k^{n-1}\,(j+k+1)!\,2^{j+k}\,\frac{(iv)^j\,(i\eta)^k}{(w-\overline{\omega})^{j+k}}.$$

Introducing

$$\begin{aligned}
\kappa_{j,k}^{n-1} &= \lambda_j^{n-1}\,\lambda_k^{n-1}\,(j+k+1)! \\
&= (-1)^{j+k}\left(\frac{n!}{j!\,k!}\right)^2 \frac{(j+k+1)!}{(n-j)!\,(n-k)!},
\end{aligned}$$

we have finally

$$\begin{aligned}
K_\Pi^{(n)}(w,\omega) &= -\frac{1}{\pi}\,\frac{1}{(w-\overline{\omega})^2}\sum_{j=0}^{n-1}\sum_{k=0}^{n-1}\kappa_{j,k}^{n-1}\left(\frac{2iv}{w-\overline{\omega}}\right)^j\left(\frac{2i\eta}{w-\overline{\omega}}\right)^k \\
&= -\frac{1}{\pi}\,\frac{1}{(w-\overline{\omega})^2}\cdot\sum_{j=0}^{n-1}\sum_{k=0}^{n-1}\kappa_{j,k}^{n-1}\left(\frac{w-\overline{w}}{w-\overline{\omega}}\right)^j\left(\frac{\omega-\overline{\omega}}{w-\overline{\omega}}\right)^k.
\end{aligned}$$

$\square$

We describe now the orthogonal projection $\widetilde{B}_\Pi^{(n)}$ of $L_2(\Pi)$ onto the space $\widetilde{\mathcal{A}}^2_{(n)}(\Pi)$ of true-n-anti-analytic functions.

Theorem 3.4.2. *The orthogonal projection $\widetilde{B}_\Pi^{(n)}$ of $L_2(\Pi)$ onto the space $\widetilde{\mathcal{A}}^2_{(n)}(\Pi)$ of true-n-anti-analytic functions is given by*

$$(\widetilde{B}_\Pi^{(n)}\varphi)(w) = \int_\Pi \widetilde{K}_\Pi^{(n)}(w,\omega)\,\varphi(\omega)\,d\mu(\omega),$$

where

$$\widetilde{K}_\Pi^{(n)}(w,\omega) = -\frac{1}{\pi}\,\frac{1}{(\overline{w}-\omega)^2}\cdot\sum_{j=0}^{n-1}\sum_{k=0}^{n-1}\kappa_{j,k}^{n-1}\left(\frac{\overline{w}-w}{\overline{w}-\omega}\right)^j\left(\frac{\overline{\omega}-\omega}{\overline{w}-\omega}\right)^k$$

and $\kappa_{j,k}^{n-1}$ are given by (3.4.2).

Proof. Follow all the steps of the proof of Theorem 3.4.1; we will mention only the key formulas. We have the following connections between the projections,

$$\begin{aligned}
\widetilde{B}_\Pi^{(n)} &= (F^{-1}\otimes I)\widetilde{B}_1^{(n)}(F\otimes I), \\
\widetilde{B}_1^{(n)} &= U_2^1\,\widetilde{B}_2^{(n)}\,U_2.
\end{aligned}$$

The projections $\widetilde{B}_2^{(n)}$ and $\widetilde{B}_1^{(n)}$ are given by

$$(\widetilde{B}_2^{(n)}\varphi)(x,y) = \chi_-(x)\,\ell_{n-1}(y)\int_{\mathbb{R}_+}\varphi(x,\nu)\,\ell_{n-1}(\nu)\,d\nu,$$

where ℓ_{n-1} is defined in (3.3.2), and by

$$\begin{aligned}
(\widetilde{B}_1^{(n)}\varphi)(u,v) &= \sqrt{2|u|}\chi_-(u)\ell_{n-1}(2|u|v)\int_{\mathbb{R}_+}\frac{1}{\sqrt{2|u|}}\varphi\left(u,\frac{\nu}{2|u|}\right)\ell_{n-1}(\nu)d\nu \\
&= -2u\chi_-(u)e^{uv}L_{n-1}(-2uv)\int_{\mathbb{R}_+}\varphi(u,\eta)e^{u\eta}L_{n-1}(-2u\eta)d\eta,
\end{aligned}$$

where $L_{n-1}(y)$ is the Laguerre polynomial (3.3.1).

Calculate

$$
\begin{aligned}
(F^{-1} \otimes I)\widetilde{B}_1^{(n)}\varphi &= \frac{1}{\sqrt{2\pi}}\int_{\mathbb{R}}\left(\int_{\mathbb{R}_+}(-2\xi)\chi_-(\xi)e^{v\xi}L_{n-1}(-2v\xi)L_{n-1}(-2\eta\xi)\right. \\
&\qquad \left.\cdot\, \varphi(\xi,\eta)e^{\eta\xi}d\eta\right)e^{iu\xi}d\xi \\
&= \sqrt{\frac{2}{\pi}}\int_{\Pi}\varphi(\xi,\eta)(-\xi)\chi_-(\xi)L_{n-1}(-2v\xi)L_{n-1}(-2\eta\xi) \\
&\qquad \cdot\, e^{(\eta+i\,\overline{w})\xi}d\xi d\eta.
\end{aligned}
$$

Because of

$$
\begin{aligned}
(F^{-1} \otimes I)\widetilde{B}_1^{(n)}\varphi &= \widetilde{B}_\Pi^{(n)}(F^{-1} \otimes I)\varphi \\
&= \langle (F^{-1} \otimes I)\varphi, \overline{\widetilde{K}_\Pi^{(n)}(w,\omega)}\rangle \\
&= \langle \varphi, (F \otimes I)\overline{\widetilde{K}_\Pi^{(n)}(w,\omega)}\rangle
\end{aligned}
$$

we have

$$
\begin{aligned}
(F \otimes I)\overline{\widetilde{K}_\Pi^{(n)}(w,\omega)} &= \sqrt{\frac{2}{\pi}}\,\chi_-(\xi)\,(-\xi)\,L_{n-1}(-2v\xi)L_{n-1}(-2\eta\xi)\,e^{(\eta-i\,w)\xi} \\
&= \sum_{j=0}^{n-1}\sum_{k=0}^{n-1}\lambda_j^{n-1}\lambda_k^{n-1}2^{j+k}v^j\eta^k \\
&\qquad \cdot\sqrt{\frac{2}{\pi}}\,\chi_-(\xi)\,(-\xi)^{j+k+1}\,e^{(\eta-i\,w)\xi},
\end{aligned}
$$

where λ_k^{n-1} was introduced in (3.4.4).

By [19] the function

$$
\sqrt{\frac{2}{\pi}}\,\chi_-(\xi)\,(-\xi)^m\,e^{(\eta-i\,w)\xi}
$$

is the Fourier transform of the function

$$
\frac{1}{\pi}\frac{m!}{(i\xi+\eta-i\,w)^{m+1}}.
$$

Thus

$$
\overline{\widetilde{K}_\Pi^{(n)}(w,\omega)} = \frac{1}{\pi}\sum_{j=0}^{n-1}\sum_{k=0}^{n-1}\lambda_j^{n-1}\,\lambda_k^{n-1}\,2^{j+k}\,v^j\,\eta^k\,\frac{(j+k+1)!}{(i\xi+\eta-i\,w)^{j+k+2}},
$$

or

$$
\begin{aligned}
\widetilde{K}_\Pi^{(n)}(w,\omega) &= -\frac{1}{\pi}\frac{1}{(\overline{w}-\omega)^2}\sum_{j=0}^{n-1}\sum_{k=0}^{n-1}\lambda_j^{n-1}\lambda_k^{n-1}(j+k+1)!\,2^{j+k}\frac{(-iv)^j(-i\eta)^k}{(\overline{w}-\omega)^{j+k}}\\[2mm]
&= -\frac{1}{\pi}\frac{1}{(\overline{w}-\omega)^2}\sum_{j=0}^{n-1}\sum_{k=0}^{n-1}\kappa_{j,k}^{n-1}\left(\frac{-2iv}{\overline{w}-\omega}\right)^j\left(\frac{-2i\eta}{\overline{w}-\omega}\right)^k\\[2mm]
&= -\frac{1}{\pi}\frac{1}{(\overline{w}-\omega)^2}\cdot\sum_{j=0}^{n-1}\sum_{k=0}^{n-1}\kappa_{j,k}^{n-1}\left(\frac{\overline{w}-w}{\overline{w}-\omega}\right)^j\left(\frac{\overline{\omega}-\omega}{\overline{w}-\omega}\right)^k,
\end{aligned}
$$

where $\kappa_{j,k}^{n-1}$ is given by (3.4.2). $\qquad\square$

To uniformize the statements of Theorem 3.4.1 and Theorem 3.4.2 we introduce the function

$$
K_\Pi^{(n)}(w_1,w_2;\omega_1,\omega_2) = -\frac{1}{\pi}\frac{1}{(w_1-\omega_2)^2}\cdot\sum_{j=0}^{n-1}\sum_{k=0}^{n-1}\kappa_{j,k}^{n-1}\left(\frac{w_1-w_2}{w_1-\omega_2}\right)^j\left(\frac{\omega_1-\omega_2}{w_1-\omega_2}\right)^k,
$$

where $\kappa_{j,k}^{n-1}$ is given by (3.4.2).

Then the kernel $K_\Pi^{(n)}$ of the projection $B_\Pi^{(n)}$ admits the representation

$$
K_\Pi^{(n)}(w,\omega) = K_\Pi^{(n)}(w,\overline{w};\omega,\overline{\omega}),
$$

analogously

$$
\widetilde{K}_\Pi^{(n)}(w,\omega) = K_\Pi^{(n)}(\overline{w},w;\overline{\omega},\omega).
$$

Note that the function $K_\Pi^{(n)}(w,\overline{w};\omega,\overline{\omega})$ is Hermitian symmetric in a sense that

$$
\overline{K_\Pi^{(n)}(w,\overline{w};\omega,\overline{\omega})} = K_\Pi^{(n)}(\overline{w},w;\overline{\omega},\omega),
$$

and for $n=1$ coincides with the Bergman kernel function: $K_\Pi^{(1)}(w,\overline{w};\omega,\overline{\omega}) = K_\Pi(w,\omega)$.

Now we have

Theorem 3.4.3. *The orthogonal projections $B_\Pi^{(n)}$ of $L_2(\Pi)$ onto the space $\mathcal{A}^2_{(n)}(\Pi)$ of true-n-analytic functions, and $\widetilde{B}_\Pi^{(n)}$ of $L_2(\Pi)$ onto the space $\widetilde{\mathcal{A}}^2_{(n)}(\Pi)$ of true-n-anti-analytic functions are given by the following formulas:*

$$
(B_\Pi^{(n)}\varphi)(w) = \int_\Pi K_\Pi^{(n)}(w,\overline{w};\omega,\overline{\omega})\,\varphi(\omega)\,d\mu(\omega),
$$

and

$$
\begin{aligned}
(\widetilde{B}_\Pi^{(n)}\varphi)(w) &= \int_\Pi \overline{K_\Pi^{(n)}(w,\overline{w};\omega,\overline{\omega})}\,\varphi(\omega)\,d\mu(\omega)\\[2mm]
&= \int_\Pi K_\Pi^{(n)}(\overline{w},w;\overline{\omega},\omega)\,\varphi(\omega)\,d\mu(\omega).
\end{aligned}
$$

Corollary 3.4.4. *The unitary operator* $W : L_2(\mathbb{R}) \otimes L_2(\mathbb{R}_+) \longrightarrow L_2(\mathbb{R}) \otimes L_2(\mathbb{R}_+)$ *gives the following connections between the true-poly-Bergman projections* $B_\Pi^{(n)}$, $\widetilde{B}_\Pi^{(n)}$ *and the Szegö projections* P_Π^+, P_Π^- :

$$
\begin{aligned}
W\, B_\Pi^{(n)}\, W^{-1} &= P_\Pi^+ \otimes P_{n-1}, \\
W\, \widetilde{B}_\Pi^{(n)}\, W^{-1} &= P_\Pi^- \otimes P_{n-1},
\end{aligned}
$$

where P_{n-1} *in the orthogonal projection* (3.4.3) *of* $L_2(\mathbb{R}_+)$ *onto the one-dimensional space* L_{n-1}, *generated by the function* $\ell_{n-1}(y)$, *defined in* (3.3.2).

Proof. Follows directly from Corollaries 3.3.2 and 3.3.4. $\square$

3.5   Poly-Bergman spaces and two-dimensional singular integral operators

We start with the two-dimensional singular integral operators (2.4.1) and (2.4.2) for the upper half-plane case, which are defined as follows,

$$
\begin{aligned}
(S_\Pi \varphi)(z) &= -\frac{1}{\pi} \int_\Pi \frac{\varphi(\zeta)}{(\zeta - z)^2}\, dv(\zeta), \\
(S_\Pi^* \varphi)(z) &= -\frac{1}{\pi} \int_\Pi \frac{\varphi(\zeta)}{(\overline{\zeta} - \overline{z})^2}\, dv(\zeta),
\end{aligned}
$$

and which are obviously bounded on $L_2(\Pi) = L_2(\mathbb{R}) \otimes L_2(\mathbb{R}_+)$.

By (2.4.3) these operators admit the following representations:

$$
\begin{aligned}
S_\Pi &= (I \otimes \chi_+ I) S_{\mathbb{R}^2}(I \otimes \chi_+ I) \\
&= (I \otimes \chi_+ I)(F^{-1} \otimes F^{-1})\frac{\xi - i\eta}{\xi + i\eta}(F \otimes F)(I \otimes \chi_+ I) \qquad (3.5.1)
\end{aligned}
$$

and

$$
\begin{aligned}
S_\Pi^* &= (I \otimes \chi_+ I) S_{\mathbb{R}^2}^*(I \otimes \chi_+ I) \\
&= (I \otimes \chi_+ I)(F^{-1} \otimes F^{-1})\frac{\xi + i\eta}{\xi - i\eta}(F \otimes F)(I \otimes \chi_+ I),
\end{aligned}
$$

where $\xi, \eta \in \mathbb{R}$, and the one-dimensional Fourier transform F and its inverse F^{-1} are given by

$$
\begin{aligned}
(F\varphi)(\xi) &= \frac{1}{\sqrt{2\pi}} \int_\mathbb{R} e^{-ix\xi} f(x)\, dx, \\
(F^{-1}\varphi)(x) &= \frac{1}{\sqrt{2\pi}} \int_\mathbb{R} e^{ix\xi} f(\xi)\, d\xi.
\end{aligned}
$$

We introduce as well the integral operators

$$(S_+ f)(y) \;=\; -f(y) + e^{-\frac{y}{2}} \int_0^y e^{\frac{t}{2}} f(t)\, dt,$$

$$(S_- f)(y) \;=\; -f(y) + e^{\frac{y}{2}} \int_y^\infty e^{-\frac{t}{2}} f(t)\, dt,$$

which, as we will see soon, are bounded on $L_2(\mathbb{R}_+)$ and are mutually adjoint.

As in Section 3.1 we will use the unitary operator

$$U = U_2 U_1 : L_2(\mathbb{R}) \otimes L_2(\mathbb{R}_+) \longrightarrow L_2(\mathbb{R}) \otimes L_2(\mathbb{R}_+),$$

where the operators U_1 and U_2 are given by (3.1.1) and (3.1.4) respectively.

Theorem 3.5.1. *The unitary operator $U = U_2 U_1$ gives an isometric isomorphism of the space $L_2(\Pi) = [L_2(\mathbb{R}_+) \otimes L_2(\mathbb{R}_+)] \oplus [L_2(\mathbb{R}_-) \otimes L_2(\mathbb{R}_+)]$ under which the two-dimensional singular integral operators S_Π and S_Π^* are unitary equivalent to the operators*

$$U\, S_\Pi\, U^{-1} \;=\; (I \otimes S_+) \oplus (I \otimes S_-),$$
$$U\, S_\Pi^*\, U^{-1} \;=\; (I \otimes S_-) \oplus (I \otimes S_+).$$

Proof. By the representation (3.5.1) we have

$$\begin{aligned}
S_1 \;&=\; U_1 S_\Pi U_1^{-1} = (F \otimes I) S_\Pi (F^{-1} \otimes I)\\
&=\; (I \otimes \chi_+ I)(I \otimes F^{-1}) \frac{\xi - i\eta}{\xi + i\eta} (I \otimes F)(I \otimes \chi_+ I).
\end{aligned}$$

The operator U_2 is unitary on both $L_2(\mathbb{R}_+)$ and $L_2(\mathbb{R})$, and furthermore it commutes with $\chi_{\mathbb{R}_+} I$. Direct calculation shows that

$$U_2(I \otimes F^{-1}) \frac{\xi - i\eta}{\xi + i\eta} (I \otimes F) U_2^{-1} = (I \otimes F^{-1}) \frac{\frac{1}{2}\operatorname{sign}x - i\eta}{\frac{1}{2}\operatorname{sign}x + i\eta} (I \otimes F).$$

Thus

$$\begin{aligned}
S_2 \;&=\; U S_\Pi U^{-1} = U_2 S_1 U_2^{-1}\\
&=\; (\chi_+ I \otimes \chi_+ I)(I \otimes F^{-1}) \frac{\frac{1}{2} - i\eta}{\frac{1}{2} + i\eta} (I \otimes F)(\chi_+ I \otimes \chi_+ I)\\
&\quad + (\chi_- I \otimes \chi_+ I)(I \otimes F^{-1}) \frac{\frac{1}{2} + i\eta}{\frac{1}{2} - i\eta} (I \otimes F)(\chi_- I \otimes \chi_+ I)
\end{aligned}$$

and

$$\begin{aligned}
S_2^* \;&=\; U S_\Pi^* U^{-1}\\
&=\; (\chi_+ I \otimes \chi_+ I)(I \otimes F^{-1}) \frac{\frac{1}{2} + i\eta}{\frac{1}{2} - i\eta} (I \otimes F)(\chi_+ I \otimes \chi_+ I)\\
&\quad + (\chi_- I \otimes \chi_+ I)(I \otimes F^{-1}) \frac{\frac{1}{2} - i\eta}{\frac{1}{2} + i\eta} (I \otimes F)(\chi_- I \otimes \chi_+ I).
\end{aligned}$$

The defining symbols of two involved convolution operators

$$\widetilde{S}_+ = F^{-1}\frac{\frac{1}{2} - i\eta}{\frac{1}{2} + i\eta}F \quad \text{and} \quad \widetilde{S}_- = F^{-1}\frac{\frac{1}{2} + i\eta}{\frac{1}{2} - i\eta}F,$$

which are obviously bounded on $L_2(\mathbb{R})$, admit the representations

$$\frac{\frac{1}{2} - i\eta}{\frac{1}{2} + i\eta} = -1 - \frac{i\eta}{\frac{1}{4} + \eta^2} + \frac{\frac{1}{2}}{\frac{1}{4} + \eta^2} \quad \text{and} \quad \frac{\frac{1}{2} + i\eta}{\frac{1}{2} - i\eta} = -1 + \frac{i\eta}{\frac{1}{4} + \eta^2} + \frac{\frac{1}{2}}{\frac{1}{4} + \eta^2},$$

respectively.

Using the formulas 17.23.14 and 17.23.15 of [86] we have

$$F\left(\frac{\frac{1}{2}}{\frac{1}{4} + \eta^2}\right) = \sqrt{\frac{\pi}{2}}\,e^{-\frac{|y|}{2}}, \qquad F\left(\frac{i\eta}{\frac{1}{4} + \eta^2}\right) = \sqrt{\frac{\pi}{2}}\,\text{sign}\,y\,e^{-\frac{|y|}{2}},$$

and thus

$$\begin{aligned}
(\widetilde{S}_+ f)(y) &= -f(y) + \frac{1}{\sqrt{2\pi}}\int_{\mathbb{R}}\sqrt{\frac{\pi}{2}}\,e^{-\frac{|t-y|}{2}}(1 - \text{sign}(t-y))f(t)\,dt \\
&= -f(t) + \int_{\mathbb{R}} e^{-\frac{|t-y|}{2}}\chi_-(t-y)f(t)\,dt
\end{aligned}$$

and

$$\begin{aligned}
(\widetilde{S}_- f)(y) &= -f(y) + \frac{1}{\sqrt{2\pi}}\int_{\mathbb{R}}\sqrt{\frac{\pi}{2}}\,e^{-\frac{|t-y|}{2}}(1 + \text{sign}(t-y))f(t)\,dt \\
&= -f(t) + \int_{\mathbb{R}} e^{-\frac{|t-y|}{2}}\chi_+(t-y)f(t)\,dt.
\end{aligned}$$

Then the operators $S_+ = \chi_+\widetilde{S}_+\chi_+ I_{|L_2(\mathbb{R}_+)}$ and $S_- = \chi_+\widetilde{S}_-\chi_+ I_{|L_2(\mathbb{R}_+)}$, acting on $L_2(\mathbb{R}_+)$, are as follows:

$$\begin{aligned}
(S_+ f)(y) &= -f(t) + \int_{\mathbb{R}_+} e^{-\frac{|t-y|}{2}}\chi_-(t-y)f(t)\,dt \\
&= -f(y) + e^{-\frac{y}{2}}\int_0^y e^{\frac{t}{2}}f(t)\,dt
\end{aligned}$$

and

$$\begin{aligned}
(S_- f)(y) &= -f(t) + \int_{\mathbb{R}_+} e^{-\frac{|t-y|}{2}}\chi_+(t-y)f(t)\,dt \\
&= -f(y) + e^{\frac{y}{2}}\int_y^\infty e^{-\frac{t}{2}}f(t)\,dt.
\end{aligned}$$

Thus finally

$$
\begin{aligned}
US_\Pi U^{-1} &= (\chi_+ I \otimes \chi_+ I)(I \otimes \widetilde{S}_+)(\chi_+ I \otimes \chi_+ I)\\
&\quad + (\chi_- I \otimes \chi_+ I)(I \otimes \widetilde{S}_-)(\chi_- I \otimes \chi_+ I)\\
&= \chi_+ I \otimes S_+ + \chi_- I \otimes S_-\\
&= (I \otimes S_+) \oplus (I \otimes S_-)
\end{aligned}
$$

and

$$
\begin{aligned}
US_\Pi^* U^{-1} &= (\chi_+ I \otimes \chi_+ I)(I \otimes \widetilde{S}_-)(\chi_+ I \otimes \chi_+ I)\\
&\quad + (\chi_- I \otimes \chi_+ I)(I \otimes \widetilde{S}_+)(\chi_- I \otimes \chi_+ I)\\
&= \chi_+ I \otimes S_- + \chi_- I \otimes S_+\\
&= (I \otimes S_-) \oplus (I \otimes S_+),
\end{aligned}
$$

where the last lines in both representations are written according to the splitting

$$
L_2(\Pi) = [L_2(\mathbb{R}_+) \otimes L_2(\mathbb{R}_+)] \oplus [L_2(\mathbb{R}_-) \otimes L_2(\mathbb{R}_+)] . \qquad \square
$$

Recall that the system of functions

$$
\ell_n(y) = e^{-y/2} L_n(y), \qquad n = 0, 1, 2, \ldots,
$$

where the Laguerre polynomials $L_n(y)$ are given by (3.3.1), forms an orthonormal base in the space $L_2(\mathbb{R}_+)$.

Theorem 3.5.2. *For each admissible n, the following equalities hold,*

$$
(S_+\ell_n)(y) = -\ell_{n+1}(y), \qquad (S_-\ell_n)(y) = -\ell_{n-1}(y), \qquad \text{and} \qquad (S_-\ell_0)(y) = 0.
$$

Proof. By [86], formula 8.971.1, we have

$$
L_n'(y) - L_{n+1}'(y) = L_n(y). \tag{3.5.2}
$$

Taking into account that $L_n(0) = 1$, for all n, the integral form of the above formula is

$$
L_n(y) - L_{n+1}(y) = \int_0^y L_n(t)\,dt.
$$

Calculate now

$$
\begin{aligned}
(S_+\ell_n)(y) &= -e^{-\frac{y}{2}} L_n(y) + e^{-\frac{y}{2}} \int_0^y L_n(t)\,dt\\
&= e^{-\frac{y}{2}} \left(-L_n(y) + L_n(y) - L_{n+1}(y) \right) = -\ell_{n+1}(y).
\end{aligned}
$$

Integrating by parts twice and using (3.5.2), we have

$$
\begin{aligned}
\int_y^\infty e^{-t} L_n(t)dt &= e^{-y} L_n(y) + \int_y^\infty e^{-t} L'_{n-1}(t)dt - \int_y^\infty e^{-t} L_{n-1}(t)dt \\
&= e^{-y} L_n(y) - \int_y^\infty e^{-t} L_{n-1}(t)dt \\
&\quad -e^{-y} L_{n-1}(y) + \int_y^\infty e^{-t} L_{n-1}(t)dt \\
&= e^{-y} L_n(y) - e^{-y} L_{n-1}(y).
\end{aligned}
$$

Thus

$$
\begin{aligned}
(S_- \ell_n)(y) &= -e^{-\frac{y}{2}} L_n(y) + e^{\frac{y}{2}} \int_y^\infty e^{-t} L_n(t)dt \\
&= -e^{-\frac{y}{2}} L_n(y) + e^{\frac{y}{2}} \left(e^{-y} L_n(y) - e^{-y} L_{n-1}(y) \right) = -\ell_{n-1}(y).
\end{aligned}
$$

Finally,

$$
(S_- \ell_0)(y) = -e^{-\frac{y}{2}} + e^{\frac{y}{2}} \int_y^\infty e^{-t} dt = 0. \qquad \square
$$

Remark 3.5.3. As the previous theorem shows, the operator $-S_+$ is an isometric operator on $L_2(\mathbb{R}_+)$ and is nothing but the unilateral forward shift with respect to the base $\{\ell_n(y)\}_{n=0}^\infty$. Its adjoint operator $-S_-$ is the unilateral backward shift with respect to the same base, and its kernel coincides with the one-dimensional space L_0 generated by $\ell_0(y) = e^{-\frac{y}{2}}$.

As in Section 3.3, we denote by L_n, $n = 0, 1, 2, \ldots$, the one-dimensional subspace of $L_2(\mathbb{R}_+)$ generated by the function $\ell_n(y)$. Let

$$
L_n^\oplus = \bigoplus_{k=0}^n L_k
$$

be the direct sum of the first $(n+1)$ spaces. We denote by P_n and $P_n^\oplus$ the orthogonal projections of $L_2(\mathbb{R}_+)$ onto L_n and $L_n^\oplus$, respectively.

Corollary 3.5.4. *For all admissible indices, we have*

$$
\begin{aligned}
P_0 &= I - S_- S_+, \\
P_n &= S_-^n P_0 S_+^n, \\
P_n^\oplus &= I - S_-^{n+1} S_+^{n+1}, \\
(-S_+)^k \big|_{L_n} &: \quad L_n \longrightarrow L_{n+k}, \\
(-S_-)^k \big|_{L_n} &: \quad L_n \longrightarrow L_{n-k}.
\end{aligned}
$$

The next result was obtained in [115] (see Theorem 2.4 and Corollary 2.6 therein) and shows that the action of both operators S_Π and S_Π^* is extremely transparent according to the decomposition

$$L_2(\Pi) = \bigoplus_{k=1}^\infty \mathcal{A}_{(k)}^2(\Pi) \oplus \bigoplus_{k=1}^\infty \widetilde{\mathcal{A}}_{(k)}^2(\Pi).$$

In our approach it is just a straightforward corollary of Theorems 3.3.5, 3.5.1, and Corollary 3.5.4.

Theorem 3.5.5. *For all admissible indices, we have*

$$(S_\Pi)^k \big|_{\mathcal{A}_{(n)}^2(\Pi)} \quad : \quad \mathcal{A}_{(n)}^2(\Pi) \longrightarrow \mathcal{A}_{(n+k)}^2(\Pi),$$

$$(S_\Pi)^k \big|_{\widetilde{\mathcal{A}}_{(n)}^2(\Pi)} \quad : \quad \widetilde{\mathcal{A}}_{(n)}^2(\Pi) \longrightarrow \widetilde{\mathcal{A}}_{(n-k)}^2(\Pi),$$

$$(S_\Pi^*)^k \big|_{\widetilde{\mathcal{A}}_{(n)}^2(\Pi)} \quad : \quad \widetilde{\mathcal{A}}_{(n)}^2(\Pi) \longrightarrow \widetilde{\mathcal{A}}_{(n+k)}^2(\Pi),$$

$$(S_\Pi^*)^k \big|_{\mathcal{A}_{(n)}^2(\Pi)} \quad : \quad \mathcal{A}_{(n)}^2(\Pi) \longrightarrow \mathcal{A}_{(n-k)}^2(\Pi),$$

$$\ker(S_\Pi)^n = \widetilde{\mathcal{A}}_n^2(\Pi), \qquad (\operatorname{Im}(S_\Pi)^n)^\perp = \mathcal{A}_n^2(\Pi),$$

$$\ker(S_\Pi^*)^n = \mathcal{A}_n^2(\Pi), \qquad (\operatorname{Im}(S_\Pi^*)^n)^\perp = \widetilde{\mathcal{A}}_n^2(\Pi).$$

Corollary 3.5.6. *Each true-n-analytic function ψ admits the representation*

$$\psi = (S_\Pi)^{n-1}\varphi,$$

where $\varphi \in \mathcal{A}^2(\Pi)$.

Each true-n-anti-analytic function g admits the representation

$$g = (S_\Pi^*)^{n-1}f,$$

where $f \in \widetilde{\mathcal{A}}^2(\Pi)$.

We continue to denote by $B_\Pi^{(n)}$ and $\widetilde{B}_\Pi^{(n)}$ the orthogonal projections of $L_2(\Pi)$ onto the spaces $\mathcal{A}_{(n)}^2$ and $\widetilde{\mathcal{A}}_{(n)}^2$ consisting of true-n-analytic and true-n-anti-analytic functions respectively. Let $B_{\Pi,n}$ and $\widetilde{B}_{\Pi,n}$ be the orthogonal projections of $L_2(\Pi)$ onto the spaces $\mathcal{A}_n^2$ and $\widetilde{\mathcal{A}}_n^2$, consisting of n-analytic and n-anti-analytic functions respectively.

We summarize now some important properties of the above projections in terms of singular operators.

Theorem 3.5.7. *For all admissible indices, we have*

$$
\begin{aligned}
B_\Pi &= I - S_\Pi S_\Pi^*, \\
\widetilde{B}_\Pi &= I - S_\Pi^* S_\Pi, \\
B_{\Pi,n} &= I - (S_\Pi)^n (S_\Pi^*)^n, \\
\widetilde{B}_{\Pi,n} &= I - (S_\Pi^*)^n (S_\Pi)^n, \\
B_{\Pi,(n)} = (S_\Pi)^{n-1} B_\Pi (S_\Pi^*)^{n-1} &= (S_\Pi)^{n-1}(S_\Pi^*)^{n-1} - (S_\Pi)^n (S_\Pi^*)^n, \\
\widetilde{B}_{\Pi,(n)} = (S_\Pi^*)^{n-1} \widetilde{B}_\Pi (S_\Pi)^{n-1} &= (S_\Pi^*)^{n-1}(S_\Pi)^{n-1} - (S_\Pi^*)^n (S_\Pi)^n, \\
B_{\Pi,(n+1)} &= S_\Pi B_{\Pi,(n)} S_\Pi^*, \\
\widetilde{B}_{\Pi,(n+1)} &= S_\Pi^* \widetilde{B}_{\Pi,(n)} S_\Pi.
\end{aligned}
$$

Proof. Follows directly from Theorems 3.3.5, 3.5.1, and Corollary 3.5.4. $\qquad\square$

As direct corollaries of the above results we mention as well the following statements.

Theorem 3.5.8. *For all $n \in \mathbb{N}$, we have*

$$
\begin{aligned}
(S_\Pi)^n (S_\Pi^*)^n (S_\Pi)^n &= (S_\Pi)^n, \\
(S_\Pi^*)^n (S_\Pi)^n (S_\Pi^*)^n &= (S_\Pi^*)^n.
\end{aligned}
$$

Introduce the one-dimensional singular integral operator

$$
(S_\mathbb{R}\varphi)(x) = \frac{1}{\pi i} \int_\mathbb{R} \frac{\varphi(\xi)}{\xi - x}\, d\xi,
$$

acting on $L_2(\mathbb{R})$. It is well known that the operators

$$
P_\pm = \frac{1}{2}(I \pm S_\mathbb{R})
$$

are the Szegö projections of $L_2(\mathbb{R})$ onto the Hardy spaces on the upper and lower half-planes, respectively.

Theorem 3.5.9. *We have*

$$
\begin{aligned}
\mathrm{s-}\lim_{n\to\infty} (S_\Pi^*)^n (S_\Pi^*)^n &= P_+ \otimes I, \\
\mathrm{s-}\lim_{n\to\infty} (S_\Pi)^n (S_\Pi^*)^n &= P_- \otimes I, \\
\mathrm{s-}\lim_{n\to\infty} [(S_\Pi^*)^n (S_\Pi^*)^n, (S_\Pi)^n (S_\Pi^*)^n] &= S_\mathbb{R} \otimes I, \\
\mathrm{s-}\lim_{n\to\infty} ((S_\Pi^*)^n (S_\Pi^*)^n + (S_\Pi)^n (S_\Pi^*)^n) &= I.
\end{aligned}
$$

Corollary 3.5.10. *We have*

$$
\begin{aligned}
\mathrm{s-}\lim_{n\to\infty} B_{\Pi,n} &= P_+ \otimes I, \\
\mathrm{s-}\lim_{n\to\infty} \widetilde{B}_{\Pi,n} &= P_- \otimes I.
\end{aligned}
$$

Chapter 4

Bergman Type Spaces on the Unit Disk

Let $\mathbb{D}$ be the unit disk in $\mathbb{C}$. Rearranging the basis of $L_2(\mathbb{D})$, L. Peng, R. Rochberg and Z. Wu [154] proved that the space $L_2(\mathbb{D})$ can be decomposed onto a direct sum of the Bergman type spaces

$$L_2(\mathbb{D}) = \bigoplus_{n=1}^{\infty} \mathcal{A}^2_{(n)} \oplus \bigoplus_{n=1}^{\infty} \widetilde{\mathcal{A}}^2_{(n)}, \tag{4.0.1}$$

where

$$\mathcal{A}^2_{(n)} = \ker(\bar{z}\,\partial/\partial\bar{z})^n \ominus \ker(\bar{z}\,\partial/\partial\bar{z})^{n-1}, \quad \widetilde{\mathcal{A}}^2_{(n)} = \ker(z\,\partial/\partial z)^n \ominus \ker(z\,\partial/\partial z)^{n-1}.$$

Studying this question we follow the ideas of Chapter 3, which allows us to obtain more information.

4.1 Bergman space and Bergman projection

We consider the space $L_2(\mathbb{D})$ with usual Lebesgue plane measure $dv(z) = dxdy$, $z = x + iy$, and its Bergman subspace $\mathcal{A}^2(\mathbb{D})$. The Bergman projection $B_{\mathbb{D}}$ of $L_2(\mathbb{D})$ onto $\mathcal{A}^2(\mathbb{D})$ has the form (see Example 2.2.4)

$$(B_{\mathbb{D}}\varphi)(z) = \frac{1}{\pi} \int_{\mathbb{D}} \frac{\varphi(\zeta)\,dv(\zeta)}{(1 - z\bar{\zeta})^2}.$$

The Bergman space $\mathcal{A}^2(\mathbb{D})$ can be described alternatively as the (closed) subspace of $L_2(\mathbb{D})$, which consists of all functions satisfying the equation

$$\frac{\partial}{\partial\bar{z}}\varphi = \frac{1}{2}\left(\frac{\partial}{\partial x} + i\frac{\partial}{\partial y}\right)\varphi = 0,$$

where $z = x + iy$.

Passing to the polar coordinates we have

$$
\begin{aligned}
L_2(\mathbb{D}) &= L_2([0,1), r\,dr) \otimes L_2([0, 2\pi), d\alpha) \\
&= L_2([0,1), r\,dr) \otimes L_2(S^1, \frac{dt}{it}) = L_2([0,1), r\,dr) \otimes L_2(S^1),
\end{aligned}
$$

where S^1 is the unit circle, and

$$
\frac{dt}{it} = |dt| = d\alpha
$$

is the element of length; in addition

$$
\frac{\partial}{\partial \bar{z}} = \frac{\cos\alpha + i\sin\alpha}{2}\left(\frac{\partial}{\partial r} + i\frac{1}{r}\frac{\partial}{\partial \alpha}\right) = \frac{t}{2}\left(\frac{\partial}{\partial r} - \frac{t}{r}\frac{\partial}{\partial t}\right).
$$

Introduce the unitary operator

$$
U_1 = I \otimes \mathcal{F} : L_2([0,1), r\,dr) \otimes L_2(S^1) \longrightarrow L_2([0,1), r\,dr) \otimes l_2 = l_2(L_2([0,1), r\,dr)),
$$

where the discrete Fourier transform $\mathcal{F} : L_2(S^1) \to l_2$ is given by

$$
\mathcal{F} : f \longmapsto c_n = \frac{1}{\sqrt{2\pi}} \int_{S^1} f(t)\, t^{-n}\, \frac{dt}{it}, \quad n \in \mathbb{Z}, \tag{4.1.1}
$$

and its inverse $\mathcal{F}^{-1} = \mathcal{F}^* : l_2 \to L_2(S^1)$ is given by

$$
\mathcal{F}^{-1} : \{c_n\}_{n\in\mathbb{Z}} \longmapsto f = \frac{1}{\sqrt{2\pi}} \sum_{n\in\mathbb{Z}} c_n\, t^n.
$$

Calculate

$$
(I \otimes \mathcal{F})\frac{t}{2}\left(\frac{\partial}{\partial r} - \frac{t}{r}\frac{\partial}{\partial t}\right)(I \otimes \mathcal{F}^{-1}) \quad : \quad \{c_n(r)\}_{n\in\mathbb{Z}} \longmapsto \frac{1}{\sqrt{2\pi}} \sum_{n\in\mathbb{Z}} c_n(r)\, t^n
$$

$$
\longmapsto \frac{1}{\sqrt{2\pi}} \sum_{n\in\mathbb{Z}} \frac{t}{2}\left(\frac{\partial}{\partial r} - \frac{n}{r}\right) c_n(r)\, t^n
$$

$$
\longmapsto \{d_n\} = \left\{\frac{1}{2}\left(\frac{\partial}{\partial r} - \frac{n-1}{r}\right) c_{n-1}(r)\right\}_{n\in\mathbb{Z}},
$$

or

$$
(I \otimes \mathcal{F})\frac{t}{2}\left(\frac{\partial}{\partial r} - \frac{t}{r}\frac{\partial}{\partial t}\right)(I \otimes \mathcal{F}^{-1})\{c_n(r)\}_{n\in\mathbb{Z}} = \left\{\frac{1}{2}\left(\frac{\partial}{\partial r} - \frac{n-1}{r}\right) c_{n-1}(r)\right\}_{n\in\mathbb{Z}}.
$$

Thus the image of the Bergman space $\mathcal{A}_1^2 = U_1(\mathcal{A}^2(\mathbb{D}))$ can be described as the (closed) subspace of $L_2([0,1), r\,dr) \otimes l_2 = l_2(L_2([0,1), r\,dr))$ which consists of all sequences $\{c_n(r)\}_{n\in\mathbb{Z}}$ satisfying the equations

$$
\frac{1}{2}\left(\frac{\partial}{\partial r} - \frac{n}{r}\right) c_n(r) = 0, \quad n \in \mathbb{Z}. \tag{4.1.2}
$$

The equations (4.1.2) are easy to solve, and their general solutions have the form

$$c_n(r) = c'_n \, r^n = \sqrt{2(|n|+1)}\, c_n \, r^n, \quad n \in \mathbb{Z}.$$

But each function $c_n(r) = \sqrt{2(|n|+1)}\, c_n \, r^n$ has to be in $L_2([0,1), r\,dr)$, which implies that $c_n(r) \equiv 0$, for each $n < 0$. Thus the space $\mathcal{A}_1^2$ $(\subset L_2([0,1), r\,dr) \otimes l_2 = l_2(L_2([0,1), r\,dr)))$ coincides with the space of all two-sided sequences $\{c_n(r)\}_{n\in\mathbb{Z}}$ with the entries

$$c_n(r) = \begin{cases} \sqrt{2(n+1)}\, c_n \, r^n, & \text{if } n \in \mathbb{Z}_+ \\ 0, & \text{if } n \in \mathbb{Z}_- \end{cases},$$

where $\mathbb{Z}_+ = \{0\} \cup \mathbb{N}$, $\mathbb{Z}_- = \mathbb{Z} \setminus \mathbb{Z}_+$, and

$$\|\{c_n(r)\}_{n\in\mathbb{Z}}\| = \left(\sum_{n\in\mathbb{Z}_+} |c_n|^2 \right)^{1/2} = \|\{c_n\}_{n\in\mathbb{Z}_+}\|_{l_2}.$$

For each $n \in \mathbb{Z}_+$ introduce the unitary operator

$$u_n : L_2([0,1), r\,dr) \longrightarrow L_2([0,1), r\,dr)$$

by the rule

$$(u_n f)(r) = \frac{1}{\sqrt{n+1}}\, r^{-\frac{n}{n+1}}\, f(r^{\frac{1}{n+1}}),$$

then the inverse operator $u_n^{-1} = u_n^* : L_2([0,1), r\,dr) \longrightarrow L_2([0,1), r\,dr)$ is given by

$$(u_n^{-1} f)(r) = \sqrt{n+1}\, r^n\, f(r^{n+1}).$$

Finally, define the unitary operator

$$U_2 : l_2(L_2([0,1), r\,dr)) \longrightarrow l_2(L_2([0,1), r\,dr)) = L_2([0,1), r\,dr) \otimes l_2$$

as

$$U_2 : \{c_n(r)\}_{n\in\mathbb{Z}} \longmapsto \{(u_{|n|} c_n)(r)\}_{n\in\mathbb{Z}}.$$

Then the space $\mathcal{A}_2^2 = U_2(\mathcal{A}_1^2)$ coincides with the space of all sequences $\{d_n(r)\}_{n\in\mathbb{Z}}$, where

$$d_n = u_n(\sqrt{2(n+1)}\, c_n r^n) = \sqrt{2}\, c_n,$$

for $n \in \mathbb{Z}_+$, and $d_n(r) \equiv 0$, for $n \in \mathbb{Z}_-$.

We introduce some notation. Let $\ell_0(r) = \sqrt{2}$; we have $\ell_0(r) \in L_2([0,1), r\,dr)$ and $\|\ell_0(r)\| = 1$. Denote by L_0 the one-dimensional subspace of $L_2([0,1), r\,dr)$ generated by $\ell_0(r)$, then the one-dimensional projection P_0 of $L_2([0,1), r\,dr)$ onto L_0 has the form

$$(P_0 f)(r) = \langle f, \ell_0 \rangle \cdot \ell_0 = \sqrt{2} \int_0^1 f(\rho)\, \sqrt{2}\rho \, d\rho. \tag{4.1.3}$$

Denote by l_2^+ (l_2^-) the subspace of (two-sided) l_2, consisting of all sequences $\{c_n\}_{n\in\mathbb{Z}}$ such that $c_n = 0$ for all $n \in \mathbb{Z}_-$ ($n \in \mathbb{Z}_+$). Then $l_2 = l_2^+ \oplus l_2^-$, and denote by p^+ (p^-) the orthogonal projection of l_2 onto l_2^+ (l_2^-). Introduce the sequences $\chi_\pm = \{\chi_\pm(n)\}_{n\in\mathbb{Z}} \in l_\infty$, where $\chi_\pm(n) = 1$ for $n \in \mathbb{Z}_\pm$, and $\chi_\pm(n) = 0$ for $n \in \mathbb{Z}_\mp$. Then obviously $p^\pm = \chi_\pm I$.

Now $\mathcal{A}_2^2 = L_0 \otimes l_2^+$, so the orthogonal projection B_2 of $l_2(L_2([0,1), r\,dr)) = L_2([0,1), r\,dr) \otimes l_2$ onto $\mathcal{A}_2^2$ obviously has the form

$$B_2 = P_0 \otimes p^+.$$

This leads to the following theorem.

Theorem 4.1.1. *The unitary operator $U = U_2 U_1$ gives an isometric isomorphism of the space $L_2(\mathbb{D})$ onto $L_2([0,1), r\,dr) \otimes l_2$ under which*

1. *the Bergman space $\mathcal{A}^2(\mathbb{D})$ is mapped onto $L_0 \otimes l_2^+$,*

$$U : \mathcal{A}^2(\mathbb{D}) \longrightarrow L_0 \otimes l_2^+,$$

  where L_0 is the one-dimensional subspace of $L_2([0,1), r\,dr)$, generated by $\ell_0(r) = \sqrt{2}$,

2. *the Bergman projection $B_\mathbb{D}$ is unitary equivalent to*

$$U\, B_\mathbb{D}\, U^{-1} = P_0 \otimes p^+,$$

  where P_0 is the one-dimensional projection (4.1.3) of $L_2([0,1), r\,dr)$ onto L_0.

Introduce the isometric imbedding

$$R_0 : l_2^+ \longrightarrow L_2([0,1), r\,dr) \otimes l_2$$

by the rule

$$R_0 : \{c_n\}_{n\in\mathbb{Z}_+} \longmapsto \ell_0(r)\{\chi_+(n)c_n\}_{n\in\mathbb{Z}},$$

where we extend the sequence $\{c_n\}_{n\in\mathbb{Z}_+}$ to an element of l_2 setting $c_n = 0$ for negative indices $n < 0$. The image of R_0 obviously coincides with the space $\mathcal{A}_2^2$. The adjoint operator $R_0^* : L_2([0,1), r\,dr) \otimes l_2 \to l_2^+$ is given by

$$R_0^* : \{c_n(r)\}_{n\in\mathbb{Z}} \longmapsto \left\{ \int_0^1 c_n(\rho)\sqrt{2}\,\rho\,d\rho \right\}_{n\in\mathbb{Z}_+},$$

and

$$R_0^* R_0 = I \quad : \quad l_2^+ \longrightarrow l_2^+,$$
$$R_0 R_0^* = B_2 \quad : \quad L_2([0,1), r\,dr) \otimes l_2 \longrightarrow \mathcal{A}_2^2 = L_0 \otimes l_2^+.$$

Now the operator $R = R_0^* U$ maps the space $L_2(\mathbb{D})$ onto l_2^+, and the restriction

$$R|_{\mathcal{A}^2(\mathbb{D})} : \mathcal{A}^2(\mathbb{D}) \longrightarrow l_2^+$$

is an isometric isomorphism. The adjoint operator

$$R^* = U^* R_0 : l_2^+ \longrightarrow \mathcal{A}^2(\mathbb{D}) \subset L_2(\mathbb{D})$$

is an isometric isomorphism of l_2^+ onto the subspace $\mathcal{A}^2(\mathbb{D})$ of the space $L_2(\mathbb{D})$.

Remark 4.1.2. We have

$$\begin{aligned} R R^* = I &: l_2^+ \longrightarrow l_2^+, \\ R^* R = B_{\mathbb{D}} &: L_2(\mathbb{D}) \longrightarrow \mathcal{A}^2(\mathbb{D}). \end{aligned}$$

Theorem 4.1.3. *The isometric isomorphism*

$$R^* = U^* R_0 : l_2^+ \longrightarrow \mathcal{A}^2(\mathbb{D})$$

is given by

$$R^* : \{c_n\}_{n\in\mathbb{Z}_+} \longmapsto \frac{1}{\sqrt{2\pi}} \sum_{n\in\mathbb{Z}_+} \sqrt{2(n+1)}\, c_n\, z^n. \tag{4.1.4}$$

Proof. Calculate

$$\begin{aligned} R^* = U_1^* U_2^* R_0 \quad &: \quad \{c_n\}_{n\in\mathbb{Z}_+} \longmapsto U_1^* U_2^* (\{\sqrt{2}\, c_n\}_{n\in\mathbb{Z}_+}) \\ &= \quad U_1^* (\{\sqrt{2(n+1)}\, c_n\, r^n\}_{n\in\mathbb{Z}_+}) \\ &= \quad \frac{1}{\sqrt{2\pi}} \sum_{n\in\mathbb{Z}_+} \sqrt{2(n+1)}\, c_n\, (rt)^n = \frac{1}{\sqrt{2\pi}} \sum_{n\in\mathbb{Z}_+} \sqrt{2(n+1)}\, c_n\, z^n. \end{aligned}$$

$\square$

Corollary 4.1.4. *The inverse isomorphism*

$$R : \mathcal{A}^2(\mathbb{D}) \longrightarrow l_2^+$$

is given by

$$R : \varphi(z) \longmapsto \left\{ \frac{\sqrt{2(n+1)}}{\sqrt{2\pi}} \int_{\mathbb{D}} \varphi(z)\, \overline{z}^n\, d\mu(z) \right\}_{n\in\mathbb{Z}_+}. \tag{4.1.5}$$

Remark 4.1.5. The above operator R is defined on the whole of $L_2(\mathbb{D})$ and its restriction to $\mathcal{A}^2(\mathbb{D})$ gives an isomorphism onto l_2^+. For the restriction one has an alternative formula

$$R : \varphi(z) \longmapsto \left\{ \frac{\sqrt{2\pi}}{\sqrt{2(n+1)}} \frac{\varphi^{(n)}(0)}{n!} \right\}_{n\in\mathbb{Z}_+}.$$

Introduce the operator $\widetilde{R} : L_2(\mathbb{D}) \to L_2(S^1)$ as follows,

$$\widetilde{R} = \mathcal{F}^{-1} R.$$

We denote by $H_+^2(S^1)$ the classical Hardy space on the unit disk. It can be characterized, for example, as the space of all $L_2(S^1)$-functions whose discrete Fourier transform (4.1.1) vanishes for all negative indices.

Corollary 4.1.6. *We have the following isometric isomorphisms between the Bergman $\mathcal{A}^2(\mathbb{D})$ and the Hardy $H^2_+(S^1)$ spaces:*

$$\widetilde{R}|_{\mathcal{A}^2(\mathbb{D})} \;:\; \mathcal{A}^2(\mathbb{D}) \longrightarrow H^2_+(S^1),$$
$$\widetilde{R}^*|_{H^2_+(S^1)} \;:\; H^2_+(S^1) \longrightarrow \mathcal{A}^2(\mathbb{D}).$$

The operators $\widetilde{R}$ and $\widetilde{R}^*$ provide the following decomposition of the Bergman $B_{\mathbb{D}}$ and the Szegö P_{S^1} projections:

$$\widetilde{R}^*\widetilde{R} = B_{\mathbb{D}} \;:\; L_2(\mathbb{D}) \longrightarrow \mathcal{A}^2(\mathbb{D}),$$
$$\widetilde{R}\,\widetilde{R}^* = P_{S^1} \;:\; L_2(S^1) \longrightarrow H^2_+(S^1).$$

Another connection between the Bergman and the Hardy spaces, and between the corresponding projections, gives the next theorem.

Theorem 4.1.7. *The unitary operator $W = (I \otimes \mathcal{F}^{-1})U_2(I \otimes \mathcal{F})$ gives an isometric isomorphism of the space $L_2(\mathbb{D}) = L_2([0,1), rdr) \otimes L_2(S^1, \frac{dt}{it})$ under which*

1. *the Bergman $\mathcal{A}^2(\mathbb{D})$ and the Hardy $H^2_+(S^1)$ spaces are connected by the formula*

$$W(\mathcal{A}^2(\mathbb{D})) = L_0 \otimes H^2_+(S^1),$$

2. *the Bergman $B_{\mathbb{D}}$ and the Szegö $P^+_{S^1}$ projections are connected by the formula*

$$W B_{\mathbb{D}} W^{-1} = P_0 \otimes P^+_{S^1},$$

where P_0 is the one-dimensional projection (4.1.3) of $L_2([0,1), rdr)$ onto the one-dimensional space L_0 generated by $\ell_0(r) = \sqrt{2} \in L_2([0,1), rdr)$.

Proof. Follows directly from Theorem 4.1.1. $\qquad\qquad\square$

In addition to the Bergman space $\mathcal{A}^2(\mathbb{D})$ of analytic functions in $\mathbb{D}$, introduce the space $\widetilde{\mathcal{A}}^2(\mathbb{D})$ as the (closed) subspace of $L_2(\mathbb{D})$ consisting of all functions which are anti-analytic in $\mathbb{D}$ and take the zero value at the point $0 \in \mathbb{D}$ (otherwise the spaces $\mathcal{A}^2(\mathbb{D})$ and $\widetilde{\mathcal{A}}^2(\mathbb{D})$ would intersect in the constants). Denote by $\widetilde{B}_{\mathbb{D}}$ the orthogonal Bergman projection of $L_2(\mathbb{D})$ onto $\widetilde{\mathcal{A}}^2(\mathbb{D})$.

Theorem 4.1.8. *The unitary operator $U = U_2U_1$ gives an isometric isomorphism of the space $L_2(\mathbb{D})$ onto $L_2([0,1), rdr) \otimes l_2$ under which*

1. *the space $\widetilde{\mathcal{A}}^2(\mathbb{D})$ is mapped onto $L_0 \otimes l^-_2$,*

$$U : \widetilde{\mathcal{A}}^2(\mathbb{D}) \longrightarrow L_0 \otimes l^-_2,$$

where L_0 is the one-dimensional subspace of $L_2([0,1), rdr)$, generated by $\ell_0(r) = \sqrt{2}$,

2. *the projection $\widetilde{B}_{\mathbb{D}}$ is unitary equivalent to*

$$U\,\widetilde{B}_{\mathbb{D}}\,U^{-1} = P_0 \otimes p^-,$$

where $p^- = \chi_- I$ is the orthogonal projection of (two-sided) l_2 onto l_2^-, and P_0 is the one-dimensional projection (4.1.3) of $L_2([0,1), r\,dr)$ onto L_0.

Proof. The space $\widetilde{\mathcal{A}}^2(\mathbb{D})$ can be described alternatively as the (closed) subspace of $L_2(\mathbb{D})$ which consists of all functions satisfying the equation

$$\mathcal{D}\,\varphi = z\frac{\partial}{\partial z}\varphi = \frac{z}{2}\left(\frac{\partial}{\partial x} - i\frac{\partial}{\partial y}\right)\varphi = 0,$$

where $z = x + iy$. Passing to polar coordinates we have

$$\mathcal{D} = \frac{r}{2}\left(\frac{\partial}{\partial r} + \frac{t}{r}\frac{\partial}{\partial t}\right).$$

Calculate

$$(I \otimes \mathcal{F})\mathcal{D}(I \otimes \mathcal{F}^{-1}) \quad : \quad \{c_n(r)\}_{n\in\mathbb{Z}} \longmapsto \frac{1}{\sqrt{2\pi}}\sum_{n\in\mathbb{Z}} c_n(r)\,t^n$$

$$\longmapsto \frac{1}{\sqrt{2\pi}}\sum_{n\in\mathbb{Z}}\frac{r}{2}\left(\frac{\partial}{\partial r} + \frac{n}{r}\right)c_n(r)\,t^n$$

$$\longmapsto \left\{\frac{r}{2}\left(\frac{\partial}{\partial r} + \frac{n}{r}\right)c_n(r)\right\}_{n\in\mathbb{Z}}.$$

Now

$$U_2 U_1 \mathcal{D} U_1^{-1} U_2^{-1} = U_2\left\{\frac{r}{2}\left(\frac{\partial}{\partial r} + \frac{n}{r}\right)c_n(r)\right\}_{n\in\mathbb{Z}} U_2^{-1} = \{\mathcal{D}_n\}_{n\in\mathbb{Z}},$$

where

$$\mathcal{D}_n = \begin{cases} \frac{n+1}{2}\left(\frac{2n}{n+1} + r\frac{d}{dr}\right), & \text{if } n \in \mathbb{Z}_+ \\ \frac{|n|+1}{2}\,r\frac{d}{dr}, & \text{if } n \in \mathbb{Z}_- \end{cases} \tag{4.1.6}$$

The general solution of the equations $\mathcal{D}_n c_n(r) = 0$, $n \in \mathbb{Z}$, has the form

$$c_n(r) = c_n\, r^{-\frac{2n}{n+1}}, \quad \text{for } n \in \mathbb{Z}_+,$$
$$c_n(r) = \sqrt{2}\,c_n, \quad \text{for } n \in \mathbb{Z}_-.$$

All functions in $\widetilde{\mathcal{A}}^2(\mathbb{D})$ have the value zero at the origin, which implies that $c_0(r) = c_0 = 0$. Again, each $c_n(r)$, $n \in \mathbb{N}$, has to be in $L_2([0,1), r\,dr)$, which implies that $c_n = 0$, for all $n \in \mathbb{N}$.

Then the space $\widetilde{\mathcal{A}}_2^2 = U(\widetilde{\mathcal{A}}^2(\mathbb{D}))$ coincides with the space of all sequences $\{c_n(r)\}_{n \in \mathbb{Z}}$, where

$$c_n(r) = \begin{cases} 0, & \text{if } n \in \mathbb{Z}_+ \\ \sqrt{2}\, c_n, & \text{if } n \in \mathbb{Z}_- \end{cases},$$

and $\|\{c_n(r)\}\|_{l_2(L_2([0,1),rdr))} = \|\{c_n\}\|_{l_2^-}$. Thus $\widetilde{\mathcal{A}}_2^2 = L_0 \otimes l_2^-$, and the orthogonal projection $\widetilde{B}_2$ of $l_2(L_2([0,1),rdr)) = L_2([0,1),rdr) \otimes l_2$ onto $\widetilde{\mathcal{A}}_2^2$ has the form $\widetilde{B}_2 = P_0 \otimes p^-$. $\qquad\qquad\qquad\qquad\qquad\qquad\qquad\qquad\qquad\qquad\qquad\square$

Introduce now the Hardy space $H_-^2(S^1)$ on the exterior of the unit disk. It can be characterized, for example, as the space of all $L_2(S^1)$-functions whose discrete Fourier transform (4.1.1) vanishes for all non-negative indices. Denote by $P_{S^1}^-$ the orthogonal (Szegö) projection of $L_2(S^1)$ onto $H_-^2(S^1)$.

Corollary 4.1.9. *The unitary operator* $W = (I \otimes \mathcal{F}^{-1})U_2(I \otimes \mathcal{F})$ *gives an isometric isomorphism of the space* $L_2(\mathbb{D}) = L_2([0,1),rdr) \otimes L_2(S^1, \frac{dt}{it})$ *under which*

1. *the spaces* $\widetilde{\mathcal{A}}^2(\mathbb{D})$ *and* $H_-^2(S^1)$ *are connected by the formula*

$$W(\widetilde{\mathcal{A}}^2(\mathbb{D})) = L_0 \otimes H_-^2(S^1),$$

2. *the projections* $\widetilde{B}_{\mathbb{D}}$ *and* $P_{S^1}^-$ *are connected by the formula*

$$W\, \widetilde{B}_{\mathbb{D}}\, W^{-1} = P_0 \otimes P_{S^1}^-,$$

where P_0 *is the one-dimensional projection* (4.1.3) *of* $L_2([0,1),rdr)$ *onto one-dimensional space* L_0 *generated by* $\ell_0(r) = \sqrt{2} \in L_2([0,1),rdr)$.

4.2   Poly-Bergman type spaces, decomposition of $L_2(\mathbb{D})$

Besides the operator

$$D = z\frac{\partial}{\partial z} = \frac{r}{2}\left(\frac{\partial}{\partial r} + \frac{t}{r}\frac{\partial}{\partial t}\right)$$

introduce

$$\overline{D} = \overline{z}\frac{\partial}{\partial \overline{z}} = \frac{r}{2}\left(\frac{\partial}{\partial r} - \frac{t}{r}\frac{\partial}{\partial t}\right).$$

Analogously to the Bergman spaces $\mathcal{A}^2(\mathbb{D})$ and $\widetilde{\mathcal{A}}^2(\mathbb{D})$, which can be treated as the L_2-kernels of $\overline{D}$ and D respectively, following [154] introduce the spaces of poly-$\overline{D}$-analytic and poly-D-analytic functions, the poly-Bergman type spaces.

Define the space $\mathcal{A}_n^2(\mathbb{D})$ of n-$\overline{D}$-analytic functions as the (closed) subspace of $L_2(\mathbb{D})$ of all functions $\varphi = \varphi(z,\overline{z}) = \varphi(x,y)$, which satisfy the equation

$$\overline{D}^n\varphi = \left(\overline{z}\frac{\partial}{\partial \overline{z}}\right)^n \varphi = 0.$$

Similarly, define the space $\widetilde{\mathcal{A}}_n^2(\mathbb{D})$ of n-$\mathcal{D}$-analytic functions as the (closed) subspace of $L_2(\mathbb{D})$ of all functions $\varphi = \varphi(z,\bar{z}) = \varphi(x,y)$, which satisfy the equation

$$\mathcal{D}^n \varphi = \left(z \frac{\partial}{\partial z} \right)^n \varphi = 0,$$

and take the value zero at the origin.

Of course, we have $\mathcal{A}_1^2(\mathbb{D}) = \mathcal{A}^2(\mathbb{D})$ and $\widetilde{\mathcal{A}}_1^2(\mathbb{D}) = \widetilde{\mathcal{A}}^2(\mathbb{D})$, for $n = 1$, as well as $\mathcal{A}_n^2(\mathbb{D}) \subset \mathcal{A}_{n+1}^2(\mathbb{D})$ and $\widetilde{\mathcal{A}}_n^2(\mathbb{D}) \subset \widetilde{\mathcal{A}}_{n+1}^2(\mathbb{D})$, for each $n \in \mathbb{N}$.

Recall, that the system of functions $\{e^{-x/2} L_n(x)\}_{n \in \mathbb{Z}_+}$, where $L_n(x)$, $n \in \mathbb{Z}_+$, are the Laguerre polynomials (3.3.1), forms an orthonormal base in $L_2(\mathbb{R}_+)$; that is

$$\int_0^\infty L_n(x)\, L_m(x)\, e^{-x}\, dx = \delta_{n,m}. \tag{4.2.1}$$

Following [154] we introduce

$$\ell_n(r) = \sqrt{2}\, L_n(\log r^{-2}) = \sqrt{2} \sum_{k=0}^n \frac{n!}{k!(n-k)!} \frac{2^k}{k!} (\log r)^k, \tag{4.2.2}$$

where $r \in (0,1]$ and $n \in \mathbb{Z}_+$. Changing variables in (4.2.1) we have

$$\int_0^1 \ell_n(r)\, \ell_m(r)\, r dr = \delta_{n,m},$$

that is, the system of functions $\{\ell_n(r)\}_{n \in \mathbb{Z}_+}$ forms an orthonormal basis in the space $L_2([0,1), rdr)$.

Denote by L_n, $n \in \mathbb{Z}_+$, the one-dimensional subspace of $L_2([0,1), rdr)$, generated by the function $\ell_n(r)$. Note, that for $n = 0$ this definition gives exactly the previously defined space L_0. And let

$$L_n^\oplus = \bigoplus_{k=0}^n L_k$$

be the direct sum of the first $(n+1)$ spaces.

Theorem 4.2.1. *The unitary operator $U : L_2(\mathbb{D}) \to L_2([0,1), rdr) \otimes l_2$ maps the space $\mathcal{A}_n^2(\mathbb{D})$ of n-$\overline{\mathcal{D}}$-analytic functions onto the space $L_{n-1}^\oplus \otimes l_2^+$.*

Proof. The space $U(\mathcal{A}_n^2(\mathbb{D}))$ obviously coincides with the set of all sequences from $l_2(L_2([0,1), rdr)) = L_2([0,1), rdr) \otimes l_2$, which satisfy the equation

$$U\overline{\mathcal{D}}^n U^{-1} \{c_k(r)\}_{k \in \mathbb{Z}} = U_2 \left\{ \left[\frac{r}{2} \left(\frac{\partial}{\partial r} - \frac{k}{r} \right) \right]^n \right\}_{k \in \mathbb{Z}} U_2^{-1} \{c_k(r)\}_{k \in \mathbb{Z}}$$
$$= \{\overline{\mathcal{D}}_k^n c_k(r)\}_{k \in \mathbb{Z}} = 0,$$

where

$$\overline{\mathcal{D}}_k = \begin{cases} \frac{k+1}{2}\, r\, \frac{d}{dr}, & \text{if } k \in \mathbb{Z}_+ \\ \frac{|k|+1}{2}\left(\frac{2|k|}{|k|+1} + r\,\frac{d}{dr}\right), & \text{if } k \in \mathbb{Z}_- \end{cases}.$$

It is easy to see that the intersection of the general solution of this equation with the space $l_2(L_2([0,1), rdr)) = L_2([0,1), rdr) \otimes l_2$ coincides with the set of all sequences of the form

$$\sum_{m=0}^{n-1} (\log r)^m \{\chi_+(k)\, d_k^{(m)}\}_{k\in\mathbb{Z}},$$

where $\{d_k^{(m)}\}_{k\in\mathbb{Z}} \in l_2$, for all $m = \overline{0, n-1}$, or rearranging polynomials on $\log r$, with the set of all sequences

$$\sum_{m=0}^{n-1} \ell_m(r)\{c_k^{(m)}\}_{k\in\mathbb{Z}_+},$$

where $\{c_k^{(m)}\}_{k\in\mathbb{Z}_+} \in l_2^+$, for all $m = \overline{0, n-1}$. $\qquad\qquad\square$

Introduce the space $\mathcal{A}^2_{(n)}$ of true-n-$\overline{\mathcal{D}}$-analytic functions by

$$\mathcal{A}^2_{(n)} = \mathcal{A}^2_n \ominus \mathcal{A}^2_{n-1},$$

for $n > 1$, and by $\mathcal{A}^2_{(1)} = \mathcal{A}^2_1$, for $n = 1$, then, of course,

$$\mathcal{A}^2_n = \bigoplus_{k=1}^{n} \mathcal{A}^2_{(k)}.$$

Corollary 4.2.2. *The unitary operator $U : L_2(\mathbb{D}) \to L_2([0,1), rdr) \otimes l_2$ maps the space $\mathcal{A}^2_{(n)}(\mathbb{D})$ of true-n-$\overline{\mathcal{D}}$-analytic functions onto the space $L_{n-1} \otimes l_2^+$.*

Analogously in the $\mathcal{D}$-analytic situations we have the following assertions.

Theorem 4.2.3. *The unitary operator $U : L_2(\mathbb{D}) \to L_2([0,1), rdr) \otimes l_2$ maps the space $\widetilde{\mathcal{A}}^2_n(\mathbb{D})$ of n-$\mathcal{D}$-analytic functions onto the space $L^{\oplus}_{n-1} \otimes l_2^-$.*

Proof. We follow all the steps of the proof of Theorem 4.2.1. The space $U(\widetilde{\mathcal{A}}^2_n(\mathbb{D}))$ obviously coincides with the set of all sequences from the space $l_2(L_2([0,1), rdr)) = L_2([0,1), rdr) \otimes l_2$, which satisfy the equation

$$\begin{aligned} U\mathcal{D}^n U^{-1}\{c_k(r)\}_{k\in\mathbb{Z}} &= U_2\left\{\left[\frac{r}{2}\left(\frac{\partial}{\partial r} + \frac{k}{r}\right)\right]^n\right\}_{k\in\mathbb{Z}} U_2^{-1}\{c_k(r)\}_{k\in\mathbb{Z}} \\ &= \{\mathcal{D}_k^n c_k(r)\}_{k\in\mathbb{Z}} = 0, \end{aligned}$$

where the operator $\{\mathcal{D}_k\}_{k\in\mathbb{Z}}$ is given by (4.1.6).

Now the intersection of the general solution of this equation with the space $l_2(L_2([0,1), r\,dr)) = L_2([0,1), r\,dr) \otimes l_2$ coincides with the set of all sequences of the form

$$\sum_{m=0}^{n-1} (\log r)^m \, \{\chi_-(k) \, d_k^{(m)}\}_{k\in\mathbb{Z}},$$

where $\{d_k^{(m)}\}_{k\in\mathbb{Z}} \in l_2$, for all $m = \overline{0, n-1}$, or rearranging polynomials on $\log r$, with the set of all sequences

$$\sum_{m=0}^{n-1} \ell_m(r)\{c_k^{(m)}\}_{k\in\mathbb{Z}_-},$$

where $\{c_k^{(m)}\}_{k\in\mathbb{Z}_-} \in l_2^-$, for all $m = \overline{0, n-1}$. $\square$

Symmetrically, introduce the space $\widetilde{\mathcal{A}}_{(n)}^2$ of true-n-$\mathcal{D}$-analytic functions by

$$\widetilde{\mathcal{A}}_{(n)}^2 = \widetilde{\mathcal{A}}_n^2 \ominus \widetilde{\mathcal{A}}_{n-1}^2,$$

for $n > 1$, and by $\widetilde{\mathcal{A}}_{(1)}^2 = \widetilde{\mathcal{A}}_1^2$, for $n = 1$, analogously,

$$\widetilde{\mathcal{A}}_n^2 = \bigoplus_{k=1}^{n} \widetilde{\mathcal{A}}_{(k)}^2.$$

Corollary 4.2.4. *The unitary operator* $U : L_2(\mathbb{D}) \to L_2([0,1), r\,dr) \otimes l_2$ *maps the space* $\widetilde{\mathcal{A}}_{(n)}^2(\mathbb{D})$ *of true-n-$\mathcal{D}$-analytic functions onto the space* $L_{n-1} \otimes l_2^-$.

The above results lead up to the following theorem.

Theorem 4.2.5. *We have the following isometric isomorphisms and decompositions of spaces:*

1. *Isomorphic images of poly-$\overline{\mathcal{D}}$-analytic spaces,*

$$W \;\; : \;\; \mathcal{A}_{(n)}^2(\mathbb{D}) \longrightarrow L_{n-1} \otimes H_+^2(S^1),$$

$$W \;\; : \;\; \mathcal{A}_n^2(\mathbb{D}) \longrightarrow \bigoplus_{k=0}^{n-1} L_k \otimes H_+^2(S^1),$$

$$W \;\; : \;\; \bigoplus_{k=1}^{\infty} \mathcal{A}_{(k)}^2(\mathbb{D}) \longrightarrow L_2([0,1), r\,dr) \otimes H_+^2(S^1).$$

2. *Isomorphic images of poly-$\mathcal{D}$-analytic spaces,*

$$W \quad : \quad \widetilde{\mathcal{A}}^2_{(n)}(\mathbb{D}) \longrightarrow L_{n-1} \otimes H^2_-(S^1),$$

$$W \quad : \quad \widetilde{\mathcal{A}}^2_n(\mathbb{D}) \longrightarrow \bigoplus_{k=0}^{n-1} L_k \otimes H^2_-(S^1),$$

$$W \quad : \quad \bigoplus_{k=1}^{\infty} \widetilde{\mathcal{A}}^2_{(k)}(\mathbb{D}) \longrightarrow L_2([0,1), r\,dr) \otimes H^2_-(S^1).$$

3. *Decomposition of the space $L_2(\mathbb{D})$,*

$$L_2(\mathbb{D}) \quad = \quad \bigoplus_{k=1}^{\infty} (\mathcal{A}^2_{(k)}(\mathbb{D}) \oplus \widetilde{\mathcal{A}}^2_{(k)}(\mathbb{D}))$$

$$= \quad \bigoplus_{k=1}^{\infty} \mathcal{A}^2_{(k)}(\mathbb{D}) \oplus \bigoplus_{k=1}^{\infty} \widetilde{\mathcal{A}}^2_{(k)}(\mathbb{D}).$$

Here L_n is the one-dimensional subspace of $L_2([0,1), r\,dr)$ generated by the function $\ell_n(r)$ of the form (4.2.2).

Chapter 5

Toeplitz Operators with Commutative Symbol Algebras

Theorems 2.4.5, 2.8.3, 2.8.6, and 2.8.7 show a certain difference between the compactness properties of commutators and semi-commutators. In order to understand this difference we start with the following setting.

Let $\mathcal{A}(\mathbb{D})$ be a C^*-subalgebra of $L_\infty(\mathbb{D})$. Consider the following statements (which may be false or true depending on $\mathcal{A}(\mathbb{D})$):

1) For each $a \in \mathcal{A}(\mathbb{D})$ the commutator $[B_\mathbb{D}, aI] = B_\mathbb{D}aI - aB_\mathbb{D}$ is compact.

2) For each pair $a, b \in \mathcal{A}(\mathbb{D})$ the semi-commutator $[T_a, T_b) = T_aT_b - T_{ab}$ is compact.

3) For each pair $a, b \in \mathcal{A}(\mathbb{D})$ the commutator $[T_a, T_b] = T_aT_b - T_bT_a$ is compact.

In the next section we show that the first two statements are equivalent; that is, they can be true only simultaneously, and each of them implies the third statement. At the same time, as Theorems 2.8.6, and 2.8.7 show, the third statement does not imply in general either the first or the second statement. In this stage it is important and interesting to understand the gap between the third and the first two properties. In fact this gap turns out to be quite substantial. To show this we construct a number of algebras $\mathcal{A} = \mathcal{A}(\mathbb{D})$ which have (only) property 3), and moreover even have a stronger property: *for each $a, b \in \mathcal{A}$, $[T_a, T_b] = 0$, while $[T_a, T_b)$ is not compact.*

Another related concept is as follows. Given a C^*-subalgebra $\mathcal{A}(\mathbb{D})$ of $L_\infty(\mathbb{D})$, we introduce two operator C^*-algebras: the algebra $\mathcal{T}(\mathcal{A}(\mathbb{D}))$, which is generated by all Toeplitz operators

$$T_a : \varphi \in \mathcal{A}^2(\mathbb{D}) \longmapsto B_\mathbb{D}a\varphi \in \mathcal{A}^2(\mathbb{D})$$

with defining symbols $a \in \mathcal{A}(\mathbb{D})$, and the algebra $\mathcal{R}(\mathcal{A}(\mathbb{D}), B_\mathbb{D})$, which is generated by all operators of the form

$$A = aI + bB_{\mathbb{D}},$$

where a, $b \in \mathcal{A}(\mathbb{D})$, acting on $L_2(\mathbb{D})$.

One of the main features of algebras $\mathcal{A}(\mathbb{D})$ having property 2) is that the corresponding Toeplitz operator algebra $\mathcal{T}(\mathcal{A}(\mathbb{D}))$ admits a commutative symbolic calculus, i.e., the Fredholm symbol algebra $\mathrm{Sym}\,\mathcal{T}(\mathcal{A}(\mathbb{D})) = \mathcal{T}(\mathcal{A}(\mathbb{D}))/\mathcal{K}$, where $\mathcal{K}$ is the ideal of compact operators, is commutative.

Note, that under the condition 2) the Fredholm symbol algebra

$$\mathrm{Sym}\,\mathcal{R}(\mathcal{A}(\mathbb{D}), B_{\mathbb{D}}) = \mathcal{R}(\mathcal{A}(\mathbb{D}), B_{\mathbb{D}})/\mathcal{K}$$

of the algebra $\mathcal{R}(\mathcal{A}(\mathbb{D}), B_{\mathbb{D}})$ is commutative as well.

At the same time, under (only) condition 3) the Fredholm symbol algebra $\mathrm{Sym}\,\mathcal{R}(\mathcal{A}(\mathbb{D}), B_{\mathbb{D}}) = \mathcal{R}(\mathcal{A}(\mathbb{D}), B_{\mathbb{D}})/\mathcal{K}$ of the algebra $\mathcal{R}(\mathcal{A}(\mathbb{D}), B_{\mathbb{D}})$ is *non-commutative*, while the Fredholm symbol algebra $\mathrm{Sym}\,\mathcal{T}(\mathcal{A}(\mathbb{D})) = \mathcal{T}(\mathcal{A}(\mathbb{D}))/\mathcal{K}$ still remains *commutative*.

Another way to understand the difference between properties 2) and 3) is by comparing the representations of these algebras. In other words, it is important to understand how complicated the Fredholm symbol algebra of the algebra $\mathcal{R}(\mathcal{A}(\mathbb{D}), B_{\mathbb{D}})$ can be, while the Toeplitz operator algebra $\mathcal{T}(\mathcal{A}(\mathbb{D}))$ still admits the commutative symbolic calculus.

Answering this question we show that for each finite set of integers $\Lambda = \langle n_0, n_1, \ldots, n_m \rangle$, where $1 = n_0 < n_1 < \ldots < n_m \leq \infty$, and $n_k \in \mathbb{N} \cup \{\infty\}$, there is an algebra $\mathcal{A}_\Lambda$ (with only property 3)), such that the Fredholm symbol algebra $\mathrm{Sym}\,\mathcal{T}(\mathcal{A}_\Lambda)$ of the algebra $\mathcal{T}(\mathcal{A}_\Lambda)$ is *commutative*, while the Fredholm symbol algebra $\mathrm{Sym}\,\mathcal{R}(\mathcal{A}_\Lambda, B_{\mathbb{D}})$ of the algebra $\mathcal{R}(\mathcal{A}_\Lambda, B_{\mathbb{D}})$ has *irreducible representations of exactly the predefined dimensions* $n_0, n_1, \ldots, n_m$.

It is worth mentioning that for Toeplitz operators on Hardy spaces, the functional algebras with the property 2) were known a long time ago [172, 173], and that the Gohberg-Krupnik results of the late 1960s [82, 84] gave examples of algebras with only property 3). Nevertheless the above questions still remain open for the case of Toeplitz operators on the Hardy space.

5.1   Semi-commutator versus commutator

In this section we clarify the general interrelations among the above mentioned three properties. We note that these interrelations do not depend in fact on the specific nature of the Bergman space, or on the fact that $\mathcal{A}(\mathbb{D})$ is a functional algebra. We consider them thus in the following abstract setting.

Let H be a separable Hilbert space and let H_0 be its closed subspace. We denote by P the orthogonal projection from H onto H_0.

Given a bounded linear operator A acting on H, introduce the "abstract" Toeplitz operator with defining symbol A,

$$T_A = PAP = PA : H_0 \longrightarrow H_0,$$

and the "abstract" Hankel operator with defining symbol A,

$$H_A = (I - P)AP,$$

acting either from H_0 to $H_0^\perp$, or on H.

We start now with a C^*-algebra $\mathcal{C}$ of bounded linear operators acting on H and having the property that the commutator $[A, B]$ is compact for each $A,\, B \in \mathcal{C}$. Consider the following three statements (which again may be false or true depending on $\mathcal{C}$):

1) For each $A \in \mathcal{C}$ the commutator $[P, A] = PA - AP$ is compact.

2) For each pair $A,\, B \in \mathcal{C}$ the semi-commutator $[T_A, T_B) = T_A T_B - T_{AB}$ is compact.

3) For each pair $A,\, B \in \mathcal{C}$ the commutator $[T_A, T_B] = T_A T_B - T_B T_A$ is compact.

From the equality

$$[T_A, T_B] = [T_A, T_B) - [T_B, T_A) - T_{[A,B]}$$

we have that the property 2) always implies the property 3). At the same time we know that the inverse is not valid in general.

Further we have

Theorem 5.1.1. *The following three statements are equivalent.*

(i) *For each $A \in \mathcal{C}$ the commutator $[P, A] = PA - AP$ is compact.*

(ii) *For each pair $A,\, B \in \mathcal{C}$ the semi-commutator $[T_A, T_B) = T_A T_B - T_{AB}$ is compact.*

(iii) *For each $A \in \mathcal{C}$ the Hankel operator H_A is compact.*

Proof. (i) $\Rightarrow$ (ii) follows from the equality

$$[T_A, T_B) = PAPBP - PABP = PA[P, B]P.$$

(ii) $\Rightarrow$ (iii) follows from the equality

$$(H_A)^* H_A = PA^*(I - P)AP = P_{A^*A} - PA^*PAP = -[T_{A^*}, T_A)$$

and the general fact that an operator X acting on H is compact if and only if the operator X^*X is compact.

(iii) $\Rightarrow$ (i) follows from the equality

$$[P, A] = (H_{A^*})^* - H_A. \qquad \square$$

where

$$\gamma_a(x) = \int_{\mathbb{R}_+} a\left(\frac{\eta}{2x}\right) e^{-\eta}\, d\eta, \quad x \in \mathbb{R}_+.$$

$\square$

Denote by $\mathcal{A}_\infty$ the C^*-algebra of all $L_\infty(\Pi)$-functions which depend on $v = \operatorname{Im} w$ only, and consider two operator algebras, the Toeplitz operator algebra $\mathcal{T}(\mathcal{A}_\infty)$, generated by all the operators of the form

$$T_a : \varphi \in \mathcal{A}^2(\Pi) \longmapsto B_\Pi a\varphi \in \mathcal{A}^2(\Pi),$$

where $a = a(v) \in \mathcal{A}_\infty$, and the algebra $\mathcal{R}(\mathcal{A}_\infty, B_\Pi)$, generated by all the operators of the form

$$A = aI + bB_\Pi,$$

where $a = a(v)$, $b = b(v) \in \mathcal{A}_\infty$, acting on $L_2(\Pi)$.

Corollary 5.2.2. *The algebra $\mathcal{T}(\mathcal{A}_\infty)$ is commutative. The isomorphic imbedding*

$$\tau_\infty : \mathcal{T}(\mathcal{A}_\infty) \longrightarrow C_b(\mathbb{R}_+)$$

is generated by the following mapping of generators of the algebra $\mathcal{T}(\mathcal{A}_\infty)$,

$$\tau_\infty : T_a \longmapsto \gamma_a(x) = \int_{\mathbb{R}_+} a\left(\frac{\eta}{2x}\right) e^{-\eta}\, d\eta, \quad x \in \mathbb{R}_+;$$

here $C_b(\mathbb{R}_+)$ is the algebra of all functions bounded and continuous in $\mathbb{R}_+$.

Remark 5.2.3. Given two functions $a,\ b \in \mathcal{A}_\infty$ we have obviously that $\gamma_a(x)\,\gamma_b(x) - \gamma_{ab}(x) \neq 0$, in general. Thus the C^*-algebra $\mathcal{A}_\infty$ provides us with an example of the algebra $\mathcal{A}$ with the property that *for each $a,\ b \in \mathcal{A}$, $[T_a, T_b] = 0$, while $[T_a, T_b)$ is not compact*, in general.

Return now to the algebra $\mathcal{R}(\mathcal{A}_\infty, B_\Pi)$. The algebra $\mathcal{R}(\mathcal{A}_\infty, B_\Pi)$ is naturally isomorphic to the algebra

$$\mathcal{R}_0 = U\,\mathcal{R}(\mathcal{A}_\infty, B_\Pi)\,U^{-1} = U_2(F \otimes I)\mathcal{R}(\mathcal{A}_\infty, B_\Pi)(F^{-1} \otimes I)U_2^{-1},$$

where $U = U_2(F \otimes I)$, and the operator U_2 is given by (3.1.4). Under this isomorphism the generators of the algebra $\mathcal{R}(\mathcal{A}_\infty, B_\Pi)$ are mapped to the following generators of the algebra $\mathcal{R}_0$:

$$\begin{aligned}
U\,B_\Pi\,U^{-1} &= \chi_+(x)I \otimes P_0, \\
U\,a(v)\,U^{-1} &= a\left(\frac{y}{2|x|}\right)I,
\end{aligned}$$

which act on the space $L_2(\mathbb{R}) \otimes L_2(\mathbb{R}_+)$. We recall that the one-dimensional projection P_0 is given by (3.1.5) and is the orthogonal projection onto the one-dimensional subspace of $L_2(\mathbb{R}_+)$ generated by $\ell_0(y) = e^{-y/2}$.

Thus the algebra $\mathcal{R}_0$ splits onto the direct integral of the algebras $\mathcal{R}_0(x)$, $x \in \mathbb{R} \setminus \{0\}$, where each algebra $\mathcal{R}_0(x)$ is generated by the following operators, acting on $L_2(\mathbb{R}_+)$:

for $x \in \mathbb{R}_+$,

$$P_0 \quad \text{and} \quad a(\frac{y}{2x})I,$$

for $x \in \mathbb{R}_-$,

$$0 \quad \text{and} \quad a(-\frac{y}{2x})I.$$

The proof of the following lemma is trivial.

Lemma 5.2.4. *For $x \in \mathbb{R}_-$ all the algebras $\mathcal{R}_0(x)$ are isomorphic to $L_\infty(\mathbb{R}_+)$. The homomorphism*

$$\nu_x : \mathcal{R}(\mathcal{A}_\infty, B_\Pi) \cong \mathcal{R}_0 \longrightarrow \mathcal{R}_0(x) \cong L_\infty(\mathbb{R}_+) \tag{5.2.2}$$

is generated by the following mapping of generators of the algebra $\mathcal{R}(\mathcal{A}_\infty, B_\Pi)$:

$$\nu_x \; : \; aI \; \longmapsto \; a(y),$$
$$\nu_x \; : \; B_\Pi \; \longmapsto \; 0.$$

Lemma 5.2.5. *For each $x \in \mathbb{R}_+$ the algebra $\mathcal{R}_0(x)$ is irreducible, contains the ideal $\mathcal{K}$ of all compact in $L_2(\mathbb{R}_+)$ operators, and its Calkin algebra $\widehat{\mathcal{R}}_0(x) = \mathcal{R}_0(x)/\mathcal{K}$ is isomorphic to $L_\infty(\mathbb{R}_+)$.*

Proof. To prove the first statement we show that the algebra $\mathcal{R}_0(x)$ does not have any non-trivial invariant subspace. First, observe that all the invariant subspaces of the subalgebra

$$\{a(\frac{y}{2x})I : a \in L_\infty(\mathbb{R}_+)\} \cong L_\infty(\mathbb{R}_+)$$

of the algebra $\mathcal{R}_0(x)$ have the form

$$X_M = \{\psi_M(y) = \chi_M(y)\psi(y) : \psi(y) \in L_2(\mathbb{R}_+)\},$$

where M is a measurable subset of $\mathbb{R}_+$ having positive measure. Thus the algebra $\mathcal{R}_0(x)$ has an invariant subspace if and only if there exists a measurable set M of a positive measure, such that for each $\psi_M(y) \in X_M$ we have $P_0\psi_M(y) \in X_M$. Further

$$P_0\psi_M = \langle \chi_M\psi, \ell_0 \rangle \cdot \ell_0.$$

The function ℓ_0 is always non-zero on $\mathbb{R}_+$, thus $\langle \chi_M\psi, \ell_0 \rangle$ has to be zero on $\mathbb{R}_+ \setminus M$ for all $\psi \in L_2(\mathbb{R}_+)$, but for $\psi = \ell_0$ we have

$$\langle \chi_M\ell_0, \ell_0 \rangle = \langle \chi_M\ell_0, \chi_M\ell_0 \rangle = \|\chi_M\ell_0\|^2.$$

Thus χ_M must be zero a.e. on $\mathbb{R}_+$, or M must have measure zero. Therefore the algebra $\mathcal{R}_0(x)$ does not have any non-trivial invariant subspace, and thus is irreducible. Now, being irreducible, the algebra $\mathcal{R}_0(x)$ contains the one-dimensional operator P_0, and thus (see, for example, [148] Theorem 2.4.9) it contains the whole ideal $\mathcal{K}$ of compact operators.

The last statement of the lemma is obvious. $\qquad\square$

Note that all algebras $\mathcal{R}_0(x)$, $x \in \mathbb{R}_+$, are the same, and equal to the algebra $\mathcal{R}(L_\infty(\mathbb{R}_+), P_0)$, which is generated by P_0 and all multiplication operators aI, $a \in L_\infty(\mathbb{R}_+)$, acting on $L_2(\mathbb{R}_+)$.

Lemma 5.2.6. *For each $x \in \mathbb{R}_+$ the homomorphism*

$$\nu_x : \mathcal{R}(\mathcal{A}_\infty, B_\Pi) \cong \mathcal{R}_0 \longrightarrow \mathcal{R}_0(x) \cong \mathcal{R}(L_\infty(\mathbb{R}_+), P_0)$$

gives the (infinite dimensional) irreducible representation of $\mathcal{R}(\mathcal{A}_\infty, B_\Pi)$, and for different values of $x \in \mathbb{R}_+$ the representations ν_x are not unitary equivalent.

Proof. The first statement follows directly from Lemma 5.2.5. Let now $0 < x_1 < x_2 < \infty$, and let $a(y)$ be a function from L_∞, such that $\gamma_a(x_1) \neq \gamma_a(x_2)$, where $\gamma_a(x)$ is given by (5.2.1). Then for the operator $A = B_\Pi(\gamma_a(x_1) - a(y))B_\Pi \in \mathcal{R}(\mathcal{A}_\infty, B_\Pi)$ we have

$$\begin{aligned} v_{x_1} &: \quad A \longmapsto (\gamma_a(x_1) - \gamma_a(x_1))P_0 = 0, \\ v_{x_2} &: \quad A \longmapsto (\gamma_a(x_1) - \gamma_a(x_2))P_0 \neq 0. \end{aligned}$$

Thus the representations v_{x_1} and v_{x_2} are not unitary equivalent. $\qquad\square$

Theorem 5.2.7. *The algebra $\mathcal{R}(\mathcal{A}_\infty, B_\Pi)$ does not contain any non-zero compact operator. All its infinite dimensional representations are parameterized by the points $x \in \mathbb{R}_+$ and are of the form*

$$\nu_x : \mathcal{R}(\mathcal{A}_\infty, B_\Pi) \cong \mathcal{R}_0 \longrightarrow \mathcal{R}(L_\infty(\mathbb{R}_+), P_0),$$

where

$$\begin{aligned} v_x &: \quad a(y)I \in \mathcal{R}(\mathcal{A}_\infty, B_\Pi) \longmapsto a\left(\frac{y}{2x}\right)I \in \mathcal{R}(L_\infty(\mathbb{R}_+), P_0), \\ v_x &: \quad B_\Pi \in \mathcal{R}(\mathcal{A}_\infty, B_\Pi) \longmapsto P_0 \in \mathcal{R}(L_\infty(\mathbb{R}_+), P_0). \end{aligned}$$

All finite dimensional irreducible representations of the algebra $\mathcal{R}(\mathcal{A}_\infty, B_\Pi)$ are one-dimensional, and are the compositions of the homomorphism (5.2.2) of Lemma 5.2.4 with one-dimensional representations of $L_\infty(\mathbb{R}_+)$.

Proof. Follows directly from Lemmas 5.2.4, 5.2.5 and 5.2.6. $\qquad\square$

We pass now from the upper half-plane to the unit disk $\mathbb{D}$. The Möbius transformation

$$w = \frac{z + i}{1 + iz} \tag{5.2.3}$$

maps the unit disk $\mathbb{D}$ onto the upper half-plane Π, and generates the unitary operator $U_0 : L_2(\mathbb{D}) \to L_2(\Pi)$ acting by the rule

$$(U_0\varphi)(w) = \frac{2}{(1-iw)^2}\, \varphi\left(\frac{w-i}{1-iw}\right). \tag{5.2.4}$$

Then the inverse (and adjoint) operator $U_0^{-1} : L_2(\Pi) \to L_2(\mathbb{D})$ is given by

$$(U_0^{-1}\varphi)(z) = \frac{2}{(1+iz)^2}\, \varphi\left(\frac{z+i}{1+iz}\right).$$

Denote by $\widetilde{\mathcal{A}}_\infty$ the image of the algebra $\mathcal{A}_\infty$ on the disk, i.e.,

$$\widetilde{\mathcal{A}}_\infty = U_0^{-1}\mathcal{A}_\infty U_0 = \{a(\omega(z)) : a \in \mathcal{A}_\infty\},$$

where $\omega(z)$ is the Möbius transformation (5.2.3).

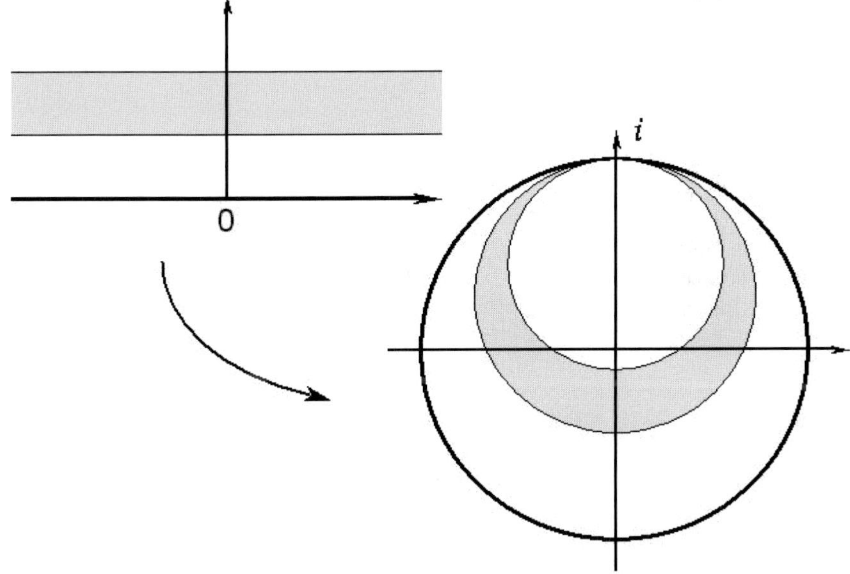

Figure 5.1: From the upper half-plane to the unit disk.

Each horizontal line $\operatorname{Im}\omega = \eta_0 = \text{const}$ goes under the mapping $z = z(\omega)$ (inverse to the mapping (5.2.3)) to the circle $C(\eta_0)$ (*horocycle*, see Section 9.4) having center at the point $i\frac{\eta_0}{1+\eta_0}$ and radius $\frac{1}{1+\eta_0}$, and which is tangent to $\partial\mathbb{D}$ at the point i. Figure 5.1 shows the image of a horizontal strip in the upper half-plane under the mapping $z = z(\omega)$. Note that the functions from $\widetilde{\mathcal{A}}_\infty$ are constant on each horocycle $C(\eta_0)$, $\eta_0 \in \mathbb{R}_+$.

The algebras $\mathcal{R}(\widetilde{\mathcal{A}}_\infty, B_{\mathbb{D}})$ and $\mathcal{R}(\mathcal{A}_\infty, B_\Pi)$, as well as the algebras $\mathcal{T}(\widetilde{\mathcal{A}}_\infty)$ and $\mathcal{T}(\mathcal{A}_\infty)$, are unitary equivalent. Thus Corollary 5.2.2 and Theorem 5.2.7 lead directly to the following result.

Theorem 5.2.8. *The algebra $\mathcal{T}(\widetilde{\mathcal{A}}_\infty)$ is commutative. The isomorphic imbedding*

$$\tilde{\tau}_\infty : \mathcal{T}(\widetilde{\mathcal{A}}_\infty) \longrightarrow C_b(\mathbb{R}_+)$$

is generated by the following mapping of generators of the algebra $\mathcal{T}(\widetilde{\mathcal{A}}_\infty)$,

$$\tilde{\tau}_\infty : T_{a(\omega(z))} \longmapsto \gamma_a(x) = \int_{\mathbb{R}_+} a(\frac{\eta}{2x})\, \ell_0^2(\eta)\, d\eta, \quad x \in \mathbb{R}_+,$$

where $a = a(\omega(z)) \in \widetilde{\mathcal{A}}_\infty$, and $\ell_0(\eta) = e^{-\eta/2}$.

The algebra $\mathcal{R}(\widetilde{\mathcal{A}}_\infty, B_{\mathbb{D}})$ does not contain any non-zero compact operator. All its infinite dimensional representations are parameterized by the points $x \in \mathbb{R}_+$ and are of the form

$$\tilde{\nu}_x : \mathcal{R}(\widetilde{\mathcal{A}}_\infty, B_{\mathbb{D}}) \cong \mathcal{R}_0 \longrightarrow \mathcal{R}(L_\infty(\mathbb{R}_+), P_0),$$

where

$$\tilde{\nu}_x \quad : \quad a(\omega(z))I \in \mathcal{R}(\widetilde{\mathcal{A}}_\infty, B_{\mathbb{D}}) \longmapsto a(\frac{y}{2x})I \in \mathcal{R}(L_\infty(\mathbb{R}_+), P_0),$$

$$\tilde{\nu}_x \quad : \quad B_{\mathbb{D}} \in \mathcal{R}(\widetilde{\mathcal{A}}_\infty, B_{\mathbb{D}}) \longmapsto P_0 \in \mathcal{R}(L_\infty(\mathbb{R}_+), P_0).$$

All finite dimensional irreducible representations of the algebra $\mathcal{R}(\widetilde{\mathcal{A}}_\infty, B_{\mathbb{D}})$ are one-dimensional, and are the composition of the homomorphism (5.2.2) of Lemma 5.2.4 with one-dimensional representations of $L_\infty(\mathbb{R}_+)$.

5.3   Spectra and compactness

We give here some applications of the results obtained in the previous section to spectral and compactness properties of Toeplitz operators.

The next lemma describes the spectrum of Toeplitz operators with $\mathcal{A}_\infty$ defining symbols, extending the results of [142, 237, 240] on connectedness of the spectrum to this class of symbols.

Lemma 5.3.1. *For each $a \in \mathcal{A}_\infty$ the spectrum of the Toeplitz operator $T_a \in \mathcal{T}(\mathcal{A}_\infty)$ is connected and coincides with the closure of the image of the function $\gamma_a(x)$.*

For a real-valued function $a \in \mathcal{A}_\infty$ we have

$$\operatorname{sp} T_a = [\,\inf_{x \in \mathbb{R}_+} \gamma_a(x),\ \sup_{x \in \mathbb{R}_+} \gamma_a(x)].$$

Proof. Direct corollary of Theorem 5.2.1. $\qquad\qquad\qquad\qquad\qquad\square$

Let M be a measurable set in $\mathbb{R}_+$, and $M' = \mathbb{R} + iM$ be the corresponding set in Π. Introduce the characteristic functions χ_M and $\chi_{M'}$ of the sets M and M' respectively.

Consider now one special form of a set $M \subset \mathbb{R}_+$. For $\sigma \in (0,1]$ and $s \in (0,\sigma)$ introduce the sets $\Delta_0 = [1, 1+s]$ $(\subset [1, 1+\sigma))$, $\Delta_n = (1+\sigma)^n \cdot \Delta_0$, where $n \in \mathbb{Z}$, and

$$M(s) = \frac{1}{2} \bigcup_{n \in \mathbb{Z}} \Delta_n. \tag{5.3.1}$$

The function

$$\gamma_s(x) = \int_{\mathbb{R}_+} \chi_{M(s)}\left(\frac{\eta}{2x}\right) e^{-\eta}\, d\eta = \int_{2x \cdot M(s)} e^{-\eta}\, d\eta = \int_{\bigcup_{n \in \mathbb{Z}} \Delta_n} xe^{-x\eta}\, d\eta$$

is "multiplicative-periodic", i.e., $\gamma_s((1+\sigma)x) = \gamma_s(x)$, for all $x \in \mathbb{R}_+$. Thus to analyze the values of $\gamma_s(x)$ it is sufficient to consider only $x \in [1, 1+\sigma)$.

Lemma 5.3.2. *For each $\varepsilon > 0$ there is $\sigma \in (0,1)$ such that for each set $M(s)$ of the form (5.3.1) we have*

$$\sup_{x \in \mathbb{R}_+} \gamma_s(x) - \inf_{x \in \mathbb{R}_+} \gamma_s(x) < \varepsilon.$$

Proof. Given $x_1, x_2 \in [1, 1+\sigma)$ consider

$$
\begin{aligned}
\gamma_s(x_1) - \gamma_s(x_2) &= \int_{\bigcup_{n \in \mathbb{Z}} \Delta_n} (x_1 e^{-x_1 \eta} - x_2 e^{-x_2 \eta})\, d\eta \\
&= \sum_{n \in \mathbb{Z}} \int_{\Delta_n} (x_1 e^{-x_1 \eta} - x_2 e^{-x_2 \eta})\, d\eta \\
&= s \sum_{n \in \mathbb{Z}} (1+\sigma)^n (x_1 e^{-x_1 \eta_n} - x_2 e^{-x_2 \eta_n}),
\end{aligned}
$$

for some $\eta_n \in \Delta_n$. Note, that $|\Delta_n| = s \cdot (1+\sigma)^n$.

Further

$$x_1 e^{-x_1 \eta_n} - x_2 e^{-x_2 \eta_n} = (x_1 - x_2) e^{-\xi_n \eta_n} (1 - \xi_n \eta_n)$$

for some $\xi_n \in (x_1, x_2) \subset [1, 1+\sigma)$.

Now, for $\zeta_n = \xi_n \eta_n \in (1+\sigma)^n \cdot (1, (1+\sigma)^2)$ we have

$$\gamma_s(x_1) - \gamma_s(x_2) = s(x_1 - x_2) \sum_{n \in \mathbb{Z}} (1+\sigma)^n e^{-\zeta_n} (1 - \zeta_n),$$

thus

$$|\gamma_s(x_1) - \gamma_s(x_2)| \;\leq\; \sigma^2 \sum_{n\in\mathbb{Z}} (1+\sigma)^n e^{-\zeta_n} |1-\zeta_n|$$

$$= \;\sigma^2 \sum_{n=1}^{\infty} (1+\sigma)^n e^{-\zeta_n} |1-\zeta_n|$$

$$+\sigma^2 \sum_{k=0}^{\infty} \frac{1}{(1+\sigma)^k} e^{-\zeta_k} |1-\zeta_k|.$$

Finally we estimate each of these sums:

$$\sigma^2 \sum_{k=0}^{\infty} \frac{1}{(1+\sigma)^k} e^{-\zeta_k} |1-\zeta_k| < \sigma^2 \sum_{k=0}^{\infty} \frac{1}{(1+\sigma)^k} \cdot 1 \cdot 3 < 6\sigma,$$

$$\sigma^2 \sum_{n=1}^{\infty} (1+\sigma)^n e^{-\zeta_n} |1-\zeta_n| \;<\; \sigma^2 \sum_{n=1}^{\infty} (1+\sigma)^n e^{-\zeta_n} \zeta_n$$

$$<\; \sigma^2 \sum_{n=1}^{\infty} (1+\sigma)^{2(n+1)} e^{-(1+\sigma)^n}$$

$$<\; (1+\sigma)^3 \sigma \int_1^{\infty} y\, e^{-y}\, dy$$

$$<\; 8\sigma \int_{\mathbb{R}_+} y\, e^{-y}\, dy = 8\sigma.$$

$\square$

We consider now the Toeplitz operator with a specific defining symbol, the characteristic function of some subset in Π.

Theorem 5.3.3. *The spectrum of the Toeplitz operator $T_{\chi_{M'}}$ is given by the formula*

$$\operatorname{sp} T_{\chi_{M'}} = [\alpha, \beta],$$

where $\alpha = \inf_{x\in\mathbb{R}_+} \gamma_{\chi_{M'}}(x) \geq 0$, $\beta = \sup_{x\in\mathbb{R}_+} \gamma_{\chi_{M'}}(x) \leq 1$.

Moreover, for every α, β with $0 \leq \alpha < \beta \leq 1$, there exists a set M such that the spectrum of the Toeplitz operator $T_{\chi_{M'}}$ is equal to $[\alpha, \beta]$.

Proof. The first statement follows directly from Lemma 5.3.1. Consider now arbitrary α and β with $0 \leq \alpha < \beta \leq 1$. For $\varepsilon = (\beta - \alpha)/2$ fix σ according to Lemma 5.3.2. It is clear that for each $\lambda \in [0, 1]$ there exists s_λ ($\in (0, \sigma)$) such that for the corresponding set $M_\lambda = M(s_\lambda)$ of the form (5.3.1) we have

$$\sup_{x\in[1,1+\sigma)} \int_{2x\cdot M_\lambda} |\ell_0(y)|^2 dy = \lambda,$$

similarly, there exists a set m_λ of the above form, such that

$$\inf_{x \in [1,1+\sigma)} \int_{2x \cdot m_\lambda} |\ell_0(y)|^2 dy = \lambda.$$

Now for the set $M = (m_\alpha \cap (0,1)) \cup (M_\beta \cap (1,+\infty))$ and

$$\gamma_{\chi_{M'}}(x) = \int_{2x \cdot M} |\ell_0(\eta)|^2 \, d\eta$$

we have

$$\inf_{x \in \mathbb{R}_+} \gamma_{\chi_{M'}}(x) = \liminf_{x \to +\infty} \gamma_{\chi_{M'}}(x) = \alpha,$$

$$\sup_{x \in \mathbb{R}_+} \gamma_{\chi_{M'}}(x) = \limsup_{x \to 0} \gamma_{\chi_{M'}}(x) = \beta.$$

$\square$

Remark 5.3.4. All Toeplitz operators T_a of the above theorem have as their defining symbols the functions a of the same type – characteristic functions with the same range, the two point set $\{0,1\}$, but the spectra of T_a are *quite different*. Recall that for Toeplitz operators on the Hardy space the situation is drastically different: one *always* has

$$\operatorname{sp} T_\chi = [0,1] \ (= \text{closure of convex hull of Im } \chi),$$

for *any* characteristic function χ.

Using the unitary operator U_0 of the form (5.2.4), the above spectrum characterizations allow immediate reformulation for a Toeplitz operator with $\widetilde{\mathcal{A}}_\infty$ defining symbols, acting on $L_2(\mathbb{D})$.

Toeplitz operators on the unit disk with defining symbols constant on horocycles as well as Toeplitz operators on the upper half-plane with defining symbols depending only on $y = \operatorname{Im} z$, being unitary equivalent to multiplication operators, can never be compact. This fact permits us to construct a fine example illustrating the sharpness of the hypothesis of the first statement of Theorem 2.8.3.

All horocycles $C(\eta)$, $\eta \in \mathbb{R}_+$, are tangent to the unit circle $\partial \mathbb{D} = C(0)$ at the point i, and near this point $C(\eta)$ are the graphs of the functions

$$y_\eta(x) = \frac{\eta}{1+\eta} + \sqrt{\frac{1}{(1+\eta)^2} - x^2}.$$

The "degree of adhesion" of $C(\eta)$ to $\partial \mathbb{D}$ at the point i can be naturally measured by the quantity

$$\delta(\eta) = y_0''(0) - y_\eta''(0) = \frac{\eta(2+\eta)}{(1+\eta)^2},$$

as a function, $\delta(\eta)$, $\eta \in \mathbb{R}_+$, is continuous, strictly increasing and $\delta(0) = 0$.

We will say that a function $a(z)$, defined on $\overline{\mathbb{D}}$, has an ε-limit equal to A at the point i:

$$\varepsilon\text{-}\lim_{z \to i} a(z) = A,$$

if

$$\lim_{z \to i} a(z)|_{\overline{\mathbb{D}}(\eta)} = A,$$

where $\overline{\mathbb{D}}(\eta)$ is the closed disk with boundary $C(\eta)$, and such that $\delta(\eta) = \varepsilon$.

The ε-limit has the following obvious properties:

1. if for any $\varepsilon > 0$ we have $\varepsilon\text{-}\lim_{z \to i} a(z) = 0$, then all non-tangent limit values of the function a at the point i exist and are equal to zero;

2. if $\varepsilon_1 < \varepsilon_2$ and $\varepsilon_1\text{-}\lim_{z \to i} a(z) = 0$, then $\varepsilon_2\text{-}\lim_{z \to i} a(z) = 0$;

3. $0\text{-}\lim_{z \to i} a(z) = 0$ if and only if $\lim_{z \to i} a(z) = 0$.

Lemma 5.3.5. *For each $\varepsilon > 0$ there exists a function $a \in L_\infty(\mathbb{D})$ such that*

1. *a is continuous at each point of $\partial \mathbb{D} \setminus \{i\}$, and $a|_{\partial \mathbb{D} \setminus \{i\}} \equiv 0$;*

2. *$\varepsilon\text{-}\lim_{z \to i} a(z) = 0$;*

3. *the Toeplitz operator T_a is not compact on $\mathcal{A}^2(\mathbb{D})$.*

Proof. Fix positive ε and consider η having the property $\delta(\eta) = \varepsilon$. Take a function $b \in \mathcal{A}_\infty$ with $\mathbb{R} \times [\eta/3, 2\eta/3] \subset \operatorname{supp} b$ (for example, $b = b(x,y)$ is the characteristic function of the strip $\mathbb{R} \times [\eta/3, 2\eta/3]$, see Figure 5.1). Then the function $a(z) = (b \cdot \chi_{\mathbb{R} \times [\eta/3, 2\eta/3]})(\omega(z))$ satisfies the first two conditions of the lemma. The operator T_a is unitary equivalent to a (non-zero) multiplication operator, and thus is not compact. $\qquad\square$

5.4   Finite dimensional representations

Fix a number $n \in \mathbb{N}$. Let Y_k, $k = \overline{1,n}$, be disjoint measurable sets in $\mathbb{R}_+$ having positive measure, and such that $\bigcup_{k=1}^{n} Y_k = \mathbb{R}_+$, and let

$$\Pi_k = \mathbb{R} + iY_k, \quad k = \overline{1,n},$$

be the corresponding sets in the upper half-plane $\Pi = \mathbb{R} + i\mathbb{R}_+$. Denote by $\chi_k(y)$ $(\in L_\infty(\mathbb{R}_+))$ the characteristic function of the set Y_k, and by χ_{Π_k} $(\in L_\infty(\Pi))$ the characteristic function of the set Π_k $k = \overline{1,n}$.

Introduce the algebra

$$\mathcal{A}_n = \{a_1 \chi_{\Pi_1} + \ldots + a_n \chi_{\Pi_n} : a_k \in \mathbb{C}, \quad k = \overline{1,n}\} \cong \mathbb{C}^n.$$

We will study the algebras $\mathcal{T}(\mathcal{A}_n)$ and $\mathcal{R}(\mathcal{A}_n, B_\Pi)$.

Theorem 5.4.1. *The algebra $\mathcal{T}(\mathcal{A}_n)$ is isomorphic and isometric to the algebra $C(\Delta)$, where*

$$\Delta = \Delta(Y_1, \ldots, Y_n) = \mathrm{clos}\left\{\, (\gamma_{\chi_1}(x), \ldots, \gamma_{\chi_n}(x)) : x \in \mathbb{R}_+ \,\right\} \quad \subset \Delta_{n-1}, \quad (5.4.1)$$

and $\gamma_{\chi_k}(x)$, $k = \overline{1,n}$, are given by

$$\gamma_{\chi_k}(x) = \int_{\mathbb{R}_+} \chi_k\left(\frac{\eta}{2x}\right) |\ell_0(\eta)|^2 \, d\eta = \int_{2x \cdot Y_k} |\ell_0(\eta)|^2 \, d\eta.$$

The isomorphism

$$\tau_n : \mathcal{T}(\mathcal{A}_n) \longrightarrow C(\Delta)$$

is generated by the following mapping of the generators T_a, where $a = a_1\chi_{\Pi_1} + \ldots + a_n\chi_{\Pi_n} \in \mathcal{A}_n$, of the algebra $\mathcal{T}(\mathcal{A}_n)$

$$\tau_n : T_a \longmapsto a_1 t_1 + \ldots + a_n t_n, \quad t = (t_1, \ldots, t_n) \in \Delta.$$

Proof. Follows directly from Theorem 5.2.1. $\qquad\qquad\qquad\qquad\qquad\qquad\square$

The algebra $\mathcal{R}(\mathcal{A}_n, B_\Pi)$ can be defined by another set of generators. Namely, the orthogonal projections

$$P = B_\Pi, \quad \text{and} \quad Q_k = \chi_{\Pi_k} I, \quad k = \overline{1,n}, \quad (5.4.2)$$

acting on $L_2(\Pi)$, obviously generate this algebra. To describe the C^*-algebra $\mathcal{R}(\mathcal{A}_n, B_\Pi) = \mathcal{R}(P, Q_1, \ldots, Q_n)$ we use the results of Section 1.2. These projections are all-but-one and satisfy the property (1.2.13). At the same time, the analysis of the projections $UB_\Pi U^{-1} = \chi_+(x)I \otimes P_0$ and $U\chi_{\Pi_k}U^{-1} = \chi_{\Pi_k}\left(\frac{y}{2|x|}\right)I$, $k = \overline{1,n}$, shows that the property (1.2.14) does not hold for the projections (5.4.2) for all $k = \overline{1,n}$. That is we are in the situation of Theorem 1.2.33 with $m = n$. The joint spectrum of the operators $C_k = PQ_kP = T_{\chi_{\Pi_k}}$, $k = \overline{1,n}$, is obviously equal to $\Delta = \Delta(Y_1, \ldots, Y_n)$. Thus as a direct corollary from Theorem 1.2.33 we have

Theorem 5.4.2. *The C^*-algebra $\mathcal{R}(\mathcal{A}_n, B_\Pi)$ is isomorphic and isometric to a subalgebra of the algebra $\mathfrak{S}(\Delta) \oplus \mathbb{C}^n$, where $\Delta = \Delta(Y_1, \ldots, Y_n)$ is given by (5.4.1). The isomorphic imbedding*

$$\nu : \mathcal{R}(\mathcal{A}_n, B_\Pi) \longrightarrow \mathfrak{S}(\Delta) \oplus \mathbb{C}^n$$

is generated by the following mapping of the generators of the algebra $\mathcal{R}(\mathcal{A}_n, B_\Pi)$:

$$\nu : B_\Pi \longmapsto (p(t), (0, 0, \ldots, 0)),$$
$$\nu : \chi_{\Pi_k} I \longmapsto (q_k(t), (0, \ldots, 0, \underset{k\text{-place}}{1}, 0, \ldots, 0)), \quad k = \overline{1,n},$$

where

$$p(t) = \left(\sqrt{t_j t_k}\right)_{j,k=1}^n,$$
$$q_k(t) = \mathrm{diag}\,(0, \ldots, 0, \underset{k\text{-place}}{1}, 0, \ldots, 0),$$

$$t = (t_1, \ldots, t_n) \in \Delta = \Delta(Y_1, \ldots, Y_n).$$

Remark 5.4.3. It is easy to construct measurable sets Y_k, $k = \overline{1,n}$, for which the corresponding set $\Delta = \Delta(Y_1, \ldots, Y_n)$ belongs to the interior of the simplex Δ_{n-1}, see, for example, the detailed study of the joint spectrum in Section 7.5. In such a case the algebra $\mathfrak{S}(\Delta)$ in Theorem 5.4.2 is just the algebra $C(\Delta, \mathrm{Mat}_n(\mathbb{C}))$.

5.5   General case

Denote by L_∞^0 the subalgebra of L_∞, which consists of all functions having zero limit at the point $0 \in R_+$. Let $\mathcal{A}_\infty^0 = \mathbb{C} \otimes L_\infty^0 \subset \mathcal{A}_\infty$, and

$$\widetilde{\mathcal{A}}_\infty^0 = U_0^{-1}\mathcal{A}_\infty^0 U_0 = \{a(\omega(z)) : a \in \mathcal{A}_\infty^0\} \subset \widetilde{\mathcal{A}}_\infty,$$

where the operator U_0 is given by (5.2.4).

Besides the algebra $\widetilde{\mathcal{A}}_\infty^0$ we will consider its rotated variants. To do this we introduce the unitary (rotation) operator on $L_2(\mathbb{D})$: given $t_1, t_2 \in \partial\mathbb{D}$, let

$$(U_{(t_1,t_2)}\varphi)(z) = \varphi(\frac{t_2}{t_1}z);$$

then of course $U_{(t_1,t_2)}^{-1} = U_{(t_2,t_1)}$. Denote now $\widetilde{\mathcal{A}}_\infty^0(i) = \widetilde{\mathcal{A}}_\infty^0$, and for each $t \in \partial\mathbb{D}$ introduce

$$\widetilde{\mathcal{A}}_\infty^0(t) = U_{(i,t)}^{-1}\widetilde{\mathcal{A}}_\infty^0(i)U_{(i,t)} = \{a(\frac{i}{t}z) : a \in \widetilde{\mathcal{A}}_\infty^0(i)\}.$$

All the functions from $\widetilde{\mathcal{A}}_\infty^0(t)$ are continuous at each point of $\partial\mathbb{D} \setminus \{t\}$, and their restrictions onto $\partial\mathbb{D} \setminus \{t\}$ are identically zero.

Lemma 5.5.1. *Given two distinct points* $t_1, t_2 \in \partial\mathbb{D}$, *the* C^*-*algebras* $\mathcal{R}(\mathbb{C} + \widetilde{\mathcal{A}}_\infty^0(t_1), B_\mathbb{D})$ *and* $\mathcal{R}(\mathbb{C} + \widetilde{\mathcal{A}}_\infty^0(t_2), B_\mathbb{D})$, *as well as the algebras* $\mathcal{T}(\mathbb{C} + \widetilde{\mathcal{A}}_\infty^0(t_1))$ *and* $\mathcal{T}(\mathbb{C} + \widetilde{\mathcal{A}}_\infty^0(t_2))$, *are unitary equivalent.*

For all functions $a_k \in \widetilde{\mathcal{A}}_\infty^0(t_k)$, $k = 1, 2$, *the operators* $T_{a_1 \cdot a_2}$ *and* $T_{a_1}T_{a_2}$ *are compact.*

Proof. The first statement follows from

$$\widetilde{\mathcal{A}}_\infty^0(t_2) = U_{(t_1,t_2)}^{-1}\widetilde{\mathcal{A}}_\infty^0(t_1)U_{(t_1,t_2)}, \qquad B_\mathbb{D} = U_{(t_1,t_2)}^{-1}B_\mathbb{D}U_{(t_1,t_2)}.$$

The operator $T_{a_1 \cdot a_2}$ is compact by Theorem 2.8.3. Let now $c_{(t_1,t_2)}(z)$ be a continuous function on $\overline{\mathbb{D}}$, equal to 0 on a small neighborhood of the point t_1, and equal to 1 on a small neighborhood of the point t_2. Then the compactness of the operator $T_{a_1}T_{a_2}$ follows from

$$\begin{aligned}
T_{a_1}T_{a_2} &= B_\mathbb{D}a_1 B_\mathbb{D}(c_{(t_1,t_2)} + 1 - c_{(t_1,t_2)})B_\mathbb{D}a_2 B_\mathbb{D} \\
&= B_\mathbb{D}a_1 c_{(t_1,t_2)}B_\mathbb{D}a_2 B_\mathbb{D} + B_\mathbb{D}a_1 B_\mathbb{D}(1 - c_{(t_1,t_2)})a_2 B_\mathbb{D} + K \\
&= T_{a_1 c_{(t_1,t_2)}} \cdot T_{a_2} + T_{a_1} \cdot T_{(1-c_{(t_1,t_2)})a_2} + K,
\end{aligned}$$

where K is compact, and from the compactness of $T_{a_1 c_{(t_1,t_2)}}$ and $T_{(1-c_{(t_1,t_2)})a_2}$. $\quad\square$

The next theorem characterizes, in a sense, the gap between compactness commutator and compactness semi-commutator properties. It shows that, for an appropriate choice of the coefficient algebra $\mathcal{A}$, the Fredholm symbol algebra $\operatorname{Sym}\mathcal{R}(\mathcal{A}, B_{\mathbb{D}})$ can have irreducible representations of any predefined dimensions, while the Fredholm symbol algebra $\operatorname{Sym}\mathcal{T}(\mathcal{A})$ still remains commutative.

Theorem 5.5.2. *For each finite (ordered) set $\Lambda = \langle n_0, n_1, \ldots, n_m \rangle$, where $1 = n_0 < n_1 < \ldots < n_m \leq \infty$, and $n_k \in \mathbb{N} \cup \{\infty\}$, there is an algebra $\mathcal{A}_\Lambda$, the subalgebra of $L_\infty(\mathbb{D})$, such that the Fredholm symbol algebra $\operatorname{Sym}\mathcal{T}(\mathcal{A}_\Lambda)$ of the algebra $\mathcal{T}(\mathcal{A}_\Lambda)$ is commutative, while the Fredholm symbol algebra $\operatorname{Sym}\mathcal{R}(\mathcal{A}_\Lambda, B_{\mathbb{D}})$ of the algebra $\mathcal{R}(\mathcal{A}_\Lambda, B_{\mathbb{D}})$ has irreducible representations exactly of the predefined dimensions $n_0, n_1, \ldots, n_m$.*

Proof. Fix a set $\Lambda = \langle n_0, n_1, \ldots, n_m \rangle$, and denote by $t_1, t_2, \ldots, t_m$ any m different points on the unit circle $\partial\mathbb{D}$. Now for each k ($k = \overline{1, m}$, if $n_m < \infty$, and $k = \overline{1, m-1}$, if $n_m = \infty$) select the subalgebra $\widetilde{\mathcal{A}}^0_{n_k}(t_k)$ of the algebra $\widetilde{\mathcal{A}}^0_\infty(t_k)$ as follows.

We start with a special case of the algebra $\mathcal{A}_n$ of Section 5.4 considering the algebra $\mathcal{A}_{n_k}$, which is generated by the sets $Y_{k,1}, \ldots, Y_{k,n_k}$, such that $Y_{k,n_k} = [0, \eta_k]$, for some positive η_k, and such that the corresponding set $\Delta = \Delta(Y_{k,1}, \ldots, Y_{k,n_k})$ (5.4.1) intersects the boundary of the simplex $\Delta_{n_k - 1}$ only by the point $(0, \ldots, 0, 1)$.

Then the algebra $\mathcal{A}^0_{n_k} = \mathcal{A}^0_\infty \cap \mathcal{A}_{n_k} \cong \mathbb{C}^{n_k - 1}$ is generated by the characteristic functions $\chi_{\Pi_{k,j}}$, where $\Pi_{k,j} = \mathbb{R} + iY_{k,j}$, $j = \overline{1, n_k - 1}$, and, of course, $\mathbb{C} + \mathcal{A}^0_{n_k} = \mathcal{A}_{n_k}$. We define now

$$\widetilde{\mathcal{A}}^0_{n_k}(t_k) = U^{-1}_{(i,t_k)} U^{-1}_0 \mathcal{A}^0_{n_k} U_0 U_{(i,t_k)} = \{ a(\omega(\frac{i}{t_k} z)) : a \in \mathcal{A}^0_{n_k} \} \subset \mathcal{A}^0_\infty(t_k).$$

If $n_m = \infty$, then we define $\widetilde{\mathcal{A}}^0_{n_m}(t_m) = \widetilde{\mathcal{A}}^0_\infty(t_m)$. Finally we define the algebra $\mathcal{A}_\Lambda$ as

$$\mathcal{A}_\Lambda = C(\overline{\mathbb{D}}) + \widetilde{\mathcal{A}}^0_{n_1}(t_1) + \ldots + \widetilde{\mathcal{A}}^0_{n_m}(t_m).$$

Now the commutativity of the Fredholm symbol algebra $\operatorname{Sym}\mathcal{T}(\mathcal{A}_\Lambda) = \mathcal{T}(\mathcal{A}_\Lambda)/\mathcal{K}$ follows from the commutativity of $\operatorname{Sym}\mathcal{T}(C(\overline{\mathbb{D}}))$, Theorem 5.2.1 and Lemma 5.5.1.

To analyze the Fredholm symbol algebra $\operatorname{Sym}\mathcal{R}(\mathcal{A}_\Lambda, B_{\mathbb{D}})$ of the algebra $\mathcal{R}(\mathcal{A}_\Lambda, B_{\mathbb{D}})$ we use the Douglas-Varela local principle. As the algebra $C(\overline{\mathbb{D}})$ is obviously a central commutative subalgebra of $\widehat{\mathcal{R}} = \operatorname{Sym}\mathcal{R}(\mathcal{A}_\Lambda, B_{\mathbb{D}})$, we localize by points $z_0 \in \overline{\mathbb{D}}$.

We denote by J_{z_0} the maximal ideal of $C(\overline{\mathbb{D}})$ corresponding to the point $z_0 \in \overline{\mathbb{D}}$, and by $J(z_0)$ the closed two-sided ideal of the algebra $\widehat{\mathcal{R}}$, generated by J_{z_0}. Then the local algebra at the point z_0 is defined as $\mathcal{R}(z_0) = \widehat{\mathcal{R}}/J(z_0)$, and the natural projection

$$\pi_{z_0} : \mathcal{R}(\mathcal{A}_\Lambda, B_{\mathbb{D}}) \longrightarrow \operatorname{Sym}\mathcal{R}(\mathcal{A}_\Lambda, B_{\mathbb{D}}) \longrightarrow \mathcal{R}(z_0)$$

identifies elements of the algebra $\mathcal{R}(\mathcal{A}_\Lambda, B_{\mathbb{D}})$ locally equivalent at the point z_0.

We recall (see Theorem 1.1.17) that the set of all irreducible representations of the algebra $\operatorname{Sym}\mathcal{R}(\mathcal{A}_\Lambda, B_{\mathbb{D}})$ coincides with the collection of irreducible representations of all local algebras $\mathcal{R}(z_0)$, $z_0 \in \overline{\mathbb{D}}$. Thus we need to analyze the irreducible representations of each local algebra.

For each $z_0 \in \mathbb{D} = \operatorname{Int}\overline{\mathbb{D}}$ the Bergman projection $B_{\mathbb{D}}$ is locally equivalent to 0. Thus the local algebra $\mathcal{R}(z_0)$ is commutative, and all its irreducible representations are just the one-dimensional representations of the algebra $\mathcal{A}_\Lambda|_{\mathbb{D}}$.

Recall that every function $a(z)$ which is continuous at some point z_0 is locally equivalent at this point to its value $a(z_0)$, and that all functions from $\widetilde{\mathcal{A}}^0_{n_k}(t_k)$ are continuous in each point of $\partial\mathbb{D} \setminus \{t_k\}$, and are identically zero in $\partial\mathbb{D} \setminus \{t_k\}$.

At the points $z_0 \in \partial\mathbb{D} \setminus \{t_1, \ldots, t_m\}$ the functions $a(z) \in C(\overline{\mathbb{D}})$ are locally equivalent to $a(z_0) \in \mathbb{C}$, while all functions from each $\widetilde{\mathcal{A}}^0_{n_k}(t_k)$ are locally equivalent to 0. The local algebra $\mathcal{R}(z_0)$ in this case is generated by $\mathbb{C}$ and the projection $\pi_{z_0}(B_{\mathbb{D}})$, and thus the algebra $\mathcal{R}(z_0)$ is commutative and isomorphic to $\mathbb{C}^2$.

Let finally $z_0 = t_k$ for some $k = \overline{1, m}$. Again, the functions $a(z) \in C(\overline{\mathbb{D}})$ are locally equivalent to $a(z_0) \in \mathbb{C}$, but now only functions from $\widetilde{\mathcal{A}}^0_{n_j}(t_j)$, with $j \neq k$, are locally equivalent to 0. Thus the local algebra $\mathcal{R}(t_k)$ is generated by the images (under the mapping π_{t_k}) of $B_{\mathbb{D}}$ and of the functions from $\mathbb{C} + \widetilde{\mathcal{A}}^0_{n_k}(t_k)$.

Further, the algebra $\mathcal{R}(\mathbb{C} + \widetilde{\mathcal{A}}^0_{n_k}(t_k), B_{\mathbb{D}})$ is unitary equivalent to $\mathcal{R}(\mathcal{A}_{n_k}, B_\Pi)$, which does not contain any non-zero compact operator. Thus its image on the Fredholm symbol algebra (which can be described via continuous functions on $\overline{\mathbb{D}}$, valued in local algebras $\mathcal{R}(z_o)$ for each z_0) is isomorphic to the initial algebra $\mathcal{R}(\mathbb{C} + \widetilde{\mathcal{A}}^0_{n_k}(t_k), B_{\mathbb{D}})$. But $\pi_{z_0}(B_{\mathbb{D}}) = 0$ for each $z_0 \in \mathbb{D}$, and $\pi_{z_0}(\widetilde{\mathcal{A}}^0_{n_k}(t_k)) = 0$ for each $z_0 \in \partial\mathbb{D} \setminus \{t_k\}$. Thus

$$\mathcal{R}(t_k) \cong \mathcal{R}(\mathbb{C} + \widetilde{\mathcal{A}}^0_{n_k}(t_k), B_{\mathbb{D}}) \cong \mathcal{R}(\mathcal{A}_{n_k}, B_\Pi),$$

and by Theorem 5.2.7 (for $n_k = \infty$), or Theorem 5.4.2 (for $n_k < \infty$) the algebra $\mathcal{R}(t_k)$ has irreducible representations only of dimensions n_k and 1. $\square$

Remark 5.5.3. Given Λ, the algebra $\mathcal{A}_\Lambda$ constructed is not unique. In the proof of the theorem we rather demonstrate key steps for constructing such algebras. Similar results can be given starting with defining symbols which are constant on cycles for the other two possible types of pencils of hyperbolic geodesics (see Section 9.4). In fact, the results of Section 7.4 give another example of algebras $\mathcal{A}_\Lambda$ constructed using hyperbolic pencils. It is very interesting to understand whether there exist algebras with the above properties but having a fundamentally different structure.

Functions a from the algebra $\mathcal{A}_\Lambda = C(\overline{\mathbb{D}}) + \widetilde{\mathcal{A}}^0_{n_1}(t_1) + \ldots + \widetilde{\mathcal{A}}^0_{n_m}(t_m)$ are obviously continuous at each point of $\partial\mathbb{D} \setminus \{t_1, \ldots, t_m\}$, and have well-defined limit values at the points t_k, $k = \overline{1, m}$, over the unit circle $\partial\mathbb{D}$:

$$a(t_k) = \lim_{\partial\mathbb{D}\ni t\to t_k} a(t). \tag{5.5.1}$$

Routine verification shows that for each $a \in \mathcal{A}_\Lambda$ and each $k = \overline{1,m}$ there is a unique function $\tilde{a}_k \in \widetilde{\mathcal{A}}^0_{n_k}(t_k)$ such that

$$\lim_{\mathbb{D} \ni z \to t_k} \left(a(z) - a(t_k) - \tilde{a}_k(z) \right) = 0.$$

For the functions $\tilde{a}_k(z) \in \widetilde{\mathcal{A}}^0_{n_k}(t_k)$ we will use their "origins" $a_k \in \mathcal{A}^0_{n_k} \subset \mathbb{C} \otimes L^0_\infty(\mathbb{R}_+)$, i.e., the functions

$$a_k(\omega) = \tilde{a}_k\left(\frac{t_k}{i} z(\omega) \right), \quad \omega \in \Pi. \tag{5.5.2}$$

Denote by Γ the disjoint union of the unit circle $\partial\mathbb{D}$ and m copies of $(0, +\infty] = (0, +\infty]_k$, $k = \overline{1,m}$, with the points $t_k \in \partial\mathbb{D}$ and $+\infty \in (0, +\infty]_k$ identified, for each $k = \overline{1,m}$.

Corollary 5.5.4. *The Fredholm symbol algebra* $\operatorname{Sym} \mathcal{T}(\mathcal{A}_\Lambda)$ *is isomorphic and isometric to a subalgebra of* $C_b(\Gamma)$. *The homomorphism*

$$\operatorname{sym} : \mathcal{T}(\mathcal{A}_\Lambda) \longrightarrow \operatorname{Sym} \mathcal{T}(\mathcal{A}_\Lambda) \subset C_b(\Gamma)$$

is generated by the following mapping of generators of the algebra $\mathcal{T}(\mathcal{A}_\Lambda)$:

$$\operatorname{sym} : T_a \longmapsto \begin{cases} a(t), & \text{if } t \in \partial\mathbb{D} \setminus \{t_1, \ldots, t_m\} \\ a(t_k), & \text{if } t = t_k \\ \gamma_{a_k}(x), & \text{if } x \in (0, +\infty]_k \end{cases},$$

where $a \in \mathcal{A}_\Lambda$, *and for the function* a_k, *defined in* (5.5.2), *the function* $\gamma_{a_k}(x)$ *is given by* (5.2.1).

Corollary 5.5.5. *The essential spectrum of each Toeplitz operator* T_a, $a \in \mathcal{A}_\Lambda$, *is connected and coincides with the closure of the range of its Fredholm symbol*

$$\operatorname{ess-sp} T_a = \operatorname{clos} \{ (\operatorname{sym} T_a)(u) : u \in \Gamma \}.$$

Chapter 6

Toeplitz Operators on the Unit Disk with Radial Symbols

As follows, for example, from Theorem 2.8.3, the Toeplitz operator with radial defining symbols $a(r)$, which is continuous at the boundary point 1, has a trivial structure, nothing but a compact perturbation of a scalar operator, $T_{a(r)} = a(1)I + K$.

One of the principal differences between Toeplitz operators on the Bergman and Hardy spaces is that in the first case there is an additional direction: "inside the domain". In particular, the defining symbols with quite a nice behaviour with respect to the circular direction may have very complicated irregular behaviour with respect to the radial direction. As a consequence the Toeplitz operators with radial defining symbols may have and, as we will see, do have interesting and rich structure.

To analyze the impact of the radial component itself we study the Toeplitz operators having pure radial defining symbols. In [120], studying the Toeplitz operators with *bounded* radial symbols, B. Korenblum and K. Zhu found two of their important properties: the diagonal form of Toeplitz operators with respect to the standard polynomial basis in $\mathcal{A}^2(\mathbb{D})$, and the criterion for compactness of such operators. The methods used in [120] did not permit them to consider *unbounded* defining symbols, which was left as an open problem. Let us mention as well the papers by J. Miao [144], Sangadji and K. Stroethoff [171], and K. Stroethoff [181, 182], where the compactness of Toeplitz operators with bounded radial defining symbols were studied in different settings.

It turns out that Toeplitz operators with radial symbols possess many interesting properties. In particular there exist *compact* Toeplitz operators whose (radial) defining symbols are *unbounded* near the unit circle $\partial\mathbb{D}$. The essential spectra of Toeplitz operators with pure radial symbols have a very rich structure, and can even be massive (i.e., have positive plane measure). For bounded operators

T_a and T_b whose defining symbols are unbounded, the operator $T_{a \cdot b}$ may not be bounded at all. That is, contrary to commonly known cases, the set of defining symbols for which corresponding Toeplitz operators are bounded neither forms an algebra (under the pointwise multiplication), nor admits any natural norm.

6.1 Toeplitz operators with radial symbols

Theorem 6.1.1. *Let $a = a(r)$ be a measurable function on the segment $[0,1]$. Then the Toeplitz operator T_a acting on $\mathcal{A}^2(\mathbb{D})$ is unitary equivalent to the multiplication operator $\gamma_a I = R T_a R^*$, where R and R^* are given by (4.1.5) and (4.1.4) respectively, acting on l_2^+. The sequence $\gamma_a = \{\gamma_a(n)\}_{n \in \mathbb{Z}_+}$ is given by*

$$\gamma_a(n) = \int_0^1 a(r^{\frac{1}{2(n+1)}})\, dr$$

$$= (n+1) \int_0^1 a(\sqrt{r})\, r^n\, dr, \quad n \in \mathbb{Z}_+. \tag{6.1.1}$$

Proof. The operator T_a is obviously unitary equivalent to the operator

$$\begin{aligned}
R T_a R^* &= R B_{\mathbb{D}} a B_{\mathbb{D}} R^* = R(R^* R) a (R^* R) R^* \\
&= (RR^*) R a R^* (RR^*) = R a R^* \\
&= R_0^* U_2 (I \otimes \mathcal{F}) a(r) (I \otimes \mathcal{F}^{-1}) U_2^{-1} R_0 \\
&= R_0^* U_2 \{a(r)\} U_2^{-1} R_0 \\
&= R_0^* \{a(r^{\frac{1}{n+1}})\} R_0.
\end{aligned}$$

Now

$$R_0^* \{a(r^{\frac{1}{n+1}})\} R_0 \{c_n\}_{n \in \mathbb{Z}_+} = \left\{ \int_0^1 a(r^{\frac{1}{n+1}})\, 2\, c_n\, r dr \right\}_{n \in \mathbb{Z}_+} = \{\gamma_a(n) \cdot c_n\}_{n \in \mathbb{Z}_+},$$

where

$$\gamma_a(n) = 2 \int_0^1 a(r^{\frac{1}{n+1}})\, r dr = \int_0^1 a(r^{\frac{1}{2(n+1)}})\, dr. \qquad \square$$

Corollary 6.1.2. *The Toeplitz operator T_a with measurable radial defining symbol $a = a(r)$ is bounded on $\mathcal{A}^2(\mathbb{D})$ if and only if*

$$\gamma_a = \{\gamma_a(n)\}_{n \in \mathbb{Z}_+} \in l_\infty$$

and

$$\|T_a\| = \sup_{n \in \mathbb{Z}_+} |\gamma_a(n)|.$$

The Toeplitz operator T_a is compact if and only if $\gamma_a \in c_0$, that is,

$$\lim_{n \to \infty} \gamma_a(n) = 0.$$

The spectrum of the bounded Toeplitz operator T_a is given by

$$\operatorname{sp} T_a = \overline{\{\gamma_a(n) : n \in \mathbb{Z}_+\}},$$

and its essential spectrum ess-sp T_a coincides with the set of all limit points of the sequence $\{\gamma_a(n)\}_{n\in\mathbb{Z}_+}$.

As was proved in [142], the spectrum of a Toeplitz operator with a defining symbol real harmonic on $\mathbb{D}$ is always *connected*. Nevertheless for defining symbols harmonic in $\mathbb{D}$ except for even a single point the situation can be quite different.

Example 6.1.3. The general form of a radial function which is harmonic in $\mathbb{D}$ (in $\mathbb{D} \setminus \{0\}$, to be more precise) is

$$h(r) = c_1 \ln \frac{1}{r} + c_2, \qquad c_1, c_2 \in \mathbb{C}.$$

We have

$$\gamma_h(n) = \int_0^1 h(r^{\frac{1}{2(n+1)}})\, dr = \frac{c_1}{2(n+1)} + c_2,$$

that is, the Toeplitz operator T_h is *bounded* on $\mathcal{A}^2(\mathbb{D})$, and its *discrete* spectrum is given by

$$\operatorname{sp} T_h = \{\frac{c_1}{2(n+1)} + c_2\}_{n\in\mathbb{Z}_+} \cup \{c_2\}.$$

The Toeplitz operator T_h is compact if and only if $c_2 = 0$.

To study Toeplitz operators with radial symbols it is useful first to understand the behaviour of sequences of the type (6.1.1). We have

$$\gamma_a(n) = \int_0^1 a(r^{\frac{1}{2(n+1)}})\, dr = (n+1) \int_0^1 b(u)u^n\, du, \qquad n \in \mathbb{Z}_+,$$

with $b(u) = a(\sqrt{u})$. It is natural to assume that

$$\int_0^1 |b(u)|du < \infty,$$

or equivalently

$$\int_0^1 |a(r)|rdr < \infty.$$

We note that the sequence

$$\eta_b(n) = \frac{1}{n+1}\gamma_a(n)$$

forms the sequence of the power momenta of the function $b(u)$.

The following uniqueness result is standard in momentum theory, and will be important for us.

Theorem 6.1.4. *Let*

$$\eta_b(n_k) = 0, \quad k \in \mathbb{Z}_+,$$

where $n_k = n_0 + dk$, $n_0 \in \mathbb{Z}_+$, $d \in \mathbb{N}$. Then $b(u) = 0$ almost everywhere.

Proof. We have

$$\int_0^1 b(u) u^{n_0} u^{dk} \, du = 0.$$

Changing the variable $u^d = s$, we obtain

$$\int_0^1 \left[b(s^{1/d}) s^{\frac{n_0+1-d}{d}} \right] s^k \, ds = 0, \quad k = 0, 1, 2, \ldots$$

Now the function in square brackets belongs to $L_1(0,1)$ and is orthogonal to all polynomials. Thus this function is equal to zero almost everywhere, so $b(u) = 0$ almost everywhere as well. $\qquad\square$

Corollary 6.1.5. *There is no function $b(u) \in L_1(0,1)$ for which $\eta_b(n) \neq 0$ only at a finite number of points.*

The behaviour of a sequence $\gamma_a(n)$ when $n \to \infty$ is mainly determined by the behaviour of a function $a(r)$ (or a function b) in a neighborhood of the point $r = 1$. Given $b \in L_1(0,1)$, introduce the function

$$B(s) = \int_s^1 b(u)\, du.$$

Theorem 6.1.6. *If the function $B(s)$ when $s \to 1$ has the form*

$$|B(s)| = O(1 - s), \tag{6.1.2}$$

then

$$\sup_{n \in \mathbb{Z}_+} |\gamma_a(n)| < \infty.$$

If

$$|B(s)| = o(1 - s), \tag{6.1.3}$$

then

$$\lim_{n \to \infty} \gamma_a(n) = 0.$$

Proof. Let $b(u) = a(\sqrt{u}) \in L_1(0,1)$. Integrating by parts we have for $n \geq 1$,

$$\gamma_a(n) = (n+1)n \int_0^1 B(s) s^{n-1} \, ds.$$

Let $\varepsilon = \varepsilon(n) = n^{-2/3}$. Then assuming (6.1.2), we estimate

$$
\begin{aligned}
|\gamma_a(n)| \;\leq\;& (n+1)n \int_{1-\varepsilon}^{1} |B(s)| s^{n-1} ds + (n+1)n \int_{0}^{1-\varepsilon} |B(s)| s^{n-1} ds \\
\leq\;& (n+1)n\, c_\varepsilon \int_{1-\varepsilon}^{1} (1-s) s^{n-1} ds + (n+1)n(1-\varepsilon)^{n-1} \int_{0}^{1} |B(s)| ds \\
\leq\;& (n+1)n\, c_\varepsilon \left(\frac{s^n}{n} - \frac{s^{n+1}}{n+1} \right) \Bigg|_{1-\varepsilon}^{1} \\
& + \text{const}\,(n+1)n \exp((n-1)\ln(1-\varepsilon)) \\
\leq\;& c_\varepsilon (n+1)n \left(\frac{1}{(n+1)n} + \frac{\exp((n+1)\ln(1-\varepsilon))}{n+1} \right) \\
& + \text{const}\,(n+1)n(\exp(-(n-1)\varepsilon + (n-1)O(\varepsilon^2))) \\
\leq\;& c_\varepsilon (1 + n \exp(-(n+1)\varepsilon + O(\varepsilon^2))) + \text{const}\, n^2 \exp(-n^{1/3}) \\
\leq\;& c_\varepsilon + \text{const}\, n^2 \exp(n^{-1/3}),
\end{aligned}
$$

where "const" denotes a quantity uniformly bounded in ε. Having (6.1.2), the quantity c_ε is uniformly bounded on ε, and thus $\gamma_a \in l_\infty$.

Having (6.1.3), the quantity c_ε can be chosen in such a way that

$$
\lim_{\varepsilon \to 0} c_\varepsilon = \lim_{n \to \infty} c_{\varepsilon(n)} = 0,
$$

and thus $\gamma_a \in c_0$. $\qquad \square$

In fact Theorem 6.1.6 says that the behaviour near the boundary of a certain average of defining symbols, rather than the behaviour of the symbols themselves, is responsible for the boundedness and compactness properties of the corresponding Toeplitz operators. That is, in spite of bad behaviour of a defining symbol, which can even be unbounded near the boundary, the corresponding Toeplitz operator can be bounded and even compact.

Example 6.1.7. Let

$$
a(r) = (1 - r^2)^{-\beta} \sin(1 - r^2)^{-\alpha}, \tag{6.1.4}
$$

where $\alpha > 0$ and $\beta < 1$. Consider the corresponding function

$$
B(v) = \int_{v}^{1} (1 - u)^{-\beta} \sin(1 - u)^{-\alpha} du.
$$

Changing variables

$$
s = (1 - u)^{-\alpha}, \quad u = 1 - s^{-1/\alpha},
$$

we have

$$
B(v) = \frac{1}{\alpha} \int_{(1-v)^{-\alpha}}^{\infty} s^{-\delta} \sin s\, ds, \quad \delta = \frac{1 - \beta}{\alpha} + 1.
$$

Integrate by parts twice:

$$\begin{aligned}
B(v) &= \frac{1}{\alpha}(\cos(1-v)^{-\alpha})(1-v)^{\alpha\delta} + \frac{\delta}{\alpha}\int_{(1-v)^{-\alpha}}^{\infty} s^{-\delta-1}\cos s\,ds \\
&= \frac{1}{\alpha}(\cos(1-v)^{-\alpha})(1-v)^{\alpha\delta} - \frac{\delta}{\alpha}(\sin(1-v)^{-\alpha})(1-v)^{\alpha(\delta+1)} \\
&\quad - \frac{\delta(\delta+1)}{\alpha}\int_{(1-v)^{-\alpha}}^{\infty} s^{-\delta-2}\sin s\,ds.
\end{aligned}$$

This implies that

$$B(v) = \frac{\cos(1-v)^{-\alpha}}{\alpha}(1-v)^{\alpha-\beta+1} + O((1-v)^{2\alpha-\beta+1}). \qquad (6.1.5)$$

Thus considering the Toeplitz operator T_a with the radial symbol a of the form (6.1.4) we have

- for $\alpha \geq \beta$ the sequence $\gamma_a(n)$ is bounded and thus the Toeplitz operator is bounded on $\mathcal{A}^2(\mathbb{D})$;

- for $\alpha > \beta$ the sequence $\gamma_a(n)$ belongs to c_0 and thus the Toeplitz operator is compact on $\mathcal{A}^2(\mathbb{D})$.

Moreover for $\beta \leq 0$ the symbol (6.1.4) is bounded, while for $\beta > 0$ the symbol (6.1.4) is unbounded near the boundary $\partial\mathbb{D}$.

The conditions (6.1.2) and (6.1.3) are sufficient for boundedness and compactness of an operator T_a, in general. It is known [120], that for bounded symbols $a(r) \in L_\infty(0,1)$ the condition (6.1.3) is necessary and sufficient for compactness of T_a on $\mathcal{A}^2(\mathbb{D})$.

Another case when conditions (6.1.2) and (6.1.3) are necessary for L_1 defining symbols is described by the following theorem.

Theorem 6.1.8. *Let $b(u) \in L_1(0,1)$, and $b(u) \geq 0$ almost everywhere. Then the conditions (6.1.2) and (6.1.3) are necessary and sufficient for $\gamma_a \in l_\infty$ and $\gamma_a \in c_0$, respectively.*

Proof. Let $n = [(1-s)^{-1}]$, then

$$\gamma_a(n) \geq (n+1)\int_s^1 b(u)u^n du \geq \text{const}\,(n+1)\int_s^1 b(u)du = \text{const}\,(n+1)B(s).$$

Thus

$$B(s) \leq \text{const}\,(1-s)\gamma_a(n). \qquad \square$$

Example 6.1.9. Consider the following family of radial defining symbols

$$a_\alpha(r) = (1-r)^{\alpha-1}, \quad \text{where} \quad \alpha > 0,$$

which scales the (polynomial) growth of symbols near the boundary. We have

$$\gamma_{a_\alpha}(n) = (n+1) \int_0^1 (1 - \sqrt{r})^{\alpha-1} r^n dr$$

and

$$B_\alpha(s) = \int_s^1 a_\alpha(\sqrt{r})\, dr = \frac{2}{\alpha} s(1-s)^\alpha + \frac{2}{\alpha(\alpha+1)}(1-s)^{\alpha+1}.$$

By Theorem 6.1.8 the operator T_{a_α} is bounded if and only if $\alpha \geq 1$, and compact if and only if $\alpha > 1$. That is, in this scale unbounded defining symbols generate a unbounded Toeplitz operators. Moreover, as it will follow from Corollary 6.1.10, to generate bounded or compact Toeplitz operator its unbounded defining symbol must necessarily have sufficiently sophisticated oscillating behaviour near the unit circle $\partial \mathbb{D}$.

For a non-negative symbol $a(r)$ introduce the function

$$m_a(u) = \inf_{r \in [u,1)} a(r),$$

which is obviously always monotone.

Corollary 6.1.10. *If* $\lim_{u \to 1} m_a(u) = +\infty$ *(which is equivalent to* $\lim_{r \to 1} a(r) = +\infty$*), then the Toeplitz operator* T_a *is unbounded.*

Proof. Estimate

$$B(s) = \int_s^1 b(u)du \geq \inf_{r \in [s^2,1]} a(r) \cdot (1-s),$$

then $\lim_{t \to \infty} (B(s)/(1-s)) = +\infty$. Thus according to Theorem 6.1.8 the operator T_a is unbounded. $\square$

Note that for general L_1 defining symbols the conditions (6.1.2) and (6.1.3) fail to be necessary.

Example 6.1.11. Let

$$a(r) = -(1-\gamma)(1-r^2)^{-\gamma} \sin(1-r^2)^{-\alpha} + \alpha(1-r^2)^{-\alpha-\gamma} \cos(1-r^2)^{-\alpha}. \quad (6.1.6)$$

Then

$$B(u) = (1-u)^{1-\gamma} \sin(1-u)^{-\alpha}.$$

Suppose that

$$0 < \gamma < \alpha \qquad\qquad (6.1.7)$$

and

$$\alpha + \gamma < 1.$$

Then $b(r)(= a(\sqrt{r})) \in L_1(0,1)$ but conditions (6.1.2) and (6.1.3) are not realized. However we can show that operator T_a is bounded and compact on $\mathcal{A}^2(\mathbb{D})$. Indeed

$$\gamma_a(n) = (n+1)\int_0^1 b(r)r^n dr = (n+1)n\int_0^1 B(r)r^{n-1} dr.$$

Integrating by parts once more we have

$$\gamma_a(n) = (n+1)n(n-1)\int_0^1 C(r)r^{n-2} dr,$$

where

$$C(r) = \int_r^1 B(s)ds.$$

Setting $\beta = \gamma - 1$ from (6.1.5) it follows that

$$C(r) = \frac{\cos(1-r)^{-\alpha}}{\alpha}(1-r)^{\alpha-\gamma+2} + O((1-r)^{2\alpha-\gamma+2}).$$

Thus

$$|\gamma_a(n)| \le \text{const} \cdot n^3 \int_0^1 (1-r)^{\alpha-\gamma+2} r^{n-2} dr.$$

Integrating by parts twice we have

$$|\gamma_a(n)| \le \text{const} \cdot (n+1)\int_0^1 (1-r)^{\alpha-\gamma} r^n dr,$$

and due to (6.1.7) from Theorem 6.1.6 we have

$$\lim_{n\to\infty} \gamma_a(n) = 0.$$

In the above example we have used the following fact: the second "antiderivative" of the function $b(r)$ in a neighborhood of the point $r = 1$ has the asymptotic

$$C(r) = o((1-r)^2).$$

This observation hints at the following generalization of Theorem 6.1.6.

Given $b(u)$, introduce the functions

$$B^{(j)}(u) = \int_u^1 B^{(j-1)}(s)ds, \quad j = 1, 2, \ldots,$$

where $B^{(0)}(s) = b(s)$.

Note, that $B(u) = B^{(1)}(u)$, and $C(u) = B^{(2)}(u)$.

Theorem 6.1.12. *Let a function $B^{(j)}(s)$, $j \in \mathbb{N}$, have the following asymptotic, when $s \to 1$,*

$$|B^{(j)}(s)| = O((1-s)^j),$$

then

$$\sup_{n \in \mathbb{Z}_+} |\gamma_a(n)| < \infty.$$

If

$$|B^{(j)}(s)| = o((1-s)^j),$$

then

$$\lim_{n \to \infty} \gamma_a(n) = 0.$$

Proof. The proof is analogous to that of Theorem 6.1.6. Note that the function (6.1.6) of Example 6.1.11 satisfies the hypothesis of Theorem 6.1.12 for $j = 2$. $\square$

The above result permits us to get more complete information on boundedness and compactness of the Toeplitz operator of Example 6.1.7.

Corollary 6.1.13. *Let $a(r)$ be of the form (6.1.4),*

$$a(r) = (1 - r^2)^{-\beta} \sin(1 - r^2)^{-\alpha},$$

where $\alpha > 0$ and $\beta < 1$. Then the Toeplitz operator T_a is bounded and compact for all $\alpha > 0$.

Proof. It is sufficient to consider the case $0 < \beta < 1$ only. Setting $\beta_1 = \beta - \alpha - 1$ in (6.1.5) and integrating $B^{(2)}(r)$ by parts twice we get, analogously to (6.1.5), the representation

$$B^{(2)}(r) = -\frac{\sin(1-r)^{-\alpha}}{\alpha^2}(1-r)^{\alpha-\beta_1+1} + O((1-r)^{2\alpha-\beta_1+1}).$$

Then by Theorem 6.1.12 with $j = 2$, we have that for $\alpha - \beta_1 + 1 > 2$, or for $\alpha > \frac{\beta}{2}$ the operator T_a is compact. Repeating the above procedure consequently, we get that for $\alpha > \frac{\beta}{j}$ the operator T_a is compact for every $j \in \mathbb{N}$. $\square$

Another useful characterization of a sequence $\gamma_a(n)$ is given by the following theorem.

Theorem 6.1.14. *Let $b(u) \in L_1(0,1)$. Then*

$$\lim_{n \to \infty} (\gamma_a(n) - \gamma_a(n+1)) = 0. \tag{6.1.8}$$

Proof. Consider

$$\gamma_a(n) - \gamma_a(n+1) = (n+1) \int_0^1 (1-u)u^n b(u)\,du - \int_0^1 u^{n+1} b(u)\,du$$
$$= I_1(n) + I_2(n).$$

To estimate the first summand we find first the maximum of the function $s(u) = (1 - u)u^n$. This is obviously at $u_0 = 1 - \frac{1}{n+1}$. Thus

$$\sup_{u\in[0,1]} s(u) = s(u_0) = \frac{\left(1 - \frac{1}{n+1}\right)^n}{n + 1} \leq \frac{\text{const}}{n + 1}.$$

Let $\varepsilon = \varepsilon(n) = \frac{1}{\sqrt{n}}$, then

$$\begin{aligned}
|I_1(n)| &\leq (n+1)\int_{1-\varepsilon}^{1}(1 - u)u^n|b(u)|du + (n+1)\int_{0}^{1-\varepsilon}(1 - u)u^n|b(u)|du \\
&\leq \text{const}\int_{1-\varepsilon}^{1}|b(u)|du + (n+1)(1 - \varepsilon)^n\int_{0}^{1}|b(u)|du.
\end{aligned}$$

Now from

$$\lim_{n\to\infty}\int_{1-\varepsilon}^{1}|b(u)|du = 0$$

and

$$(n+1)\left(1 - \frac{1}{\sqrt{n}}\right)^n = (n+1)\exp\left(-n\left(\frac{1}{\sqrt{n}} - \frac{1}{2n} + O\left(\frac{1}{n^{3/2}}\right)\right)\right) \to 0$$

it follows that

$$\lim_{n\to\infty} I_1(n) = 0.$$

Analogously splitting the integral $I_2(n)$ on segments $[1-1/\sqrt{n}, 1]$ and $[0, 1-1/\sqrt{n}]$ one can show that

$$\lim_{n\to\infty} I_2(n) = 0,$$

which finishes the proof of the theorem. $\qquad\square$

Corollary 6.1.15. *Let $b(u) \in L_1(0,1)$. Then the set of all limit points of the sequence $\gamma_a(n)$ is a closed connected subset of $\mathbb{C}$. In particular the sequence $\gamma_a(n)$ cannot have a finite or countable set of distinct limit points.*

Proof. Suppose the set K of limit points is not connected. Then there exist two closed subsets K_1 and K_2 (intersecting K) with a positive distance between them such that $K \subset K_1 \cup K_2$. Without loss of generality we can assume that $\gamma_a(n) \in K_1 \cup K_2$ for each n starting from some N. Thus there exist infinitely many $n_j \in \mathbb{N}$ such that $\gamma_a(n_j) \in K_1$ but, at that time, either $\gamma_a(n_j+1) \in K_2$, or $\gamma_a(n_j-1) \in K_2$, which contradicts (6.1.8). $\qquad\square$

Corollary 6.1.16. *The essential spectrum of a bounded Toeplitz operator with a radial defining symbol is always connected.*

That is, if an l_∞ sequence $\gamma_a(n)$ does not have a limit, then the essential spectrum of the corresponding Toeplitz operator may be either a compact connected curve, or a compact connected subset of $\mathbb{C}$ having positive planar measure. Let us show that both these cases can be realized.

Example 6.1.17 (Unit circle and unit interval). Let $a_p(r) = \alpha_p(\ln r^{-2})^{ip}$, with $\alpha_p \in \mathbb{C}$, and $p \in \mathbb{R}$. Then

$$
\begin{aligned}
\gamma_{a_p}(n) &= \alpha_p(n+1) \int_0^1 (\ln u^{-1})^{ip} u^n du \\
&= \alpha_p \int_0^1 (\ln(s^{-1/(n+1)}))^{ip} ds \\
&= (n+1)^{-ip} \left[\alpha_p \int_0^1 (\ln s^{-1})^{ip} ds \right].
\end{aligned}
$$

Select now α_p in such a way that the multiple in the square brackets is equal to 1. It is easy to see that $\alpha_p = 1/\Gamma(ip+1)$. Then

$$
\gamma_{a_p}(n) = (n+1)^{-ip} = \exp(-ip\ln(n+1)).
$$

Thus

$$
\operatorname{sp} T_{a_p} = \operatorname{ess-sp} T_{a_p} = S^1.
$$

If $c_p(r) = \operatorname{Im} \alpha_p(\ln r^{-2})^{ip}$, then

$$
\gamma_{c_p}(n) = -\sin(p\ln(n+1)), \tag{6.1.9}
$$

and

$$
\operatorname{sp} T_{c_p} = \operatorname{ess-sp} T_{c_p} = [-1, 1].
$$

Example 6.1.18 (Square). Let $a(r) = c_1(r) + ic_{\sqrt{2}}(r)$, then by (6.1.9) we have

$$
\gamma_a(n) = -(\sin\ln(n+1) + i\sin\sqrt{2}\ln(n+1)).
$$

Since the number $\sqrt{2}$ is irrational, the points $\{\gamma_a(n)\}_{n\in\mathbb{Z}_+}$ form a dense set in the square, and thus

$$
\operatorname{sp} T_a = \operatorname{ess-sp} T_a = [-1, 1] \times [-1, 1].
$$

Example 6.1.19 (A more complicated curve). Let $a(r) = c_1(r) + ic_2(r)$, then by (6.1.9) we have

$$
\gamma_a(n) = -(\sin\ln(n+1) + i\sin 2\ln(n+1)).
$$

Now the points of this sequence are located on the curve

$$
y^2 - 4x^2 + 4x^4 = 0.
$$

6.2   Algebras of Toeplitz operators

We are going to consider now the C^*-algebra generated by bounded Toeplitz operators with radial defining symbols. First observe that our class of symbols as well as the corresponding Toeplitz C^*-algebra will have certain peculiarities. In particular, contrary to commonly known and studied cases (see, for example [240]), the Toeplitz operator algebra is commutative, but the semi-commutators $[T_{a_1}, T_{a_2}) = T_{a_1} \cdot T_{a_2} - T_{a_1 \cdot a_2}$ are not compact in general. Moreover, the symbols under study do not form an algebra (under the pointwise multiplication). That is, having two radial defining symbols $a_1(r)$ and $a_2(r)$, for which the corresponding Toeplitz operators $T_{a_1(r)}$ and $T_{a_2(r)}$ are bounded, the Toeplitz operator $T_{a_1 \cdot a_2}$, which corresponds to the product of these symbols, is not necessarily bounded. The natural structure on the set of defining symbols under consideration is a linear space (in the algebraic sense, i.e., no norm structure assumed).

Example 6.2.1. Let
$$a_1(r) = \sin(1 - r^2)^{-\alpha}$$
and
$$a_2(r) = (1 - r^2)^{-\beta} \sin(1 - r^2)^{-\alpha}$$
where $0 \le \beta < 1$ and $\beta < \alpha$. Then according to the Example 6.1.7, both operators T_{a_1} and T_{a_2} are bounded and compact.

The product $a_1 \cdot a_2$ has the form
$$a_1(r) \cdot a_2(r) = \frac{(1 - r^2)^{-\beta}}{2} - \frac{(1 - r^2)^{-\beta} \cos 2(1 - r^2)^{-\alpha}}{2}.$$

Let $\beta = 0$, then $T_{a_1 \cdot a_2} = \frac{1}{2}I - T_{a_3}$, where the operator T_{a_3} with the defining symbol
$$a_3(r) = \frac{1}{2} \cos 2(1 - r^2)^{-\alpha}$$
is compact. The compactness of T_{a_3} can be shown as in Example 6.1.7.

Now let $\beta > 0$. Then, according to Example 6.1.9, the operator T_{a_4} with the defining symbol
$$a_4(r) = \frac{1}{2}(1 - r^2)^{-\beta}$$
is unbounded. At the same time the operator T_{a_5} with the defining symbol
$$a_5(r) = \frac{1}{2}(1 - r^2)^{-\beta} \cos 2(1 - r^2)^{-\alpha}$$

is compact (again analogously to Example 6.1.7).

That is

(i) for $\beta = 0$ the operators T_{a_1} and T_{a_2} are compact, while the operator $T_{a_1 \cdot a_2}$ is bounded but not compact; that is, the semi-commutator $[T_{a_1}, T_{a_2}) = T_{a_1} \cdot T_{a_2} - T_{a_1 \cdot a_2}$ is *not compact*;

(ii) for $\beta > 0$ the operators T_{a_1} and T_{a_2} are bounded, but the operator $T_{a_1 \cdot a_2}$ is *not bounded* at all.

Denote by $\mathcal{M}$ the linear space of measurable functions such that for each $a(r) \in \mathcal{M}$ the Toeplitz operator $T_{a(r)}$ is bounded on $L_2(\mathbb{D})$, and denote by $\mathcal{T}(\mathcal{M})$ the C^*-algebra generated by all Toeplitz operators T_a with defining symbols $a \in \mathcal{M}$.

Theorem 6.2.2. *The C^*-algebra $\mathcal{T}(\mathcal{M})$ is commutative and isomorphically embedded into the algebra l_∞. The embedding*

$$\nu : \mathcal{T}(\mathcal{M}) \longrightarrow l_\infty$$

is generated by the mapping

$$\nu : T_a \longmapsto \gamma_a,$$

where $a(r) \in \mathcal{M}$, and the sequence γ_a is given by (6.1.1).

Proof. Follows directly from Theorem 6.1.1 and Corollary 6.1.2. $\qquad\square$

In the next statements we deal with the Wick function and the Wick symbol of a Toeplitz operator. Their definitions are given in Appendix A, formulas (A.1.2)–(A.1.3), and (A.1.6)

Theorem 6.2.3. *Let T_a be the Toeplitz operator with a radial defining symbol $a = a(r)$. Then the corresponding Wick function has the form*

$$\widetilde{a}(z, \zeta) = (1 - z\overline{\zeta})^2 \sum_{n=0}^{\infty} (n + 1)(z\overline{\zeta})^n \, \gamma_a(n).$$

Proof. Calculate

$$
\begin{aligned}
\widetilde{a}(z, \zeta) &= \frac{\langle a k_\zeta, k_z \rangle}{\langle k_\zeta, k_z \rangle} = k_\zeta^{-1}(z) \, \langle a k_\zeta, k_z \rangle \\
&= \pi\,(1 - z\overline{\zeta})^2 \, \frac{1}{\pi^2} \int_{\mathbb{D}} \frac{a(|\tau|)\, dv(\tau)}{(1 - \tau\overline{\zeta})^2\,(1 - z\overline{\tau})^2} \\
&= \frac{1}{\pi}\,(1 - z\overline{\zeta})^2 \int_{\mathbb{D}} a(|\tau|) \sum_{k=0}^{\infty} (k + 1)(\tau\overline{\zeta})^k \sum_{n=0}^{\infty} (n + 1)(z\overline{\tau})^n \, dv(\tau) \\
&= \frac{1}{\pi}\,(1 - z\overline{\zeta})^2 \sum_{k=0}^{\infty} \sum_{n=1}^{\infty} (k + 1)(n + 1)\overline{\zeta}^k z^n \int_0^1 a(r) r^{k+n+1} dr \\
&\qquad \cdot \frac{1}{i} \int_{S^1} t^{k-n-1} dt \\
&= (1 - z\overline{\zeta})^2 \sum_{k=0}^{\infty} (n + 1)(z\overline{\zeta})^n \, 2(n + 1) \int_0^1 a(r) r^{2n+1} dr \\
&= (1 - z\overline{\zeta})^2 \sum_{n=0}^{\infty} (n + 1)(z\overline{\zeta})^n \, \gamma_a(n).
\end{aligned}
$$

$\qquad\square$

Denote by L_n the one-dimensional subspace of $\mathcal{A}^2(\mathbb{D})$ generated by the base element $e_n(z) = \sqrt{\frac{n+1}{\pi}}\, z^n$, $n \in \mathbb{Z}_+$. Then the one-dimensional projection P_n of $\mathcal{A}^2(\mathbb{D})$ onto L_n obviously has the form

$$P_n f = \langle f, e_n \rangle\, e_n = \frac{n+1}{\pi}\, z^n \int_{\mathbb{D}} f(\zeta)\, \overline{\zeta}^n\, d\mu(\zeta).$$

A bounded Toeplitz operator with a radial defining symbol is obviously diagonal with respect to the basis $\{e_n(z)\}_{n \in \mathbb{Z}_+}$. It is important to mention that its representation in the form with the Wick symbol (A.1.4) leads exactly to this diagonal form.

Corollary 6.2.4. *Let T_a be a bounded Toeplitz operator having radial defining symbol $a(r)$. Then the representation of the operator T_a using the Wick symbol gives the following spectral decomposition of the operator T_a:*

$$T_a = \sum_{n=0}^{\infty} \gamma_a(n) P_n.$$

Corollary 6.2.5. *Let T_a be a bounded Toeplitz operator with radial defining symbol $a(r)$. Then the Wick symbol of the operator T_a is radial as well, and is given by the formula*

$$\widetilde{a}(r) = (1 - r^2)^2 \sum_{n=0}^{\infty} (n+1) r^{2n}\, \gamma_a(n).$$

Corollary 6.2.6. *For any $n \in \mathbb{Z}_+$ the one-dimensional space L_n is an eigenspace for any operator $T \in \mathcal{T}(\mathcal{M})$, and the corresponding eigenvalue $\gamma(n)$ is given by the formula*

$$\gamma(n) = \frac{d^n}{dr^n}\left(\frac{1}{(1-r)^2(n+1)!}\, \widetilde{a}(\sqrt{r}) \right)_{r=0},$$

where $\widetilde{a}$ is the Wick symbol of the operator T.

Chapter 7

Toeplitz Operators on the Upper Half-Plane with Homogeneous Symbols

7.1 Representation of the Bergman space

In this chapter we return to the upper half-plane Π, the space $L_2(\Pi)$ and its Bergman subspace $\mathcal{A}^2(\Pi)$. Passing to polar coordinates we have

$$L_2(\Pi) = L_2(\mathbb{R}_+, r\,dr) \otimes L_2([0, \pi], d\theta) = L_2(\mathbb{R}_+, r\,dr) \otimes L_2(0, \pi),$$

and

$$\frac{\partial}{\partial \overline{z}} = \frac{\cos\theta + i\sin\theta}{2}\left(\frac{\partial}{\partial r} + i\frac{1}{r}\frac{\partial}{\partial \theta}\right) = \frac{\cos\theta + i\sin\theta}{2r}\left(r\frac{\partial}{\partial r} + i\frac{\partial}{\partial \theta}\right).$$

The Bergman space $\mathcal{A}^2(\Pi)$ can be described alternatively as the (closed) subspace of $L_2(\Pi)$ which consists of all functions satisfying the equation

$$\left(r\frac{\partial}{\partial r} + i\frac{\partial}{\partial \theta}\right)\varphi(r, \theta) = 0.$$

Introduce the unitary operator

$$U_1 = M \otimes I : L_2(\mathbb{R}_+, r\,dr) \otimes L_2(0, \pi) \longrightarrow L_2(\mathbb{R}) \otimes L_2(0, \pi),$$

where the Mellin transform $M : L_2(\mathbb{R}_+, r\,dr) \longrightarrow L_2(\mathbb{R})$ is given by

$$(M\psi)(\lambda) = \frac{1}{\sqrt{2\pi}} \int_{\mathbb{R}_+} r^{-i\lambda}\,\psi(r)\,dr. \tag{7.1.1}$$

The inverse Mellin transform $M^{-1} : L_2(\mathbb{R}) \longrightarrow L_2(\mathbb{R}_+, r\,dr)$ has the form

$$(M^{-1}\psi)(r) = \frac{1}{\sqrt{2\pi}} \int_{\mathbb{R}} r^{i\lambda-1}\,\psi(\lambda)\,d\lambda.$$

Then, as is easy to see,

$$(M \otimes I)\left(r\frac{\partial}{\partial r} + i\frac{\partial}{\partial \theta}\right)(M^{-1} \otimes I) = (i\lambda - 1) + i\frac{\partial}{\partial \theta} = i\left((\lambda + i) + \frac{\partial}{\partial \theta}\right).$$

Thus the image of the Bergman space $\mathcal{A}_1^2 = U_1(\mathcal{A}^2(\Pi))$ can be described as the (closed) subspace of $L_2(\mathbb{R}) \otimes L_2(0, \pi)$ which consists of all functions $\psi(\lambda, \theta)$ satisfying the equation

$$\left((\lambda + i) + \frac{\partial}{\partial \theta}\right)\psi(\lambda, \theta) = 0.$$

This equation is easy to solve and its general L_2 solution has the form

$$\psi(\lambda, \theta) = f(\lambda) \cdot \sqrt{\frac{2\lambda}{1 - e^{-2\pi\lambda}}}\, e^{-(\lambda+i)\theta},$$

where $f(\lambda)$ is an arbitrary $L_2(\mathbb{R})$ function, and furthermore

$$\|\psi(\lambda, \theta)\|_{L_2(\mathbb{R}\times[0,\pi])} = \|f(\lambda)\|_{L_2(\mathbb{R})}.$$

Thus we arrive at the following lemma.

Lemma 7.1.1. *The unitary operator U_1 is an isometric isomorphism of the space $L_2(\Pi)$ onto $L_2(\mathbb{R}) \otimes L_2(0, \pi)$ under which the Bergman space $\mathcal{A}^2(\Pi)$ is mapped onto*

$$\mathcal{A}_1^2 = \left\{ f(\lambda) \cdot \sqrt{\frac{2\lambda}{1 - e^{-2\pi\lambda}}}\, e^{-(\lambda+i)\theta} \ : \ f(\lambda) \in L_2(\mathbb{R}) \right\}.$$

Corollary 7.1.2. *The orthogonal projection $B_1 = (M \otimes I)B_\Pi(M^{-1} \otimes I)$ of $L_2(\Pi)$ onto $L_2(\mathbb{R}) \otimes L_2(0, \pi)$ onto the image of the Bergman space $\mathcal{A}_1^2$ has the form*

$$(B_1\psi)(\lambda, \theta) = \sqrt{\frac{2\lambda}{1 - e^{-2\pi\lambda}}}\, e^{-(\lambda+i)\theta} \int_0^\pi \psi(\lambda, \eta)\sqrt{\frac{2\lambda}{1 - e^{-2\pi\lambda}}}\, e^{-(\lambda-i)\eta}d\eta.$$

For each $\lambda \in \mathbb{R}$, introduce the function

$$
\begin{aligned}
\ell_\lambda(\theta) &= \sqrt{\frac{2\lambda}{1 - e^{-2\pi\lambda}}}\, e^{-(\lambda+i)\theta} \\[2mm]
&= \sqrt{\frac{2\lambda}{1 - e^{-2\pi\lambda}}}\, e^{i\theta(i\lambda-1)}, \qquad \theta \in [0, \pi],
\end{aligned}
\tag{7.1.2}
$$

which obviously belongs to $L_2(0, \pi)$, and $\|\ell_\lambda\| = 1$.

For each $\lambda \in \mathbb{R}$ we denote by L_λ the one-dimensional subspace of $L_2(0, \pi)$ generated by ℓ_λ. The one-dimensional orthogonal projection $B(\lambda)$ of $L_2(0, \pi)$ onto L_λ obviously has the form

$$B(\lambda)f = \langle f, \ell_\lambda \rangle \ell_\lambda. \tag{7.1.3}$$

The next corollary shows that the projection B_1 splits onto the direct integral (on $\lambda \in \mathbb{R}$) of one-dimensional projections $B(\lambda)$ acting on $L_2(0, \pi)$.

Corollary 7.1.3. *We have*

$$B_1 = I \otimes B(\lambda),$$

where the operator $I \otimes B(\lambda)$ acts on $L_2(\mathbb{R}) \otimes L_2(0, \pi)$ by

$$(I \otimes B(\lambda)\psi)(\lambda, \theta) = (B(\lambda)\psi(\lambda, \cdot))(\theta).$$

Introduce the isometric imbedding $R_0 : L_2(\mathbb{R}) \longrightarrow \mathcal{A}_1^2 \subset L_2(\mathbb{R} \times [0, \pi])$ by the rule

$$(R_0 f)(\lambda, \theta) = f(\lambda) \cdot \sqrt{\frac{2\lambda}{1 - e^{-2\pi\lambda}}}\, e^{-(\lambda+i)\theta}.$$

The adjoint operator $R_0^* : L_2(\mathbb{R} \times [0, \pi]) \longrightarrow L_2(\mathbb{R})$ is given by

$$(R_0^*\psi)(\lambda) = \sqrt{\frac{2\lambda}{1 - e^{-2\pi\lambda}}} \int_0^\pi \psi(\lambda, \theta)\, e^{-(\lambda-i)\theta}\, d\theta,$$

and

$$\begin{aligned}
R_0^* R_0 &= I &&: \quad L_2(\mathbb{R}) \longrightarrow L_2(\mathbb{R}), \\
R_0 R_0^* &= B_1 &&: \quad L_2(\mathbb{R} \times [0, \pi]) \longrightarrow \mathcal{A}_1^2,
\end{aligned}$$

where B_1 is the orthogonal projection of $L_2(\mathbb{R} \times [0, \pi])$ onto $\mathcal{A}_1^2$.

Now the operator $R = R_0^* U_1$ maps the space $L_2(\Pi)$ onto $L_2(\mathbb{R})$, and its restriction

$$R|_{\mathcal{A}^2(\Pi)} : \mathcal{A}^2(\Pi) \longrightarrow L_2(\mathbb{R})$$

is an isometric isomorphism. The adjoint operator

$$R^* = U_1^* R_0 : L_2(\mathbb{R}) \longrightarrow \mathcal{A}^2(\Pi) \subset L_2(\Pi)$$

is an isometric isomorphism of $L_2(\mathbb{R})$ onto the Bergman subspace $\mathcal{A}^2(\Pi)$ of the space $L_2(\Pi)$.

Remark 7.1.4. We have

$$\begin{aligned}
RR^* &= I &&: \quad L_2(\mathbb{R}) \longrightarrow L_2(\mathbb{R}), \\
R^* R &= B_\Pi &&: \quad L_2(\Pi) \longrightarrow \mathcal{A}^2(\Pi).
\end{aligned}$$

Let us mention the differences between this case and the two cases studied in Chapters 3 and 4. First of all, here there is no second unitary operator (like U_2 in the previous section) to permit a complete separation of variables in the Bergman space representation, and thus unlike Theorem 3.1.1 and Theorem 4.1.1 we must stop here with Lemma 7.1.1.

Further, the coordinate systems used (and the systems of cycles later on) in the first two cases contain as a coordinate line the boundary of the domain, and is not compatible with the boundary in the third case. As a corollary, in the first two cases there exist transparent connections between the Bergman and Hardy spaces given by Theorems 3.2.1 and 4.1.7.

However, there is no natural connection between the Bergman and Hardy spaces compatible with the third type of description of the Bergman space.

7.2　Toeplitz operators with homogeneous symbols

Denote by $\mathcal{A}_\infty$ the C^*-algebra of bounded measurable homogeneous functions on Π of order zero, or functions depending only on the polar coordinate θ. Introduce the Toeplitz operator algebra $\mathcal{T}(\mathcal{A}_\infty)$ generated by all Toeplitz operators

$$T_a : \varphi \in \mathcal{A}^2(\Pi) \longmapsto B_\Pi a\varphi \in \mathcal{A}^2(\Pi)$$

with defining symbols $a = a(\theta) \in \mathcal{A}_\infty$.

Theorem 7.2.1. *Let* $a = a(\theta) \in \mathcal{A}_\infty$. *Then the Toeplitz operator* T_a *acting on* $\mathcal{A}^2(\Pi)$ *is unitary equivalent to the multiplication operator* $\gamma_a I = R T_a R^*$ *acting on* $L_2(\mathbb{R})$. *The function* $\gamma_a(\lambda)$ *is given by*

$$\gamma_a(\lambda) = \frac{2\lambda}{1 - e^{-2\pi\lambda}} \int_0^\pi a(\theta)\, e^{-2\lambda\theta}\, d\theta, \qquad \lambda \in \mathbb{R}. \tag{7.2.1}$$

Proof. Given $a = a(\theta) \in \mathcal{A}_\infty$, calculate

$$\begin{aligned}
R T_a R^* &= R B_\Pi a B_\Pi R^* = R(R^* R)a(R^* R)R^* \\
&= (RR^*)RaR^*(RR^*) = RaR^* \\
&= R_0^*(M \otimes I)a(\theta)(M^{-1} \otimes I)R_0 \\
&= R_0^* a(\theta) R_0.
\end{aligned}$$

Finally

$$(R_0^* a(\theta) R_0 f)(\lambda) = \frac{2\lambda}{1 - e^{-2\pi\lambda}} \int_0^\pi a(\theta)\, f(\lambda)\, e^{-2\lambda\theta}\, d\theta = \gamma_a(\lambda) \cdot f(\lambda),$$

where

$$\gamma_a(\lambda) = \frac{2\lambda}{1 - e^{-2\pi\lambda}} \int_0^\pi a(\theta)\, e^{-2\lambda\theta}\, d\theta, \qquad \lambda \in \mathbb{R}. \qquad \square$$

Corollary 7.2.2. *The algebra $T(\mathcal{A}_\infty)$ is commutative. The isomorphic imbedding*

$$\tau_\infty : T(\mathcal{A}_\infty) \longrightarrow C_b(\mathbb{R})$$

is generated by the mapping

$$\tau_\infty : T_a \longmapsto \gamma_a(\lambda)$$

of generators of the algebra $T(\mathcal{A}_\infty)$, where $a = a(\theta) \in \mathcal{A}_\infty$.

We consider now a number of important special cases.

Note first that for each $a(\theta) \in L_\infty(0, \pi)$ the function $\gamma_a(\lambda)$ is continuous in all finite points $\lambda \in \mathbb{R}$. For a "very large λ" ($\lambda \to +\infty$) the exponent $e^{-2\lambda\theta}$ has a very sharp maximum at the point $\theta = 0$, and thus the major contribution to the integral in (7.2.1) for these "very large λ" is determined by values of $a(\theta)$ at a neighborhood of the point 0. The major contribution for a "very large negative λ" ($\lambda \to -\infty$) is determined by values of $a(\theta)$ at a neighborhood of π, due to a very sharp maximum of $e^{-2\lambda\theta}$ at $\theta = \pi$ for these values of λ. In particular, as the following lemma states, if $a(\theta)$ has limits at the points 0 and π, then

$$\lim_{\lambda \to +\infty} \gamma_a(\lambda) = \lim_{\theta \to 0} a(\theta) \qquad \text{and} \qquad \lim_{\lambda \to -\infty} \gamma_a(\lambda) = \lim_{\theta \to \pi} a(\theta).$$

Lemma 7.2.3. *Let $a(\theta) \in L_\infty(0, \pi)$ and let the following limits exist:*

$$\lim_{\theta \to 0} a(\theta) = a_0, \qquad\qquad \lim_{\theta \to \pi} a(\theta) = a_\pi.$$

Then $\gamma_a(\lambda) \in C(\overline{\mathbb{R}})$, and

$$\gamma_a(+\infty) = a_0, \qquad\qquad \gamma_a(-\infty) = a_\pi. \tag{7.2.2}$$

Proof. Let $\lambda \to +\infty$. Then for a sufficiently small $\delta > 0$ we represent the function $\gamma_a(\lambda)$ as

$$\gamma_a(\lambda) \;=\; \frac{2\lambda}{1 - e^{-2\pi\lambda}} \left(\int_0^\delta a(\theta) e^{-2\lambda\theta} d\theta + \int_\delta^\pi a(\theta) e^{-2\lambda\theta} d\theta \right)$$

$$:=\; I_1^{(0)}(\lambda) + I_2^{(0)}(\lambda).$$

Consider first $I_1^{(0)}(\lambda)$,

$$I_1^{(0)}(\lambda) \;=\; \frac{2\lambda}{1 - e^{-2\pi\lambda}} \left(a_0 \int_0^\delta a_0 e^{-2\lambda\theta} d\theta + \int_0^\delta (a(\theta) - a_0) e^{-2\lambda\theta} d\theta \right)$$

$$:=\; I_{1,1}^{(0)}(\lambda) + I_{1,2}^{(0)}(\lambda).$$

It is obvious that

$$I_{1,1}^{(0)}(\lambda) = a_0 \frac{1 - e^{-2\delta\lambda}}{1 - e^{-2\pi\lambda}} = a_0 + \alpha(\delta, \lambda),$$

where for a sufficiently large λ,

$$|\alpha(\delta, \lambda)| < \varepsilon.$$

Then

$$|I_{1,2}^{(0)}(\lambda)| \leq \sup_{\theta \in (0,\delta)} |a(\theta) - a_0| \frac{2\lambda}{1 - e^{-2\pi\lambda}} \int_0^\delta e^{-2\lambda\theta} d\theta = \sup_{\theta \in (0,\delta)} |a(\theta) - a_0| \frac{1 - e^{-2\delta\lambda}}{1 - e^{-2\pi\lambda}};$$

that is, for an appropriate choice of δ and a sufficiently large λ, we have

$$|I_{1,2}^{(0)}(\lambda)| < \varepsilon.$$

Consider now $I_2^{(0)}(\lambda)$:

$$|I_2^{(0)}(\lambda)| \leq \frac{\lambda}{1 - e^{-2\pi\lambda}} e^{-2\delta\lambda} \int_\delta^\pi |a(\theta)| \, d\theta \leq 2\pi\lambda \, e^{-2\delta\lambda} \, \|a\|_{L_\infty(0,\pi)}.$$

That is, for a sufficiently large λ, we have

$$|I_2^{(0)}(\lambda)| < \varepsilon.$$

Thus, summarizing the above, we have that for any ε and an appropriate choice of δ there is $\lambda_0 > 0$ such that for each $\lambda \geq \lambda_0$ one has

$$|\gamma_a(\lambda) - a_0| < 3\varepsilon;$$

which proves the first equality in (7.2.2).

The second equality in (7.2.2) follows then from the directly verified property

$$\gamma_{a(\theta)}(-\lambda) = \gamma_{a(\pi-\theta)}(\lambda).$$

The continuity of $\gamma_a(\lambda)$ is obvious. $\square$

Given a linear subset $\mathcal{A}$ of $L_\infty(0,\pi)$, denote by $H(\mathcal{A})$ the subset of $\mathcal{A}_\infty$ which consists of all homogeneous functions of zero order functions on the upper half-plane whose restrictions onto the upper half of the unit circle (parameterized by $\theta \in [0,\pi]$) belong to $\mathcal{A}$. And let $\mathcal{T}(H(\mathcal{A}))$ be the the C^*-algebra generated by all Toeplitz operators T_a with defining symbols $a \in H(\mathcal{A})$.

Denote by $L_\infty^{(0,\pi)}([0,\pi])$ the C^*-subalgebra of $L_\infty([0,\pi])$ which consists of all functions having limits at the points 0 and π. Let $C([0,\pi])$ be, as usual, the algebra of all continuous functions on $[0,\pi]$; denote by $PC([0,\pi], \Lambda)$ the algebra of all piece-wise continuous functions on $[0,\pi]$, continuous in $[0,\pi] \setminus \Lambda$ and having one-sided limit values at the points of $\Lambda = \{\theta_1, \theta_2, \ldots, \theta_{n-1}\}$. Let $PCo([0,\pi]), \Lambda)$ be the subalgebra of $PC([0,\pi], \Lambda)$ consisting of all piece-wise constant functions. Given a function $a_0(\theta)$, denote by $L(1, a_0)$ the linear two-dimensional space generated by 1 and the function a_0.

Note that $PCo([0,\pi],\{\theta_1\}) = L(1, \chi_{[0,\theta_1]})$, where $\chi_{[0,\theta_1]}(\theta)$ is the characteristic function of $[0,\theta_1]$.

For a continuous function a_0, a set Λ, and an arbitrary point $\theta_k \in \Lambda$, we have a chain of proper inclusions

$$\begin{array}{c} L(1,a_0) \subset C([0,\pi]) \\ PCo([0,\pi],\{\theta_k\}) \subset PCo([0,\pi]), \Lambda \end{array} \subset PC([0,\pi],\Lambda) \subset L_\infty^{(0,\pi)}([0,\pi]). \quad (7.2.3)$$

Let $\mathcal{A}$ be any of the above sets. We are interested in the C^*-algebra $\mathcal{T}(H(\mathcal{A}))$. Note that for any real non-constant function a_0, the algebra $\mathcal{T}(H(L(1,a_0)))$ is a C^*-algebra with identity generated by a *single* element, the Toeplitz operator T_{a_0}.

For the largest set (algebra) $L_\infty^{(0,\pi)}([0,\pi])$ we have

Theorem 7.2.4. *The C^*-algebra $\mathcal{T}(H(L_\infty^{(0,\pi)}([0,\pi])))$ is isomorphic and isometric to $C(\overline{\mathbb{R}})$, where $\overline{\mathbb{R}} = \mathbb{R} \cup \{-\infty\} \cup \{+\infty\}$ is the two-point compactification of $\mathbb{R}$. The isomorphic isomorphism*

$$\tau_\infty : \mathcal{T}(H(L_\infty^{(0,\pi)}([0,\pi]))) \longrightarrow C(\overline{\mathbb{R}})$$

is generated by the following mapping of generators of $\mathcal{T}(H(L_\infty^{(0,\pi)}([0,\pi])))$,

$$\tau_\infty : T_a \longmapsto \gamma_a(\lambda), \qquad (7.2.4)$$

where $a = a(\theta) \in H(L_\infty^{(0,\pi)}([0,\pi]))$.

Proof. We need to show only that the mapping (7.2.4) is onto. The inclusion

$$\tau_\infty(\mathcal{T}(H(L_\infty^{(0,\pi)}([0,\pi])))) \subset C(\overline{\mathbb{R}})$$

is trivial. The inverse inclusion will follow from the next theorem. $\square$

Passing to the other extreme, the smallest possible set, we have

Theorem 7.2.5. *Let $a_0(\theta) \in L_\infty^{(0,\pi)}([0,\pi])$ be a real-valued function such that the function $\gamma_{a_0}(\lambda)$ separates the points of $\overline{\mathbb{R}}$. Then the C^*-algebra $\mathcal{T}(H(L(1,a_0)))$ is isomorphic and isometric to $C(\overline{\mathbb{R}})$. The isomorphic isomorphism*

$$\tau_\infty : \mathcal{T}(H(L(1,a_0))) \longrightarrow C(\overline{\mathbb{R}})$$

is generated by the same mapping

$$\tau_\infty : T_a \longmapsto \gamma_a(\lambda)$$

of generators of the algebra $\mathcal{T}(H(L(1,a_0)))$.

Proof. Follows directly from the Stone-Weierstrass theorem. $\square$

Corollary 7.2.6. *Given a point $\theta_0 \in (0,\pi)$, the C^*-algebra $\mathcal{T}(H(PCo([0,\pi],\{\theta_0\})))$ is isomorphically isometric to $C(\overline{\mathbb{R}})$.*

Proof. As was already mentioned $PCo([0,\pi],\{\theta_0\}) = L(1,\chi_{[0,\theta_0]})$. We need to prove that the real-valued function

$$\gamma_{\chi_{[0,\theta_0]}}(\lambda) = \frac{2\lambda}{1 - e^{-2\pi\lambda}} \int_0^{\theta_0} e^{-2\lambda\theta}\, d\theta = \frac{e^{-2\theta_0\lambda} - 1}{e^{-2\pi\lambda} - 1}$$

separates the points of $\overline{\mathbb{R}}$. We show instead that the function $\gamma_{\chi_{[0,\theta_0]}}$ is strictly increasing by a simple but a bit lengthy procedure.

After the scaling $t = 2\pi\lambda$, $\theta_0 = \alpha\pi$, with $\alpha \in (0,1)$, we have

$$\gamma(t) = \frac{e^{-\alpha t} - 1}{e^{-t} - 1}, \quad t \in \mathbb{R}.$$

Let first $t > 0$, calculate

$$\gamma'(t) = \frac{\alpha e^{-\alpha t}(1 - e^{-t}) - e^{-t}(1 - e^{-\alpha t})}{(1 - e^{-t})^2}.$$

To show that $\gamma'(t) > 0$ is equivalent to showing that

$$\alpha e^{-\alpha t}\frac{e^t - 1}{e^t} - e^{-t}\frac{e^t - 1}{e^{\alpha t}} > 0,$$

or that

$$\alpha(e^t - 1) - (e^{\alpha t} - 1) > 0,$$

or that

$$\sum_{k=1}^{\infty}(\alpha - \alpha^k)\frac{t^k}{k!} > 0.$$

The last inequality is absolutely evident because of $\alpha \in (0,1)$.

Pass now to $t < 0$. Substituting $x = -t$, $x \in \mathbb{R}_+$, we have

$$\gamma(t(x)) = \frac{e^{\alpha x} - 1}{e^x - 1}$$

and

$$\gamma'(t(x)) = \frac{\alpha e^{\alpha x}(e^x - 1) - e^x(e^{\alpha x} - 1)}{(e^x - 1)^2}.$$

Now we need to show that the function $\gamma(t(x))$ is strictly decreasing, or that $\gamma'(t(x)) < 0$. This is equivalent to

$$e^x(e^{\alpha x} - 1) - \alpha e^{\alpha x}(e^x - 1) > 0$$

or to

$$(1 - \alpha)(e^x - 1) - (e^{(1-\alpha)x} - 1) > 0,$$

or to

$$\sum_{k=1}^{\infty}[(1 - \alpha) - (1 - \alpha)^k]\frac{x^k}{k!} > 0.$$

Again the last inequality is absolutely evident because of $\alpha \in (0,1)$. $\qquad\square$

Let now $a_0(\theta) = \frac{\theta}{\pi}$. This function $a_0(\theta)$ is obviously real-valued and continuous on $[0, \pi]$.

Corollary 7.2.7. *The C^*-algebra $\mathcal{T}(H(L(1, a_0)))$ is isomorphic and isometric to $C(\overline{\mathbb{R}})$.*

Proof. We have

$$
\begin{aligned}
\gamma_{a_0}(\lambda) &= \frac{2\lambda}{\pi(1 - e^{-2\pi\lambda})} \int_0^\pi \theta e^{-2\lambda\theta}\, d\theta \\
&= \frac{1}{\pi(1 - e^{-2\pi\lambda})} \left[-\pi e^{-2\pi\lambda} - \frac{1}{2\lambda}(e^{-2\pi\lambda} - 1) \right] = \frac{1}{2\pi\lambda} - \frac{1}{e^{2\pi\lambda} - 1}.
\end{aligned}
$$

The function $\gamma_{a_0}(\lambda)$ is continuous on $\overline{\mathbb{R}}$, and

$$
\begin{aligned}
\lim_{\lambda \to 0} \gamma_{a_0}(\lambda) &= \gamma_{a_0}(0) = \frac{1}{2}, \\
\lim_{\lambda \to -\infty} \gamma_{a_0}(\lambda) &= \gamma_{a_0}(-\infty) = 1, \\
\lim_{\lambda \to +\infty} \gamma_{a_0}(\lambda) &= \gamma_{a_0}(+\infty) = 0.
\end{aligned}
$$

To finish the proof we need to show that the function $\gamma_{a_0}(\lambda)$ separates the points of $\overline{\mathbb{R}}$. To do this we show that the function

$$
\gamma(t) = \frac{1}{t} - \frac{1}{e^t - 1}, \qquad t \in \mathbb{R}
$$

is strictly decreasing, or that $\gamma'(t) < 0$ for all $t \neq 0$.

The function

$$
\gamma'(t) = -\frac{1}{t^2} + \frac{e^t}{(e^t - 1)^2}
$$

is even. Thus it is sufficient to prove that

$$
\frac{e^t}{(e^t - 1)^2} < \frac{1}{t^2},
$$

or that

$$
t^2 e^t < (e^t - 1)^2,
$$

for each $t > 0$. The last inequality is easy to check comparing corresponding coefficients of the power series

$$
\begin{aligned}
t^2 e^t &= \sum_{n=2}^\infty \frac{1}{(n-2)!} t^n, \\
(e^t - 1)^2 &= \sum_{n=2}^\infty \frac{2(2^{n-1} - 1)}{n!} t^n.
\end{aligned}
$$

$\square$

Remark 7.2.8. The above statements show that in spite of the fact that the generating sets of defining symbols in (7.2.3) are quite different, the resulting Toeplitz C^*-algebras are the same. Moreover, this (common) C^*-algebra with identity can be generated by a *single* Toeplitz operator with either *continuous*, or *piece-wise constant* defining symbol. Further, although the algebraic operations with Toeplitz operators do not give a Toeplitz operator, in general, the resulting (single-generated) algebra is extremely rich on Toeplitz operators: *each Toeplitz operator with defining symbol from $H(L_\infty^{(0,\pi)}([0,\pi]))$ belongs to this algebra.*

Given a finite number of different points on $[0,\pi]$,

$$0 = \theta_0 < \theta_1 < \theta_2 < \ldots < \theta_{n-1} < \theta_n = \pi,$$

we consider the algebra $PCo([0,\pi],\Lambda)$, with $\Lambda = \{\theta_1,\theta_2,\ldots,\theta_{n-1}\}$, and the algebra $H(PCo([0,\pi],\Lambda))$. Each (piece-wise constant) function $a \in H(PCo([0,\pi],\Lambda))$ has obviously n limit values at the origin.

Denote by V_k, $k = 1,2,\ldots,n$, the cone (angle) on the upper half-plane Π, supported on $(\theta_{k-1},\theta_k]$. Then the n-dimensional algebra $H(PCo([0,\pi],\Lambda))$ consists of all functions of the form

$$a(z) = a_1\chi_{V_1}(z) + a_2\chi_{V_2}(z) + \ldots + a_n\chi_{V_n}(z),$$

where $(a_1,a_2,\ldots,a_n) \in \mathbb{C}^n$, and $\chi_k(z)$ are the characteristic functions of the cones V_k, $k = 1,2,\ldots,n$.

The Toeplitz C^*-algebra $\mathcal{T}(H(PCo([0,\pi],\Lambda)))$ is obviously generated by n *commuting* Toeplitz operators $T_{\chi_{V_k}}$, $k = 1,2,\ldots,n$. We have

$$\gamma_{\chi_{V_k}}(\lambda) = \frac{2\lambda}{1 - e^{-2\pi\lambda}} \int_{\theta_{k-1}}^{\theta_k} e^{-2\lambda\theta}\, d\theta = \frac{e^{-2\theta_k\lambda} - e^{-2\theta_{k-1}\lambda}}{e^{-2\pi\lambda} - 1}, \qquad \lambda \in \mathbb{R}. \qquad (7.2.5)$$

Each function $\gamma_{\chi_{V_k}}$ is continuous on $\overline{\mathbb{R}}$ and

$$\lim_{\lambda\to-\infty} \gamma_{\chi_{V_1}}(\lambda) = 0, \qquad \lim_{\lambda\to+\infty} \gamma_{\chi_{V_1}}(\lambda) = 1,$$

$$\lim_{\lambda\to-\infty} \gamma_{\chi_{V_k}}(\lambda) = 0, \qquad \lim_{\lambda\to+\infty} \gamma_{\chi_{V_k}}(\lambda) = 0, \qquad k = 2,3,\ldots,n-1,$$

$$\lim_{\lambda\to-\infty} \gamma_{\chi_{V_n}}(\lambda) = 1, \qquad \lim_{\lambda\to+\infty} \gamma_{\chi_{V_n}}(\lambda) = 0.$$

Furthermore, each function $\gamma_{\chi_{V_k}}$ is non-negative and

$$\sum_{k=0}^{n} \gamma_{\chi_{V_k}}(\lambda) \equiv 1.$$

Thus the set

$$\Delta(\Lambda) = \{t = (t_1,t_2,\ldots,t_n) : t_k = \gamma_{\chi_{V_k}}(\lambda), \ \lambda \in \overline{\mathbb{R}}, \ k = 1,\ldots,n\} \qquad (7.2.6)$$

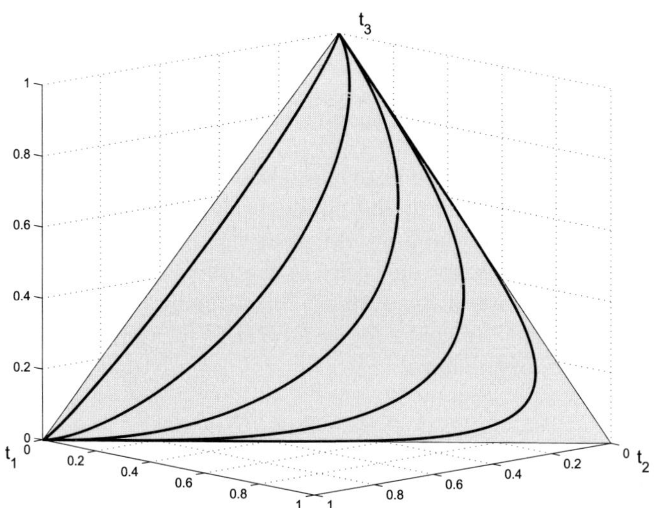

Figure 7.1: The angles (θ_1, θ_2) left to right: $(0.48\pi, 0.52\pi)$, $(0.4\pi, 0.6\pi)$, $(0.3\pi, 0.7\pi)$, $(0.2\pi, 0.8\pi)$, $(0.1\pi, 0.9\pi)$.

is a continuous curve lying on the standard $(n-1)$-dimensional simplex, and connecting the vertices $(1, 0, \ldots, 0)$ and $(0, \ldots, 0, 1)$.

On Figure 7.1 we present the behaviour of the joint spectrum $\Delta(\Lambda)$ for the case n=3 in dependence of the angles (θ_1, θ_2).

Theorem 7.2.9. *Given a set* $\Lambda = \{\theta_1, \theta_2, \ldots, \theta_{n-1}\}$, *the Toeplitz C^*-algebra* $\mathcal{T}(H(PCo([0, \pi], \Lambda)))$ *is isomorphic and isometric to* $C(\Delta(\Lambda))$. *The isomorphism*

$$\tau : \mathcal{T}(H(PCo([0, \pi], \Lambda))) \longrightarrow C(\Delta(\Lambda))$$

is generated by the following mapping of generators of $\mathcal{T}(H(PCo([0, \pi], \Lambda)))$: *if* $a(z) = a_1 \chi_{V_1}(z) + a_2 \chi_{V_2}(z) + \ldots + a_n \chi_{V_n}(z)$, *then*

$$\tau : T_a \longmapsto a_1 t_1 + a_2 t_2 + \ldots + a_n t_n,$$

where $t = (t_1, t_2, \ldots, t_n) \in \Delta(\Lambda)$.

Proof. The C^*-algebra $\mathcal{T}(H(PCo([0, \pi], \Lambda)))$ is commutative, and is generated by n generators $T_{\chi_{V_k}}$, $k = 1, 2, \ldots, n$. Thus it is isomorphic and isometric to the algebra of all continuous functions on the compact set of the maximal ideals of the algebra $\mathcal{T}(H(PCo([0, \pi], \Lambda)))$. This compact set is nothing but the joint spectrum of the above generators and coincides obviously with $\Delta(\Lambda)$. $\square$

7.3 Bergman projection and homogeneous functions

We start again with a finite number of different points on $[0, \pi]$:

$$0 = \theta_0 < \theta_1 < \theta_2 < \ldots < \theta_{n-1} < \theta_n = \pi.$$

Consider the algebra $PC([0, \pi], \Lambda)$ of piece-wise continuous functions, with $\Lambda = \{\theta_1, \theta_2, \ldots, \theta_{n-1}\}$, and the coresponding algebra $H(PC([0, \pi], \Lambda))$. Each function $a \in H(PC([0, \pi], \Lambda))$ has radial limit values at the origin which are defined by the restriction of a onto the upper half part of the unit circle and are parameterized by the point of the closure of the segment $[0, \pi]$ cut by points of Λ.

We introduce the C^*-algebra $\mathcal{R}_\Lambda = \mathcal{R}(H(PC([0, \pi], \Lambda)), B_\Pi)$ which is generated by all operators of the form

$$A = a(z)I + b(z)B_\Pi,$$

where $a, b \in H(PC([0, \pi], \Lambda))$, acting on the space $L_2(\Pi)$.

The algebra $\mathcal{R}_\Lambda$ is isomorphically isometric to the algebra

$$\mathcal{R}_1 = (M \otimes I)\mathcal{R}_\Lambda(M^{-1} \otimes I),$$

which is generated by the following operators, acting on $L_2(\mathbb{R}) \otimes L_2(0, \pi)$,

$$\begin{aligned}
(M \otimes I)B_\Pi(M^{-1} \otimes I) &= I \otimes B(\lambda), \\
(M \otimes I)a(z)I(M^{-1} \otimes I) &= I \otimes a(\theta)I,
\end{aligned}$$

where the one-dimensional projection $B(\lambda)$ is given by (7.1.3), $\lambda \in \mathbb{R}$, M is the Mellin transform (7.1.1), and $a(\theta) \in PC([0, \pi], \Lambda)$ is the restriction of the function $a(z) \in H(PC([0, \pi], \Lambda))$ onto the upper half part of the unit circle parameterized by angle θ.

That is, the algebra $\mathcal{R}_1$ splits onto the direct integral of C^*-algebras $\mathcal{R}_1(\lambda)$, $\lambda \in \mathbb{R}$, generated by the following operators, acting on $L_2(0, \pi)$,

$$B(\lambda) \quad \text{and} \quad aI, \quad a \in PC([0, \pi], \Lambda).$$

Lemma 7.3.1. *For each $\lambda \in \mathbb{R}$ the C^*-algebra $\mathcal{R}_1(\lambda)$ is irreducible and contains the ideal $\mathcal{K}$ of all compact operators on $L_2(0, \pi)$. The algebra $\mathcal{R}_1(\lambda)$ consists of all operators of the form*

$$A = aI + K, \quad a \in PC([0, \pi], \Lambda), \quad K \in \mathcal{K},$$

and its Calkin algebra $\widehat{\mathcal{R}_1}(\lambda) = \mathcal{R}_1(\lambda)/\mathcal{K}$ is isomorphic to $PC([0, \pi], \Lambda)$.

Proof. We prove first that the algebra $\mathcal{R}_1(\lambda)$ does not have any non-trivial invariant subspace. Each non-trivial invariant subspace of the subalgebra

$$\{aI : a \in PC([0, \pi], \Lambda)\}$$

of the algebra $\mathcal{R}_1(\lambda)$ obviously has the form

$$X_M = \{\psi_M(\theta) = \chi_M(\theta)\psi(\theta) : \psi \in L_2(0,\pi)\},$$

where χ_M is the characteristic function of a measurable subset M of $[0,\pi]$ having positive measure. That is the the algebra $\mathcal{R}_1(\lambda)$ has an invariant subspace if and only if there exists a measurable set M of positive measure such that $B(\lambda)\psi_M \in X_M$, for all $\psi_M \in X_M$. We have

$$B(\lambda)\psi_M = \langle \psi_M, \ell_\lambda \rangle \ell_\lambda.$$

The function ℓ_λ (7.1.2) is always non-zero on $[0,\pi]$, thus $\langle \psi_M, \ell_\lambda \rangle$ has to be zero on $[0,1] \setminus M$ for all $\psi_M \in X_M$. But for $\psi_M = \chi_M \ell_\lambda$ we have

$$\langle \psi_M \ell_\lambda, \ell_\lambda \rangle = \langle \psi_M \ell_\lambda, \psi_M \ell_\lambda \rangle = \|\psi_M \ell_\lambda\|^2.$$

Thus χ_M has to be zero a.e. on $[0,\pi]$, or M must be a zero measure set. That is, the algebra $\mathcal{R}_1(\lambda)$ does not have any non-trivial invariant subspace, and thus is irreducible.

Further, being irreducible the algebra $\mathcal{R}_1(\lambda)$ contains non-trivial compact operators, for example $B(\lambda)$; thus (see, for example, [148] Theorem 2.4.9) it contains the whole ideal $\mathcal{K}$ of compact operators.

The last statement of the lemma is obvious. $\qquad\square$

We note that all the algebras $\mathcal{R}_1(\lambda)$, $\lambda \in \mathbb{R}$, are the same and coincide with the algebra $\mathcal{R}_{pc}$ which consists of all operators of the form $A = aI + K$, where $a \in PC([0,\pi],\Lambda)$, and $K \in \mathcal{K}$. Apart from the infinite dimensional identical irreducible representation, this common C^*-algebra $\mathcal{R}_{pc}$ has one-dimensional irreducible representations. To describe them, we mention first that the compact set of maximal ideals of the algebra $PC([0,\pi],\Lambda)$ can be identified with the closure of the segment $[0,\pi]$ cut by points of Λ,

$$\widehat{[0,\pi]} = [0,\theta_1^-] \sqcup [\theta_1^+,\theta_2^-] \sqcup \ldots \sqcup [\theta_{n-1}^+,\pi].$$

Given a point $\theta \in \widehat{[0,\pi]}$, the one-dimensional irreducible representation $\pi(\theta)$ of the algebra $PC([0,\pi],\Lambda)$ is defined by

$$\pi(\theta) : a \longmapsto \begin{cases} a(\theta) \in \mathbb{C}, & \text{for } \theta \in [0,\pi] \setminus \Lambda \subset \widehat{[0,\pi]} \\ a(\theta_k^\pm) = \lim_{\theta \to \theta_k \pm 0} a(\theta) \in \mathbb{C}, & \text{for } \theta_k^\pm \in \widehat{[0,\pi]} \end{cases},$$

where $k = 1, 2, \ldots, n-1$.

Now the one-dimensional irreducible representations π_θ of the algebra $\mathcal{R}_{pc}$ parameterized by points $\theta \in \widehat{[0,\pi]}$ are

$$\pi_\theta : \mathcal{R}_{pc} \longrightarrow \mathcal{R}_{pc}/\mathcal{K} \cong PC([0,\pi],\Lambda) \xrightarrow{\pi(\theta)} \mathbb{C}.$$

Lemma 7.3.2. *For each $\lambda \in \mathbb{R}$ the homomorphism*

$$\nu_\lambda : \mathcal{R}_\Lambda \cong \mathcal{R}_1 \longrightarrow \mathcal{R}_1(\lambda) = \mathcal{R}_{pc}$$

gives an (infinite dimensional) irreducible representation of the algebra $\mathcal{R}_\Lambda = \mathcal{R}(H(PC([0,\pi],\Lambda)), B_\Pi)$, and for different $\lambda \in \mathbb{R}$ the representations ν_λ are not unitary equivalent.

Proof. The first statement follows directly from the previous lemma. Consider now two distinct real numbers $\lambda_1 \neq \lambda_2$. Recall that for $a_0(\theta) = \frac{\theta}{\pi}$, used in Corollary 7.2.7, the corresponding function $\gamma_{a_0}(\lambda)$ separates the points of $\mathbb{R}$. That is, in particular, $\gamma_{a_0}(\lambda_1) \neq \gamma_{a_0}(\lambda_2)$.

Then for the operator $A = B_\Pi(\gamma_{a_0}(\lambda_1) - a_0(\theta))B_\Pi \in \mathcal{R}_\Lambda$ we have

$$
\begin{aligned}
\nu_{\lambda_1} &: \quad A \longmapsto (\gamma_{a_0}(\lambda_1) - \gamma_{a_0}(\lambda_1))B(\lambda_1) = 0, \\
\nu_{\lambda_2} &: \quad A \longmapsto (\gamma_{a_0}(\lambda_1) - \gamma_{a_0}(\lambda_2))B(\lambda_2) \neq 0.
\end{aligned}
$$

Thus the representations ν_{λ_1} and ν_{λ_2} are not unitary equivalent. $\qquad\square$

Theorem 7.3.3. *The algebra $\mathcal{R}_\Lambda = \mathcal{R}(H(PC([0,\pi],\Lambda)), B_\Pi)$ does not contain any non-zero compact operator. All its infinite dimensional irreducible representations are parameterized by points $\lambda \in \mathbb{R}$ and are of the form*

$$\nu_\lambda : \mathcal{R}_\Lambda \cong \mathcal{R}_1 \longrightarrow \mathcal{R}_1(\lambda) = \mathcal{R}_{pc},$$

where

$$
\begin{aligned}
\nu_\lambda &: \quad a(z)I \in \mathcal{R}_\Lambda \longmapsto a(\theta)I \in \mathcal{R}_{pc}, \\
\nu_\lambda &: \quad B_\Pi \in \mathcal{R}_\Lambda \longmapsto B(\lambda) \in \mathcal{R}_{pc}.
\end{aligned}
$$

All finite dimensional irreducible representations of the algebra $\mathcal{R}_\Lambda$ are one-dimensional, parameterized by points $\theta \in \widehat{[0,\pi]}$, and are the composition of (any) representation ν_λ with one-dimensional representation ν_θ of the algebra $\mathcal{R}_{pc}$:

$$\nu_\theta : \mathcal{R}_\Lambda \cong \mathcal{R}_1 \xrightarrow{\ \nu_\lambda\ } \mathcal{R}_1(\lambda) = \mathcal{R}_{pc} \xrightarrow{\ \pi_\theta\ } \mathbb{C}.$$

Proof. Follows directly from the two previous lemmas. $\qquad\square$

We consider now two subalgebras of $\mathcal{R}_\Lambda = \mathcal{R}(H(PC([0,\pi],\Lambda)), B_\Pi)$ which are of special interest in what follows. For the first subalgebra $\mathcal{R}(H(C([0,\pi])), B_\Pi)$ we have obviously the following statement.

Denote by $\mathcal{R}_c$ the C^*-algebra consisting of all operators of the form $a(\theta)I + K$, where $a \in C([0,\pi])$ and K is compact, acting on $L_2(0,\pi)$.

Theorem 7.3.4. *The C^*-algebra $\mathcal{R}(H(C([0,\pi])), B_\Pi)$ is isomorphic to a certain subalgebra of $C_b(\mathbb{R}, \mathcal{R}_c)$. The isomorphic imbedding*

$$\nu : \mathcal{R}(H(C([0,\pi])), B_\Pi) \longrightarrow C_b(\mathbb{R}, \mathcal{R}_c)$$

is generated by the following mapping of the generators: if $A = aI + bB_\Pi$, where $a, b \in H(C([0, \pi]))$, then

$$\nu : A \longmapsto A(\lambda) = aI + bB(\lambda) \in \mathcal{R}_c, \qquad \lambda \in \mathbb{R},$$

where a and b are considered as the restrictions of the functions from $H(C([0, \pi]))$ onto $[0, \pi]$.

All infinite dimensional irreducible representations of the C^*-algebra

$$\mathcal{R}(H(C([0, \pi])), B_\Pi)$$

are parameterized by points $\lambda \in \mathbb{R}$ and are generated by the mapping

$$\nu_\lambda : A \in \mathcal{R}(H(C([0, \pi])), B_\Pi) \longrightarrow A(\lambda) \in \mathcal{R}_c.$$

The set of all finite dimensional irreducible representations of $\mathcal{R}(H(C([0, \pi])), B_\Pi)$ is one-dimensional, parameterized by points $\theta \in [0, \pi]$, and generated by the mapping

$$\nu_\theta : A = aI + bB_\Pi \in \mathcal{R}(H(C([0, \pi])), B_\Pi) \longrightarrow a(\theta) \in \mathbb{C}.$$

The structure of the second subalgebra $\mathcal{R}(H(PCo([0, \pi], \Lambda)), B_\Pi)$ is quite different. This algebra can also be defined by other generators, which are the following $n + 1$ orthogonal projections, acting on $L_2(\Pi)$,

$$P = B_\Pi \qquad \text{and} \qquad Q_k = \chi_{V_k} I, \qquad k = 1, 2, \ldots, n,$$

where V_k is the cone in Π supported on $(\theta_{k-1}, \theta_k]$.

To describe the algebra $\mathcal{R}(H(PCo([0, \pi], \Lambda)), B_\Pi) = \mathcal{R}(P, Q_1, \ldots, Q_n)$ we use the results of Section 1.2. The above projections P, Q_1, ..., Q_n are obviously all-but-one and satisfy the property (1.2.13). At the same time the property (1.2.14) does not hold for each $k = 1, \ldots, n$. That is, for each $k = 1, \ldots, n$, the intersection $\operatorname{Im} Q_k \cap \ker P$ is not trivial. We show this by a bit lengthy but simple procedure.

We start with the representations

$$(M \otimes I)P(M^{-1} \otimes I) = I \otimes B(\lambda),$$
$$(M \otimes I)Q_k(M^{-1} \otimes I) = I \otimes \chi_{(\theta_{k-1}, \theta_k]} I,$$

where the one-dimensional projection $B(\lambda)$, $\lambda \in \mathbb{R}$, is given by (7.1.3).

First we describe the intersection

$$\operatorname{Im} \chi_{(\theta_{k-1}, \theta_k]} I \cap \ker B(\lambda).$$

The function $\ell_{\lambda,k} = \ell_\lambda|_{(\theta_{k-1}, \theta_k]} = \chi_{(\theta_{k-1}, \theta_k]} \ell_\lambda$ is obviously in $L_2(\theta_{k-1}, \theta_k)$, and, see (7.2.5), $\|\ell_{\lambda,k}\|^2 = \gamma_{\chi_{V_k}}(\lambda)$, which is continuous and bounded as a function of $\lambda \in \mathbb{R}$.

For each $g \in L_2(\theta_{k-1}, \theta_k)$ the function

$$g_\lambda = \|\ell_{\lambda,k}\|^2 g - \langle g, \ell_{\lambda,k} \rangle \ell_{\lambda,k}$$

belongs to $g \in L_2(\theta_{k-1}, \theta_k) = \operatorname{Im} \chi_{(\theta_{k-1}, \theta_k]} I$, and

$$\|g_\lambda\| \leq \|\ell_{\lambda,k}\|^2 \cdot \|g\| + |\langle g, \ell_{\lambda,k}\rangle| \cdot \|\ell_{\lambda,k}\| \leq 2\|\ell_{\lambda,k}\|^2 \cdot \|g\| \leq C \cdot \|g\|,$$

where the constant C does not depend on λ.

Then, $g_\lambda \in \ker B(\lambda)$, because of

$$\langle g_\lambda, \ell_\lambda \rangle_{L_2(0,\pi)} = \langle g_\lambda, \ell_{\lambda,k}\rangle_{L_2(\theta_{k-1}, \theta_k)} = 0.$$

Thus $g_\lambda \in \operatorname{Im} \chi_{(\theta_{k-1}, \theta_k]} I \cap \ker B(\lambda)$.

Finally, for any pair $f \in L_2(\mathbb{R})$ and $g \in L_2(\theta_{k-1}, \theta_k) \subset L_2(0, \pi)$, the function

$$\varphi(z) = \varphi(re^{i\theta}) = \left[(M^{-1} \otimes I)(f(\lambda)g_\lambda(\theta))\right](r, \theta)$$

belongs to $\operatorname{Im} Q_k \cap \ker P$.

That is, describing the algebra $\mathcal{R}(H(PCo([0, \pi], \Lambda)), B_\Pi) = \mathcal{R}(P, Q_1, \ldots, Q_n)$ we are in the situation of the last part of Section 1.2.6. The corresponding positive operators $C_k = PQ_kP = T_{\chi_{V_k}} \in \mathcal{T}(H(PCo([0, \pi], \Lambda)))$, $k = 1, \ldots, n$, commute. Thus to describe the algebra $\mathcal{R}(H(PCo([0, \pi], \Lambda)), B_\Pi) = \mathcal{R}(P, Q_1, \ldots, Q_n)$ we can use Theorem 1.2.33. The joint spectrum of the operators $C_k = T_{\chi_{V_k}}$, $k = 1, \ldots, n$, is obviously $\Delta(\Lambda)$ given by (7.2.6). The algebra $\mathfrak{S}(\Delta)$ used in Theorem 1.2.33 coincides in our case with the algebra $\mathfrak{S}(\Delta(\Lambda))$ of all $n \times n$ matrix-functions continuous on $\Delta(\Lambda)$ and diagonal at the vertices $(1, 0, \ldots, 0)$, $(0, \ldots, 0, 1)$.

Thus, as a direct corollary from Theorem 1.2.33 we have

Theorem 7.3.5. *The C^*-algebra $\mathcal{R}(H(PCo([0, \pi], \Lambda)), B_\Pi)$ is isomorphic and isometric to a subalgebra of the algebra $\mathfrak{S}(\Delta(\Lambda)) \oplus \mathbb{C}^n$, where $\Delta(\Lambda)$ is given by (7.2.6). The isomorphic imbedding*

$$\nu : \mathcal{R}(H(PCo([0, \pi], \Lambda)), B_\Pi) \longrightarrow \mathfrak{S}(\Delta) \oplus \mathbb{C}^n$$

is generated by the following mapping of the generators of the algebra $\mathcal{R}(H(PCo([0, \pi], \Lambda)), B_\Pi)$:

$$\nu : B_\Pi \longmapsto (p(t), (0, 0, \ldots, 0)),$$
$$\nu : \chi_{V_k} I \longmapsto (q_k(t), (0, \ldots, 0, \underset{k\text{-}place}{1}, 0, \ldots, 0)), \quad k = 1, \ldots, n,$$

where

$$p(t) = \left(\sqrt{t_j t_k}\right)_{j,k=1}^n,$$
$$q_k(t) = \operatorname{diag}(0, \ldots, 0, \underset{k\text{-}place}{1}, 0, \ldots, 0),$$

and $t = (t_1, \ldots, t_m) \in \Delta(\Lambda)$.

7.4 Algebra generated by the Bergman projection and discontinuous coefficients

The C^*-algebra generated by the Bergman projection and multiplication operators by piece-wise continuous functions having two limit values at boundary points of discontinuity has been described in Section 2.7. The more difficult part, the characterization of the local algebra corresponding to a boundary point of discontinuity, was based on the general description of the C^*-algebra generated by two orthogonal projections (see Sections 2.6 and 1.2.2). The next natural step, the description of the C^*-algebra generated by the Bergman projection and multiplication operators by piece-wise continuous functions having *more* than two limit values at boundary points of discontinuity remained open for a very long period. The main reason was the absence of a general description of the C^*-algebra generated by *more* than two orthogonal projections. Although the local algebra at the boundary point of discontinuity involves a *special* set of projections (P and $Q_1, \ldots, Q_n$, with $Q_1 + \ldots + Q_n = I$), which are all-but-one in terminology of Section 1.2, the C^*-algebra generated by such projections is still wild.

In spite of this, in the previous section we described in fact a model case of the algebra generated by the Bergman projection and functions having $n > 2$ limit values at a boundary point. This was made possible by the following reasons. The concrete all-but-one projections $P = B_\Pi$ and $Q_k = \chi_{V_k} I$, $k = 1, \ldots, n$, involved generate the special positive operators $C_k = T_{\chi_{V_k}}$, which happen to belong to a commutative algebra of Toeplitz operators. The C^*-algebra generated by all-but-one orthogonal projections, for which the corresponding positive operators C_k are pair-wise commuting, can be described.

Note that in fact we already described such an algebra in Theorem 5.4.2, as a model case, and in Section 5.5, as a general case. We did not underline this fact at that time because, first, the result was used for other purposes, and, second, the type of discontinuity was different from the type one would expect as a natural development of a two-limit value case of Section 2.7.

We proceed now with the description of the C^*-algebra generated by the Bergman projection and multiplication operators by piece-wise continuous functions having more than two limit values at boundary points of discontinuity.

As in Section 2.7 we start by introducing of a curve of function discontinuities. Denote by ℓ a piece-wise smooth curve on the closed unit disk $\overline{\mathbb{D}}$ satisfying the following properties: there are a finite number of points (nodes), which divide ℓ onto simple oriented smooth curves ℓ_j, $j = \overline{1, k}$. We assume that the endpoints of ℓ are among the nodes. Denote by U the set of all nodes of the curve ℓ which do not belong to the boundary γ. We will refer to nodes using symbols u_{q,r_q}, where r_q is a number of lines meeting at this node, and q corresponds to the node numbering. Denote by T the set of all nodes from $\ell \cap \gamma$, and assume that T consists of m points. For each node $t_{q,r_q-1} \in T$ there are $r_q - 1$ curves meeting at t_{q,r_q-1}, $q = 1, \ldots, m$. We assume as well that locally near t_{q,r_q-1} these curves are hypercycles (see Section 9.4), that is, there is a Möbius transformation of the unit disk to the upper half-plane under which the node t_{q,r_q-1} goes to the origin and the curves meeting at t_{q,r_q-1} are mapped to curves which locally near the origin are straight line segments meeting at the origin.

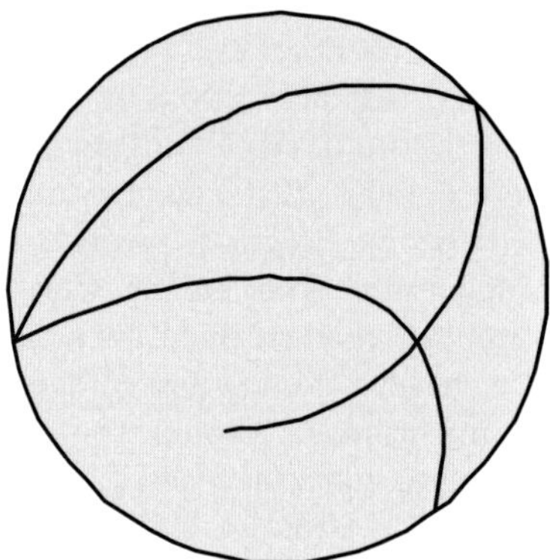

Figure 7.2: An example of a curve ℓ.

Denote by $PC(\overline{\mathbb{D}}, \ell)$ the algebra of all functions $a(z)$ continuous in $\overline{\mathbb{D}} \setminus \ell$ and having left and right limit values at all points of ℓ_j: $a^+(z)$ and $a^-(z)$. At the nodes of the type $u_{q,r_q} \in U$ the functions from $PC(\overline{\mathbb{D}}, \ell)$ have r limit values. We denote them by $a^{(1)}(u_{q,r_q})$, ..., $a^{(r)}(u_{q,r_q})$. At the nodes of the type $t_{q,r_q-1} \in T$ the functions from $PC(\overline{\mathbb{D}}, \ell)$ have r limit values, as well. We denote them by $a^{(1)}(t_{q,r_q-1})$, ..., $a^{(r)}(t_{q,r_q-1})$, counting counter-clockwise.

We study the algebra $\mathcal{R} = \mathcal{R}(PC(\overline{\mathbb{D}}, \ell), B_{\mathbb{D}})$ generated by the operators acting on the space $L_2(\mathbb{D})$ of the form

$$A = a(z)I + b(z)B_{\mathbb{D}},$$

where $a(z), b(z) \in PC(\overline{\mathbb{D}}, \ell)$ and $B_{\mathbb{D}}$ is the Bergman projection on the unit disk.

The algebra $\mathcal{R}$ contains the ideal $\mathcal{K}$ of all compact operators (its subalgebra $\mathcal{R}(C(\overline{\mathbb{D}}), B_{\mathbb{D}})$ already does), and is irreducible.

To describe the Fredholm symbol algebra $\operatorname{Sym}\mathcal{R}$ of $\mathcal{R} = \mathcal{R}(PC(\overline{\mathbb{D}}, \ell), B_{\mathbb{D}})$ we use the Douglas-Varela local principle. By Lemma 2.4.4 the C^*-algebra $\pi(C(\overline{\mathbb{D}})) \cong C(\overline{\mathbb{D}})$ is a central commutative subalgebra of $\widehat{\mathcal{R}} = \operatorname{Sym}\mathcal{R}$, thus we localize by the points of $\overline{\mathbb{D}}$. As in Section 2.7 there are five different cases of local algebras $\widehat{\mathcal{R}}(t_0)$ depending on a point $t_0 \in \overline{D}$. The first four are the same as in Section 2.7.

a) Let $t_0 \in \overline{\mathbb{D}} \setminus (\gamma \cup \ell)$. Then $\widehat{\mathcal{R}}(t_0) \cong \mathbb{C}$, and for the operator $A = a(z)I + b(z)B_{\mathbb{D}}$ we have $\pi_{t_0}(A) = a(t_0)$.

b) Let $t_0 \in \gamma \setminus \ell$. Then $\widehat{\mathcal{R}}(t_0) \cong \mathbb{C}^2$, and for the operator $A = a(z)I + b(z)B_{\mathbb{D}}$ we have $\pi_{t_0}(A) = (a(t_0), a(t_0) + b(t_0))$.

c) Let $t_0 \in \ell \setminus (U \cup T)$. Then $\widehat{B}_{\mathbb{D}}(t_0) = 0$ and $a_1(z) \overset{t_0}{\sim} a_2(z)$ if and only if $a_1^+(t_0) = a_2^+(t_0)$ and $a_1^-(t_0) = a_2^-(t_0)$. Thus $\widehat{\mathcal{R}}(t_0) \cong \mathbb{C}^2$, and for the operator $A = a(z)I + b(z)B_{\mathbb{D}}$ we have $\pi_{t_0}(A) = (a^+(t_0), a^-(t_0))$.

d) Let $t_0 \in U$ and $t_0 = u_q^r$ for some q. Then again $\widehat{B}_{\mathbb{D}}(t_0) = 0$, but a function $a(z) \in PC(\overline{D}, \ell)$ now has r limit values at the point t_0. Thus $\widehat{\mathcal{R}}(t_0) \cong \mathbb{C}^r$, and for the operator $A = a(z)I + b(z)B_{\mathbb{D}}$ we have $\pi_{t_0}(A) = (a^{(1)}(t_0), \ldots, a^{(r)}(t_0))$.

e) Let finally, $t_0 \in T = \gamma \cap \ell$.

e.1) If there is only one curve meeting at the node $t_0 \in T$, then we are again in the situation of Section 2.7. That is, the local algebra $\widehat{\mathcal{R}}(t_0)$ is isomorphic and isometric to the algebra of all 2×2 matrix-functions continuous on $[0, 1]$ and diagonal at the points 0 and 1. Identifying them, for the operator $A = a(z)I + b(z)B_{\mathbb{D}} \in \mathcal{R} = \mathcal{R}(PC(\overline{\mathbb{D}}, \ell), B_{\mathbb{D}})$ we have

$$\pi_{t_0}(A) = \begin{pmatrix} a(t_0 + 0)x + a(t_0 - 0)(1 - x) & (c(t_0 + 0) - c(t_0 - 0))\sqrt{x(1 - x)} \\ (a(t_0 + 0) - a(t_0 - 0))\sqrt{x(1 - x)} & c(t_0 + 0)(1 - x) + c(t_0 - 0)x \end{pmatrix},$$

where $x \in [0, 1]$, and $c(z) = a(z) + b(z)$.

e.2) Let now $t_0 = t_{q, r_q - 1} \in T$, that is, there are $r_q - 1$ curves meeting at this node. Recall that the functions $a(z) \in PC(\overline{\mathbb{D}}, \ell)$ have at this node r_q limit values, which we denoted by $a^{(1)}(t_{q, r_q - 1}), \ldots, a^{(r)}(t_{q, r_q - 1})$, counting counter-clockwise. Introduce the ordered set

$$\Lambda_q = \{\theta_1, \theta_2, \ldots, \theta_{r_q - 1}\}$$

of the angles formed by the above $r_q - 1$ curves meeting at the node $t_0 = t_{q, r_q - 1} \in \gamma$, counting them counter-clockwise.

By our hypotheses there exists a Möbius transformation ω_{t_0} from the unit disk onto the upper half-plane mapping the node t_0 to the origin and mapping the $r_q - 1$ curves meeting at t_0 to the $r_q - 1$ lines, which locally near the origin, are the straight line segments meeting at the origin. Note that these straight line segments have the same angles $\theta_1, \theta_2, \ldots, \theta_{r_q - 1}$ with the real axis.

Under the above Möbius transformation ω_{t_0} the local algebra $\widehat{\mathcal{R}}(t_0)$ of the initial algebra $\mathcal{R}(PC(\overline{\mathbb{D}}, \ell), B_{\mathbb{D}})$ is obviously isomorphic to the local algebra $\widehat{\mathcal{R}}'(0)$ at the origin $0 = w_{t_0}(t_0)$ of the algebra $\mathcal{R}' = \mathcal{R}'(PC(\overline{\Pi}, w_{t_0}(\ell)), B_{\Pi})$, obtained by the unitary equivalence with the initial algebra $\mathcal{R} = \mathcal{R}(PC(\overline{\mathbb{D}}, \ell), B_{\mathbb{D}})$. Observe now that the local algebra $\widehat{\mathcal{R}}'(0)$ is generated by the all-but-one projections

$$P = B_{\Pi} \quad \text{and} \quad Q_k = \chi_{V_k} I, \quad k = 1, \ldots, r_q,$$

and thus it is nothing but the algebra $\mathcal{R}(H(PCo([0, \pi], \Lambda_q)), B_{\Pi})$, which was described in the previous section.

Thus by Theorem 7.3.5 with $n = r_q$ we have

Theorem 7.4.1. *The local algebra $\widehat{\mathcal{R}}(t_0)$ is isomorphic and isometric to a subalgebra of the algebra $\mathfrak{S}(\Delta(\Lambda_q)) \oplus \mathbb{C}^{r_q}$, where $\Delta(\Lambda_q)$ is given by (7.2.6).*

For the generators $a(z)I$ and $B_{\mathbb{D}}$ of the algebra $\mathcal{R} = \mathcal{R}(PC(\overline{\mathbb{D}}, \ell), B_{\mathbb{D}})$ we have

$$\pi_{t_0}(aI) = \left(\operatorname{diag}(a^{(1)}(t_0), \ldots, a^{(r_q)}(t_0)), (a^{(1)}(t_0), \ldots, a^{(r_q)}(t_0)) \right),$$

$$\pi_{t_0}(B_{\mathbb{D}}) = \left(\left(\sqrt{t_j t_k} \right)_{j,k=1}^{r_q}, (0, 0, \ldots, 0) \right),$$

where $t = (t_1, \ldots, t_{r_q}) \in \Delta(\Lambda_q)$.

As it follows from the theorem, for the operator

$$A = aI + bB_{\mathbb{D}} = a(I - B_{\mathbb{D}}) + cB_{\mathbb{D}} \in \mathcal{R}(PC(\overline{\mathbb{D}}, \ell), B_{\mathbb{D}}),$$

where $c = a + b$, we have $\pi_{t_0}(A) =$

$$\left(\left(a^{(k)}(t_0)(\delta_{jk} - \sqrt{t_j t_k}) + c^{(k)}(t_0)\sqrt{t_j t_k} \right)_{j,k=1}^{r_q}, (a^{(1)}(t_0), \ldots, a^{(r_q)}(t_0)) \right),$$

where $t = (t_1, \ldots, t_{r_q}) \in \Delta(\Lambda_q)$.

In the sequel it is convenient to parameterize the curve $\Delta(\Lambda_q)$ by $x \in [0, 1]$ as follows (compare with (7.2.6)). Let

$$t_k(x) = \gamma_{\chi_{V_k}} \left(\frac{1 - 2x}{\sqrt{1 - (1 - 2x)^2}} \right), \tag{7.4.1}$$

where $x \in [0, 1]$ and $k = 1, \ldots, r_q$, then

$$\Delta(\Lambda_q) = \{ t = (t_1(x), t_2(x), \ldots, t_{r_q}(x)) : \ x \in [0, 1] \}. \tag{7.4.2}$$

Denote by $\mathfrak{S}(r_q)$ the algebra of all $r_q \times r_q$ matrix-functions continuous on $[0, 1]$ and diagonal at the points 0 and 1.

Corollary 7.4.2. *The local algebra $\widehat{\mathcal{R}}(t_0)$ is isomorphic and isometric to a subalgebra of the algebra $\mathfrak{S}(r_q) \oplus \mathbb{C}^{r_q}$. Identifying them, for the operator $A = aI + bB_{\mathbb{D}} = a(I - B_{\mathbb{D}}) + cB_{\mathbb{D}} \in \mathcal{R}(PC(\overline{\mathbb{D}}, \ell), B_{\mathbb{D}})$, where $c = a + b$, we have*

$$\pi_{t_0}(A) = \left(\sigma_A^r(x), (a^{(1)}(t_0), \ldots, a^{(r_q)}(t_0)) \right),$$

where

$$\sigma_A^{r_q}(x) = \left(a^{(k)}(t_0) \left(\delta_{jk} - \sqrt{t_j(x) t_k(x)} \right) + c^{(k)}(t_0)\sqrt{t_j(x) t_k(x)} \right)_{j,k=1}^{r_q} \in \mathfrak{S}(r_q),$$

with $t_k(x)$ given by (7.4.1), and $x \in [0, 1]$.

At the boundary points $\{0,1\}$ of $[0,1]$ the matrix part $\sigma_A^{r_q}(x)$ of $\pi_{t_0}(A)$ becomes diagonal:

$$
\sigma_A^{r_q}(0) \;=\; \begin{pmatrix} c^{(1)}(t_0) & & & & 0 \\ & a^{(2)}(t_0) & & & \\ & & \ddots & & \\ 0 & & & & a^{(r_q)}(t_0) \end{pmatrix},
$$

$$
\sigma_A^{r_q}(1) \;=\; \begin{pmatrix} a^{(1)}(t_0) & & & 0 \\ & \ddots & & \\ & & a^{(r_q-1)}(t_0) & \\ 0 & & & c^{(r_q)}(t_0) \end{pmatrix}.
$$

As follows from the description of the local algebras $\widehat{\mathcal{R}}(t_0)$, the algebra $\operatorname{Sym}\mathcal{R} = \mathcal{R}/\mathcal{K}$ is a C^*-algebra which has finite dimensional representations only. It has one-dimensional representations for the cases a) – d), one- and two-dimensional representations for the case e.1), and one- and r_q-dimensional representations for nodes $t_{q,r_q-1} \in T$ of the case e.2).

Let

$$
1 = n_1 < n_2 < \ldots < n_p
$$

be the ordered set of the dimensions of the representations of the Fredholm symbol algebra $\operatorname{Sym}\mathcal{R}$. To formulate the final result we introduce some notation.

Denote by $\widehat{\mathbb{D}}$ the compactification of the set $\overline{\mathbb{D}}$, cut along the line ℓ. Under that each point from $(\ell \setminus U) \cup (\ell \cap \gamma)$ will correspond to a pair of points in $\widehat{D}$; similarly each node of the form $u_{q,r_q} \in U$ will correspond to r_q points of $\widehat{\mathbb{D}}$, and each node of the form $t_{q,r_q-1} \in T$ will correspond to r_q points of $\widehat{\mathbb{D}}$, which we will denote by $t^{(1)}_{q,r_q-1}, \ldots, t^{(r_q)}_{q,r_q-1}$. For a node $t_{q,1} \in T$, we will denote by $t'_q - 0$ and $t'_q + 0$, following the positive orientation on γ, the pair of points which correspond to this node. The set $\widehat{D}$ introduced coincides obviously with the compact of maximal ideals of the algebra $PC(\overline{\mathbb{D}}, \ell)$. Analogously, denote by $\widehat{\gamma}$ the compactification of the curve γ, cut by nodes $t_{q,r_q-1} \in \ell \cap \gamma$. For a node $t_{q,r_q-1} \in T$, the pair of points which correspond to this node will be denoted by $t''_q - 0$ and $t''_q + 0$, following to the positive orientation on γ.

Let $Y = \widehat{D} \cup \widehat{\gamma}$, and let $\overline{X} = \cup_{q=1}^m \Delta_q$ be a disjoint union of segments $\Delta_q = [0,1]$. Let $\overline{X}_{n_s}$ be the disjoint union of those segments Δ_q which correspond to nodes t_{q,r_q-1} with $r_q = n_s$, $s = 1,2,\ldots,p$. Then of course $\overline{X} = \cup_{s=1}^p \overline{X}_{n_s}$.

Denote by μ the mapping which identifies the points of $\partial \overline{X} = \cup_{q=1}^m \{0_q, 1_q\}$ with a certain finite number of points of Y by the following rule:

if $t_0 = t_{q,1} \in T$, for some q, then

$$
\mu(0_q) = (t'_0 - 0, t''_0 + 0), \quad \text{where} \quad 0_q \in \Delta_q, \ t'_0 - 0 \in \widehat{\mathbb{D}}, \ t''_0 + 0 \in \widehat{\gamma},
$$
$$
\mu(1_q) = (t'_0 + 0, t''_0 - 0), \quad \text{where} \quad 1_q \in \Delta_p, \ t'_0 + 0 \in \widehat{\mathbb{D}}, \ t''_0 - 0 \in \widehat{\gamma};
$$

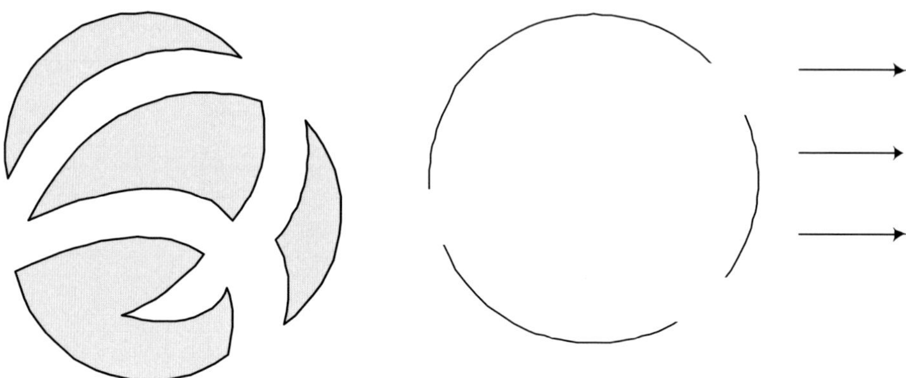

Figure 7.3: The sets $\widehat{\mathbb{D}}$, $\widehat{\gamma}$, and $\overline{X} = \cup_{q=1}^{m}\Delta_q$.

if $t_0 = t_{q,r_q-1} \in T$ with $r_q > 2$, for some q, then

$$\mu(0_q) = (t_0'' + 0, t_0^{(2)}, \ldots, t_0^{(r_q)}), \quad \text{where} \quad 0_q \in \Delta_q, \ t_0^{(2)}, \ldots, t_0^{(r_q)} \in \widehat{\mathbb{D}}, \ t_0'' + 0 \in \widehat{\gamma},$$
$$\mu(1_q) = (t_0^{(1)}, \ldots, t_0^{(r_q-1)}, t_0'' - 0), \quad \text{where} \quad 1_q \in \Delta_p, \ t_0^{(1)}, \ldots, t_0^{(r_q-1)} \in \widehat{\mathbb{D}}, \ t_0'' - 0 \in \widehat{\gamma}.$$

From the descriptions of the local algebras $\widehat{\mathcal{R}}(t_0)$ it follows that $\mathfrak{M} = \overline{X} \cup_\mu Y$ is the spectrum of the C^*-algebra $\widehat{\mathcal{R}} = \operatorname{Sym}\mathcal{R}$.

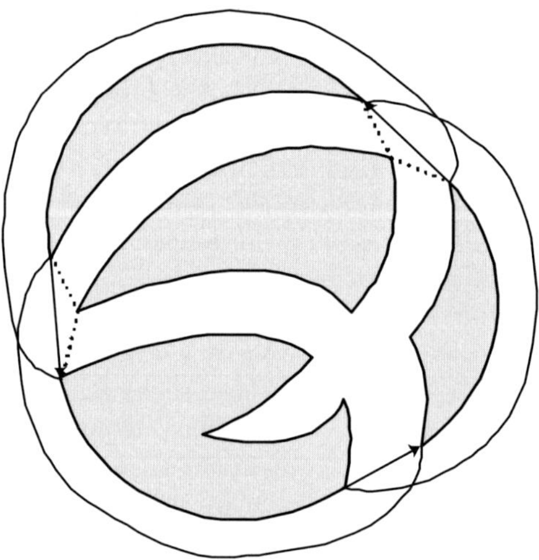

Figure 7.4: The set $\mathfrak{M} = \overline{X} \cup_\mu Y$.

Denote by $\mathfrak{S}$ the algebra of all p-tuples $\sigma = (\sigma_{n_1}, \sigma_{n_2}, \ldots, \sigma_{n_p})$, where $\sigma_{n_1} \in C(Y)$, and $\sigma_{n_s} \in C(\overline{X}_{n_s}, \mathrm{Mat}_{n_s}(\mathbb{C}))$, where $s = 1, \ldots, p$, which satisfy the condition

if $\mu(x_0) = (y_1, y_2, \ldots, y_{n_s})$, $x_0 \in \partial \overline{X}_{n_s}$, $y_1, \ldots, y_{n_s} \in Y$, then

$$\lim_{\substack{x \to x_0 \\ x \in X_{n_s}}} \sigma_{n_s}(x) = \begin{pmatrix} \sigma_1(y_1) & & 0 \\ & \ddots & \\ 0 & & \sigma_1(y_{n_s}) \end{pmatrix}; \qquad (7.4.3)$$

the norm in the algebra $\mathfrak{S}$ is given by

$$\|\sigma\| = \max\{\sup_Y |\sigma_1(y)|, \ \sup_{\overline{X}_{n_s}, s} \|\sigma_{n_s}(x)\|\},$$

where $\|\sigma_{n_s}(x)\|^2$ is the largest eigenvalue of the matrix $\sigma_{n_s}(x)\sigma^*_{n_s}(x)$.

Note, that each above p-tuple $\sigma = (\sigma_1, \sigma_{n_2}, \ldots, \sigma_{n_p})$ defines a continuous object on $\mathfrak{M} = \overline{X} \cup_\mu Y$, being a continuous function on $Y = \widehat{D} \cup \widehat{\gamma}$, being an $n_s \times n_s$ matrix-function continuous on $\overline{X}_{n_s}$, for each $s = 2, \ldots, p$, and being a diagonal $n_s \times n_s$ matrix at the points of $\partial \overline{X}_{n_s}$, whose scalar diagonal values are glued with certain values of the function on $Y = \widehat{D} \cup \widehat{\gamma}$ according to (7.4.3).

The descriptions of the local algebras $\widehat{\mathcal{R}}(t_0)$ lead up to the following theorem.

Theorem 7.4.3. *The Fredholm symbol algebra* $\mathrm{Sym}\,\mathcal{R}$ *of the C^*-algebra*

$$\mathcal{R} = \mathcal{R}(PC(\overline{\mathbb{D}}, \ell), B_{\mathbb{D}})$$

is isomorphic and isometric to the algebra $\mathfrak{S}$. Identifying them, the homomorphism

$$\mathrm{sym} : \mathcal{R} \to \mathfrak{S}$$

is generated by the following mapping of generators of the algebra $\mathcal{R}$:

$$\mathrm{sym} : A = a(z)I + b(z)B_{\mathbb{D}} + K \longmapsto$$

$$\begin{cases} a(t), & t \in \widehat{D}, \\ c(t), & t \in \widehat{\gamma}, \\ \begin{pmatrix} a(t_{q,1}+0)(1-x) + a(t_{q,1}-0)x & (c(t_{q,1}+0) - c(t_{q,1}-0))\sqrt{x(1-x)} \\ (a(t_{q,1}+0) - a(t_{q,1}-0))\sqrt{x(1-x)} & c(t_{q,1}p+0)(1-x) + c(t_{q,1}-0)x \end{pmatrix}, \\ \left(a^{(k)}(t_0)\left(\delta_{jk} - \sqrt{t_j(x)t_k(x)}\right) + c^{(k)}(t_0)\sqrt{t_j(x)t_k(x)} \right)_{j,k=1}^{r_q}, \end{cases}$$

where $x \in [0,1]$, $t_{q,1} \in T$, t_0 runs through all nodes $t_{q,r_g-1} \in T$ with $r_q > 2$, and $c(z) = a(z) + b(z)$.

Corollary 7.4.4. *An operator A from $\mathcal{R} = \mathcal{R}(PC(\overline{\mathbb{D}}, \ell), B_{\mathbb{D}})$ is Fredholm if and only if its symbol is invertible, i.e.,*

$$\mathrm{sym}\,A \neq 0 \ \text{ on } \ Y, \quad \det \mathrm{sym}\,A \neq 0 \ \text{ on } \ \overline{X}.$$

7.5   Some particular cases

It is instructive to consider a number of model algebras. Some of them or their slight variations might be used then for a local description at the points of discontinuity of the algebra $\mathcal{R}(PC(\overline{\mathbb{D}}, \ell), B_{\mathbb{D}})$ with more sophisticated types of discontinuities.

Although we could pursue our investigations for functions having any fixed number of limit values at the boundary points, we consider only the three-limit-value case which permits us to make more transparent pictures.

As in the last part of Section 7.3 we study the C^*-algebra

$$\mathcal{R}(H(PCo([0, \pi], \Lambda)), B_{\Pi})$$

which is generated now by four orthogonal projections acting on $L_2(\Pi)$,

$$P = B_{\Pi} \qquad \text{and} \qquad Q_k = \chi_{V_k} I, \qquad k = 1, 2, 3,$$

where each cone $V_k = \cup_j V_{k_j}$, $k = 1, 2, 3$, is a union of finite or countable numbers of cones of V_{k_j} supported on $(\theta_{k_j-1}, \theta_{k_j}]$, and $\Pi = \cup_{k=1}^3 V_k$.

That is, each piece-wise constant function

$$a(z) = a_1 \chi_{V_1}(z) + a_2 \chi_{V_2}(z) + a_3 \chi_{V_3}(z)$$

achieves its limit value a_k at the origin inside the cone V_k, $k = 1, 2, 3$, as in Section 7.3. At the same time, and contrary to the previous case, each cone V_k is no longer supported on only one subinterval in general.

Theorem 7.3.5 still describes our algebra; however there are some peculiarities. The curve $\Delta(\Lambda)$, lying on the two-dimensional simplex and being the joint spectrum of the corresponding commuting positive operators

$$C_k = PQ_k P = T_{V_k}, \qquad k = 1, 2, 3,$$

now does not have so regular behaviour, as in Figure 7.1, and does not necessarily connect the vertices $(1, 0, 0)$ and $(0, 0, 1)$.

To illustrate this we start with the case of a finite partition of $[0, \pi]$.

In the pictures we present four curves which correspond to the following projections and angles.

In the first picture:

$a.1:$ $\theta_1 = 0.05\pi, \quad \theta_2 = 0.07\pi, \quad \theta_3 = 0.5\pi, \quad \theta_4 = 0.65\pi, \quad \theta_5 = 0.8\pi$

            $Q_1 = \chi_{[0,\theta_1] \cup (\theta_5,\pi]} I, \quad Q_2 = \chi_{(\theta_1,\theta_2] \cup (\theta_3,\theta_4]} I, \quad Q_3 = \chi_{(\theta_2,\theta_3] \cup (\theta_4,\theta_5]} I;$

$a.2:$ $\theta_1 = 0.1\pi, \quad \theta_2 = 0.2\pi, \quad \theta_3 = 0.88\pi, \quad \theta_4 = 0.98\pi,$

            $Q_1 = \chi_{[0,\theta_1] \cup (\theta_4,\pi]} I, \quad Q_2 = \chi_{(\theta_1,\theta_2] \cup (\theta_3,\theta_4]} I, \quad Q_3 = \chi_{(\theta_2,\theta_3]} I;$

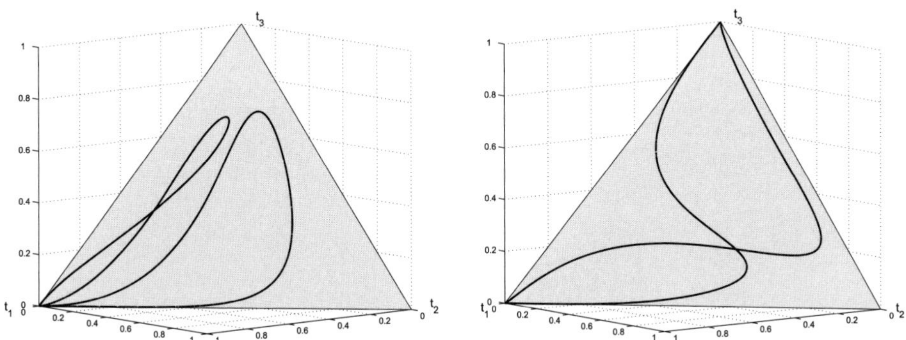

Figure 7.5: Left to right, curves: a.1, a.2, and b.1, b.2.

and in the second picture:

$$b.1: \quad \theta_1 = 0.1\pi, \quad \theta_2 = 0.4\pi, \quad \theta_3 = 0.5\pi, \quad \theta_4 = 0.7\pi, \quad \theta_5 = 0.9\pi$$

$$Q_1 = \chi_{[0,\theta_1]\cup(\theta_4,\theta_5]}I, \quad Q_2 = \chi_{(\theta_1,\theta_2]\cup(\theta_3,\theta_4]}I, \quad Q_3 = \chi_{(\theta_2,\theta_3]\cup(\theta_5,\pi]}I;$$

$$b.2: \quad \theta_1 = 0.1\pi, \quad \theta_2 = 0.11\pi, \quad \theta_3 = 0.2\pi, \quad \theta_4 = 0.89\pi, \quad \theta_5 = 0.9\pi$$

$$Q_1 = \chi_{[0,\theta_1]\cup(\theta_4,\theta_5]}I, \quad Q_2 = \chi_{(\theta_1,\theta_2]\cup(\theta_3,\theta_4]}I, \quad Q_3 = \chi_{(\theta_2,\theta_3]\cup(\theta_5,\pi]}I.$$

We consider now examples of countable numbers of subintervals dividing the segment $[0, \pi]$. In the next two pictures we present five curves which correspond to the projections $Q_k = \chi_{V_k}I$, $k = 1, 2, 3$, where the supports of the cones V_k for each picture are given respectively by

$$c.1: \ \operatorname{supp} V_1 = \bigcup_{k=1}^{\infty} \left(\frac{\pi}{2k}, \frac{\pi}{2k-1} \right],$$

$$\operatorname{supp} V_2 = \left(\frac{\pi}{3}, \frac{\pi}{2} \right],$$

$$\operatorname{supp} V_3 = \bigcup_{k=2}^{\infty} \left(\frac{\pi}{2k+1}, \frac{\pi}{2k} \right],$$

$$c.2: \ \operatorname{supp} V_1 = \bigcup_{k=1}^{\infty} \left(\frac{\pi}{4k}, \frac{\pi}{2(2k-1)} \right],$$

$$\operatorname{supp} V_2 = \bigcup_{k=1}^{\infty} \left(\frac{\pi}{2(2k+1)}, \frac{\pi}{4k} \right] \cup \bigcup_{k=1}^{\infty} \left(\pi\left(1 - \frac{1}{4k}\right), \pi\left(1 - \frac{1}{2(2k+1)}\right) \right],$$

$$\operatorname{supp} V_3 = \bigcup_{k=1}^{\infty} \left(\pi\left(1 - \frac{1}{2(2k-1)}\right), \pi\left(1 - \frac{1}{4k}\right) \right],$$

and

$$d.1: \quad \operatorname{supp} V_1 = \bigcup_{k=1}^{\infty} \left(\frac{\pi}{2(5k-2)}, \frac{\pi}{2(5k-4)} \right]$$

$$\cup \bigcup_{k=1}^{\infty} \left(\pi \left(1 - \frac{1}{2(5k-2)} \right), \pi \left(1 - \frac{1}{2(5k+1)} \right) \right],$$

$$\operatorname{supp} V_2 = \bigcup_{k=1}^{\infty} \left(\frac{\pi}{2(5k-1)}, \frac{\pi}{2(5k-2)} \right],$$

$$\operatorname{supp} V_3 = \bigcup_{k=1}^{\infty} \left(\frac{\pi}{2(5k+1)}, \frac{\pi}{2(5k-1)} \right]$$

$$\cup \bigcup_{k=1}^{\infty} \left(\pi \left(1 - \frac{1}{2(5k-4)} \right), \pi \left(1 - \frac{1}{2(5k-2)} \right) \right],$$

$$d.2: \quad \operatorname{supp} V_1 = \bigcup_{k=1}^{\infty} \left(\frac{\pi}{2(5k-3)}, \frac{\pi}{2(5k-4)} \right]$$

$$\cup \bigcup_{k=1}^{\infty} \left(\pi \left(1 - \frac{1}{2(5k-2)} \right), \pi \left(1 - \frac{1}{10k} \right) \right],$$

$$\operatorname{supp} V_2 = \bigcup_{k=1}^{\infty} \left(\frac{\pi}{2(5k-2)}, \frac{\pi}{2(5k-3)} \right]$$

$$\cup \bigcup_{k=1}^{\infty} \left(\pi \left(1 - \frac{1}{2(5k-4)} \right), \pi \left(1 - \frac{1}{2(5k-2)} \right) \right],$$

$$\operatorname{supp} V_3 = \bigcup_{k=1}^{\infty} \left(\frac{\pi}{2(5k+1)}, \frac{\pi}{2(5k-2)} \right]$$

$$\cup \bigcup_{k=1}^{\infty} \left(\pi \left(1 - \frac{1}{10k} \right), \pi \left(1 - \frac{1}{2(5k+1)} \right) \right],$$

$$d.3: \quad \operatorname{supp} V_1 = \bigcup_{k=1}^{\infty} \left(\frac{\pi}{5k}, \frac{\pi}{5k-3} \right],$$

$$\operatorname{supp} V_2 = \bigcup_{k=1}^{\infty} \left(\frac{\pi}{5k-3}, \frac{\pi}{5k-4} \right],$$

$$\operatorname{supp} V_3 = \bigcup_{k=1}^{\infty} \left(\frac{\pi}{5k+1}, \frac{\pi}{5k} \right].$$

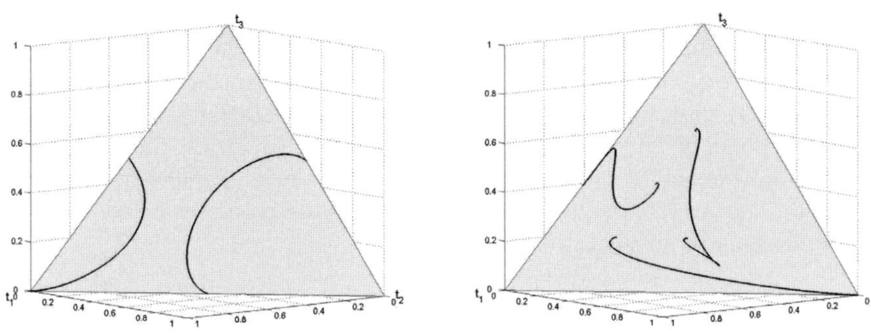

Figure 7.6: Left to right, curves: c.1, c.2, and d.1, d.2, d.3.

The behaviour of the boundary points of the curve $\Delta(\Lambda)$ is determined by the quantities

$$
\begin{aligned}
L_k(0) &= \lim_{\theta\to 0} \chi_{V_k}(\theta) \quad \text{(in case of existence, is equal to } \lim_{\lambda\to+\infty} \gamma_{\chi_{V_k}}(\lambda)), \\
L_k(\pi) &= \lim_{\theta\to\pi} \chi_{V_k}(\theta) \quad \text{(in case of existence, is equal to } \lim_{\lambda\to-\infty} \gamma_{\chi_{V_k}}(\lambda)),
\end{aligned}
$$

where $k = 1, 2, 3$, and is as follows.

Let $v_1 = (1, 0, 0)$, $v_2 = (0, 1, 0)$, and $v_3 = (0, 0, 3)$ be the vertices of the standard two-dimensional simplex, and let s_k be the boundary edge of the simplex opposite to the vertex v_k, $k = 1, 2, 3$. Then

(i) if the projection Q_k satisfies the property that either $L_k(0)$ or $L_k(\pi)$ exists and is equal to 1, then the vertex v_k belongs to $\Delta(\Lambda)$;

(ii) if for the projection Q_k either $L_k(0)$ or $L_k(\pi)$ exists and is equal to 0 and while for the other two projections the corresponding limits (at the same point) do not exist, then the curve $\Delta(\Lambda)$ ends on the edge s_k;

(iii) if for all projections Q_k the above limits do not exist either at 0 or at π (the same point for all projections), then the corresponding boundary point of $\Delta(\Lambda)$ belongs to the interior of the simplex.

The different behaviour of the boundary points of the joint spectrum $\Delta(\Lambda)$ reflects the differences in the properties of the corresponding algebra $\mathfrak{S}(\Delta(\Lambda))$. For the case of three-limit-value functions, the algebra $\mathfrak{S}(\Delta(\Lambda))$ used in Theorem 7.3.5 coincides with the algebra of all 3×3 matrix-functions continuous on $\Delta(\Lambda)$ and diagonal at the vertices $(1, 0, 0)$ and $(0, 0, 1)$, which are the endpoints of $\Delta(\Lambda)$.

The same type of the endpoint behaviour remains valid for case (i) above. In case (ii), however, at the corresponding endpoint the matrix, which is continuous on $\Delta(\Lambda)$, becomes block diagonal with 1×1 and 2×2 blocks, whose positions within the 3×3 matrix depend on which boundary edge s_k the endpoint belongs to. At the endpoints of case (iii) the matrix remains a 3×3 matrix.

Thus we see that the dimensions of irreducible representations of the algebra $\mathfrak{S}(\Delta(\Lambda))$ may vary. More precisely, the representation which corresponds to each endpoint of $\Delta(\Lambda)$ can be of the three following types: the direct sum of three one-dimensional representations; the direct sum of a one-dimensional representation and an irreducible two-dimensional representation; or an irreducible three-dimensional representation.

7.6   Toeplitz operator algebra. A first look

We start with the unit disk $\mathbb{D}$, a curve ℓ, and the C^*-algebra $PC(\overline{\mathbb{D}}, \ell)$ as defined in Section 7.4. Denote by $\mathcal{T} = \mathcal{T}(PC(\overline{\mathbb{D}}, \ell))$ the Toeplitz C^*-algebra, i.e., the C^*-algebra generated by all Toeplitz operators T_a with defining symbols $a \in PC(\overline{\mathbb{D}}, \ell)$.

We obviously have $\mathcal{T}(C(\overline{\mathbb{D}})) \subset \mathcal{T}(PC(\overline{\mathbb{D}}, \ell))$, and thus by Lemma 2.8.4 the algebra $\mathcal{T}(PC(\overline{\mathbb{D}}, \ell))$ contains the ideal $\mathcal{K}$ of all compact operators on $\mathcal{A}^2(\mathbb{D})$ and is irreducible.

Lemma 2.4.4 and Theorem 2.8.5 show, in particular, that the C^*-algebra $\widehat{\mathcal{T}}_0 = \mathcal{T}(C(\overline{\mathbb{D}})/\mathcal{K} \cong C(\gamma)$, where $\gamma = \partial\mathbb{D}$ is the unit circle, is a central commutative subalgebra of $\widehat{\mathcal{T}} = \mathcal{T}(PC(\overline{\mathbb{D}}, \ell))/\mathcal{K}$. Thus to describe the Fredholm symbol algebra $\mathrm{Sym}\,\mathcal{T} = \widehat{\mathcal{T}} = \mathcal{T}(PC(\overline{\mathbb{D}}, \ell))/\mathcal{K}$ we use the Douglas-Varela local principle (see Subsection 1.1.6), localizing by points $t_0 \in \gamma$.

There are two different types of local algebras. The first one corresponds to the boundary points of continuity, i.e., $t_0 \in \gamma \setminus T$, where $T = \ell \cap \gamma$ is the set of boundary points of discontinuity, as in Section 7.4. The second one corresponds to the boundary points of discontinuity, i.e., $t_0 \in T$.

1) Let $t_0 \in \gamma \setminus T$. Then we are essentially in the situation of continuous defining symbols. That is, as is easy to see, a Toeplitz operator T_a with defining symbol $a \in PC(\overline{\mathbb{D}}, \ell)$ is locally equivalent at the point t_0 to the constant defining symbol operator $T_{a(t_0)} = a(t_0)I$. This is because at a point of continuity a continuous function (symbol) is locally equivalent to its values at this point, and $B_{\mathbb{D}} = I$ on the space $\mathcal{A}^2(\mathbb{D})$. Thus the local algebra $\widehat{\mathcal{T}}(t_0)$, which corresponds to the point t_0, is isomorphic to $\mathbb{C}$, and the homomorphism

$$\pi_{t_0} : \mathcal{T} \longrightarrow \widehat{\mathcal{T}}(t_0) = \mathbb{C}$$

is generated by the mapping of generators

$$\pi_{t_0} : T_a \longmapsto a(t_0).$$

2.a) Let $t_0 \in T = \ell \cap \gamma$. Assume first that t_0 is a node of type $t_{q,1}$ that is, only one curve meets the boundary at the node t_0. This case in fact was covered by Theorem 2.8.8, and the result is as follows. Denote by $t_0 - 0$ and $t_0 + 0$ the points of the compact of maximal ideals of the algebra $PC(\overline{\mathbb{D}}, \ell)|_\gamma$ such that

$$a(t_0 - 0) = \lim_{t \to t_0,\, t \prec t_0} a(t) \quad \text{and} \quad a(t_0 + 0) = \lim_{t \to t_0,\, t_0 \prec t} a(t),$$

for each $a \in PC(\overline{\mathbb{D}}, \ell)$. Then the local algebra $\widehat{T}(t_0)$, which corresponds to the point t_0, is isomorphic to $C([0, 1])$, and the homomorphism

$$\pi_{t_0} : T \longrightarrow \widehat{T}(t_0) = C([0, 1])$$

is generated by the mapping of generators

$$\pi_{t_0} : T_a \longmapsto a(t_0 - 0)(1 - x) + a(t_0 + 0)x,$$

where $x \in [0, 1]$.

2.b) Let finally $t_0 = t_{q, r_q - 1} \in T$, that is, there are $r_q - 1$ curves meeting at this node. Recall that any function $a(z) \in PC(\overline{\mathbb{D}}, \ell)$ has r_q limit values at this node, which we denoted by $a^{(1)}(t_{q, r_q - 1}), \ldots, a^{(r)}(t_{q, r_q - 1})$, counting counter-clockwise.

Introduce the ordered set

$$\Lambda_q = \{\theta_1, \theta_2, \ldots, \theta_{r_q - 1}\}$$

of the angles which the above $r_q - 1$ curves form at the node $t_0 = t_{q, r_q - 1} \in \gamma$, counting them counter-clockwise. Then, as is easy to see, each Toeplitz operator T_a with defining symbol $a \in PC(\overline{\mathbb{D}}, \ell)$ is locally equivalent at the point t_0 to the Toeplitz operator $T_{a_{t_0}}$ whose defining symbol a_{t_0} is piece-wise constant in some neighborhood U of t_0 and having the following form in U,

$$a_{t_0} = a_{t_0}^{(1)} \chi_{U_1} + a_{t_0}^{(2)} \chi_{U_2} + \ldots + a_{t_0}^{(r_q)} \chi_{U_{r_q}},$$

where each χ_{U_k} is the characteristic function of the region which is the part of U lying between the $k - 1$ and k curves meeting at the node $t_0 = t_{q, r_q - 1}$.

By the hypotheses imposed on ℓ, there exists a Möbius transformation ω_{t_0} from the unit disk onto the upper half-plane mapping the node t_0 to the origin and mapping the $r_q - 1$ curves meeting at t_0 to the $r_q - 1$ lines which locally near the origin are straight line segments meeting at the origin. Note that these straight line segments have the same angles $\theta_1, \theta_2, \ldots, \theta_{r_q - 1}$ with the real axis.

Under the above Möbius transformation ω_{t_0} the local algebra $\widehat{T}(t_0)$ of the initial algebra $T(PC(\overline{\mathbb{D}}, \ell))$ is obviously isomorphic to the local algebra $\widehat{T}'(0)$ at the origin $0 = \omega_{t_0}(t_0)$ of the algebra $T' = T'(PC(\overline{\Pi}, \omega_{t_0}(\ell)))$, obtained by the unitary equivalence with the initial algebra T. Observe now that the local algebra $\widehat{T}'(0)$ is generated by the Toeplitz operators with $H(PCo([0, \pi]0, \Lambda_q))$ defining symbols, and thus is nothing but the algebra $T(H(PCo([0, \pi], \Lambda_q)))$, which was described in Theorem 7.2.9. That is, we have

Lemma 7.6.1. *The local algebra $\widehat{T}(t_0)$, which corresponds to the point t_0, is isomorphic to $T(H(PCo([0,\pi],\Lambda_q))) \cong C(\Delta(\Lambda_q))$. The homomorphism*

$$\pi_{t_0} : T \longrightarrow T(H(PCo([0,\pi],\Lambda))) = C(\Delta(\Lambda_q))$$

is generated by the mapping of generators

$$\pi_{t_0} : T_a \longmapsto a_{t_0}^{(1)} t_1 + a_{t_0}^{(2)} t_2 + \ldots + a_{t_0}^{(r_q)} t_{r_q}, \qquad t = (t_1, t_2, \ldots, t_{r_q}) \in \Delta(\Lambda_q)$$

where, see (7.2.6),

$$\Delta(\Lambda_q) = \{(t_1, t_2, \ldots, t_{r_q}) : t_k = \gamma_{\chi_{V_k}}(\lambda), \quad \lambda \in \overline{\mathbb{R}}, \quad k = 1, \ldots, r_q\} \qquad (7.6.1)$$

is a continuous curve lying on the standard $(r_q - 1)$-dimensional simplex, and connecting the vertices $(1, 0, \ldots, 0)$ and $(0, \ldots, 0, 1)$, and each $t_k = \gamma_{\chi_{V_k}}(\lambda)$, $k = 1, \ldots, r_q$, is given by, see (7.2.5),

$$t_k = \gamma_{\chi_{V_k}}(\lambda) = \frac{2\lambda}{1 - e^{-2\pi\lambda}} \int_{\theta_{k-1}}^{\theta_k} e^{-2\lambda\theta}\, d\theta = \frac{e^{-2\theta_k\lambda} - e^{-2\theta_{k-1}\lambda}}{e^{-2\pi\lambda} - 1}, \qquad \lambda \in \overline{\mathbb{R}}.$$

It is worth mentioning, that putting in the lemma $r_q = 2$, $a_{t_0}^{(1)} = a(t_0 + 0)$, $a_{t_0}^{(r_q)} = a_{t_0}^{(2)} = a(t_0 - 0)$, $t_1 = x \in [0, 1]$, $t_2 = 1 - x$, we obtain the description of case 2.a) above.

Now we paste together all the local descriptions. Denote by $\widehat{\gamma}$ the set γ, cut at the points $t_{q, r_q - 1} \in T = \ell \cap \gamma$. The pair of points which correspond to a point $t_{q, r_q - 1} \in T$ will be denoted by $t_{q, r_q - 1} - 0$ and $t_{q, r_q - 1} + 0$, following the positive orientation of γ. Let $\overline{X} = \cup_q \Delta(\Lambda_q)$ be the disjoint union of the sets (7.6.1). Denote by Γ the union $\widehat{\gamma} \cup \overline{X}$ with the point identification

$$t_{q, r_q - 1} - 0 \equiv (1, 0, \ldots, 0) \qquad t_{q, r_q - 1} + 0 \equiv (0, \ldots, 0, 1),$$

where $t_{q, r_q - 1} \pm 0 \in \widehat{\gamma}$, and the vertices $(1, 0, \ldots, 0)$ and $(0, \ldots, 0, 1)$ are the boundary points of $\Delta(\Lambda_q)$.

Then we have obviously

Theorem 7.6.2. *The C^*-algebra $T = T(PC(\overline{\mathbb{D}}, \ell))$ is irreducible and contains the ideal K of compact operators. The Fredholm symbol algebra $\operatorname{Sym} T = T/K$ is isomorphic to the algebra $C(\Gamma)$. Identifying them, the symbol homomorphism*

$$\operatorname{sym} : T \to \operatorname{Sym} T = C(\Gamma)$$

is generated by the following mapping of generators of T,

$$\operatorname{sym} : T_a \longmapsto \begin{cases} a(t), & t \in \widehat{\gamma} \\ a_{t_{q,r_q-1}}^{(1)} t_1 + a_{t_{q,r_q-1}}^{(2)} t_2 + \ldots + a_{t_{q,r_q-1}}^{(r_q)} t_{r_q}, & (t_1, t_2, \ldots, t_{r_q}) \in \Delta(\Lambda_q) \end{cases},$$

where $t_{q, r_q - 1} \in T$.

Each operator $T \in T$ is Fredholm if and only if its symbol is invertible, i.e., the function $\operatorname{sym} T$ is non-zero on Γ, and

$$\operatorname{Ind} T = -\frac{1}{2\pi} \{\operatorname{sym} T\}_\Gamma.$$

7.7 Toeplitz operator algebra. Some more analysis

The representation of the Fredholm symbol for a Toeplitz operator T_a, with $a \in PC(\overline{\mathbb{D}}, \ell)$, given by Theorem 7.6.2 is very convenient for the description of the essential spectrum of T_a and understanding of the geometric regularities of its behaviour.

Indeed, given a defining symbol $a \in PC(\overline{\mathbb{D}}, \ell)$, the essential spectrum ess-sp T_a of the operator T_a, which is obviously equal to $\operatorname{Im} \operatorname{sym} T_a$, consists of two parts. Its regular part is the image of the Fredholm symbol restricted on the boundary points of continuity, i.e., $\operatorname{sym} T_a|_{\hat{\gamma}} = a|_{\hat{\gamma}}$. Its complementary part is a finite number of additional arcs, each one of which is the restriction of $\operatorname{sym} T_a$ to the curve $\Delta(\Lambda_q)$ which corresponds to a boundary point of discontinuity $t_{q, r_q - 1}$.

We note that each such curve $\operatorname{sym} T_a|_{\Delta(\Lambda_q)}$ describes as well the spectrum of the local representative at the point $t_{q, r_q - 1}$ of the initial operator T_a.

Let us assume that t_0 is a boundary point of discontinuity for functions from $PC(\overline{\mathbb{D}}, \ell)$ in which n curves from ℓ meet. As previously, introduce the ordered set

$$\Lambda = \{\theta_1, \theta_2, \ldots, \theta_{n-1}\}$$

of the angles which the above n curves form with the boundary γ, counting them counter-clockwise. As above, we add $\theta_0 = 0$ and $\theta_n = \pi$. Given a defining symbol $a \in PC(\overline{\mathbb{D}}, \ell)$, introduce the ordered set

$$A = \{a_1, a_2, \ldots, a_n\},$$

where each a_k, $k = 1, 2, \ldots, n$, is the limit value of a at point t_0 reached from the region between the $(k-1)$-th and k-th curves.

The local representative at the point t_0 of the operator T_a can be taken as the Toeplitz operator $T_{A,\Lambda}$ with the piece-wise constant defining symbol

$$a_{A,\Lambda}(\theta) = a_1 \chi_{V_1}(\theta) + \ldots + a_n \chi_{V_n}(\theta) \in H(PCo([0, \pi], \Lambda)),$$

where each χ_{V_k} is the characteristic function of the cone V_k supported on $(\theta_{k-1}, \theta_k]$.

That is, the spectrum of $T_{a_{A,\Lambda}}$, which is the same as the corresponding portion of the essential spectrum of T_a, is governed by the sets A and Λ and is given by the formula

$$\operatorname{sp} T_{a_{A,\Lambda}} = \{a_1 t_1 + \ldots + a_n t_n : t = (t_1, t_2, \ldots, t_n) \in \Delta(\Lambda)\}. \tag{7.7.1}$$

It is instructive to understand the geometric regularities of its behaviour.

We start with the simplest case of just two limit values. Let $A = (a_1, a_2)$ and $\Lambda = \{\theta_1\}$. In this case the spectrum $\operatorname{sp} T_{a_{A,\Lambda}}$ *does not depend* at all on Λ, is *uniquely* determined by A, and is the straight line segment connecting the points a_1 and a_2. This is an effect of low dimension: each line connecting the vertices of an *one*-dimensional simplex is the simplex itself, and is the straight line segment connecting the vertices.

Passing to the case $n > 2$ we consider first the most transparent case $n = 3$. In this case the curve $\Delta(\Lambda)$ lies on a two-dimensional simplex having the same dimension as the complex plane where the spectrum lies.

As we already know (see Figure 7.1), the continuous curve $\Delta(\Lambda)$ connecting the vertices $v_1 = (1, 0, 0)$ and $v_3 = (0, 0, 1)$ *does depend* essentially on Λ. Then by (7.7.1) the spectrum, geometrically, $\operatorname{sp} T_{a_{A,\Lambda}}$ is the image of the curve $\Delta(\Lambda)$ under the projection (affine mapping) of the two-dimensional simplex to the complex plane such that each vertex v_k is projected to a_k, $k = 1, 2, 3$, and $a_k \in A$. That is, the set A determines the triangle to which the simplex is projected, while the set Λ determines the shape of the line $\Delta(\Lambda)$ whose projection to the already defined triangle gives the spectrum.

In the next two pictures we illustrate this for three different sets Λ, being the first, third, and fifth set of angles of Figure 7.1. That is, we consider the sets of angles $(0.48\pi, 0.52\pi)$, $(0.3\pi, 0.7\pi)$, and $(0.1\pi, 0.9\pi)$, ordered as generated from less to more curved lines. For the first picture the set A is given by $(0.1 + 0.1i, 0.9i, 0.9 + 0.5i)$, and $A = ((0.1 + 0.1i, 1 + 0.2i, 0.9 + 0.5i)$, for the second picture. For both sets we leave the same values of a_1 and a_3, making the pictures "one-parametric" in their dependence on a_2.

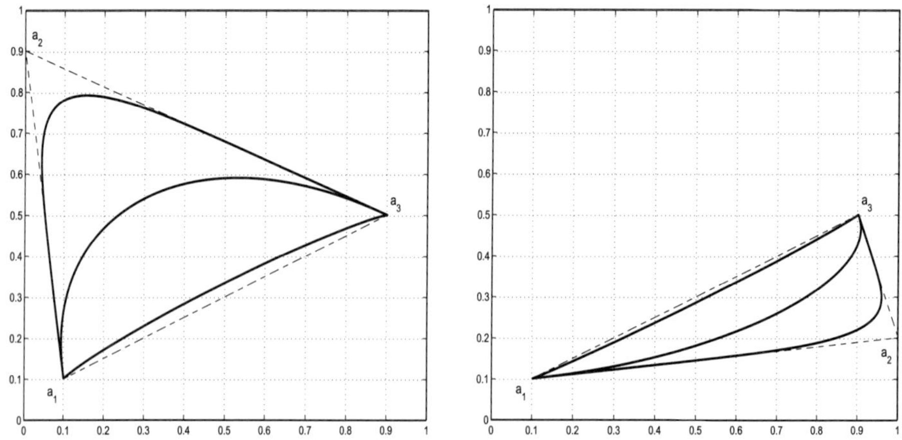

Figure 7.7: Spectra of $T_{a_{A,\Lambda}}$ for three-limit-values defining symbols.

The case $n > 3$ maintains in principle the same features. The spectrum $\operatorname{sp} T_{a_{A,\Lambda}}$ is the image of the curve $\Delta(\Lambda)$ under the projection (affine mapping) of, now, the $(n-1)$-dimensional simplex onto a certain convex polygon on the complex plane such that each vertex v_k is projected to a_k, $k = 1, 2, , \ldots, n$, and $a_k \in A$. The curve $\Delta(\Lambda)$ connecting the vertices $v_1 = (1, 0, \ldots, 0)$ and $v_n = (0, \ldots, 0, 1)$ *does depend* essentially on Λ. The set A determines the polygon to which the simplex is projected, while the set Λ determines the shape of the line $\Delta(\Lambda)$ whose projection to the already defined polygon gives the spectrum. The only difference is that now this convex polygon has n or *fewer* vertices, depending on the way, prescribed by A, in which the $(n-1)$-dimensional simplex is projected to the two-dimensional polygon. That is, the projections of some vertices can be (or not) in the interior of the polygon.

In the next two pictures we present the cases of five limit-values defining symbols for which the 4-dimensional simplex is projected onto a pentagon and a triangle, respectively. We consider the sets A,

$$(0.2 + 0.1i, 0.4 + 0.9i, 0.8 + 0.1i, 0.1 + 0.7i, 0.9 + 0.8i)$$

and

$$(0.2 + 0.1i, 0.5 + 0.6i, 0.1 + 0.9i, 0.3 + 0.4i, 0.9 + 0.8i),$$

maintaining the same values of a_1 and a_5 for both cases. Both pictures represent three spectra for which the sets Λ are

$$(0.46\pi, 0.48\pi, 0.52\pi, 0.54\pi)$$
$$(0.2\pi, 0.2\pi, 0.7\pi, 0.8\pi)$$
$$(0.0002\pi, 0.01\pi, 0.99\pi, 0.9998\pi),$$

and which again correspond to lines ordered from less to more curved.

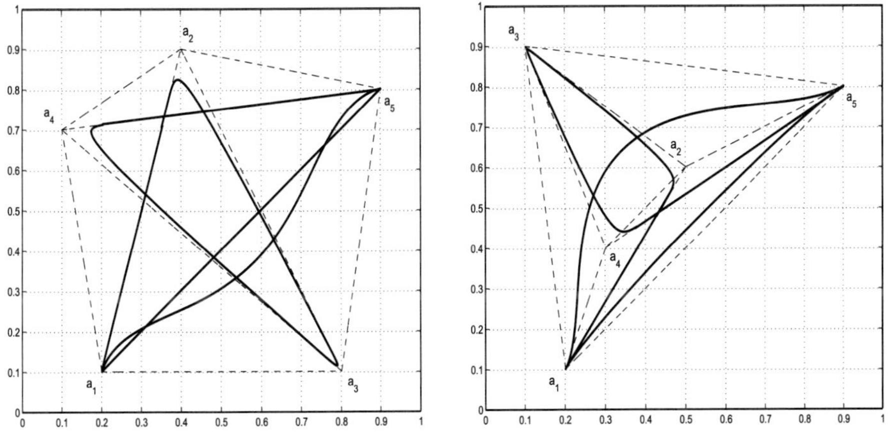

Figure 7.8: Spectra of $T_{a_{A,\Lambda}}$ for five limit-values defining symbols (pentagon and triangle).

One more illustration presents two pictures of the five limit-values case for the same as above three sets Λ and for the sets A,

$$(0.2 + 0.1i, 0.1 + 0.8i, 0.5 + 0.9i, 0.9 + 0.8i, 0.8 + 0.1i)$$
$$(0.2 + 0.1i, 0.1 + 0.8i, 0.3 + 0.7i, 0.9 + 0.8i, 0.8 + 0.1i),$$

the only difference in them being the value of a_3, which reflects the change of a pentagon to a quadrangle.

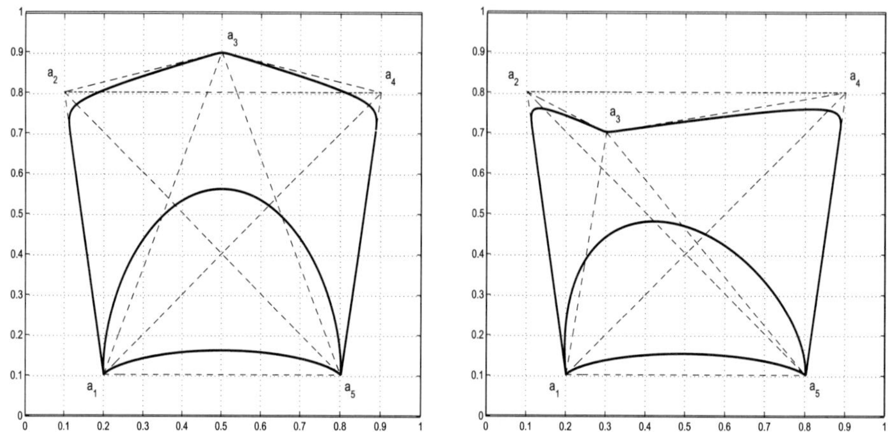

Figure 7.9: Spectra of $T_{a_{A,\Lambda}}$ for five limit-values defining symbols (pentagon and quadrangle).

We note that the spectrum $\operatorname{sp} T_{a_{A,\Lambda}}$ becomes more rectilinear and more stable under the perturbations of $a_k \in A$, $k = 2, \ldots, n-1$, for bigger values of the angles θ_1 and $\pi - \theta_{n-1}$. Further, the spectrum approaches to the straight line segment connecting the images of the vertices $(1, 0, \ldots, 0)$ and $(0, \ldots, 0, 1)$ when the sum of these angles tends to π. The opposite tendency, in a sense, appears when the angles between the curves intersecting at t_0 and the boundary of the domain tend to 0. In that case the spectrum approaches the union of straight line segments passing in series through the images of the vertices $(1, 0, \ldots, 0)$, $(0, 1, 0, \ldots, 0)$, $\ldots$, $(0, \ldots, 0, 1)$.

At the same time the description given by Theorem 7.6.2 hides some essential properties of the above Toeplitz operator algebras. It turns out that each Toeplitz operator algebra $\mathcal{T}(PC(\overline{\mathbb{D}}, \ell))$ contains, apart from of its initial generators T_a with defining symbols $a \in PC(\overline{\mathbb{D}}, \ell)$, many other Toeplitz operators with much more general defining symbols.

To show this we start with the local situation of piece-wise continuous defining symbols having only two-limit-values at the point of discontinuity. Consider $\mathcal{A}^2(\Pi)$ and the Toeplitz operator T_+ with defining symbol $a_+(z) = \chi_+(\operatorname{Re} z) = \chi_+(x)$, where χ_+ is the characteristic function of the positive half-line. We have as well that $a_+(z) = a_+(re^{i\theta}) = \chi_{[0,\pi/2]}(\theta)$, and thus $a_+ \in H(PCo([0,\pi], \{\pi/2\}))$.

The Toeplitz operator $T_+ \in \mathcal{T}(H(PCo([0,\pi], \{\pi/2\})))$ is unitary equivalent to the multiplication operator $\gamma_{a_+} I$, where, by (7.2.1),

$$
\begin{aligned}
\gamma_{a_+}(\lambda) &= \frac{2\lambda}{1 - e^{-2\pi\lambda}} \int_0^\pi \chi_{[0,\pi/2]}(\theta)\, e^{-2\lambda\theta}\, d\theta \\
&= \frac{e^{-\pi\lambda} - 1}{e^{-2\pi\lambda} - 1} = \frac{1}{e^{-\pi\lambda} + 1}, \qquad \lambda \in \mathbb{R}.
\end{aligned}
\qquad (7.7.2)
$$

The operator T_+ is obviously self-adjoint and $\operatorname{sp} T_+ = [0, 1]$. Thus for any function f continuous on $[0, 1]$ the operator $f(T_+)$ is well defined by the standard functional calculus in C^*-algebras; furthermore the operator $f(T_+)$ belongs to the same algebra $\mathcal{T}(H(PCo([0, \pi], \{\pi/2\})))$.

Example 7.7.1. Consider the family of functions f_α parameterized by $\alpha \in [0, 1]$ and given by

$$f_\alpha(x) = x^{2(1-\alpha)} \frac{(1 - x)^{2\alpha} - x^{2\alpha}}{(1 - x) - x}, \qquad x \in [0, 1]. \tag{7.7.3}$$

Each function f_α is continuous on $[0, 1]$, and $f_\alpha(0) = 0$, $f_\alpha(1) = 1$. Let us mention as well some particular cases

$$f_0(x) \equiv 0, \qquad f_{\frac{1}{2}}(x) = x, \qquad f_1(x) \equiv 1.$$

Then

$$f_\alpha(T_+) = T_{\chi_{[0,\alpha\pi]}} \in \mathcal{T}(H(PCo([0, \pi], \{\pi/2\}))),$$

where the defining symbol $\chi_{[0,\alpha\pi]}$ of $T_{\chi_{[0,\alpha\pi]}}$ belongs to $H(PCo([0, \pi], \{\alpha\pi\}))$.

Proof. We will exploit the isomorphism between the Toeplitz operator algebra and the functional algebra given in Corollary 7.2.2. Introduce

$$x = \gamma_{a_+}(\lambda) = \frac{1}{e^{-\pi\lambda} + 1} \in [0, 1],$$

which is equivalent to

$$\lambda = \lambda(x) = -\frac{1}{\pi} \ln \frac{1 - x}{x}.$$

Then for the operator $T_{\chi_{[0,\alpha\pi]}}$ the corresponding function $\gamma_{\chi_{[0,\alpha\pi]}}$ is given by

$$\gamma_{\chi_{[0,\alpha\pi]}}(\lambda) = \frac{2\lambda}{1 - e^{-2\pi\lambda}} \int_0^\pi \chi_{[0,\lambda\pi]}(\theta)\, e^{-2\lambda\theta}\, d\theta = \frac{e^{-2\alpha\pi\lambda} - 1}{e^{-2\pi\lambda} - 1}, \qquad \lambda \in \mathbb{R}.$$

Substituting $\lambda = \lambda(x)$ we have

$$\begin{aligned}
\gamma_{\chi_{[0,\alpha\pi]}}(\lambda(x)) &= \frac{e^{2\alpha\pi\frac{1}{\pi} \ln \frac{1-x}{x}} - 1}{e^{2\pi\frac{1}{\pi} \ln \frac{1-x}{x}} - 1} \\[2mm]
&= \frac{\left(\frac{1-x}{x}\right)^{2\alpha} - 1}{\left(\frac{1-x}{x}\right)^2 - 1} = x^{2(1-\alpha)} \frac{(1 - x)^{2\alpha} - x^{2\alpha}}{(1 - x) - x}.
\end{aligned}$$

$\qquad\square$

Note, that the above mentioned particular cases of f_α lead to the equalities

$$f_0(T_+) = 0, \qquad f_{\frac{1}{2}}(T_+) = T_+, \qquad f_1(T_+) = I,$$

as should be.

In the next example we present a connection between Toeplitz operators with piece-wise constant defining symbols having just two and more than two limit values at the single point of discontinuity.

Example 7.7.2. Given a finite ordered set of numbers

$$0 < \alpha_1 < \alpha_2 < \ldots < \alpha_{n-1} < 1,$$

we introduce

$$\Lambda = \{\alpha_1\pi, \alpha_2\pi, \ldots, \alpha_{n-1}\pi\},$$

for convenience we add $\alpha_0 = 0$ and $\alpha_n = 1$. Let further $A = \{a_1, a_2, \ldots, a_n\}$ be an ordered set of complex numbers.

Given both A and Λ, we define the piece-wise constant symbol

$$a_{A,\Lambda}(\theta) = \sum_{k=1}^{n} a_k \chi_{(\alpha_{k-1}\pi, \alpha_k\pi]} \in PCo([0,\pi], \Lambda)$$

and the function $f_{A,\Lambda} = f_{A,\Lambda}(x)$ continuous on $[0,1]$ defined by

$$f_{A,\Lambda}(x) = \sum_{k=1}^{n} a_k \frac{(1-x)^{2\alpha_k} x^{2(1-\alpha_k)} - (1-x)^{2\alpha_{k-1}} x^{2(1-\alpha_{k-1})}}{(1-x) - x}.$$

Then

$$f_{A,\Lambda}(T_+) = T_{a_{A,\Lambda}} \in T(H(PCo([0,\pi], \{\pi/2\}))).$$

Proof. Consider the Toeplitz operator $T_{a_{A,\Lambda}}$. Using (7.2.5), we have

$$
\begin{aligned}
\gamma_{a_{A,\Lambda}}(\lambda) &= \frac{2\lambda}{1 - e^{-2\pi\lambda}} \int_0^{\pi} a_{A,\Lambda}(\theta) e^{-2\lambda\theta} \, d\theta \\
&= \sum_{k=1}^{n} a_k \frac{e^{-2\alpha_k \pi\lambda} - e^{-2\alpha_{k-1}\pi\lambda}}{e^{-2\pi\lambda} - 1}, \qquad \lambda \in \mathbb{R}.
\end{aligned}
$$

Substitute, as in the previous example, $\lambda = \lambda(x)$. Then after a simple calculation we have

$$\gamma_{a_{A,\Lambda}}(\lambda(x)) = \sum_{k=1}^{n} a_k \frac{(1-x)^{2\alpha_k} x^{2(1-\alpha_k)} - (1-x)^{2\alpha_{k-1}} x^{2(1-\alpha_{k-1})}}{(1-x) - x}. \qquad \square$$

Theorem 7.2.5 and Corollary 7.2.6 imply, in particular, that every Toeplitz operator with $H(L_\infty^{(0,\pi)}([0,\pi]))$-symbol can be obtained in a similar way. The exact formula for the corresponding continuous function $f(x)$, though necessarily rather implicit, is given in the next example.

Example 7.7.3. Given a function $a = a(\theta) \in H(L_\infty^{(0,\pi)}([0,\pi]))$, let

$$f_a(x) = \frac{2x^2}{\pi} \frac{\ln(1-x) - \ln x}{(1-x) - x} \int_0^\pi a(\theta) \left(\frac{1-x}{x}\right)^{\frac{2\theta}{\pi}} d\theta.$$

Then

$$f_a(T_+) = T_a.$$

Remark 7.7.4. In the above examples we have considered the Toeplitz operator T_+ as the starting operator for a very simple reason. Precisely in this specific case the generically transcendental equation $x = \gamma_a(\lambda)$ admits an explicit solution.

At the same time we can start as well from any Toeplitz operator T_α having the defining symbol $\chi_{[0,\alpha\pi]}$, where $\alpha \in (0,\pi)$. Indeed, as follows from the proof of Corollary 7.2.6, the function $\gamma_{\chi_{[0,\alpha\pi]}}(\lambda)$ is strictly increasing. This implies that the function $f_\alpha(x)$ (see (7.7.3)), which maps $[0,1]$ onto $[0,1]$, is strictly increasing as well. Thus the function $f_\alpha^{-1}(x)$ is well defined and continuous on $[0,1]$.

Finally, given $\alpha, \beta \in (0,\pi)$, A, Λ, and $a = a(\theta) \in H(L_\infty^{(0,\pi)}([0,\pi]))$, we have

$$T_\alpha \in \mathcal{T}(H(PCo([0,\pi],\{\alpha\pi\})))$$

and

$$\begin{aligned}
(f_\beta \circ f_\alpha^{-1})(T_\alpha) &= T_\beta, \\
(f_{A,\Lambda} \circ f_\alpha^{-1})(T_\alpha) &= T_{a_{A,\Lambda}}, \\
(f_a \circ f_\alpha^{-1})(T_\alpha) &= T_a,
\end{aligned}$$

where all Toeplitz operators on the right-hand side of the above equalities belong to $\mathcal{T}(H(PCo([0,\pi],\{\alpha\pi\})))$.

The above examples show that studying the algebra generated by Toeplitz operators, whose defining symbols admit discontinuities at a finite number of boundary points, we can start from any symbol algebra selected from a wide variety of defining symbol classes. Moreover, the line ℓ entering in the definition of the defining symbol algebra $PC(\overline{\mathbb{D}},\ell)$ does not play in fact any significant role. In all such cases the resulting C^*-algebra will contain all Toeplitz operators whose defining symbols admit "homogeneous type discontinuity" locally described by the algebra $H(L_\infty^{(0,\pi)}([0,\pi]))$, as above, at every boundary point of discontinuity.

Thus it seems to be reasonable to include the Toeplitz operators with such defining symbols among the generators of the algebra from the very beginning. In this case the definition proceeds as follows.

Let $\Lambda = \{t_1, t_2, \ldots, t_m\}$ be a finite set of distinct points on the unit circle $\gamma = \partial\mathbb{D}$. Introduce the linear space $BPC(\mathbb{D},\Lambda)$ (*BPC* stands for *Boundary Piece-wise Continuous*) which consists of all functions $a(z)$ obeying the following properties:

(i) $a(z) \in L_\infty(\mathbb{D})$;

(ii) $a(z)$ has limit values at all boundary points $t \in \gamma \setminus \Lambda$, and the function $a(t)$, constructed by these limit values, is continuous in $\gamma \setminus \Lambda$;

(iii) at each point $t_0 \in \Lambda$ the function $a(z)$ has a "homogeneous type discontinuity", which means that there exists a Möbius transformation $z = z_{t_0}(w)$ of the upper half-plane Π to the unit disk $\mathbb{D}$ with $t_0 = z_{t_0}(0)$ and a homogeneous function $a_{t_0}(w) \in H(L_\infty^{(0,\pi)}([0,\pi]))$ of order zero such that

$$\lim_{w \to 0} [a(z_{t_0}(w)) - a_{t_0}(w)] = 0.$$

Let us comment on this definition. The set $BPC(\mathbb{D}, \Lambda)$ in fact is a C^*-algebra, although only the linear space structure is important for our purposes. The function $a(t)$, as a function of the boundary points, belongs to $PC(\gamma, \Lambda)$ that is, for each point $t_0 \in \Lambda$ the limits

$$\lim_{t \to t_0, \, t \prec t_0} a(t) = a(t_0 - 0) \qquad \text{and} \qquad \lim_{t \to t_0, \, t_0 \prec t} a(t) = a(t_0 + 0)$$

are well defined. The property (iii) of the above definition can be alternatively described in geometric terms of $\mathbb{D}$ as follows. For each point $t_0 \in \Lambda$ there is a hyperbolic pencil $\mathcal{P}_{t_0}$ (see Section 9.4 and Figure 9.3), such that t_0 is the endpoint of its axis, and a function $\tilde{a}_{t_0}(z)$ which is constant on cycles of $\mathcal{P}_{t_0}$ and whose values on (each) geodesic are given by an L_∞-function having limit values at the endpoints of the geodesic on γ (points at infinity in the hyperbolic geometry), such that

$$\lim_{z \to t_0} [a(z) - \tilde{a}_{t_0}(z)] = 0.$$

Consider now the C^*-algebra $\mathcal{T}_{BPC} = \mathcal{T}(BPC(\mathbb{D}, \Lambda))$ generated by all Toeplitz operators T_a with defining symbols $a \in BPC(\mathbb{D}, \Lambda)$.

Let, as previously, $\widehat{\gamma}$ be the set γ cut at the points $t_p \in \Lambda$. The pair of points which correspond to a point $t_p \in \Lambda$, $p = \overline{1, m}$, will be denoted by $t_p - 0$ and $t_p + 0$, following the positive orientation of γ. Let $\overline{X} = \sqcup_{p=1}^{m} \Delta_p$ be the disjoint union of segments $\Delta_p = [0, 1]$. Denote by Γ the union $\widehat{\gamma} \cup \overline{X}$ with the point identification

$$t_p - 0 \equiv 1_p, \qquad t_p + 0 \equiv 0_p,$$

where $t_p \pm 0 \in \widehat{\gamma}$, 0_p and 1_p are the boundary points of Δ_p, $p = 1, 2, \ldots, m$.

Then we obviously have

Theorem 7.7.5. *The C^*-algebra $\mathcal{T}_{BPC} = \mathcal{T}(BPC(\mathbb{D}, \Lambda))$ is irreducible and contains the ideal $\mathcal{K}$ of compact operators. The Fredholm symbol algebra $\operatorname{Sym} \mathcal{T}_{BPC} = \mathcal{T}_{BPC}/\mathcal{K}$ is isomorphic to the algebra $C(\Gamma)$. Identifying them, the symbol homomorphism*

$$\operatorname{sym} : \mathcal{T}_{BPC} \to \operatorname{Sym} \mathcal{T}_{BPC} = C(\Gamma)$$

is generated by the following mapping of generators of $\mathcal{T}_{BPC}$,

$$\text{sym} : T_a \longmapsto \begin{cases} a(t), & t \in \widehat{\gamma} \\ \gamma_{a_{t_p}}\left(\dfrac{1-2x}{\sqrt{1-(1-2x)^2}}\right), & x \in [0,1] \end{cases},$$

where a_{t_p} is the function defined by the above property (iii) for $a(z)$ at the point $t_p \in \Lambda$, $p = 1, 2, \ldots, m$, and

$$\gamma_{a_{t_p}}(\lambda) = \frac{2\lambda}{1 - e^{-2\pi\lambda}} \int_0^\pi a_{t_p}(\theta)\, e^{-2\lambda\theta}\, d\theta, \qquad \lambda \in \mathbb{R}.$$

The operator $T \in \mathcal{T}_{BPC}$ is Fredholm if and only if its symbol is invertible, i.e., the $\text{sym}\, T \neq 0$ on Γ, and

$$\text{Ind}\, T = -\frac{1}{2\pi}\{\text{sym}\, T\}_\Gamma.$$

Proof. Easily follows from the local principle, Theorem 7.2.4 and Theorem 7.6.2.
$\qquad\square$

We mention that the algebras described by Theorems 2.8.8, 7.6.2, and 7.7.5 consist of the same operators, in spite of the fact that their initial generators are quite different. As it turns out, the first algebra generated by Toeplitz operators with discontinuous defining symbols, which was described by Theorem 2.8.8, already contained all the operators with $BPC(\mathbb{D}, T)$-symbols. For about twenty years there was no way to see this. At the same time Theorem 7.7.5 gives a transparent description for all Toeplitz operators for all $BPC(\mathbb{D}, T)$-symbols.

Chapter 8

Anatomy of the Algebra Generated by Toeplitz Operators with Piece-wise Continuous Symbols

In this chapter we continue the study of the C^*-algebra generated by Toeplitz operators T_a with piece-wise continuous defining symbols a acting on the Bergman space $\mathcal{A}^2(\mathbb{D})$ on the unit disk $\mathbb{D}$. Our aim here is to describe explicitly each operator from this algebra and to characterize the Toeplitz operators which belong to the algebra.

The first structural result on Toeplitz operator algebras is due to L. Coburn [42] and goes back to early 1970s. It says, see Theorem 2.8.5, that *the C^*-algebra $\mathcal{T}(C(\overline{\mathbb{D}}))$ generated by Toeplitz operators with defining symbols continuous on $\overline{\mathbb{D}}$ is irreducible and contains the entire ideal $\mathcal{K}$ of compact operators on $\mathcal{A}^2(\mathbb{D})$. Every operator $T \in \mathcal{T}(C(\overline{\mathbb{D}}))$ is of the form*

$$T = T_a + K,$$

where $a \in C(\overline{\mathbb{D}})$ and K is a compact operator.

The key property of Toeplitz operators behind this result is that the semi-commutator of two Toeplitz operators with continuous defining symbols $[T_a, T_b) = T_a T_b - T_{ab}$ is compact.

The maximal class of defining symbols for which the above structural result remains true was introduced and studied by K. Zhu, see, for example, [237, 240]. For a brief description of this class see the end of Section 5.1. Recall that this class coincides with $Q = VMO_\partial(\mathbb{D}) \cap L_\infty(\mathbb{D})$ and is the maximal C^*-subalgebra of $L_\infty(\mathbb{D})$ having the compact semi-commutator property.

However, by Theorem 2.8.6, for piece-wise continuous defining symbols, the semi-commutators of Toeplitz operators are no longer compact in general. As was already mentioned, this immediately leads to a much more complicated structure for the C^*-algebra generated by such operators. Indeed, apart from the initial generators, the Toeplitz operators T_a with piece-wise continuous defining symbols, the algebra contains now all elements of the form

$$\sum_{k=1}^{p} \prod_{j=1}^{q_k} T_{a_{j,k}},$$

as well as the uniform limits of sequences of such elements.

The Fredholm symbol algebra for the C^*-algebra generated by Toeplitz operators T_a with piece-wise continuous defining symbols was described in Theorem 2.8.8. At the same time many important questions connected with the structure of the algebra itself remained unanswered. We list some of them in the following general setting.

Let $\mathcal{A} \in L_\infty(\mathbb{D})$ be a set (linear space or algebra) of initial defining symbols. Denote by $\mathcal{T}(\mathcal{A})$ the C^*-algebra generated by all Toeplitz operators T_a with symbols from $\mathcal{A}$. The following questions are of great importance.

(i) Describe the Fredholm symbol algebra $\operatorname{Sym} \mathcal{T}(\mathcal{A}) = \mathcal{T}(\mathcal{A})/\mathcal{T}(\mathcal{A}) \cap \mathcal{K}$ of the algebra $\mathcal{T}(\mathcal{A})$, as well as the symbol homomorphism $\operatorname{sym} : \mathcal{T}(\mathcal{A}) \to \operatorname{Sym} \mathcal{T}(\mathcal{A})$; here $\mathcal{K}$ is the ideal of compact operators on $\mathcal{A}^2(\mathbb{D})$.

(ii) Describe a canonical representation of elements forming $\mathcal{T}(\mathcal{A})$ clarifying thus the structure of the algebra $\mathcal{T}(\mathcal{A})$.

(iii) Given an element $\operatorname{sym} A$ from the Fredholm symbol algebra $\operatorname{Sym} \mathcal{T}(\mathcal{A})$, characterize an operator $A \in \mathcal{T}(\mathcal{A})$ having this image in the symbol algebra.

(iv) Characterize the Toeplitz operators T_b which belong to $\mathcal{T}(\mathcal{A})$, as well as the variety of their possible defining symbols b.

We consider here the case when $\mathcal{A}$ is the class of piece-wise continuous symbols defined in the next section. The complete answer to (i) is given by Theorem 2.8.8, while questions (ii)–(iv) remained unanswered. Our aim here is to answer these last questions.

In this connection we mention an intensive study of the question of when the product of two Toeplitz operators is a Toeplitz operator. Not pretending to be complete, we cite, for example, the papers by P. Ahern and Ž. Čučković [3, 4, 5], I. Louhichi, E. Strouse, and L. Zakariasy [133], and S. Pott and E. Strouse [160]. This interesting and important problem naturally leads to a more general question: *under what conditions will an application of any algebraic operation (summation, product, uniform limit) to Toeplitz operators produce a Toeplitz operator; which combinations of which Toeplitz operators give Toeplitz operators?* If we restrict ourselves to a specific class of initial Toeplitz operators, then this last question becomes precisely the fourth one from the above list.

8.1 Symbol class and operators

As was mentioned at the end of Section 7.7, considering Toeplitz operators with piece-wise continuous defining symbols, it turns out that neither the curves supporting the symbol discontinuities nor the number of such curves meeting at a boundary point of discontinuity play any essential role for the Toeplitz operator algebra studied. We can start from very different sets of defining symbols and obtain exactly the same operator algebra as a result. Thus, without loss of generality, we fix now a certain setup which is suitable for our needs.

We fix a finite number of distinct points $T = \{t_1, \dots, t_m\}$ on the boundary γ of the unit disk $\mathbb{D}$, and let

$$\delta = \min_{k \neq j}\{|t_k - t_j|, 1\}.$$

Denote by ℓ_k, $k = 1, \dots, m$, the part of the radius of $\mathbb{D}$ starting at t_k and having length $\delta/3$; and let $\mathcal{L} = \bigcup_{k=1}^{m} \ell_k$. We denote by $PC(\overline{\mathbb{D}}, T)$ the set (algebra) of all piece-wise continuous functions on $\mathbb{D}$ which are continuous in $\overline{\mathbb{D}} \setminus \mathcal{L}$ and have one-sided limit values at each point of $\mathcal{L}$. In particular, every function $a \in PC(\overline{\mathbb{D}}, T)$ has at each point $t_k \in T$ two (different, in general) limit values:

$$a^-(t_k) = a(t_k - 0) = \lim_{\gamma \ni t \to t_k,\, t \prec t_k} a(t) \quad \text{and} \quad a^+(t_k) = a(t_k + 0) = \lim_{\gamma \ni t \to t_k,\, t \succ t_k} a(t),$$

the signs $\pm$ here correspond to the standard orientation of the boundary γ of $\mathbb{D}$.

For each $k = 1, \dots, m$, denote by $\chi_k = \chi_k(z)$ the characteristic function of the half-disk obtained by cutting $\mathbb{D}$ by the diameter passing through $t_k \in T$, and such that $\chi_k^+(t_k) = 1$, and thus $\chi_k^-(t_k) = 0$.

For each $k = 1, \dots, m$, we introduce two neighborhoods of the point t_k:

$$V_k' = \{z \in \overline{\mathbb{D}} : |z - t_k| < \frac{\delta}{6}\} \quad \text{and} \quad V_k'' = \{z \in \overline{\mathbb{D}} : |z - t_k| < \frac{\delta}{3}\},$$

and fix a continuous function $v_k = v_k(z) : \overline{\mathbb{D}} \to [0, 1]$ such that

$$v_k|_{\overline{V_k'}} \equiv 1, \qquad v_k|_{\overline{\mathbb{D}} \setminus V_k''} \equiv 0.$$

As an easy consequence of Theorem 2.4.5 we have the following result.

Lemma 8.1.1. *The following properties hold:*

(i) *let $L_\infty^0(\mathbb{D})$ be the closure in $L_\infty(\mathbb{D})$ of the set of all L_∞-functions having compact support in $\mathbb{D}$. Then for each function $a \in L_\infty^0(\mathbb{D})$ the Toeplitz operator T_a is compact;*

(ii) *for each pair of functions $a \in L_\infty(\mathbb{D})$ and $b \in C(\overline{\mathbb{D}})$ the semi-commutator $[T_a, T_b) = T_a T_b - T_{ab}$ is compact;*

(iii) *for each pair of functions $a \in L_\infty(\mathbb{D})$ and $b \in C(\overline{\mathbb{D}})$ the commutator $[T_a, T_b] = T_a T_b - T_b T_a$ is compact.*

Using Lemma 8.1.1 it is easy to see that for any defining symbol $a \in PC(\overline{\mathbb{D}}, T)$, the Toeplitz operator T_a admits the canonical representations

$$
\begin{aligned}
T_a &= T_{s_a} + \sum_{k=1}^{m} T_{v_k} p_{a,k}(T_{\chi_k}) T_{v_k} + K \\
&= T_{s_a} + \sum_{k=1}^{m} T_{u_k} p_{a,k}(T_{\chi_k}) + K' \\
&= T_{s_a} + \sum_{k=1}^{m} p_{a,k}(T_{\chi_k}) T_{u_k} + K'',
\end{aligned}
$$

where $s_a(z)$ is a continuous function on $\overline{\mathbb{D}}$ such that the following restrictions on γ coincide:

$$
s_a(z)|_\gamma \equiv \left[a(z) - \sum_{k=1}^{m} [a^-(t_k) + (a^+(t_k) - a^-(t_k))\chi_k(z)] u_k(z) \right]_\gamma ,
$$

and where

$$
p_{a,k}(x) = a^-(t_k) + (a^+(t_k) - a^-(t_k))x = a^-(t_k)(1 - x) + a^+(t_k)x, \quad k = 1, \ldots, m,
$$

are first-order polynomials in x, $u_k(z) = v_k(z)^2$, and K, K', K'' are compact operators.

Indeed, by the second statement of the lemma, each right-hand side operator is a compact perturbation of the Toeplitz operator $T_{\widetilde{a}}$, where

$$
\widetilde{a}(z) = s_a(z) + \sum_{k=1}^{m} [a^-(t_k) + (a^+(t_k) - a^-(t_k))\chi_k(z)] u_k(z).
$$

We note that each function $\chi_k(z) u_k(z)$, $k = 1, \ldots, m$, belongs to $PC(\overline{\mathbb{D}}, T)$ and that $s_a(t_k) = 0$ for all $t_k \in T$. Then the difference $a(z) - \widetilde{a}(z)$ is continuous at every point of the boundary γ and $[a(z) - \widetilde{a}(z)]_\gamma \equiv 0$. Thus by the first statement of the lemma the difference $T_a - T_{\widetilde{a}}$ is compact.

Such representations are essentially unique in the sense that the values of $s_a(z)$ on γ are uniquely defined and, if the function $s_a(z)$ is changed for another one with the same boundary values, the result will be altered at most by a compact operator.

8.2   Algebra $\mathcal{T}(PC(\overline{\mathbb{D}}, T))$

Recall that according to Theorem 2.8.8 the Fredholm symbol algebra

$$
\operatorname{Sym} \mathcal{T}(PC(\overline{\mathbb{D}}, T)) = \mathcal{T}(PC(\overline{\mathbb{D}}, T))/\mathcal{K}
$$

of the algebra $\mathcal{T}(PC(\overline{\mathbb{D}}, T))$ admits the following description.

Let $\widehat{\gamma}$ be the boundary γ cut at the points $t_k \in T$. The pair of points of $\widehat{\gamma}$ which correspond to the point $t_k \in T$, $k = 1, \ldots, m$, will be denoted by $t_k - 0$ and $t_k + 0$, following the positive orientation of γ. Let $\overline{X} = \bigsqcup_{k=1}^{m} \Delta_k$ be the disjoint union of segments $\Delta_k = [0, 1]$. Denote by Γ the union $\widehat{\gamma} \cup \overline{X}$ with the point identification

$$t_k - 0 \equiv 0_k, \qquad t_k + 0 \equiv 1_k,$$

where $t_k \pm 0 \in \widehat{\gamma}$, 0_k and 1_k are the boundary points of Δ_k, $k = 1, \ldots, m$.

Theorem 8.2.1. *The Fredholm symbol algebra* $\operatorname{Sym}\mathcal{T}(PC(\overline{\mathbb{D}}, T)) = \mathcal{T}(PC(\overline{\mathbb{D}}, T))/\mathcal{K}$ *of the algebra* $\mathcal{T}(PC(\overline{\mathbb{D}}, T))$ *is isomorphic and isometric to the algebra* $C(\Gamma)$. *The homomorphism*

$$\operatorname{sym} : \ \mathcal{T}(PC(\overline{\mathbb{D}}, T)) \longrightarrow \operatorname{Sym}\mathcal{T}(PC(\overline{\mathbb{D}}, T)) \cong C(\Gamma)$$

is generated by the mapping of generators of $\mathcal{T}(PC(\overline{\mathbb{D}}, T))$

$$\operatorname{sym} : T_a \longmapsto \begin{cases} a(t), & t \in \widehat{\gamma} \\ a(t_k - 0)(1 - x) + a(t_k + 0)x, & x \in [0, 1] \end{cases},$$

where $t_k \in T$, $k = 1, 2, \ldots, m$.

In what follows we will use two different descriptions of the local algebras $\widehat{\mathcal{T}}(t_k)$, for $t_k \in T \subset \gamma$, which we now proceed to describe.

As a Toeplitz operator T_a with defining symbol continuous at the point t_k is locally equivalent at the point t_k to the scalar operator $a(t_k)I = T_{a(t_k)}$, the local algebra $\widehat{\mathcal{T}}(t_k)$ is the C^*-algebra with identity generated by the single self-adjoint element T_{χ_k}, and thus is isomorphic and isometric to $C(\operatorname{sp} T_{\chi_k})$. By (8.2.2) it is obvious that $\operatorname{sp} T_{\chi_k} = [0, 1]$. Thus as the first description of the local algebra $\widehat{\mathcal{T}}(t_k)$ we have

The local algebra $\widehat{\mathcal{T}}(t_k)$ *is isomorphic and isometric to* $C[0, 1]$ *and the isomorphism*

$$\pi'_{t_k} : \ \widehat{\mathcal{T}}(t_k) \longrightarrow C[0, 1]$$

is generated by the mapping $\pi'_{t_k} : T_{\chi_k} \mapsto x$, *where* $x \in \Delta_k = [0, 1]$.

For the second description we construct a unitary operator directly reducing T_{χ_k} to a multiplication operator.

We introduce the Möbius transformation

$$\alpha_k(z) = i\frac{t_k - z}{z + t_k},$$

which maps the unit disk $\mathbb{D}$ onto Π, sending the point t_k to 0 and the opposite point $-t_k$ to ∞. Then

$$(V_k \varphi)(z) = -\frac{2it_k}{(z + t_k)^2} \varphi\left(i\frac{t_k - z}{z + t_k}\right) \tag{8.2.1}$$

is obviously a unitary operator both from $L_2(\Pi)$ onto $L_2(\mathbb{D})$, and from $\mathcal{A}^2(\Pi)$ onto $\mathcal{A}^2(\mathbb{D})$, and its inverse (and adjoint) has the form

$$(V_k^{-1}\varphi)(w) = -\frac{2it_k}{(w+i)^2}\,\varphi\left(t_k\frac{i-w}{w+i}\right).$$

It is a simple calculation to check that

$$V_k T_{\chi_k} V_k^{-1} = T_{\chi_+}, \tag{8.2.2}$$

where χ_+ is the characteristic function of the right quarter-plane in Π.

Then by Theorem 7.2.4 and Corollary 7.2.6 the C^*-algebra with identity generated by T_{χ_+} coincides with the algebra $\mathcal{T}(H(L_\infty^{(0,\pi)}(0,\pi)))$ and is isomorphically isometric to $C(\overline{\mathbb{R}})$. The isomorphism

$$\pi_+ : \mathcal{T}(H(L_\infty^{(0,\pi)}(0,\pi))) \longrightarrow C(\overline{\mathbb{R}})$$

is generated by the mapping $\pi_+ : T_{\chi_+} \longmapsto \gamma_{\chi_+}(\lambda)$, where, see (7.7.2),

$$\gamma_{\chi_+}(\lambda) = \frac{1}{e^{-\pi\lambda}+1}, \qquad \lambda \in \overline{\mathbb{R}}. \tag{8.2.3}$$

Now to obtain the second description of the local algebra $\widehat{\mathcal{T}}(t_k)$ we consider the unitary operator $U_k = RV_k$, where the operator R is defined in Section 7.1, and note that

$$U_k T_{\chi_k} U_k^{-1} = \gamma_{\chi_+}(\lambda) \qquad \text{or} \qquad T_{\chi_k} = U_k^{-1}\gamma_{\chi_+}(\lambda)U_k. \tag{8.2.4}$$

Thus we have: *The local algebra $\widehat{\mathcal{T}}(t_k)$ is isomorphic and isometric to $C(\overline{\mathbb{R}})$ and the isomorphism*

$$\pi_{t_k}'' : \widehat{\mathcal{T}}(t_k) \longrightarrow C(\overline{\mathbb{R}})$$

is generated by the mapping $\pi_{t_k}'' : T_{\chi_k} \mapsto \gamma_{\chi_+}(\lambda)$, where $\lambda \in \overline{\mathbb{R}}$.

We summarize the above in the next proposition.

Proposition 8.2.2. *For each point $t_k \in T$, the local algebra $\widehat{\mathcal{T}}(t_k)$ consists of all operators of the form $f(T_{\chi_k})$, where $f \in C[0,1]$. Each such operator admits the representation*

$$f(T_{\chi_k}) = U_k^{-1}f(\gamma_{\chi_+}(\lambda))U_k.$$

8.3 Operators of the algebra $\mathcal{T}(PC(\overline{\mathbb{D}},T))$

As has been mentioned, the algebra $\mathcal{T}(PC(\overline{\mathbb{D}},T))$, apart from its initial generators T_a with $a \in PC(\overline{\mathbb{D}},T)$, contains all elements of the form

$$\sum_{k=1}^{p}\prod_{j=1}^{q_k} T_{a_{j,k}}, \tag{8.3.1}$$

as well as the uniform limits of sequences of such elements. Our aim here is to characterize each operator from the algebra $T(PC(\overline{\mathbb{D}}, T))$ up to a compact summand.

We start with the following lemma.

Lemma 8.3.1. *For each $n \in \mathbb{N}$ and each $k = 1, \ldots, m$, there is a function $s_{n,k} = s_{n,k}(z) \in C(\overline{\mathbb{D}})$ and a compact operator $K_{n,k}$ such that*

$$(T_{v_k} T_{\chi_k} T_{v_k})^n = T_{v_k} T^n_{\chi_k} T_{v_k} + T_{s_{n,k}} + K_{n,k}.$$

Proof. We have obviously

$$
\begin{aligned}
(T_{v_k} T_{\chi_k} T_{v_k})^n &= T_{v_k^n} T^n_{\chi_k} T_{v_k^n} + K', \\
T_{v_k} T^n_{\chi_k} T_{v_k} &= \left(T_{v_k^{1/n}} T_{\chi_k} T_{v_k^{1/n}} \right)^n + K'',
\end{aligned}
$$

where K' and K'' are compact operators. Thus both operators $T_{v_k^n} T^n_{\chi_k} T_{v_k^n}$ and $T_{v_k} T^n_{\chi_k} T_{v_k}$ belong to the algebra $T(PC(\overline{\mathbb{D}}, T))$. Calculating their Fredholm symbols we have that

$$
\begin{aligned}
\mathrm{sym}\left(T_{v_k^n} T^n_{\chi_k} T_{v_k^n} - T_{v_k} T^n_{\chi_k} T_{v_k} \right) &= s_{n,k}(t) \\
&= \begin{cases} [v_k^{2n}(t) - v_k^2(t)]\chi_k(t), & t \in \widehat{\gamma} \setminus (V_k' \cup (\overline{\mathbb{D}} \setminus V_k'')) \\ 0, & t \in \widehat{\gamma} \cap (V_k' \cup (\overline{\mathbb{D}} \setminus V_k'')) \end{cases},
\end{aligned}
$$

which is a continuous function on γ. Extending $s_{n,k}(t)$ to a continuous function on $\overline{\mathbb{D}}$ and returning from symbols to operators we obtain the desired property. $\square$

Corollary 8.3.2. *For every polynomial $p(x)$ and each $k = 1, \ldots, m$, the operator $A_{p,k} = T_{v_k} p(T_{\chi_k}) T_{v_k}$ belongs to the algebra $T(PC(\overline{\mathbb{D}}, T))$, and*

$$(\mathrm{sym}\, A_{p,k})|_{\Delta_k} = p(x), \qquad x \in [0, 1].$$

Corollary 8.3.3. *Each operator A of the form (8.3.1) admits the canonical representation*

$$A = \sum_{i=1}^{p} \prod_{j=1}^{q_i} T_{a_{i,j}} = T_{s_A} + \sum_{k=1}^{m} T_{v_k} p_{A,k}(T_{\chi_k}) T_{v_k} + K_A,$$

where $s_A = s_A(z) \in C(\overline{\mathbb{D}})$, $p_{A,k} = p_{A,k}(x)$, $k = 1, \ldots, m$, are polynomials, and K_A is a compact operator.

Lemma 8.3.4. *Let $f \in C[0, 1]$. Then for each $k = 1, \ldots, m$ the operator $A_{f,k} = T_{v_k} f(T_{\chi_k}) T_{v_k}$ belongs to the algebra $T(PC(\overline{\mathbb{D}}, T))$, and*

$$(\mathrm{sym}\, A_{f,k})|_{\Delta_k} = f(x), \qquad x \in [0, 1].$$

Proof. Recall that the operator T_{χ_k} is self-adjoint and its spectrum is equal to $[0,1]$. Let $\{p_n(x)\}_{n\in\mathbb{N}}$ be a sequence of polynomials which converges uniformly on $[0,1]$ to the function $f(x)$. Then by the standard functional calculus in a C^*-algebra we have

$$\|p_n(T_{\chi_k}) - f(T_{\chi_k})\| = \sup_{x\in[0,1]} |p_n(x) - f(x)|,$$

and thus $A_{f,k}$ is the uniform limit of the sequence $\{T_{v_k}p_n(T_{\chi_k})T_{v_k}\}_{n\in\mathbb{N}}$ of the elements of the algebra $\mathcal{T}(PC(\overline{\mathbb{D}},T))$.

Finally, the restriction $(\operatorname{sym} A_{f,k})|_{\Delta_k}$ coincides with the uniform limit of the restrictions $(\operatorname{sym} T_{v_k}p_n(T_{\chi_k})T_{v_k})|_{\Delta_k} = p_n(x)$, $x \in [0,1]$, thus giving the desired result. $\qquad\qquad\square$

The next theorem starts the characterization of operators from the algebra $\mathcal{T}(PC(\overline{\mathbb{D}},T))$, representing them in certain canonical forms.

Theorem 8.3.5. *Every operator $A \in \mathcal{T}(PC(\overline{\mathbb{D}},T))$ admits the canonical representations*

$$
\begin{aligned}
A &= T_{s_A} + \sum_{k=1}^{m} T_{v_k} f_{A,k}(T_{\chi_k})T_{v_k} + K \\
&= T_{s_A} + \sum_{k=1}^{m} T_{u_k} f_{A,k}(T_{\chi_k}) + K' \\
&= T_{s_A} + \sum_{k=1}^{m} f_{A,k}(T_{\chi_k})T_{u_k} + K'',
\end{aligned}
$$

where $s_A(z)$ is a continuous function on $\overline{\mathbb{D}}$, $u_k(z) = v_k(z)^2$, $f_{A,k}(x)$, $k = 1,\ldots,m$, are continuous functions on $[0,1]$, and K, K', K'' are compact operators.

Before we pass to the proof, we note that such representations have already been obtained in Section 8.1 for the generators T_a, where $a \in PC(\overline{\mathbb{D}},T)$, of the algebra $A \in \mathcal{T}(PC(\overline{\mathbb{D}},T))$ and for elements of the form (8.3.1) in Corollary 8.3.3.

As in Section 8.1, these representations are essentially unique in the sense that the values of $s_A(z)$ on γ and the functions $f_{A,k}(x)$ are uniquely defined by the operator A, and if the function $s_A(z)$ is changed for another one with the same boundary values the result will be altered at most by a compact operator.

Proof. We will show the first representation only; the other two follows from the fact that operators from the algebra $\mathcal{T}(PC(\overline{\mathbb{D}},T))$ commute modulo a compact operator.

Hence, given an operator $A \in \mathcal{T}(PC(\overline{\mathbb{D}},T))$, we introduce the functions $f_{A,k}(x) \in C[0,1]$, $k = 1,\ldots,m$, by

$$f_{A,k}(x) = (\operatorname{sym} A)|_{\Delta_k}, \qquad x \in [0,1].$$

Then the Fredholm symbol of the operator $A - \sum_{k=1}^{m} T_{v_k} f_{A,k}(T_{\chi_k}) T_{v_k}$ has the form

$$\text{sym}\left(A - \sum_{k=1}^{m} T_{v_k} f_{A,k}(T_{\chi_k}) T_{v_k}\right) = \begin{cases} s_A(t) & t \in \widehat{\gamma} \\ 0, & x \in \Delta_k, \ k = 1, \ldots, m \end{cases},$$

where

$$s_A(t) = (\text{sym}\, A)(t) - \sum_{k=1}^{m} v_k^2(t)[f_{A,k}(0)(1 - \chi_k(t)) + f_{A,k}(1)\chi_k(t)]$$

is a continuous function on γ, and such that $s_A(t_k) = 0$ for all $t_k \in T$.

To finish the proof we extend s_A to a continuous function on $\overline{\mathbb{D}}$ and return from symbols to operators. $\qquad\square$

Theorem 8.3.5 and Proposition 8.2.2 lead to the next characterization of elements of the algebra $\mathcal{T}(PC(\overline{\mathbb{D}}, T))$.

Corollary 8.3.6. *Every operator $A \in \mathcal{T}(PC(\overline{\mathbb{D}}, T))$ admits the representations*

$$\begin{aligned}
A &= T_{s_A} + \sum_{k=1}^{m} T_{v_k} U_k^{-1} f_{A,k}(\gamma_{\chi_+}(\lambda)) U_k T_{v_k} + K \\
&= T_{s_A} + \sum_{k=1}^{m} T_{u_k} U_k^{-1} f_{A,k}(\gamma_{\chi_+}(\lambda)) U_k + K' \\
&= T_{s_A} + \sum_{k=1}^{m} U_k^{-1} f_{A,k}(\gamma_{\chi_+}(\lambda)) U_k T_{u_k} + K'',
\end{aligned}$$

where $s_A(z)$ is a continuous function on $\overline{\mathbb{D}}$ whose restriction to γ is given by

$$s_A(t) = (\text{sym}\, A)(t) - \sum_{k=1}^{m} u_k(t)[f_{A,k}(0)(1 - \chi_k(t)) + f_{A,k}(1)\chi_k(t)],$$

where $f_{A,k}(x) = (\text{sym}\, A)|_{\Delta_k}$, the operators U_k are defined in (8.2.4), $u_k(x) = v_k^2(x)$, $k = 1, \ldots, m$, and K, K', K'' are compact operators.

8.4   Toeplitz operators of the algebra $\mathcal{T}(PC(\overline{\mathbb{D}}, T))$

In this section we show that, apart from the initial generators, the C^*-algebra $\mathcal{T}(PC(\overline{\mathbb{D}}, T))$ contains many other (non-compact) Toeplitz operators which are drastically different from the initial generators. By Toeplitz operator in this section we always mean a Toeplitz operator with *bounded measurable* defining symbol.

Let A be an operator of the algebra $\mathcal{T}(PC(\overline{\mathbb{D}}, T))$. By Theorem 8.3.5 it admits the canonical representation

$$A = T_{s_A} + \sum_{k=1}^{m} T_{v_k} f_{A,k}(T_{\chi_k}) T_{v_k} + K.$$

Lemma 8.4.1. *The operator A is a compact perturbation of a Toeplitz operator if and only if each operator $T_{v_k} f_{A,k}(T_{\chi_k}) T_{v_k}$, $k = 1, \ldots, m$, is a compact perturbation of a Toeplitz operator.*

Proof. The "if" part is obvious. To prove the "only if" part we assume that $A = T_a + K_1$ for some $a \in L_\infty(\mathbb{D})$. Using $v_k^{1/2}$ in place of v_k in Theorem 8.3.5 we represent the operator A in its second canonical form

$$A = T_{s'_A} + \sum_{k=1}^{m} T_{v_k} f_{A,k}(T_{\chi_k}) + K'.$$

Then, multiplying by T_{v_k} and using statement (ii) of Lemma 8.1.1, we have

$$AT_{v_k} = T_{s'_A v_k} + T_{v_k} f_{A,k}(T_{\chi_k}) T_{v_k} + K_2 = T_{av_k} + K_3,$$

or

$$T_{v_k} f_{A,k}(T_{\chi_k}) T_{v_k} = T_{av_k - s'_A v_k} + (K_3 - K_2). \qquad \square$$

The result of Lemma 8.4.1 obviously remains true if we change the operators $T_{v_k} f_{A,k}(T_{\chi_k}) T_{v_k}$ for either $T_{u_k} f_{A,k}(T_{\chi_k})$, or $f_{A,k}(T_{\chi_k}) T_{u_k}$, $k = 1, \ldots, m$.

Theorem 8.4.2. *For any $k = 1, \ldots, m$, the operator $T_{v_k} f_{A,k}(T_{\chi_k}) T_{v_k}$ is a compact perturbation of a Toeplitz operator if and only if the operator $f_{A,k}(T_{\chi_k})$ is a compact perturbation of a Toeplitz operator.*

Proof. The "if" part is again obvious. To prove the "only if" part we first reduce the problem to the real-valued function $f_{A,k}$. To this end we assume that

$$T_{v_k} f_{A,k}(T_{\chi_k}) T_{v_k} = T_a + K_1, \qquad \text{for some } a \in L_\infty(\mathbb{D}).$$

Passing to adjoint operators and taking into account that the functions v_k and χ_k are real-valued, we have

$$(T_{v_k} f_{A,k}(T_{\chi_k}) T_{v_k})^* = T_{v_k} \overline{f}_{A,k}(T_{\chi_k}) T_{v_k} = T_{\bar{a}} + K_2.$$

Summing up these equalities we have that $T_{v_k}(\mathrm{Re} f_{A,k})(T_{\chi_k}) T_{v_k}$ is a compact perturbation of a Toeplitz operator. Subtracting the equalities, we have that $T_{v_k}(\mathrm{Im} f_{A,k})(T_{\chi_k}) T_{v_k}$ is a compact perturbation of a Toeplitz operator as well.

That is, the operator $T_{v_k} f_{A,k}(T_{\chi_k}) T_{v_k}$ is a compact perturbation of a Toeplitz operator if and only if both $T_{v_k}(\mathrm{Re} f_{A,k})(T_{\chi_k}) T_{v_k}$ and $T_{v_k}(\mathrm{Im} f_{A,k})(T_{\chi_k}) T_{v_k}$ are compact perturbations of Toeplitz operators. Thus proving the part "only if" we can assume that the function $f_{A,k}$ is real-valued, moreover we can consider the operator $f_{A,k}(T_{\chi_k})T_{u_k}$ instead of $T_{v_k}f_{A,k}(T_{\chi_k})T_{v_k}$.

We note as well that without loss of generality we may assume in what follows that $t_k = i \in \gamma$, because otherwise, using an appropriate rotation, we come to the unitary equivalent operator with $t_k = i \in \gamma$.

Hence, let $t_k = i$ and let $f_{A,k}$ be a real-valued function such that $f_{A,k}(T_{\chi_k})T_{u_k} = T_a + K_1$ for some $a \in L_\infty(\mathbb{D})$.

We introduce now the operator

$$(Z\varphi)(z) = \overline{\varphi(\overline{z})},$$

which is obviously unitary on both $L_2(\mathbb{D})$ and $\mathcal{A}^2(\mathbb{D})$. Then, as is easy to see,

$$Z f_{A,k}(T_{\chi_k(z)})T_{u_k(z)}Z = f_{A,k}(T_{\chi_k(z)})T_{u_k(\overline{z})} = T_{\overline{a(\overline{z})}} + K_2,$$

and thus

$$f_{A,k}(T_{\chi_k(z)})\left(T_{u_k(z)} + T_{u_k(\overline{z})}\right) = T_b + K_3,$$

where $b(z) = a(z) + \overline{a(\overline{z})}$, or

$$f_{A,k}(T_{\chi_k(z)}) = f_{A,k}(T_{\chi_k(z)})\left(I - T_{u_k(z)} - T_{u_k(\overline{z})}\right) + T_b + K_3.$$

We note that the operator $f_{A,k}(T_{\chi_k(z)})\left(I - T_{u_k(z)} - T_{u_k(\overline{z})}\right)$ belongs to the algebra $\mathcal{T}(PC(\overline{\mathbb{D}}, T'))$ with $T' = \{i, -i\}$ (that is, we have only two points of symbol discontinuity: $t_1 = i$ and $t_2 = -i$) and its symbol is a continuous function on γ (that is, a continuous function on Γ which is constant on each Δ_j, $j = 1, 2$) and is identically equal to 0 at $\gamma \cap (V_k' \cup \overline{V_k'}) = \gamma \cap (V_1' \cup V_2')$. Thus the operator $f_{A,k}(T_{\chi_k(z)})\left(I - T_{u_k(z)} - T_{u_k(\overline{z})}\right)$ is a compact perturbation of some Toeplitz operator T_c with continuous defining symbol c, and thus we have finally

$$f_{A,k}(T_{\chi_k}) = T_{b+c} + K_3. \qquad \qquad \square$$

By Proposition 8.2.2 every operator of the form $f(T_{\chi_k})$, with $f \in C[0,1]$, is unitary equivalent to the multiplication operator $(f \circ \gamma_+)I$. That is, the C^*-algebra generated by (and consisting of) all such operators intersects the ideal $\mathcal{K}$ of compact operators in just the zero operator. This implies that an operator of the form $f(T_{\chi_k})$ is a compact perturbation of a Toeplitz operator if and only if it is a Toeplitz operator itself.

Summarizing the above we come to the main result of the section.

Theorem 8.4.3. *An operator $A \in \mathcal{T}(PC(\overline{\mathbb{D}}, T))$ is a compact perturbation of a Toeplitz operator if and only if every operator $f_{A,k}(T_{\chi_k})$ is a Toeplitz operator, where $f_{A,k} = (\mathrm{sym}\, A)|_{\Delta_k}$ and $k = 1, \ldots, m$.*

The next theorem gives the description of the defining symbol of a Toeplitz operator for the case when $A \in \mathcal{T}(PC(\overline{\mathbb{D}}, T))$ is of the form $A = T_a + K$.

Theorem 8.4.4. *Let $A = T_a + K$, thus, for each $k = 1, \ldots, m$, the operators $(\operatorname{sym} A)|_{\Delta_k}(T_{\chi_k})$ are Toeplitz, i.e., $(\operatorname{sym} A)|_{\Delta_k}(T_{\chi_k}) = T_{a_k}$ for some $a_k \in L_\infty(\mathbb{D})$. Then the defining symbol a of the operator T_a is given by*

$$a(z) = s_A(z) + \sum_{k=1}^{m} a_k(z) v_k^2(z),$$

where $s_A(z)$ is a continuous function on $\overline{\mathbb{D}}$ whose restriction to γ coincides with

$$s_A(t) = (\operatorname{sym} A)(t) - \sum_{k=1}^{m} \left[(\operatorname{sym} A)|_{\Delta_k}(0)(1 - \chi_k(t)) + (\operatorname{sym} A)|_{\Delta_k}(1)\chi_k(t) \right] v_k^2(t).$$

$$(8.4.1)$$

Proof. Follows directly from Corollary 8.3.6. $\square$

Note that the operators $f(T_{\chi_k})$ and $f(T_{\chi_+})$, being unitary equivalent, can be Toeplitz operators only simultaneously. That is, the question *whether an operator $A \in \mathcal{T}(PC(\overline{\mathbb{D}}, T))$ is a compact perturbation of a Toeplitz operator* reduces to the description of the Toeplitz operators in the algebra $\mathcal{T}(H(L_\infty^{(0,\pi)}(0, \pi)))$. By Theorem 7.2.1 this algebra can be generated by T_+ alone, and thus consists of all operators of the form $f(T_{\chi_+})$, where $f \in C[0,1]$.

Many Toeplitz operators in the algebra $\mathcal{T}(H(L_\infty^{(0,\pi)}(0, \pi)))$ and the corresponding functions $f \in C[0,1]$ are given by Examples 7.7.1–7.7.3. In particular, Example 7.7.3 states that for any defining symbol $a = a(\theta) \in H(L_\infty^{(0,\pi)}(0, \pi))$, the Toeplitz operator T_a belongs to the algebra $\mathcal{T}(H(L_\infty^{(0,\pi)}(0, \pi)))$, and is the following function of the operator T_{χ_+},

$$T_a = f_a(T_{\chi_+}),$$

where

$$f_a(x) = \frac{2x^2}{\pi} \frac{\ln(1 - x) - \ln x}{(1 - x) - x} \int_0^\pi a(\theta) \left(\frac{1 - x}{x} \right)^{\frac{2\theta}{\pi}} d\theta.$$

Thus, as a corollary, we have the following statement.

Proposition 8.4.5. *For each function $a = a(w) = a(e^{i\theta}) \in H(L_\infty^{(0,\pi)}(0, \pi))$, where $w = re^{i\theta} \in \Pi$, and each $k = 1, \ldots, m$, the Toeplitz operator*

$$T_{b_k} = V_k T_a V_k^{-1}$$

belongs to the algebra $\mathcal{T}(PC(\overline{\mathbb{D}}, \{t_k, -t_k\}))$ and has the defining symbol

$$b_k(z) = a(\alpha_k(z)) = a\left(i\frac{t_k - z}{z + t_k} \right).$$

Here the operator V_k is given by (8.2.1).

We can describe therefore the defining symbols of a wide variety of Toeplitz operators in $\mathcal{T}(PC(\overline{\mathbb{D}}, T))$ which are drastically different from the initial generators. All of them have at each point of discontinuity $t_k \in T$, in general, infinitely many limit values reached by the hypercycles starting at t_k (i.e., the images under the Möbius transformation α_k^{-1} of rays on the upper half-plane Π starting at origin) and parameterized by functions from $L_\infty^{(0,\pi)}(0, \pi)$. We note that each of these (bounded) defining symbols b have one-sided limit values at the point t_k and these limit values coincide with the values of $\operatorname{sym} T_b$ at the endpoints of Δ_k:

$$b(t_k - 0) = (\operatorname{sym} T_b)(0_k), \qquad b(t_k + 0) = (\operatorname{sym} T_b)(1_k).$$

Corollary 8.4.6. *For every function $a_k = a_k(w) = a_k(e^{i\theta}) \in H(L_\infty^{(0,\pi)}(0, \pi))$, where $w = re^{i\theta} \in \Pi$, $k = 1, \ldots, m$, and every function $s(z) \in L_\infty(\mathbb{D})$ having limits at all points of γ and such that $s|_\gamma \in C(\gamma)$, the Toeplitz operator T_b with defining symbol*

$$b(z) = s(z) + \sum_{k=1}^{m} a_k \left(i \frac{t_k - z}{z + t_k} \right) u_k(x)$$

belongs to the algebra $\mathcal{T}(PC(\overline{\mathbb{D}}, T))$.

8.5 More Toeplitz operators

In the previous section we reduced the description of Toeplitz operators in the algebra $\mathcal{T}(PC(\overline{\mathbb{D}}, T))$ to the description of Toeplitz operators in $\mathcal{T}(H(L_\infty^{(0,\pi)}(0, \pi)))$. We show now that the algebra $\mathcal{T}(H(L_\infty^{(0,\pi)}(0, \pi)))$ contains many more Toeplitz operators than described in the previous section. Indeed, as we will see, it also contains (bounded) Toeplitz operators whose generally unbounded defining symbols $a(\theta)$ may not have limits at the endpoints 0 and π of the segment $[0, \pi]$.

We recall that the Toeplitz operator T_a with defining symbol $a(\theta)$ belongs to the algebra $\mathcal{T}(H(L_\infty^{(0,\pi)}(0, \pi)))$ if and only if the corresponding function $\gamma_a(\lambda)$, defined by (7.2.1), belongs to $C(\overline{\mathbb{R}})$.

Remark 8.5.1. Given a defining symbol $a(\theta)$, in what follows we will study the behavior of the function $\gamma_a(\lambda)$ when $\lambda \to \pm\infty$. It is clear that the behavior of $a(\theta)$ near the point 0, or π, determines the behavior of $\gamma_a(\lambda)$ near the point $+\infty$, or $-\infty$, respectively. The equality

$$
\begin{aligned}
\gamma_{a(\theta)}(-\lambda) &= \frac{-2\lambda}{1 - e^{2\pi\lambda}} \int_0^\pi a(\theta) e^{2\lambda\theta} d\theta = \frac{-2\lambda}{1 - e^{2\pi\lambda}} \int_0^\pi a(\pi - \theta) e^{2\lambda(\pi - \theta)} d\theta \\
&= \frac{2\lambda}{1 - e^{-2\pi\lambda}} \int_0^\pi a(\pi - \theta) e^{-2\lambda\theta} d\theta = \gamma_{a(\pi-\theta)}(\lambda)
\end{aligned}
$$

permits us to reduce this study to only one case, say considering the symbol $a(\theta)$ in a neighborhood of 0 and $\gamma_a(\lambda)$ in a neighborhood of $+\infty$.

We continue to consider the homogeneous functions of zero order on the upper half-plane Π identifying them with functions $a(\theta)$, where $\theta \in [0, \pi]$.

For any L_1-symbol $a(\theta)$ we define the following averaging functions, which correspond to the endpoints of $[0, \pi]$,

$$C_a^{(1)}(\theta) = \int_0^\theta a(u)\,du, \qquad D_a^{(1)}(\theta) = \int_{\pi-\theta}^\pi a(u)\,du$$

and

$$C_a^{(p)}(\theta) = \int_0^\theta C_a^{(p-1)}(u)\,du, \qquad D_a^{(p)}(\theta) = \int_{\pi-\theta}^\pi D_a^{(p-1)}(u)\,du,$$

for each $p = 2, 3, \ldots$.

The next theorem gives the conditions on the behavior of L_1-symbols near endpoints 0 and π guaranteeing that the corresponding Toeplitz operators belong to the algebra $T(H(L_\infty^{(0,\pi)}(0, \pi)))$.

Theorem 8.5.2. *Let $a(\theta) \in L_1(0, \pi)$ and suppose that for some $p, q \in \mathbb{N}$,*

$$\lim_{\theta \to 0} \theta^{-p} C_a^{(p)}(\theta) = c_p \ (\in \mathbb{C}) \qquad \text{and} \qquad \lim_{\theta \to \pi} \theta^{-q} D_a^{(q)}(\theta) = d_q \ (\in \mathbb{C}). \quad (8.5.1)$$

Then $\gamma_a(\lambda) \in C(\overline{\mathbb{R}})$.

Proof. Consider first the case when $p = 1$ and $\lambda \to +\infty$. Integrating by parts we have

$$
\begin{aligned}
\gamma_a(\lambda) \quad &= \quad \frac{2\lambda}{1 - e^{-2\pi\lambda}} \int_0^\pi e^{-2\lambda\theta}\, dC_a^{(1)}(\theta) \\[2mm]
&= \quad \frac{2\lambda e^{-2\pi\lambda}}{1 - e^{-2\pi\lambda}} C_a^{(1)}(\pi) + \frac{4\lambda^2}{1 - e^{-2\pi\lambda}} \int_0^\pi C_a^{(1)}(\theta)\, e^{-2\lambda\theta}\, d\theta.
\end{aligned}
$$

Taking into account the first equality in (8.5.1), we have

$$
\begin{aligned}
\gamma_a(\lambda) \quad &= \quad \frac{2\lambda e^{-2\pi\lambda}}{1 - e^{-2\pi\lambda}} C_a^{(1)}(\pi) + \frac{4\lambda^2 c_1}{1 - e^{-2\pi\lambda}} \int_0^\pi \theta\, e^{-2\lambda\theta}\, d\theta \\[2mm]
&\quad + \frac{4\lambda^2 c_1}{1 - e^{-2\pi\lambda}} \int_0^\pi \alpha(\theta)\theta\, e^{-2\lambda\theta}\, d\theta \quad = \quad I_1(\lambda) + I_2(\lambda) + I_3(\lambda),
\end{aligned}
$$

where $\lim_{\theta \to 0} \alpha(\theta) = 0$.

It is obvious that for sufficiently large λ, $|I_1(\lambda)| < \varepsilon$. Then,

$$I_2(\lambda) = c_1 - \frac{2\pi\lambda e^{-2\pi\lambda}}{1 - e^{-2\pi\lambda}},$$

and thus for sufficiently large λ, $|I_2(\lambda) - c_1| < \varepsilon$.

To estimate I_3, we select a sufficiently small δ to quarantee that

$$\sup_{\theta \in (0,\delta)} |\alpha(\theta)| < \varepsilon.$$

Then

$$|I_3(\lambda)| \leq \text{const}\,\lambda^2 \left(\int_0^\delta \alpha(\theta)\theta\, e^{-2\lambda\theta}\, d\theta + \int_\delta^\pi \theta\, e^{-2\lambda\theta}\, d\theta \right)$$

$$\leq \text{const} \left(\lambda^2 \varepsilon \int_0^\delta \theta\, e^{-2\lambda\theta}\, d\theta + \lambda^2 e^{-2\lambda\delta} \int_\delta^\pi \theta\, d\theta \right)$$

$$\leq \text{const}\left(\varepsilon + \lambda^2 e^{-2\lambda\delta} \right).$$

That is, for sufficiently large λ we have as well that $|I_3(\lambda)| < \text{const}\,\varepsilon$, and the above three inequalities yield

$$\lim_{\lambda\to+\infty} \gamma_a(\lambda) = c_1.$$

The case when $q = 1$ and $\lambda \to -\infty$ follows from Remark 8.5.1 and the case just considered. The continuity of $\gamma_a(\lambda)$ in all interior points of $[0,\pi]$ is obvious.

The proof for the cases when $p > 1$ and $q > 1$ is quite analogous and requires repeated (p-times, or q-times) integration by parts. $\qquad\square$

We give now several examples of defining symbols bounded or unbounded near the endpoints of $[0,1]$ and which oscillate approaching the endpoints.

Example 8.5.3. Let

$$a(\theta) = \theta^{-\beta} \sin\theta^{-\alpha}, \qquad \text{where} \quad 0 \leq \beta < 1, \quad \alpha > 0. \tag{8.5.2}$$

This symbol oscillates near 0, is bounded when $\beta = 0$, is unbounded for all $\beta \in (0,1)$, and is continuous at the another endpoint π for all admissible values of the parameters. That is, we need to analyze the behavior of $a(\theta)$ near the point 0 only.

We have

$$C_a^{(1)}(\theta) = \int_0^\theta u^{-\beta} \sin u^{-\alpha}\, du = \frac{1}{\alpha} \int_{\theta^{-\alpha}}^\infty y^{\frac{\beta-1}{\alpha}-1} \sin y\, dy.$$

Integrating by parts twice we get

$$C_a^{(1)}(\theta) = \frac{\theta^{\alpha-\beta+1}}{\alpha} \cos\theta^{-\alpha} - \frac{(\beta-\alpha-1)}{\alpha^2} \theta^{2\alpha-\beta+1} \sin\theta^{-\alpha}$$
$$- \frac{(\beta-\alpha-1)(\beta-2\alpha-1)}{\alpha^3} \int_{\theta^{-\alpha}}^\infty y^{\frac{\beta-1}{\alpha}-3} \sin y\, dy.$$

Thus we have that

$$C_a^{(1)}(\theta) = \frac{\theta^{\alpha-\beta+1}}{\alpha} \cos\theta^{-\alpha} + O(\theta^{2\alpha-\beta+1}), \qquad \text{when} \quad \theta \to 0. \tag{8.5.3}$$

Let $\alpha > \beta$, then

$$\lim_{\theta\to 0} \theta^{-1} C_a^{(1)}(\theta) = 0,$$

and the first condition in (8.5.1) is satisfied for $p = 1$.

Further, if $\alpha \leq \beta$ we need to consider the averages of the higher order. Indeed, formula (8.5.3) implies that

$$C_a^{(2)}(\theta) = O(\theta^{2\alpha - \beta + 2}), \qquad \text{when} \quad \theta \to 0$$

and, more generally, that

$$C_a^{(p)}(\theta) = O(\theta^{p\alpha - \beta + p}), \qquad \text{when} \quad \theta \to 0.$$

Thus for each $\alpha \leq \beta$ there is $p_0 \in \mathbb{N}$ such that $p_0 \alpha > \beta$, and thus the first condition in (8.5.1) is satisfied for $p = p_0$.

That is, the Toeplitz operator T_a with defining symbol (8.5.2) does belong to the algebra $\mathcal{T}(H(L_\infty^{(0,\pi)}(0,\pi)))$ for all admissible values of the parameters.

Example 8.5.4. Let

$$a(\theta) = (\sin \theta)^{-\beta} \sin (\sin \theta)^{-\alpha}, \qquad \text{where} \quad 0 \leq \beta < 1, \quad \alpha > 0. \qquad (8.5.4)$$

This symbol oscillates near both endpoints of $[0,1]$, is bounded when $\beta = 0$, and is unbounded for all $\beta \in (0,1)$.

Analogously to Example 8.5.3 one can show that if $p_0 \alpha > \beta$ then both conditions in (8.5.1) are satisfied for $p = p_0$, and thus the Toeplitz operator T_a with defining symbol (8.5.4) belongs to the algebra $\mathcal{T}(H(L_\infty^{(0,\pi)}(0,\pi)))$ as well.

We show now that not all oscillating functions, even bounded and continuous, generate the Toeplitz operators from $\mathcal{T}(H(L_\infty^{(0,\pi)}(0,\pi)))$.

Example 8.5.5. Let

$$a(\theta) = \theta^i = e^{i \ln \theta}. \qquad (8.5.5)$$

As the symbol oscillates near the endpoint 0, we examine the behavior of $\gamma_a(\lambda)$ when $\lambda \to +\infty$. Changing the variable $t = 2\lambda\theta$, we have

$$
\begin{aligned}
\gamma_a(\lambda) &= \frac{2\lambda}{1 - e^{-2\pi\lambda}} \int_0^\pi \theta^i \, e^{-2\lambda\theta} \, d\theta \\
&= \frac{(2\lambda)^{-i}}{1 - e^{-2\pi\lambda}} \int_0^{2\pi\lambda} t \, e^{-t} \, dt \\
&= \frac{(2\lambda)^{-i}}{1 - e^{-2\pi\lambda}} \left((1 - e^{-2\pi\lambda}) - 2\pi\lambda \, e^{-2\pi\lambda} \right) \\
&= (2\lambda)^{-i} (1 + o(\lambda)),
\end{aligned}
$$

where $\lim_{\lambda \to +\infty} o(\lambda) = 0$.

That is, the function $\gamma_a(\lambda)$ oscillates and has no limit when $\lambda \to +\infty$, and thus the Toeplitz operator T_a with defining symbol (8.5.5) does not belong to the algebra $\mathcal{T}(H(L_\infty^{\{0,\pi\}}(0,\pi)))$.

The symbol

$$a(\theta) = (\sin\theta)^i$$

provides us with an example for which the corresponding function $\gamma_a(\lambda)$ has no limits both when $\lambda \to +\infty$ and $\lambda \to -\infty$.

To give a characterization of Toeplitz operators in $T(H(L_\infty^{(0,\pi)}(0,\pi)))$ which have L_∞-symbols we need the following auxiliary result.

Theorem 8.5.6. *Let $a(\theta) \in L_\infty(0,\pi)$. Then for each real-valued monotone function $q(\lambda)$ such that*

$$\lim_{\lambda\to\pm\infty} q(\lambda) = \pm\infty \qquad \text{and} \qquad \lim_{\lambda\to\pm\infty} \frac{q(\lambda)}{\lambda} = 0, \qquad (8.5.6)$$

we have

$$\lim_{\lambda\to\pm\infty} (\gamma_a(\lambda + q(\lambda)) - \gamma_a(\lambda)) = 0.$$

Proof. We calculate

$$
\begin{aligned}
\gamma_a(\lambda + q(\lambda)) - \gamma_a(\lambda) &= \frac{2(\lambda + q(\lambda))}{1 - e^{-2\pi(\lambda+q(\lambda))}} \int_0^\pi a(\theta)e^{-2(\lambda+q(\lambda))\theta}\,d\theta \\
&\quad - \frac{2\lambda}{1 - e^{-2\pi\lambda}} \int_0^\pi a(\theta)e^{-2\lambda\theta}\,d\theta \\
&= -\frac{2(\lambda + q(\lambda))}{1 - e^{-2\pi(\lambda+q(\lambda))}} \int_0^\pi a(\theta)e^{-2\lambda\theta}\left(1 - e^{-2q(\lambda)\theta}\right)\,d\theta \\
&\quad + \left(\frac{2(\lambda + q(\lambda))}{1 - e^{-2\pi(\lambda+q(\lambda))}} - \frac{2\lambda}{1 - e^{-2\pi\lambda}}\right) \int_0^\pi a(\theta)e^{-2\lambda\theta}\,d\theta \\
&= I_1(\lambda) + I_2(\lambda).
\end{aligned}
$$

Let $\lambda \to +\infty$; we introduce $\sigma(\lambda) = (\lambda q(\lambda))^{-1/2}$ and start estimating $I_1(\lambda)$,

$$
\begin{aligned}
|I_1(\lambda)| &\leq \text{const}\left(\lambda \int_0^{\sigma(\lambda)} |a(\theta)|e^{-2\lambda\theta}\left|1 - e^{-2q(\lambda)\theta}\right|\,d\theta + \lambda \int_{\sigma(\lambda)}^\pi |a(\theta)|e^{-2\lambda\theta}\,d\theta\right) \\
&\leq \text{const}\left(\lambda|q(\lambda)| \int_0^{\sigma(\lambda)} \theta\, e^{-2\lambda\theta}\,d\theta + \lambda \int_{\sigma(\lambda)}^\pi e^{-2\lambda\theta}\,d\theta\right) \\
&\leq \text{const}\left(\lambda|q(\lambda)|\left(-\frac{\sigma(\lambda)e^{-2\lambda\sigma(\lambda)}}{2\lambda} + \frac{1 - e^{-2\lambda\sigma(\lambda)}}{4\lambda^2}\right)\right. \\
&\qquad\qquad \left. -\lambda\left(\frac{e^{-2\lambda\pi} - e^{-2\lambda\sigma(\lambda)}}{2\lambda}\right)\right) \\
&\leq \text{const}\left(\left(\frac{q(\lambda)}{\lambda}\right)^{1/2} \cdot e^{-2\left(\frac{\lambda}{q(\lambda)}\right)^{1/2}} + \frac{q(\lambda)}{\lambda} + e^{-2\pi\lambda} + e^{-2\left(\frac{\lambda}{q(\lambda)}\right)^{1/2}}\right).
\end{aligned}
$$

By the second condition in (8.5.6), for each $\varepsilon > 0$ and corresponding sufficiently large λ, we have

$$|I_1(\lambda)| < \varepsilon.$$

We estimate now $I_2(\lambda)$:

$$
\begin{aligned}
|I_2(\lambda)| &\leq \text{const} \, |q(\lambda)| \int_0^\pi |a(\theta)| \, e^{-2\lambda\theta} \, d\theta \\
&\leq \text{const} \, |q(\lambda)| \int_0^\pi e^{-2\lambda\theta} \, d\theta \leq \text{const} \, \frac{q(\lambda)}{\lambda}.
\end{aligned}
$$

That is, for sufficiently large λ we have as well that

$$|I_2(\lambda)| < \varepsilon,$$

and thus

$$\lim_{\lambda \to +\infty} (\gamma_a(\lambda + q(\lambda)) - \gamma_a(\lambda)) = 0.$$

The case when $\lambda \to -\infty$ follows from Remark 8.5.1 and the above arguments. $\qquad\square$

Given any $a(\theta) \in L_\infty(0, \pi)$, we introduce now two modified averaging functions which correspond to the endpoints of $[0, \pi]$,

$$C_a'(\theta) = \frac{2}{1 - e^{-2\theta}} \int_0^\theta a(u) \, du \qquad \text{and} \qquad D_a'(\theta) = \frac{2}{1 - e^{-2\theta}} \int_{\pi-\theta}^\pi a(u) \, du. \quad (8.5.7)$$

We note that these functions are connected with the old averages by

$$C_a'(\theta) = \frac{2}{1 - e^{-2\theta}} C_a^{(1)}(\theta) \qquad \text{and} \qquad D_a'(\theta) = \frac{2}{1 - e^{-2\theta}} D_a^{(1)}(\theta).$$

Both functions $C_a'(\theta)$ and $D_a'(\theta)$ are bounded, moreover $C_a'(\theta) \in C(0, \pi]$ and $D_a'(\theta) \in C[0, \pi)$. That is, to check whether Toeplitz operators with defining symbols $C_a'(\theta)$ and $D_a'(\theta)$ belong to the algebra $\mathcal{T}(H(L_\infty^{(0,\pi)}(0, \pi)))$, one needs to study the behavior of these functions near a single point only, 0 for $C_a'(\theta)$ and π for $D_a'(\theta)$.

The next theorem shows that the study of general L_∞-symbols is equivalent to the study of these two much more easily treatable functions.

Theorem 8.5.7. *Let $a(\theta) \in L_\infty(0, \pi)$. Then $\gamma_a(\lambda) \in C(\overline{\mathbb{R}})$ if and only if*

$$\gamma_{C_a'}(\lambda) \in C(\overline{\mathbb{R}}) \qquad \text{and} \qquad \gamma_{D_a'}(\lambda) \in C(\overline{\mathbb{R}}). \quad (8.5.8)$$

Proof. Let $\gamma_a(\lambda) \in C(\overline{\mathbb{R}})$. Consider first the case when $\lambda \to +\infty$. We will integrate by parts the integral

$$\gamma_{C_a'}(\lambda) = \frac{2\lambda}{1 - e^{2\pi\lambda}} \int_0^\pi \left(\frac{2}{1 - e^{-2\theta}} \int_0^\theta a(u) \, du \right) e^{-2\lambda\theta} \, d\theta.$$

Before doing so we mention that

$$A_\lambda(\theta) := \int_\theta^\infty \frac{e^{-2\lambda u}}{1 - e^{-2u}} \, du = \int_\theta^\infty \sum_{n=0}^\infty e^{-2(\lambda+n)u} \, du = \sum_{n=0}^\infty \frac{e^{-2(\lambda+n)\theta}}{2(\lambda+n)}.$$

Thus we have

$$
\begin{aligned}
\gamma_{C_a'}(\lambda) &= \frac{2\lambda}{1 - e^{2\pi\lambda}} \left(\left(-2A_\lambda(\theta) \int_0^\theta a(u)\, du \right)\Big|_0^\pi + 2\int_0^\pi a(\theta) A_\lambda(\theta)\, d\theta \right) \\
&= -\frac{4\lambda A_\lambda(\pi)}{1 - e^{-2\pi\lambda}} \int_0^\pi a(u)\, du + \frac{2\lambda}{1 - e^{-2\pi\lambda}} \sum_{n=0}^\infty \frac{1}{\lambda+n} \int_0^\pi a(\theta)\, e^{-2(\lambda+n)\theta}\, d\theta \\
&= -\frac{4\lambda A_\lambda(\pi)}{1 - e^{-2\pi\lambda}} \int_0^\pi a(u)\, du + \frac{\lambda}{1 - e^{-2\pi\lambda}} \sum_{n=0}^\infty \frac{1 - e^{-2\pi(\lambda+n)}}{(\lambda+n)^2} \gamma_a(\lambda+n).
\end{aligned}
$$

Using the uniform boundedness $|\gamma_a(\lambda+n)| \le \|\gamma_a\|_{L_\infty(\mathbb{R})}$ and separating the leading term we come to the equality

$$\gamma_{C_a'}(\lambda) = \sum_{n=0}^\infty \frac{\lambda}{(\lambda+n)^2} \gamma_a(\lambda+n) + o(1). \qquad (8.5.9)$$

It is obvious that

$$\frac{\lambda}{(\lambda+n)^2} = \frac{\lambda}{\lambda+n} - \frac{\lambda}{\lambda+n+1} + \frac{\lambda}{(\lambda+n)^2(\lambda+n+1)}.$$

Thus taking into account that

$$\sum_{n=0}^\infty \frac{\lambda}{(\lambda+n)^2(\lambda+n+1)} = O\left(\frac{1}{\lambda}\right)$$

we obtain

$$\sum_{n=0}^\infty \frac{\lambda}{(\lambda+n)^2} = \left(1 - \frac{\lambda}{\lambda+1} + \frac{\lambda}{\lambda+1} - \frac{\lambda}{\lambda+2} + \frac{\lambda}{\lambda+2} - \ldots \right) + O\left(\frac{1}{\lambda}\right) = 1 + o(1).$$

That is from (8.5.9) we have

$$\gamma_{C_a'}(\lambda) = \gamma_a(+\infty) + \sum_{n=0}^\infty \frac{\lambda}{(\lambda+n)^2} \left(\gamma_a(\lambda+n) - \gamma_a(+\infty) \right) + o(1).$$

As $\gamma_a(\lambda) \in C(\overline{\mathbb{R}})$, for any $\varepsilon > 0$ there is $\lambda_0 > 0$ such that for each $\lambda > \lambda_0$ and each $n \in \mathbb{Z}_+$ we have

$$|\gamma_a(\lambda+n) - \gamma_a(+\infty)| < \varepsilon.$$

Thus for $\lambda > \lambda_0$ we have

$$|\gamma_{C_a'}(\lambda) - \gamma_a(+\infty)| < \varepsilon \cdot \sum_{n=0}^{\infty} \frac{\lambda}{(\lambda + n)^2} + o(1) = \varepsilon + o(1),$$

or

$$\lim_{\gamma \to +\infty} \gamma_{C_a'}(\lambda) = \gamma_a(+\infty).$$

The proof that

$$\lim_{\gamma \to -\infty} \gamma_{D_a'}(\lambda) = \gamma_a(-\infty)$$

follows now from Remark 8.5.1.

Let now $\gamma_{C_a'}(\lambda) \in C(\mathbb{R})$. Assuming that $\lambda \to +\infty$, we have

$$
\begin{aligned}
\gamma_a(\lambda) &= \frac{2\lambda}{1 - e^{-2\pi\lambda}} \int_0^\pi e^{-2\lambda\theta} \, d \int_0^\theta a(u) \, du \\
&= \frac{2\lambda}{1 - e^{-2\pi\lambda}} \left(e^{-2\pi\lambda} \int_0^\pi a(u) \, du + 2\lambda \int_0^\pi \left(\int_0^\theta a(u) \, du \right) e^{-2\lambda\theta} \, d\theta \right) \\
&= \frac{2\lambda}{1 - e^{-2\pi\lambda}} \left(\lambda \int_0^\pi C_a'(\theta)(1 - e^{-2\theta}) \, e^{-2\lambda\theta} \, d\theta \right. \\
&\qquad\qquad \left. + \frac{1}{2} e^{-2\pi\lambda}(1 - e^{-2\lambda\pi}) C_a'(\pi) \right) \\
&= \lambda \gamma_{C_a'}(\lambda) - \frac{2\lambda^2}{2(\lambda+1)} \cdot \frac{1 - e^{-2\pi(\lambda+1)}}{1 - e^{-2\pi\lambda}} \gamma_{C_a'}(\lambda + 1) + O(\lambda e^{-2\pi\lambda}) \\
&= \lambda \gamma_{C_a'}(\lambda) - \frac{\lambda^2}{\lambda+1} \gamma_{C_a'}(\lambda + 1) + O(\lambda e^{-2\pi\lambda}).
\end{aligned}
$$

That is, we come to the equality

$$\gamma_{C_a'}(\lambda) - \gamma_{C_a'}(\lambda + 1) + \frac{\gamma_{C_a'}(\lambda + 1)}{\lambda + 1} + O(e^{-2\pi\lambda}) = \frac{\gamma_a(\lambda)}{\lambda}. \tag{8.5.10}$$

Changing, if necessary, the initial symbol $a(\theta)$ by adding a constant, we may assume without loss of generality that

$$\lim_{\lambda \to +\infty} \gamma_{C_a'}(\lambda) = 0.$$

Introduce the function

$$\alpha(\lambda) = \sup_{\xi \geq \lambda} |\gamma_{C_a'}(\xi)|,$$

which is non-increasing and satisfies

$$\lim_{\lambda \to +\infty} \alpha(\lambda) = 0.$$

Substitute

$$\lambda + 1, \quad \lambda + 2, \quad \ldots, \quad \lambda + [\lambda \cdot \alpha^{1/2}(\lambda)]$$

for λ in (8.5.10); here $[\cdot]$ is the entire part of a number; summing up the equalities obtained, we have

$$\gamma_{C_a'}(\lambda) - \gamma_{C_a'}(\lambda + n_0 + 1) + \sum_{n=0}^{n_0} \frac{\gamma_{C_a'}(\lambda + n + 1)}{\lambda + n + 1} + O(n_0 e^{-2\pi\lambda}) = \sum_{n=0}^{n_0} \frac{\gamma_a(\lambda + n)}{\lambda + n},$$

$$(8.5.11)$$

where $n_0 = n_0(\lambda) = [\lambda \cdot \alpha^{1/2}(\lambda)]$.

We assume now that

$$\lim_{\lambda \to +\infty} \gamma_a(\lambda) \neq 0.$$

That is, there exists a sequence $\{\lambda_k\}_{k=1}^{\infty}$ which tends to $+\infty$ and such that for some $\sigma > 0$,

$$|\gamma_a(\lambda_k)| \geq \sigma, \qquad \text{for all} \quad k = 1, 2, \ldots \, .$$

We denote by $E_1(\lambda)$ the left-hand side of the equality in (8.5.11) and estimate it:

$$
\begin{aligned}
|E_1(\lambda_k)| &\leq 2\alpha(\lambda_k) + \alpha(\lambda_k) \sum_{n=0}^{n_0} \frac{1}{\lambda_k + n + 1} + O((n_0 + 1)e^{-2\pi\lambda_k}) \\
&\leq 2\alpha(\lambda_k) + \alpha(\lambda_k) \ln \frac{\lambda_k + n_0 + 1}{\lambda_k + 1} + O((n_0 + 1)e^{-2\pi\lambda_k}) \\
&= 2\alpha(\lambda_k) + \alpha(\lambda_k) \ln \left(1 + \frac{n_0}{\lambda_k + 1}\right) + O((n_0 + 1)e^{-2\pi\lambda_k}) \\
&\leq \text{const} \left(\alpha(\lambda_k)(1 + \alpha^{1/2}(\lambda_k)) + O((n_0 + 1)e^{-2\pi\lambda_k})\right) \\
&\leq \text{const} \left(\alpha(\lambda_k) + O((n_0 + 1)e^{-2\pi\lambda_k})\right).
\end{aligned}
$$

We denote now the right-hand side of the equality in (8.5.11) by $E_2(\lambda)$ and estimate it:

$$
\begin{aligned}
E_2(\lambda_k) &= \sum_{n=0}^{n_0} \frac{\gamma_a(\lambda_k)}{\lambda_k + n} + \sum_{n=0}^{n_0} \frac{\gamma_a(\lambda_k + n) - \gamma_a(\lambda_k)}{\lambda_k + n} \\
&= \gamma_a(\lambda_k) \ln \frac{\lambda_k + n_0}{\lambda_k} + o(1) + E_{2,2}(\lambda_k),
\end{aligned}
$$

where

$$E_{2,2}(\lambda_k) = \sum_{n=0}^{n_0} \frac{\gamma_a(\lambda_k + n) - \gamma_a(\lambda_k)}{\lambda_k + n}.$$

As the function $n_0 = n_0(\lambda)$ satisfies (8.5.6), we make use of Theorem 8.5.6. That is, for each $k \in \mathbb{N}$ there is $\sigma_k > 0$ such that for each $n \in [1, n_0] \cap \mathbb{N}$,

$$|\gamma_a(\lambda_k) - \gamma_a(\lambda_k + n)| < \sigma_k \qquad \text{and} \qquad \lim_{k \to \infty} \sigma_k = 0.$$

We have

$$|E_{2,2}(\lambda_k)| \leq \sigma_k \ln \frac{\lambda_k + n_0(\lambda_k)}{\lambda_k} \leq \frac{n_0(\lambda_k)}{\lambda_k},$$

and thus

$$\left| E_2(\lambda_k) - \gamma_a(\lambda_k) \cdot \frac{n_0(\lambda_k)}{\lambda_k} \right| \leq O\left(\left(\frac{n_0(\lambda_k)}{\lambda_k} \right)^2 + \sigma_k \frac{n_0(\lambda_k)}{\lambda_k} \right).$$

This yields

$$|E_2(\lambda_k)| \geq \frac{|\gamma_a(\lambda_k)|}{2} \cdot \frac{n_0(\lambda_k)}{\lambda_k} \geq \frac{|\gamma_a(\lambda_k)|}{2} \alpha^{1/2}(\lambda_k),$$

or

$$|E_2(\lambda_k)| \geq \frac{\sigma}{4} \alpha^{1/2}(\lambda_k).$$

Substituting $E_1(\lambda_k)$ and $E_2(\lambda_k)$ in (8.5.11) by their estimates we have

$$\text{const} \left(\alpha(\lambda_k) + \lambda_k \, \alpha^{1/2}(\lambda_k) \, e^{-2\pi\lambda_k} + e^{-2\pi\lambda_k} \right) \geq \frac{\sigma}{4} \alpha^{1/2}(\lambda_k).$$

Now, if

$$\lim_{k \to \infty} \frac{e^{-2\pi\lambda_k}}{\alpha^{1/2}(\lambda_k)} = 0, \tag{8.5.12}$$

then we come to a contradiction, and thus

$$\lim_{\lambda \to +\infty} \gamma_a(\lambda) = 0.$$

If (8.5.12) does not hold, then there exist $\sigma_1 > 0$ and a subsequence $\{\lambda_{k_l}\}_{l=1}^{\infty}$ of the sequence $\{\lambda_k\}_{k=1}^{\infty}$ such that for each $l \in \mathbb{N}$,

$$e^{-2\pi\lambda_{k_l}} \geq \sigma_1 \alpha^{1/2}(\lambda_{k_l}).$$

Then substituting $\lambda = \lambda_{k_l}$ in (8.5.10) we come to

$$\frac{e^{-4\pi\lambda_{k_l}}}{\sigma_1^2} + \frac{e^{-4\pi\lambda_{k_l}}}{\sigma_1^2} \left(1 + \frac{1}{\lambda_{k_l} + 1} \right) + \text{const}\, e^{-2\pi\lambda_{k_l}} \geq \frac{\sigma}{\lambda_{k_l}},$$

which again leads to a contradiction.

The proof that $\gamma_a(\lambda)$ is continuous at the point $-\infty$ again follows from Remark 8.5.1. $\qquad\qquad\square$

We note that the above results uncover a variety of Toeplitz operators in $\mathcal{T}(H(L_\infty^{(0,\pi)}(0,\pi)))$ whose bounded defining symbols may not have limit values at the endpoints 0 and π, extending thereby the descriptions of Toeplitz operators in $\mathcal{T}(PC(\overline{\mathbb{D}}, T))$ of Section 8.4.

At the same time we have shown that the algebra $\mathcal{T}(H(L_\infty^{(0,\pi)}(0,\pi)))$ also contains bounded Toeplitz operators with *unbounded* defining symbols. Anticipating and motivating their detailed study we now give an example showing how monstrous the defining symbols of Toeplitz operators from $\mathcal{T}(PC(\overline{\mathbb{D}}, T))$ can be.

Example 8.5.8. Consider the algebra $\mathcal{T}(PC(\overline{\mathbb{D}}, T_0))$ for the special case of the discontinuity set $T_0 = \{t_1, t_2\}$, where $t_2 = -t_1$. Then the Toeplitz operator

$$T_{\chi_1} = T_s + T_{\chi_1 v_1^2} + T_{(1-\chi_2)v_2^2} + K,$$

obviously belongs to the algebra $\mathcal{T}(PC(\overline{\mathbb{D}}, T_0))$. Here $s(z)$ is a continuous function on $\mathbb{D}$ whose restriction on γ coincides with

$$\chi_1(t) - \chi_1(t)v_1^2(t) - (1 - \chi_2(t))v_2^2(t) = \chi_1(t)(1 - v_1^2(t) - v_2^2(t))$$

and K is a compact operator. Recall that the characteristic functions χ_k and the functions v_k, $k = 1, 2$, are as defined in Section 8.1.

Thus for each function $f(x) \in C[0, 1]$ the operator $f(T_{\chi_1})$ belongs to the algebra $\mathcal{T}(PC(\overline{\mathbb{D}}, T_0))$ as well.

By (8.2.2) we have

$$V_k T_{\chi_k} V_k^{-1} = T_{\chi_+},$$

where χ_+ is the characteristic function of the right quarter-plane in Π and the operator V_k is given by (8.2.1). This unitary equivalence implies that

$$f(T_{\chi_k}) = V_k^{-1} f(T_{\chi_+}) V_k. \tag{8.5.13}$$

Now for $t_1 \in T_0$, let $a_0(z)$ be a function on the unit disk such that

$$\widehat{a}_0(\theta) = a_0(\alpha_1^{-1}(e^{i\theta})) = (\sin\theta)^{-\beta} \sin(\sin\theta)^{-\alpha},$$

where $0 \leq \beta < 1$ and $\alpha > 0$.

By Example 8.5.4 the Toeplitz operator $T_{\widehat{a}_0}$ is bounded on $\mathcal{A}^2(\Pi)$ and belongs to the algebra generated by T_{χ_+} Moreover for the function

$$f_0(x) = \frac{2x^2}{\pi} \frac{\ln(1-x) - \ln x}{(1-x) - x} \int_0^\pi (\sin\theta)^{-\beta} \sin(\sin\theta)^{-\alpha} \left(\frac{1-x}{x}\right)^{\frac{2\theta}{\pi}} d\theta,$$

which is continuous on $[0, 1]$ and has zero values at the points 0 and 1, $f_0(0) = f_0(1) = 0$, we have that $T_{\widehat{a}_0} = f_0(T_{\chi_+})$. Thus the Toeplitz operator

$$T_{a_0} = f_0(T_{\chi_1}) = V_1^{-1} f_0(T_{\chi_+}) V_1 = V_1^{-1} T_{\widehat{a}_0} V_1$$

belongs to the algebra $\mathcal{T}(PC(\overline{\mathbb{D}}, T_0))$.

We note that the defining symbol $a_0(z)$ is quite horrible, being *unbounded and oscillating* near every point of $\gamma \setminus T$ and having quite a complicated angular behavior approaching the points of T. At the same time the Fredholm symbol of the operator T_{a_0} has quite a respectable form:

$$\operatorname{sym} T_{a_0} = \begin{cases} 0, & t \in \widehat{\gamma} \\ f_0(x), & x \in \Delta_1 = [0, 1] \\ f_0(1-x), & x \in \Delta_2 = [0, 1] \end{cases}.$$

8.6   Semi-commutators involving unbounded symbols

The classical semi-commutator property

$$[T_a, T_b) = T_a T_b - T_{ab} \in \mathcal{K}, \qquad \text{for all} \quad a \in L_\infty(\mathbb{D}), \; b \in C(\overline{\mathbb{D}}),$$

played an essential role in the description of Toeplitz operators with bounded defining symbols in the algebra $\mathcal{T}(PC(\overline{\mathbb{D}}, T))$. To study Toeplitz operators with unbounded defining symbols belonging to this algebra we need to understand first the properties of semi-commutators involving unbounded symbols. As it turns out, the question of compactness of the semi-commutator for *unbounded a* is quite delicate and does not have any universal answer. We will give two examples showing that the compactness result is not valid for general, and even special, unbounded defining symbols a and arbitrary $b \in C(\overline{\mathbb{D}})$. At the same time we will prove it for a certain special, and important for us, case of unbounded defining symbols a and a special choice of $b \in C(\overline{\mathbb{D}})$.

Before passing to the examples we note that studying semi-commutators we will consider Toeplitz operators with more general defining symbols, which depend on both variables θ and r, where $z = re^{i\theta} \in \Pi$. For such general symbols $c = c(r, \theta)$ the Toeplitz operator T_c is no longer unitary equivalent to a multiplication operator. The operator RT_cR^*, where the operators R and R^* are defined at the end of Section 7.1, now has a more complicated structure: it turns out to be a pseudodifferential operator with a certain compound (or double) symbol. The next theorem clarifies this statement for bounded defining symbols of a special and important product form: $c = c(r, \theta) = a(\theta)v(r)$. The case of unbounded $a(\theta)$ will be treated in Theorem 8.6.5.

Theorem 8.6.1. *Given a bounded measurable defining symbol $a(\theta)v(r)$, the Toeplitz operator T_{av} acting on $\mathcal{A}^2(\Pi)$ is unitary equivalent to the pseudodifferential operator $A_1 = RT_{av}R^*$ acting on $L_2(\mathbb{R})$. The operator A_1 is given by*

$$(A_1 f)(\lambda) = \frac{1}{2\pi} \int_{\mathbb{R}} d\xi \int_{\mathbb{R}} a_1(x, y, \xi) e^{i(x-y)\xi} f(y) dy, \qquad x \in \mathbb{R}, \tag{8.6.1}$$

where its compound symbol $a_1(x, y, \xi)$ has the form

$$a_1(x, y, \xi) = c(x, y)\, \gamma_a\left(\frac{x+y}{2}\right) \widetilde{v}(\xi)$$

with

$$c(x, y) = \frac{1 - e^{-\pi(x+y)}}{x + y} \sqrt{\frac{2x}{1 - e^{-2\pi x}}} \sqrt{\frac{2y}{1 - e^{-2\pi y}}}, \tag{8.6.2}$$

and $\widetilde{v}(\xi) = v(e^{-\xi})$.

Proof. We have

$$
\begin{aligned}
(A_1 f)(\lambda) &= (RT_{a(\theta)v(r)}R^* f)(\lambda) = (R(R^*R)a(\theta)v(r)(R^*R)R^* f)(\lambda) \\
&= ((RR^*)Ra(\theta)v(r)R^*(RR^*)f)(\lambda) = (Ra(\theta)v(r)R^* f)(\lambda) \\
&= (R_0^* a(\theta)(M \otimes I)v(r)(M^{-1} \otimes I)R_0 f)(\lambda) \\
&= \sqrt{\frac{2\lambda}{1 - e^{-2\pi\lambda}}} \int_0^\pi e^{-(\lambda - i)\theta} a(\theta)d\theta \, \frac{1}{\sqrt{2\pi}} \int_{\mathbb{R}_+} r^{-i\lambda} v(r)dr \\
&\quad \cdot \frac{1}{\sqrt{2\pi}} \int_{\mathbb{R}} r^{i\alpha - 1} f(\alpha) \sqrt{\frac{2\alpha}{1 - e^{-2\pi\alpha}}} \, e^{-(\alpha + i)\theta} d\alpha \\
&= \frac{1}{\sqrt{2\pi}} \int_{\mathbb{R}_+} r^{-i\lambda} v(r)dr \, \frac{1}{\sqrt{2\pi}} \int_{\mathbb{R}} r^{i\alpha - 1} f(\alpha) d\alpha \\
&\quad \cdot \sqrt{\frac{2\lambda}{1 - e^{-2\pi\lambda}}} \sqrt{\frac{2\alpha}{1 - e^{-2\pi\alpha}}} \int_0^\pi e^{-(\lambda + \alpha)\theta} a(\theta)d\theta.
\end{aligned}
$$

The last integral gives

$$
\int_0^\pi e^{-(\lambda + \alpha)\theta} a(\theta)d\theta = \frac{1 - e^{-\pi(\lambda + \alpha)}}{\lambda + \alpha} \, \gamma_a \left(\frac{\lambda + \alpha}{2} \right),
$$

and thus we have

$$
\begin{aligned}
(A_1 f)(\lambda) &= (RT_{a(\theta)v(r)}R^* f)(\lambda) \\
&= \frac{1}{2\pi} \int_{\mathbb{R}_+} dr \int_{\mathbb{R}} c(\lambda, \alpha) \gamma_a \left(\frac{\lambda + \alpha}{2} \right) v(r) \, r^{-i(\lambda - \alpha) - 1} f(\alpha) d\alpha,
\end{aligned}
$$

where

$$
c(\lambda, \alpha) = \frac{1 - e^{-\pi(\lambda + \alpha)}}{\lambda + \alpha} \sqrt{\frac{2\lambda}{1 - e^{-2\pi\lambda}}} \sqrt{\frac{2\alpha}{1 - e^{-2\pi\alpha}}}. \tag{8.6.3}
$$

Changing variables, $\lambda = x$, $\alpha = y$, and $r = e^{-\xi}$, we finally have

$$
(A_1 f)(x) = \frac{1}{2\pi} \int_{\mathbb{R}} d\xi \int_{\mathbb{R}} a_1(x, y, \xi) e^{i(x - y)\xi} f(y)dy, \quad x \in \mathbb{R},
$$

with

$$
a_1(x, y, \xi) = c(x, y) \gamma_a \left(\frac{x + y}{2} \right) \widetilde{v}(\xi),
$$

where $c(x, y)$ is given by (8.6.3), and $\widetilde{v}(\xi) = v(e^{-\xi})$. $\qquad \square$

Returning now to the examples we mention that the first one is just a minor modification of Example 6.2.1.

Example 8.6.2. Let

$$
a(z) = a(r) = (1 - r^2)^{-\beta} \sin(1 - r^2)^{-\alpha} \in L_1(\mathbb{D})
$$

and

$$b(z) = b(r) = (1 - r^2)^\varepsilon \sin(1 - r^2)^{-\alpha} \in C(\overline{\mathbb{D}})$$

where $z = re^{i\theta}$, $0 < \varepsilon < \beta < 1$. Then both T_a and T_b are bounded and compact.
The product ab has the form

$$a(r)b(r) = \frac{(1 - r^2)^{-(\beta-\varepsilon)}}{2} - \frac{(1 - r^2)^{-(\beta-\varepsilon)} \cos 2(1 - r^2)^{-\alpha}}{2} = c_1(r) - c_2(r).$$

Then the operator T_{c_1} is unbounded, while the operator T_{c_2} is compact. That is, the operator T_{ab} is not bounded, and the (unbounded) semi-commutator is not compact.

In what follows we will deal with the class of unbounded defining symbols which, considered in the upper half-plane setting, are the functions $a(\theta) \in H(L_1(0, \pi))$, where $z = re^{i\theta} \in \Pi$, for which the corresponding Toeplitz operators T_a are bounded.

The second example shows that even for such specific symbols $a(\theta)$ the semi-commutator is not compact for each $b(z) \in C(\overline{\Pi})$.

Example 8.6.3. Let

$$a(z) = a(\theta) = \theta^{-\beta} \sin \theta^{-\alpha}$$

and

$$b(z) = w(r)\, \theta^\varepsilon \sin \theta^{-\alpha},$$

where $z = re^{i\theta}$, $0 < \varepsilon < \beta < 1$, $\alpha > 0$, and $w(r)$ is a $[0, 1]$-valued C^∞- function such that

$$w(r) \equiv \begin{cases} 0, & r \in [0, \delta_1] \\ 1, & r \in [\delta_2, \delta_3] \\ 0, & r \in [\delta_4, +\infty] \end{cases},$$

and $0 < \delta_1 < \delta_2 < \delta_3 < \delta_4 < +\infty$.

The operator T_a is bounded by the results of Example 8.5.3; the operator T_b is bounded as well because of $b(z) \in C(\overline{\Pi})$. The product ab has the form

$$a(\theta)b(z) = \frac{w(r)\theta^{-\delta}}{2} - \frac{w(r)\theta^{-\delta} \cos 2\theta^{-\alpha}}{2} = c_1(z) - c_2(z),$$

where $\delta = \beta - \varepsilon \in (0, 1)$.

The Toeplitz operator T_{c_2} is bounded by Theorem 8.6.5. To prove that the semi-commutator $[T_a, T_b)$ is not compact, it is sufficient to show, for example, that the operator T_{c_1} is unbounded. Let $a_\delta(\theta) = \theta^{-\delta}$, then

$$\gamma_{a_\delta}(\lambda) = \frac{2\lambda}{1 - e^{-2\pi\lambda}} \int_0^\pi \theta^{-\delta} e^{-2\lambda\theta}\, d\theta = \frac{(2\lambda)^\delta}{1 - e^{-2\pi\lambda}} \int_0^{2\pi\lambda} u^{-\delta} e^{-u}\, du.$$

It is clear that if $\lambda \to +\infty$, then we have the asymptotics

$$\gamma_{a_\delta}(\lambda) = c_0 \lambda^\delta + o(1), \tag{8.6.4}$$

$$\frac{\partial \gamma_{a_\delta}(\lambda)}{\partial \lambda} = \delta c_0 \lambda^{\delta-1} + o(1), \tag{8.6.5}$$

where $c_0 = 2^\delta \Gamma(1 - \delta)$.

We will use now the representation (8.6.1) for the operator $A_1 = R T_{c_1} R^*$. Setting

$$\widehat{w}(x - y) = \frac{1}{2\pi} \int_{\mathbb{R}} \widetilde{w}(\xi) e^{i(x-y)\xi} \, d\xi,$$

where $\widetilde{w}(\xi) = w(e^{-\xi})$, we have

$$(A_1 f)(x) = \int_{\mathbb{R}} c(x, y) \gamma_{a_\delta}\left(\frac{x + y}{2}\right) \widehat{w}(x - y) f(y) dy,$$

where the function $c(x, y)$ is given by (8.6.2).

We show now that the operator A_1 is unbounded on $L_2(\mathbb{R})$. Introduce the family of functions

$$f_{x_0}(y) = \begin{cases} \varepsilon^{-1/2}, & y \in I_\varepsilon = [x_0 - \varepsilon/2, x_0 + \varepsilon/2] \\ 0, & y \in \mathbb{R} \setminus I_\varepsilon \end{cases},$$

where $\varepsilon = \varepsilon(x_0) = x_0^{-\delta/2}$. It is clear that $\|f_{x_0}\|_{L_2(\mathbb{R})} = 1$.

Let $x \in I_\varepsilon$; setting

$$K(x, y) = c(x, y) \gamma_{a_\delta}\left(\frac{x + y}{2}\right) \widehat{w}(x - y)$$

we have

$$\begin{aligned}
(A_1 f_{x_0})(x) &= \varepsilon^{-1/2} \int_{x_0-\varepsilon/2}^{x_0+\varepsilon/2} K(x, y) dy \\
&= \varepsilon^{1/2} K(x, x) + \varepsilon^{-1/2} \int_{x_0-\varepsilon/2}^{x_0+\varepsilon/2} (K(x, y) - K(x, x)) dy \\
&= I_1(x) + I_2(x).
\end{aligned}$$

When $x_0 \to +\infty$, for the first summand we have

$$\begin{aligned}
I_1(x) &= 1 \cdot \gamma_{a_\delta}(x) \cdot \widehat{w}(0) \cdot \varepsilon^{1/2}(x_0) \\
&= \widehat{w}(0) c_0 \left(x^\delta \cdot x_0^{-\delta/4} + o(1)\right) = \widehat{w}(0) c_0 \left(x_0^{3\delta/4} + o(1)\right). \tag{8.6.6}
\end{aligned}$$

As $\widetilde{w}(\xi) \geq 0$, we have that $\widehat{w}(0) > 0$.

Now for the second summand we have

$$|I_2(x)| \leq \varepsilon^{3/2} \sup_{y \in I_\varepsilon} \left| \frac{\partial K}{\partial y}(x, y) \right|.$$

Both functions $\frac{\partial c}{\partial y}(x, y)$ and $\frac{\partial \widehat{w}}{\partial y}(x - y)$ are uniformly bounded on x. The former is bounded by Theorem 8.8.2, while the latter is bounded as the Fourier transform of a function with a compact support. Thus we have that

$$|I_2(x)| \leq \mathrm{const}\, \varepsilon^{3/2} \sup_{y \in I_\varepsilon} \left(\left| \frac{\partial \gamma_{a_\delta}}{\partial y}\left(\frac{x+y}{2} \right) \right| + \left| \gamma_{a_\delta}\left(\frac{x+y}{2} \right) \right| \right).$$

Asymptotics (8.6.4) and (8.6.5) imply that for $x_0 \to +\infty$ we have

$$|I_2(x)| \leq \mathrm{const}\, \varepsilon^{3/2} x^\delta \leq \mathrm{const}\, \left(x_0^{-\delta/2} \right)^{3/2} x_0^\delta = \mathrm{const}\, x_0^{\delta/4}. \tag{8.6.7}$$

Comparing (8.6.6) and (8.6.7), for sufficiently large x_0 and $x \in I_\varepsilon$, we have that

$$|(A_1 f_{x_0})(x)| \geq \frac{|\widehat{w}(0)|\, c_0}{2} x_0^{3\delta/4}.$$

Thus

$$
\begin{aligned}
\|A_1 f_{x_0}\|_{L_2(\mathbb{R})} &\geq \left(\left(\frac{|\widehat{w}(0)|\, c_0}{2} x_0^{3\delta/4} \right)^2 \int_{x_0 - \varepsilon/2}^{x_0 + \varepsilon/2} dx \right)^{1/2} \\
&\geq \mathrm{const}\, \left(x_0^{3\delta/2} \cdot \varepsilon(x_0) \right)^{1/2} = \mathrm{const}\, x_0^{\delta/2}.
\end{aligned}
$$

This obviously yields unboundedness of the operator A_1, which in turn implies unboundedness of $T_a b$.

Now as a special choice of continuous functions on $\overline{\mathbb{D}}$ we select any $v_k(z)$, $k = 1, 2, \ldots, m$, considered in the upper half-plane setting as a function $v = v(r)$, where $z = r e^{i\theta} \in \Pi$, as introduced in Section 8.1 but having the additional property that $v(r) \in C^\infty(0, \infty)$. That is, v is a $[0, 1]$-valued C^∞-function such that

$$v(r) \equiv \begin{cases} 1, & r \in [0, \delta_1] \\ 0, & r \in [\delta_2, +\infty] \end{cases}, \tag{8.6.8}$$

for some $0 < \delta_1 < \delta_2 < +\infty$.

Our aim is to prove that for each $a(\theta) \in H(L_1(0, \pi))$, for which the corresponding Toeplitz operator T_a is bounded, the semi-commutator $T_a T_v - T_{av}$ is compact. To do this we first represent the operators $T_a T_v$ and T_{av} in the form of pseudodifferential operators with certain compound (or double) symbols and then use the next result, which can be found, for example, in [113, Theorem 4.2 and Theorem 4.4].

Denote by $V(\mathbb{R})$ the set of all absolutely continuous functions on $\mathbb{R}$ of bounded total variation, and by $C_b(\mathbb{R}^2, V(\mathbb{R}))$ the set of all functions $a : \mathbb{R}^2 \times \mathbb{R} \to \mathbb{C}$ such that $u \mapsto a(u, \cdot)$ is a bounded continuous $V(\mathbb{R})$-valued function on $\mathbb{R}^2$. Then, for $a \in C_b(\mathbb{R}^2, V(\mathbb{R}))$, we define

$$cm_u^C(a) = \max \left\{ \|a(u + \Delta u, \cdot) - a(u, \cdot)\|_C : \Delta u \in \mathbb{R}^2,\ \|\Delta u\| \leq 1 \right\},$$

and denote by $\mathcal{E}_2^C$ the subset of all functions in $C_b(\mathbb{R}^2, V(\mathbb{R}))$ such that the $V(\mathbb{R})$-valued function $u \mapsto a(u, \cdot)$ is uniformly continuous on $\mathbb{R}^2$ and the following conditions hold,

$$\lim_{\|u\| \to \infty} cm_u^C(a) = 0 \quad \text{and} \quad \lim_{|h| \to 0} \sup_{u \in \mathbb{R}^2} \|a(u, \cdot) - a^h(u, \cdot)\|_V = 0, \tag{8.6.9}$$

where $a^h(u, \cdot) = a(u, \xi + h)$, for all $(u, \xi) \in \mathbb{R}^2 \times \mathbb{R}$.

Theorem 8.6.4 ([113]). *If $\partial_\xi^j \partial_y^k a(x, y, \xi) \in C_b(\mathbb{R} \times \mathbb{R}, V(\mathbb{R}))$ for all $k, j = 0, 1, 2$, then the pseudodifferential operator A with compound symbol $a(x, y, \xi)$ defined on functions $f \in C_0^\infty(\mathbb{R})$ by the iterated integral*

$$(Af)(x) = \frac{1}{2\pi} \int_{\mathbb{R}} d\xi \int_{\mathbb{R}} a(x, y, \xi) e^{i(x-y)\xi} f(y) dy, \qquad x \in \mathbb{R}, \tag{8.6.10}$$

extends to a bounded linear operator on every Lebesgue space $L_p(\mathbb{R})$, $p \in (1, \infty)$.

If $\partial_\xi^j \partial_y^k a(x, y, \xi) \in \mathcal{E}_2^C$ for all $k, j = 0, 1, 2$, then the pseudodifferential operator (8.6.10) with compound symbol

$$r(x, y, \xi) = a(x, y, \xi) - a(x, x, \xi)$$

is compact on every Lebesgue space $L_p(\mathbb{R})$, $p \in (1, \infty)$.

Considering semi-commutators, we prove first that for our selection of defining symbols $a(\theta)$ and $v(r)$ the Toeplitz operator T_{av} is bounded.

Theorem 8.6.5. *For every $a(\theta) \in H(L_1(0, \pi))$ such that the Toeplitz operator T_a is bounded and a $[0, 1]$-valued C^∞-function $v = v(r)$ of the form (8.6.8), the Toeplitz operator T_{av} is bounded on $\mathcal{A}^2(\Pi)$.*

Proof. We mention first that the boundedness of T_a is equivalent (by Theorem 7.2.1) to the boundedness of the corresponding function

$$\gamma_a(\lambda) = \frac{2\lambda}{1 - e^{-2\pi\lambda}} \int_0^\pi a(\theta)\, e^{-2\lambda\theta}\, d\theta, \qquad \lambda \in \mathbb{R}.$$

The C^∞-functions with compact support in $\mathbb{R}_+$ obviously form a dense set in $L_2(\mathbb{R}_+)$. Taking any such function f we consider

$$(A_1 f)(\lambda) = \frac{1}{2\pi} \int_{\mathbb{R}} d\xi \int_{\mathbb{R}} a_1(x, y, \xi) e^{i(x-y)\xi} f(y) dy, \qquad x \in \mathbb{R},$$

where the compound symbol $a_1(x, y, \xi)$ has the form

$$a_1(x, y, \xi) = c(x, y)\, \gamma_a\left(\frac{x+y}{2}\right)\, \widetilde{v}(\xi)$$

with

$$c(x, y) = \frac{1 - e^{-\pi(x+y)}}{x + y} \sqrt{\frac{2x}{1 - e^{-2\pi x}}} \sqrt{\frac{2y}{1 - e^{-2\pi y}}},$$

and $\widetilde{v}(\xi) = v(e^{-\xi})$. We note that $c(x, x) \equiv 1$.

The boundedness of the operator A_1 follows from Theorem 8.6.4, Theorems 8.8.1–8.8.4, and the fact that $\widetilde{v}(\xi)$ is a C^∞-function with a compact support.

By the calculations of Theorem 8.6.1 we have that $T_{av} = R^* A_1 R$. Thus the Toeplitz operator T_{av} is bounded on $\mathcal{A}^2(\Pi)$. $\qquad\qquad\square$

Now we are ready to prove that the semi-commutator $T_a T_v - T_{av}$ is compact.

Theorem 8.6.6. *For each $a(\theta) \in H(L_1(0, \pi))$ such that the Toeplitz operator T_a is bounded and the $[0, 1]$-valued C^∞-function $v = v(r)$ of the form (8.6.8), the semi-commutator $T_a T_v - T_{av}$ is compact.*

Proof. A calculation analogous to that of Theorem 8.6.1 yields

$$
\begin{aligned}
(A_2 f)(x) \;&=\; R T_a T_v R f = (RaR^*)(RvR^*)f \\
&=\; \gamma_a(x)\,(RvR^*)f = \frac{1}{2\pi} \int_{\mathbb{R}} d\xi \int_{\mathbb{R}} a_2(x, y, \xi) e^{i(x-y)\xi} f(y)\,dy, \quad x \in \mathbb{R},
\end{aligned}
$$

with

$$a_2(x, y, \xi) = c(x, y)\, \gamma_a(x)\, \widetilde{v}(\xi),$$

where $c(x, y)$ is given by (8.6.2), and $\widetilde{v}(\xi) = v(e^{-\xi})$.

Thus the operator $R^*(T_{av} - T_a T_v)R = A_1 - A_2$ can be represented as a difference of two pseudodifferential operators having the compound symbols

$$
\begin{aligned}
r_1(x, y, \xi) \;&=\; a_1(x, y, \xi) - a_1(x, x, \xi) \\
&=\; c(x, y)\, \gamma_a\left(\frac{x+y}{2}\right)\, \widetilde{v}(\xi) - \gamma_a(x)\, \widetilde{v}(\xi)
\end{aligned}
$$

and

$$
\begin{aligned}
r_2(x, y, \xi) \;&=\; a_2(x, y, \xi) - a_2(x, x, \xi) \\
&=\; c(x, y)\, \gamma_a(x)\, \widetilde{v}(\xi) - \gamma_a(x)\, \widetilde{v}(\xi).
\end{aligned}
$$

The compactness of each of the last pseudodifferential operators easily follows from Theorem 8.6.4, Theorems 8.8.1–8.8.4, and the fact that $\widetilde{v}(\xi)$ is a C^∞-function with a compact support. Indeed, the above property of $\widetilde{v}(\xi)$ guarantees that both

$a_1(x, y, \xi)$ and $a_2(x, y, \xi)$, as well as their two consecutive derivatives on ξ satisfy the second property in (8.6.9); while the properties

$$\lim_{(x,y)\to\infty} \frac{\partial^k d_{1,2}}{\partial y^k}(x, y) = 0, \qquad \text{for} \quad k = 1, 2,$$

where $d_1(x, y) = c(x, y)\,\gamma_a\left(\frac{x+y}{2}\right)$ and $d_2(x, y) = c(x, y)\,\gamma_a(x)$ imply the first equality in (8.6.9).

$\square$

The above result leads directly to the following extension (of the sufficient part) of Theorem 8.4.2.

Corollary 8.6.7. *Let the operator $A \in \mathcal{T}(PC(\overline{\mathbb{D}}, T))$ be such that in its canonical representation*

$$A = T_{s_A} + \sum_{k=1}^{m} T_{v_k} f_{A,k}(T_{\chi_k}) T_{v_k} + K,$$

all operators $f_{A,k}(T_{\chi_k})$ are Toeplitz with possibly unbounded defining symbols a_k, $k = 1, \ldots, m$, correspondingly. Then $A = T_a + K_A$ is a compact perturbation of the Toeplitz operator T_a, where

$$a(z) = s_A(z) + \sum_{k=1}^{m} a_k(z) v_k^2(z),$$

where $s_A(z)$ is given by (see (8.4.1))

$$s_A(t) = (\text{sym } A)(t) - \sum_{k=1}^{m} [f_{A,k}(0)(1 - \chi_k(t)) + f_{A,k}(1)\chi_k(t)]\, v_k^2(t).$$

We note that Corollary 8.6.7 immediately reveals via property (8.5.13) and Theorem 8.5.2 many Toeplitz operators in $\mathcal{T}(PC(\overline{\mathbb{D}}, T))$ having unbounded defining symbols. Indeed, the conditions (8.5.1) are obviously satisfied, with $p = q = 1$, for example, for any function $a(\theta) \in H(L_1(0, \pi))$ which has limits at the endpoints of $[0, \pi]$. Of course, the existence of symbol limits at the endpoints by no means is necessary for the Toeplitz operator T_a to be an element of $\mathcal{T}_+$. As Example 8.5.8 shows, the corresponding symbol can even be unbounded near each of the endpoints 0 and π. Many further particular symbols can be given, for example, by combining polynomial growth with logarithmic and iterated logarithmic growth, then by considering linear combinations of different symbols, etc. The following defining symbol may serve as an illustrative example:

$$a(\theta) = \sum_{k=1}^{n} c_k \theta^{-\beta_k} \log^{\lambda_k} \theta^{-1} \sin\left(\theta^{-\alpha_k} \log^{\mu_k} \theta^{-1}\right),$$

where $c_k \in \mathbb{C}$, $0 < \beta_k < 1$, $\alpha_k > 0$, $\lambda_k \in \mathbb{R}$, $\mu_k \in \mathbb{R}$, $k = 1, \ldots, n$.

We mention especially that when speaking about a compact perturbation of a Toeplitz operator, say T_a, one should always remember that the coset $T_a + \mathcal{K}$ contains many Toeplitz operators of the form T_{a+k}, for which the Toeplitz operator T_k is compact; and that all such operators have the same image $\operatorname{sym} T_{a+k} = \operatorname{sym} T_a$ in the Fredholm symbol algebra $\operatorname{Sym} \mathcal{T}(PC(\overline{\mathbb{D}}, T))$. At the same time the properties of the functions a and $a + k$ can be extremely different. Indeed, even having as nice as possible a, say $a \in C(\overline{\mathbb{D}})$, one can always add, for example, the function

$$k(z) = (1 - r^2)^{-\beta} \sin(1 - r^2)^{-\alpha} + (1 - r)\chi_Q(z), \qquad z = re^{i\theta},$$

where the first summand is taken from Example 8.6.2 and Q is the set of all points $z = r_1 + ir_2 \in \overline{\mathbb{D}}$ with rational r_1 and r_2. This converts the initial defining symbol a to the symbol $a + k$ which does not have a limit at every point of $\overline{\mathbb{D}}$, and moreover is unbounded near every point of the boundary.

That is, when speaking about the representation $A = T_a + K$ it is preferable to have a defining symbol a with less unnecessary singularities. It seems that the option given by Theorem 8.4.4 and Corollary 8.6.7 is quite optimal in this respect.

8.7   Toeplitz or not Toeplitz

The key question in the description of the Toeplitz operators in $\mathcal{T}(PC(\overline{\mathbb{D}}, T))$ is whether the operators of the form $f(T_{\chi_k})$, where $f(x) \in C[0, 1]$ and $k = 1, 2, \ldots, m$, are Toeplitz or not. By (8.5.13) this question reduces to the following question in the upper half-plane setting: given $f(x) \in C[0, 1]$, whether the operator $f(T_{\chi_+})$ is Toeplitz or not. The last questions is in turn equivalent to: whether the function $\gamma(\lambda) \in C(\overline{\mathbb{R}})$, which is connected with $f(x) \in C[0, 1]$ by (see (8.2.3))

$$\gamma(\lambda) = f\left(\frac{1}{e^{-\pi\lambda} + 1}\right),$$

admits the representation (7.2.1) for some $a(\theta) \in L_1(0, \pi)$, i.e.,

$$\gamma(\lambda) = \gamma_a(\lambda) = \frac{2\lambda}{1 - e^{-2\pi\lambda}} \int_0^\pi a(\theta)\, e^{-2\lambda\theta}\, d\theta, \qquad \lambda \in \mathbb{R}. \tag{8.7.1}$$

The statements of the next theorem are necessary for the existence of the above representation for a given function $\gamma(\lambda) \in C(\overline{\mathbb{R}})$.

Theorem 8.7.1. *Let $a(\theta) \in L_1(0, \pi)$. Then the function $\gamma_a(\lambda)$ is analytic in the whole complex plane with the exception of the points $\lambda_n = in$, where $n = \pm 1, \pm 2, \ldots$, where $\gamma_a(\lambda)$ has simple poles. Moreover, for any fixed and sufficiently small δ the function $\gamma_a(\lambda)$ admits on the set*

$$\mathbb{C} \setminus \bigcup_{\mathbb{Z}\setminus\{0\}} K_n(\delta), \qquad \text{where} \qquad K_n(\delta) = \{\lambda \in \mathbb{C} : |\lambda - in| < \delta\},$$

the estimate

$$|\gamma_a(\lambda)| \le \text{const}\,|\lambda|,$$

where const *depends on* δ.

Proof. The function

$$\beta_a(\lambda) = \int_0^\pi a(\theta)\,e^{-2\lambda\theta}\,d\theta, \qquad \lambda = x + iy,$$

is obviously analytic in $\mathbb{C}$, and for large $|\lambda|$ admits the estimate

$$|\beta_a(\lambda)| \le \int_0^\pi |a(\theta)|\,e^{-2x\theta}\,d\theta.$$

Thus for $x > 0$ we have

$$|\beta_a(\lambda)| \le \|a(\theta)\|_{L_1},$$

while for $x < 0$ we have

$$\begin{aligned}
|\beta_a(\lambda)| &\le e^{-2\pi x} \int_0^\pi |a(\theta)|\,e^{2x(\pi-\theta)}\,d\theta \\
&= e^{-2\pi x} \int_0^\pi |a(\theta)|\,d\theta \le e^{-2\pi x}\,\|a(\theta)\|_{L_1}.
\end{aligned}$$

The theorem statements now follow from

$$\gamma_a(\lambda) = \frac{2\lambda}{1 - e^{-2\pi\lambda}}\,\beta_a(\lambda). \qquad \square$$

To give a sufficient condition for the representation (8.7.1) we start with some definitions (see, for example, [61] for details).

An entire function $\varphi(\lambda)$ is called a function of exponential type if it obeys an estimate

$$|\varphi(\lambda)| \le Ae^{B|\lambda|},$$

where the positive constants A and B do not depend on $\lambda \in \mathbb{C}$. The infimum of all constants B for which this estimate holds is called the type of the function $\varphi(\lambda)$.

We denote by $\mathcal{L}_2^\sigma$ the set of all functions of exponential type less than or equal to σ whose restrictions to $\mathbb{R}$ belong to $L_2(\mathbb{R})$.

Recall that an analytic function on the upper half-plane $\varphi(\lambda)$ is said to belong to the Hardy space $H_+^2(\mathbb{R})$ if

$$\sup_{y>0} \int_{\mathbb{R}} |\varphi(x+iy)|^2\,dx < \infty.$$

The proof of the next theorem can be found, for example, in [61, Theorem 1.4].

Theorem 8.7.2. *Let $\varphi(z) \in \mathcal{L}_2^{2\pi} \cap H_+^2(\mathbb{R})$. Then there exists a function $a(\theta) \in L_2(0, 2\pi)$ such that*

$$\varphi(z) = \int_0^{2\pi} a(\theta)\, e^{iz\theta}\, d\theta, \qquad \lambda \in \mathbb{C}.$$

As $L_2(0, 2\pi) \subset L_1(0, 2\pi)$, the theorem can be used as a sufficient condition for the existence of representation (8.7.1). Indeed, given a function $\gamma(\lambda)$, introduce

$$\varphi(z) = i\,\frac{1 - e^{i\pi z}}{z}\, \gamma\left(-\frac{iz}{2}\right).$$

If this function $\varphi(z)$ belongs to $\mathcal{L}_2^{2\pi} \cap H_+^2(\mathbb{R})$ then $\gamma(\lambda)$ does admit representation (8.7.1). That is, there is a function $a(\theta) \in L_1(0, 2\pi)$ such that $\gamma(\lambda) = \gamma_a(\lambda)$ and

$$T_a = R^* \gamma(\lambda) R = f(T_{\chi_+}),$$

where

$$f(x) = \gamma\left(\gamma_{\chi_+}^{-1}(x)\right) = \gamma\left(-\frac{1}{\pi}\ln\frac{1 - x}{x}\right).$$

Theorem 8.7.3. *Let*

$$p(x) = \sum_{k=1}^{n} a_k x^k, \qquad a_n \neq 0,$$

be a polynomial of degree $n \geq 2$ with complex coefficients. Then the bounded operator $p(T_{\chi_+})$ is not a Toeplitz operator.

Proof. The operator $p(T_{\chi_+})$ belongs to the algebra generated by all Toeplitz operators on the upper half-plane with homogeneous L_∞-symbols $a(\theta)$ of zero order. Thus the operator $p(T_{\chi_+})$ being Toeplitz must have a defining symbol from $H(L_1(0, \pi))$. The corresponding function $\gamma(\lambda)$, that is, such that $p(T_{\chi_+}) = R^* \gamma(\lambda) R$, obviously has the form

$$\gamma(\lambda) = p\left(\gamma_{\chi_+}(\lambda)\right) = p\left(\frac{1}{e^{-\pi\lambda} + 1}\right).$$

But this function has poles of order n at the points $\lambda_n = i(2n - 1)$, where $n \in \mathbb{Z}$. Thus by Theorem 8.7.1 there is no function $a(\theta) \in H(L_1(0, \pi))$ for which the representation (8.7.1) holds. $\qquad\square$

Corollary 8.7.4. *Let A be an operator of the algebra $\mathcal{T}(PC(\overline{\mathbb{D}}, T))$ having the form*

$$A = \sum_{i=1}^{p} \prod_{j=1}^{q_i} T_{a_{i,j}},$$

where all $a_{i,j} \in PC(\overline{\mathbb{D}}, T)$. Then A is a compact perturbation of a Toeplitz operator if and only if A is a compact perturbation of one of the initial generators of $\mathcal{T}(PC(\overline{\mathbb{D}}, T))$, which is a Toeplitz operator T_a with $a \in PC(\overline{\mathbb{D}}, T)$.

Proof. By Corollary 8.3.3 the operator A admits the canonical representation

$$A = \sum_{i=1}^{p} \prod_{j=1}^{q_i} T_{a_{i,j}} = T_{s_A} + \sum_{k=1}^{m} T_{v_k} p_{A,k}(T_{\chi_k}) T_{v_k} + K_A,$$

where $s_A = s_A(z) \in C(\overline{\mathbb{D}})$, $p_{A,k} = p_{A,k}(x)$, $k = 1, \ldots, m$, are some polynomials, and K_A is a compact operator. Thus by Theorem 8.4.2, A is a compact perturbation of a Toeplitz operator if and only if each $p_{A,k}(T_{\chi_k})$, $k = 1, \ldots, m$, is a Toeplitz operator, or by (8.5.13) if and only if each $p_{A,k}(T_{\chi_+})$, $k = 1, \ldots, m$, is a Toeplitz operator. By Theorem 8.7.3 the last statement is equivalent to the fact that the degree of each polynomial $p_{A,k}(x)$, $k = 1, \ldots, m$, must be less than or equal to 1, which in turn is equivalent to the fact that A is a compact perturbation of a Toeplitz operator T_a with $a \in PC(\overline{\mathbb{D}}, T)$. $\square$

We summarize now the results obtained on Toeplitz operators of the algebra $\mathcal{T}(PC(\overline{\mathbb{D}}, T))$. By its construction, the C^*-algebra $\mathcal{T}(PC(\overline{\mathbb{D}}, T))$ consists of its initial generators, Toeplitz operators T_a with defining symbols $a \in PC(\overline{\mathbb{D}}, T)$, then of all elements of the form

$$\sum_{i=1}^{p} \prod_{j=1}^{q_i} T_{a_{i,j}},$$

forming thus a non-closed algebra, and finally of all elements of the uniform closure of the non-closed algebra. The information on Toeplitz operators is as follows.

- All initial generators are Toeplitz operators.

- None of the elements of the non-closed algebra which does not reduce to a compact perturbation of an initial generator can be (a compact perturbation of) a Toeplitz operator. Thus at this stage we have not increased the quantity of Toeplitz operators.

- The uniform closure of the non-closed algebra contains a huge amount of Toeplitz operators, with bounded and even unbounded defining symbols, which are drastically different from the initial generators. All of these Toeplitz operators are uniform limits of sequences of non-Toeplitz operators.

- The uniform closure, apart from the Toeplitz operators, contains many more non-Toeplitz operators (this is a consequence of Theorem 8.7.1).

At the same time each operator in the C^*-algebra $\mathcal{T}(PC(\overline{\mathbb{D}}, T))$ admits a very transparent canonical representation (given in Theorem 8.3.5).

8.8 Technical statements

We end the chapter by proving several statements whose results were used in Theorems 8.6.5 and 8.6.6.

We start with some properties of the function (see (8.6.2))

$$c(x,y) = \frac{1 - e^{-\pi(x+y)}}{x+y} \sqrt{\frac{2x}{1 - e^{-2\pi x}}} \sqrt{\frac{2y}{1 - e^{-2\pi y}}}, \qquad x, y \in \mathbb{R}.$$

Theorem 8.8.1. *The function $c(x,y)$ is bounded in $\mathbb{R}^2$; i.e.,*

$$\sup_{(x,y)\in\mathbb{R}^2} |c(x,y)| < \infty.$$

Proof. Introduce the function

$$f(u) = \sqrt{\frac{u}{1 - e^{-u}}}.$$

Then

$$c(x,y) = \frac{f(2\pi x)\, f(2\pi y)}{f^2(\pi(x+y))}.$$

Let $D_1 = [1, +\infty)$ and $D_{-1} = (-\infty, -1]$. We obviously have the following asymptotics in the above domains:

$$
\begin{aligned}
f(u) &= u^{1/2}\left(1 + O(e^{-u})\right), & u &\in D_1, & (8.8.1)\\
f(u) &= |u|^{1/2} e^{u/2}\left(1 + O(e^{u})\right), & u &\in D_{-1}, & (8.8.2)\\
f^{-2}(u) &= u^{-1}\left(1 + O(e^{-u})\right), & u &\in D_1, & (8.8.3)\\
f^{-2}(u) &= |u|^{-1} e^{-u}\left(1 + O(e^{u})\right), & u &\in D_{-1}. & (8.8.4)
\end{aligned}
$$

In what follows the relation $\varphi(u) \sim \psi(u)$ means that

$$0 < c \le \frac{\varphi(u)}{\psi(u)} \le C < \infty,$$

for all u in the domain under consideration. We note as well that if u belongs to any bounded domain in $\mathbb{R}$, then

$$f(u) \sim 1 \qquad \text{and} \qquad f^{-2}(u) \sim 1. \qquad (8.8.5)$$

We will prove the statement of the theorem considering successively all possible locations of x and y on $\mathbb{R}$. The symmetry of $c(x,y)$ with respect to its arguments implies that it is sufficient to consider only the following cases:

1. $x, y \in [-1, 1]$. Then $x + y \in [-2, 2]$, and by (8.8.5) we have that $c(x,y) \sim 1$.

2. $x \in [-1, 1]$, $y \in D_1$. Then either $x + y \in D_1$ and thus by (8.8.2) and (8.8.4) we have

$$c(x,y) \sim \frac{1 \cdot (2\pi y)^{1/2}}{\pi(x+y)} \sim y^{-1/2},$$

   or $y \in [1, 2]$ and thus, as in the first case, $c(x,y) \sim 1$.

3. $x \in [-1,1]$, $y \in D_{-1}$. Then either $x+y \in D_{-1}$ and thus by (8.8.1) and (8.8.3) we have
$$c(x,y) \sim \frac{1 \cdot |2\pi y|^{1/2} e^{\pi y}}{|\pi y| e^{\pi(x+y)}} \sim y^{-1/2},$$
or $y \in [-2,-1]$ and again $c(x,y) \sim 1$.

4. $x \in D_1$, $y \in D_{-1}$. Then we have the following three possibilities for $x+y$:

   (a) $x+y \in [-1,1]$. Then by (8.8.1), (8.8.2), and (8.8.5), assuming that $x+y = \delta \in [-1,1]$, we have
   $$c(x,y) \sim (2\pi x)^{1/2} \cdot (2\pi|y|)^{1/2} e^{\pi y} \cdot 1 \sim (\delta+y)^{1/2} |y|^{1/2} e^{\pi y} \sim |y| e^{\pi y}.$$

   (b) $x+y \in D_1$. Then by (8.8.1), (8.8.2), and (8.8.3) we have
   $$c(x,y) \sim \frac{(2\pi x)^{1/2} \cdot (2\pi|y|)^{1/2} e^{\pi y}}{\pi(x+y)} \sim \begin{cases} \frac{x^{1/2}|y|^{1/2} e^{\pi y}}{x}, & x \geq 2|y| \\ \frac{|y|^{1/2}|y|^{1/2} e^{\pi y}}{1}, & x < 2|y| \end{cases}$$
   $$\sim \begin{cases} \frac{|y|^{1/2} e^{\pi y}}{x^{1/2}}, & x \geq 2|y| \\ |y| e^{\pi y}, & x < 2|y| \end{cases}.$$

   (c) $x+y \in D_{-1}$. Then by (8.8.1), (8.8.2), and (8.8.4) we have
   $$c(x,y) \sim \frac{(2\pi x)^{1/2} \cdot (2\pi|y|)^{1/2} e^{\pi y}}{\pi|x+y| e^{\pi(x+y)}} \sim \frac{x^{1/2}|y|^{1/2} e^{-\pi x}}{|x+y|}$$
   $$\sim \begin{cases} \frac{|x|^{1/2} e^{-\pi x}}{y^{1/2}}, & |y| \geq 2x \\ x e^{-\pi x}, & |y| < 2x \end{cases}.$$

5. $x \in D_1$, $y \in D_1$. Then $x+y \in D_1$ and thus by (8.8.1) and (8.8.3) we have
$$c(x,y) = \frac{(2\pi x)^{1/2}(2\pi y)^{1/2}}{\pi(x+y)} \left(1 + O(e^{-x}) + O(e^{-y})\right)$$
$$= \frac{2x^{1/2}y^{1/2}}{x+y} \left(1 + O(e^{-x}) + O(e^{-y})\right).$$

As $2x^{1/2}y^{1/2} \leq x+y$, the boundedness of $c(x,y)$ is obvious.

6. $x \in D_{-1}$, $y \in D_{-1}$. Then $x+y \in D_{-1}$ and thus by (8.8.2) and (8.8.4) we have
$$c(x,y) = \frac{(2\pi|x|)^{1/2} e^{\pi x}(2\pi|y|)^{1/2} e^{\pi y}}{\pi|x+y| e^{\pi(x+y)}} \left(1 + O(e^{-|x|}) + O(e^{-|y|})\right)$$
$$= \frac{2|x|^{1/2}|y|^{1/2}}{|x|+|y|} \left(1 + O(e^{-|x|}) + O(e^{-|y|})\right).$$

The theorem is proved. $\square$

Theorem 8.8.2. *Both functions $\frac{\partial c}{\partial x}(x,y)$ and $\frac{\partial c}{\partial y}(x,y)$ are bounded in $\mathbb{R}^2$, and moreover*

$$\lim_{(x,y)\to\infty}\frac{\partial c}{\partial x}(x,y)=0 \qquad\text{and}\qquad \lim_{(x,y)\to\infty}\frac{\partial c}{\partial y}(x,y). \tag{8.8.6}$$

Proof. We start with the asymptotics of the derivatives of $f(u)$. We have

$$f'(u)=\frac{1}{2}\sqrt{\frac{1-e^{-u}}{u}}\cdot\frac{1-e^{-u}+ue^{-u}}{(1-e^{-u})^2},$$

thus, as is easy to see,

$$\begin{aligned}
f'(u) &= u^{-1/2}\left(1+O(ue^{-u})\right), & u\in D_1, & \tag{8.8.7}\\
f'(u) &= |u|^{1/2}e^{u/2}\left(1+O(e^u)\right), & u\in D_{-1}. & \tag{8.8.8}
\end{aligned}$$

Then

$$\begin{aligned}
\frac{\partial c}{\partial y}(x,y) &= f(2\pi x)\frac{2\pi f'(2\pi y)f^2(\pi(x+y))-2\pi f(2\pi y)f(\pi(x+y))f'(\pi(x+y))}{f^4(\pi(x+y))}\\
&= 2\pi\,c(x,y)\left(\frac{f'(2\pi y)}{f(2\pi y)}-\frac{f'(\pi(x+y))}{f(\pi(x+y))}\right). \tag{8.8.9}
\end{aligned}$$

We check now the boundedness of the logarithmic derivative of f. By (8.8.1) and (8.8.7) we have

$$\begin{aligned}
\frac{f'(2\pi y)}{f(2\pi y)} &= \frac{(2\pi y)^{-1/2}}{(2\pi y)^{1/2}}\left(1+O(ye^{-2\pi y})\right)\\
&= (2\pi y)^{-1}\left(1+O(ye^{-2\pi y})\right), & y\in D_1,
\end{aligned}$$

and by (8.8.2) and (8.8.8),

$$\begin{aligned}
\frac{f'(2\pi y)}{f(2\pi y)} &= \frac{(2\pi|y|)^{1/2}e^{\pi y}}{(2\pi|y|)^{1/2}e^{\pi y}}\left(1+O(e^{2\pi y})\right)\\
&= 1+O(e^{2\pi y}), & y\in D_{-1}.
\end{aligned}$$

Thus the function $\frac{\partial c}{\partial y}(x,y)$ is bounded in $\mathbb{R}^2$.

To prove the second equality in (8.8.6) we note that if both $|y|\to\infty$, and $|x+y|\to\infty$, and moreover $\operatorname{sign}y=\operatorname{sign}(x+y)$, then the result follows from (8.8.9) and boundedness of $c(x,y)$. If both $|y|\to\infty$, and $|x+y|\to\infty$, but $\operatorname{sign}y=-\operatorname{sign}(x+y)$, then by case 4.b of Theorem 8.8.1 we have that $c(x,y)\to 0$. If $(x,y)\to\infty$ while y belongs to a bounded domain, then $x+y$ is unbounded and we are in the situation of the cases 2 or 3 of Theorem 8.8.1, when $c(x,y)\to 0$. Finally, if $(x,y)\to\infty$ but $x+y$ is bounded, then as in the case 4.a of Theorem

8.8.1 we have that $c(x, y) \to 0$. In the last three cases the result follows from (8.8.9), boundedness of the logarithmic derivatives, and $c(x, y) \to 0$.

Boundedness of $\frac{\partial c}{\partial x}(x, y)$ and the first equality in (8.8.6) follow from the above and the symmetry of $c(x, y)$ with respect to x and y. $\qquad\square$

Theorem 8.8.3. *The function $\frac{\partial^2 c}{\partial y^2}(x, y)$ is bounded in $\mathbb{R}^2$, and moreover*

$$\lim_{(x,y)\to\infty} \frac{\partial^2 c}{\partial y^2}(x, y) = 0.$$

Proof. For the second derivative of f, after elementary calculations, we have

$$f''(u) = u^{-3/2}\left(1 + O(ue^{-u})\right), \qquad u \in D_1, \tag{8.8.10}$$

$$f''(u) = |u|^{1/2}e^{u/2}\left(1 + O(e^u)\right), \qquad u \in D_{-1}. \tag{8.8.11}$$

Differentiating (8.8.9) we have

$$
\begin{aligned}
\frac{\partial^2 c}{\partial y^2}(x, y) = {}& 2\pi \frac{\partial c}{\partial y}(x, y)\left(\frac{f'(2\pi y)}{f(2\pi y)} - \frac{f'(\pi(x + y))}{f(\pi(x + y))}\right) \\
& - 2\pi^2 c(x, y)\left[2\left(\frac{f''(2\pi y)}{f(2\pi y)} - \left(\frac{f'(2\pi y)}{f(2\pi y)}\right)^2\right)\right. \\
& \left. - \left(\frac{f''(\pi(x + y))}{f(\pi(x + y))} - \left(\frac{f'(\pi(x + y))}{f(\pi(x + y))}\right)^2\right)\right].
\end{aligned}
$$

We note that by Theorem 8.8.2 the first summand is bounded and tends to 0 as $(x, y) \to \infty$. Considering the second summand we have that if $|x + y| \to \infty$ and $y \to \infty$, then formulas (8.8.1), (8.8.2), (8.8.7), (8.8.8), (8.8.10), and (8.8.11), for $u = 2\pi y$ or $u = \pi(x + y)$, yield

$$
\begin{aligned}
\frac{f''(u)}{f(u)} - \left(\frac{f'(u)}{f(u)}\right)^2 &= \begin{cases} u^{-2}\left(1 + O(ue^{-u})\right) - u^{-2}\left(1 + O(ue^{-u})\right), & u \in D_1 \\ \left(1 + O(e^u)\right) - \left(1 + O(e^u)\right), & u \in D_{-1} \end{cases} \\
&= \begin{cases} O(u^{-1}e^{-u}), & u \in D_1 \\ O(e^u), & u \in D_{-1} \end{cases}.
\end{aligned}
$$

If $(x, y) \to \infty$, but $|x + y|$ is bounded, then we are in the situation of the case 4.a of Theorem 8.8.1, and thus $c(x, y) \to 0$ as $(x, y) \to \infty$. Finally, if $(x, y) \to \infty$, but y is bounded, then we are in the situation of the cases 2 or 3 of Theorem 8.8.1, and thus $c(x, y) \to 0$ as $(x, y) \to \infty$. $\qquad\square$

Theorem 8.8.4. *Let $a(\theta) \in L_1(0, \pi)$ be such that*

$$\gamma_a(\lambda) = \frac{2\lambda}{1 - e^{-2\pi\lambda}} \int_0^\pi a(\theta)e^{-2\lambda\theta}\, d\theta \in L_\infty(\mathbb{R}).$$

Then, for each $k = 1, 2, \ldots,$

$$\lim_{\lambda \to \pm\infty} \frac{d^j \gamma_a}{d\lambda^j}(\lambda) = 0.$$

Proof. Let $k = 1$. Then

$$\frac{d\gamma_a}{d\lambda}(\lambda) = \frac{2}{1 - e^{-2\pi\lambda}} \int_0^\pi a(\theta) e^{-2\lambda\theta}\, d\theta + \frac{4\pi\lambda e^{-2\pi\lambda}}{(1 - e^{-2\pi\lambda})^2} \int_0^\pi a(\theta) e^{-2\lambda\theta}\, d\theta$$

$$- \frac{4\lambda}{1 - e^{-2\pi\lambda}} \int_0^\pi \theta a(\theta) e^{-2\lambda\theta}\, d\theta = I_1(\lambda) + I_2(\lambda) + I_3(\lambda).$$

We consider first the behaviour of the derivative when $\lambda \to +\infty$. We have

$$I_1(\lambda) = \frac{2}{\lambda} \gamma_a(\lambda) \qquad \text{and} \qquad I_2(\lambda) = \frac{2\pi e^{-2\pi\lambda}}{1 - e^{-2\pi\lambda}} \gamma_a(\lambda),$$

and thus the first two summands tend to 0 as $\lambda \to +\infty$.
Consider now the last summand,

$$|I_3(\lambda)| \leq 2 \int_0^\delta |a(\theta)|(2\lambda\theta) e^{-2\lambda\theta} d\lambda + 4\pi\lambda e^{-2\lambda\delta} \int_\delta^\pi |a(\theta)| d\theta$$

$$\leq \int_0^\delta |a(\theta)| d\theta + 4\pi\lambda e^{-2\lambda\delta} \int_0^\pi |a(\theta)| d\theta.$$

Then for any $\varepsilon > 0$ we can select both δ small enough and $\lambda_0 = \lambda_0(\delta)$ large enough, such that

$$\int_0^\delta |a(\theta)| d\theta < \varepsilon/2$$

and

$$4\pi\lambda e^{-2\lambda\delta} \int_0^\pi |a(\theta)| d\theta < \varepsilon/2,$$

for all $\lambda \geq \lambda_0(\delta)$. That is, $\lim_{\lambda \to +\infty} I_3(\lambda) = 0$.
The case when $\lambda \to -\infty$ follows from the above and the equality

$$\gamma_{a(\theta)}(\lambda) = \gamma_{a(\pi-\theta)}(-\lambda).$$

The cases when $k > 1$ are considered analogously. $\qquad\square$

Chapter 9

Commuting Toeplitz Operators and Hyperbolic Geometry

The C^*-algebras of Toeplitz operators considered in Chapters 5, 6, and 7 are commutative. This property is impossible (except for the trivial case of scalar operators $\equiv$ constant defining symbols) for the C^*-algebras of Toeplitz operators acting on the Hardy space. The special features of the defining symbols which make this phenomenon possible were *symbols depending only on the imaginary part of a variable* for Toeplitz operators on the upper half-plane, *radial symbols* for Toeplitz operators on the unit disk, and *symbols depending only on the angular part of a variable* for Toeplitz operators on the upper half-plane, respectively. In this stage a natural question appears: *whether there exist other classes of defining symbols which generate commutative Toeplitz operator C^*-algebras, and how they can be classified.*

The invariance of the Bergman projection under biholomorphic automorphisms of a domain suggests immediately other classes of defining symbols, which are obtained from the initial ones by means of biholomorphic automorphisms. Surprisingly, the key to understanding the nature of such classes of symbols and to classifying them lies in the hyperbolic geometry of the unit disk endowed with the Poincaré metric.

It turns out that the sets of defining symbols which generate commutative C^*-algebras of Toeplitz operators are classified as follows. Each pencil of hyperbolic geodesics determines the set of symbols consisting of functions which are constant on corresponding cycles, the orthogonal trajectories to the lines forming a pencil. The C^*-algebra generated by Toeplitz operators with such defining symbols is commutative.

In Chapter 11 we show that the above classes are the only ones which generate commutative C^*-algebras of Toeplitz operators on each weighted Bergman space.

9.1　Bergman metric

We recall necessary facts about the Bergman metric. For further details see, for example [26, 176, 240].

Let D be a (bounded or unbounded) simply connected domain with smooth boundary ∂D on the complex plane $\mathbb{C}$. The Bergman metric of the domain D is defined by

$$ds^2 = \frac{1}{2}\,\frac{\partial^2 \ln K_D(z,\overline{z})}{\partial z\,\partial\overline{z}}\,(dx^2 + dy^2),$$

where $K_D(z,\zeta)$ is the Bergman kernel function of the domain D and $z = x + iy$. The element of length on the Bergman metric is given by

$$d\ell = \left(\frac{1}{2}\,\frac{\partial^2 \ln K_D(z,\overline{z})}{\partial z\,\partial\overline{z}}\right)^{\frac{1}{2}} dl,$$

where dl is the Euclidean length element. Denote by $\beta_D(z_1, z_2)$ the distance in the Bergman metric between two points z_1 and z_2 of the domain D. A very important feature of the Bergman metric is its invariance under the biholomorphic mappings: *if the function $\omega = \alpha(z)$ gives a biholomorphic mapping of the domain D onto a domain G, then*

$$\beta_D(z_1, z_2) = \beta_G(\alpha(z_1), \alpha(z_2)).$$

Denote by $\Delta_D(z_0, r)$ the (open) Bergman metric disk with center $z_0 \in D$ and radius $r > 0$, i.e.,

$$\Delta_D(z_0, r) = \{z \in D : \beta_D(z_0, z) < r\}.$$

The boundary

$$\mathcal{S}_D(z_0, r) = \{z \in D : \beta_D(z_0, z) = r\}$$

of the disk $\Delta_D(z_0, r)$ is called the Bergman circle with center z_0 and radius r.

For the case of the unit disk $\mathbb{D}$ the Bergman metric is defined by

$$ds^2 = \frac{dx^2 + dy^2}{(1 - |z|^2)^2}, \tag{9.1.1}$$

and coincides with the hyperbolic or the Poincaré metric, realizing the non-Euclidean hyperbolic geometry on the disk $\mathbb{D}$. The Bergman distance between two points z_1 and z_2 in $\mathbb{D}$ is given by

$$\beta_{\mathbb{D}}(z_1, z_2) = \frac{1}{2}\ln\frac{|1 - z_1\overline{z}_2| + |z_1 - z_2|}{|1 - z_1\overline{z}_2| - |z_1 - z_2|}, \tag{9.1.2}$$

and in particular, for $z \in \mathbb{D}$,

$$\beta_{\mathbb{D}}(0, z) = \frac{1}{2}\ln\frac{1 + |z|}{1 - |z|}.$$

Note that the Bergman disk $\Delta_{\mathbb{D}}(z_0, r)$ is actually a Euclidean disk whose (Euclidean) center and radius are

$$C = \frac{1 - s^2}{1 - s^2 |z_0|^2} z_0, \qquad R = \frac{1 - |z_0|^2}{1 - s^2 |z_0|^2} s,$$

where

$$s = \frac{e^r - e^{-r}}{e^r + e^{-r}} = \tanh r \in (0, 1).$$

Note also that the Bergman and the Euclidean centers of $\Delta_{\mathbb{D}}(z_0, r)$ coincide only for disks centered at the origin.

9.2 Basic properties of Möbius transformations

We recall here some known facts about the group of Möbius transformations $\text{Möb}(\mathbb{C})$ which consists of all mappings $g : \mathbb{C} \to \mathbb{C}$ having the form

$$g(z) = \frac{az + b}{cz + d},$$

where $a, b, c, d \in \mathbb{C}$ and $ad - bc = 1$. For further details see, for example, [21, 79, 126].

Each matrix $A \in \text{SL}(2, \mathbb{C})$ induces the mapping $g_A \in \text{Möb}(\mathbb{C})$ by the formula $A \mapsto g_A$, where

$$A = \begin{pmatrix} a & b \\ c & d \end{pmatrix}, \quad \text{and} \quad g_A(z) = \frac{az + b}{cz + d};$$

this is a group homomorphism whose kernel is equal to $\mathbb{Z}_2 = \{I, -I\}$, and thus

$$\text{Möb}(\mathbb{C}) \cong \text{SL}(2, \mathbb{C})/\mathbb{Z}_2.$$

Recall the following well-known properties of the Möbius transformations:

1) Each Möbius transformation g is a 1-1 conformal mapping of $\dot{\mathbb{C}} = \mathbb{C} \cup \{\infty\}$ onto itself.

2) Each Möbius transformation g maps a circle or a straight line onto a circle or a straight line.

3) Let z_1, z_2, z_3 be a triple of distinct points in $\dot{\mathbb{C}}$, and let w_1, w_2, w_3 be another such triple. Then there exists a unique Möbius transformation which maps z_1, z_2, z_3 to w_1, w_2, w_3 respectively.

4) If z_1 and z_2 are inverse points with respect to a circle (or a straight line) C and $g \in \text{Möb}(\mathbb{C})$, then the points $g(z_1)$ and $g(z_2)$ are inverse with respect to the circle (or a straight line) $g(C)$.

For the unit disk $\mathbb{D}$ and the upper half-plane Π one has

1. every Möbuis transformation $g : \mathbb{D} \to \mathbb{D}$ has the form

$$g(z) = \frac{az + b}{\bar{b}z + \bar{a}},$$

where $a, b \in \mathbb{C}$ and $a\bar{a} - b\bar{b} = 1$, furthermore

$$\text{Möb}(\mathbb{D}) \cong \text{SU}(1,1)/\mathbb{Z}_2;$$

2. every Möbuis transformation $g : \Pi \to \Pi$ has the form

$$g(z) = \frac{az + b}{cz + d},$$

where $a, b, c, d \in \mathbb{R}$ and $ad - bc = 1$, furthermore

$$\text{Möb}(\Pi) \cong \text{SL}(2, \mathbb{R})/\mathbb{Z}_2.$$

The function $\text{tr}^2 A$, for $A \in \text{SL}(2, \mathbb{C})$, is invariant under the transformation $A \mapsto \pm A$, and thus induces the corresponding function on Möb:

$$\text{tr}^2 g_A = \text{tr}^2 A = (a + d)^2.$$

Note that the function $\text{tr}^2 g$ is invariant under any conjugation $g \mapsto hgh^{-1}$, $h \in \text{Möb}(\mathbb{C})$, i.e.,

$$\text{tr}^2 g = \text{tr}^2 hgh^{-1}. \tag{9.2.1}$$

We will denote the conjugacy equivalence relation by $\sim$.

Introduce the following standard normalized Möbius transformations:

$$m_k(z) = kz,$$

for each $k \in \mathbb{C} \setminus \{0, 1\}$, and

$$m_1(z) = z + 1.$$

Note that in all cases

$$\text{tr}^2 m_k = k + \frac{1}{k} + 2. \tag{9.2.2}$$

Lemma 9.2.1. *Each non-identical Möbius transformation is conjugate to one of the standard form m_k.*

Proof. Each non-identical Möbius transformation g has obviously either two fixed points (denote them by α and β), or a unique fixed point (denote it by α, and let now β be any point different from α). Consider a Möbius transformation h with the properties

$$h(\alpha) = \infty, \qquad h(\beta) = 0, \qquad \text{and} \qquad h(g(\beta)) = 1 \quad \text{if} \quad g(\beta) \neq \beta.$$

Then

$$hgh^{-1}(\infty) = \infty \qquad \text{and} \qquad hgh^{-1}(0) = \begin{cases} 0, & \text{if } g(\beta) = \beta \\ 1, & \text{if } g(\beta) \neq \beta \end{cases}.$$

That is, if g has two fixed points, then the fixed points of hgh^{-1} are 0 and ∞, and thus necessarily $hgh^{-1} = m_k$ for some $k \neq 1$. If g has only one fixed point, then the unique fixed point of hgh^{-1} is ∞, and thus necessarily $hgh^{-1} = m_1$. $\square$

Theorem 9.2.2. *Two non-identical Möbius transformations f and g are conjugate if and only if* $\text{tr}^2 f = tr^2 g$.

Proof. If $f \sim g$, then by (9.2.1) we have $\text{tr}^2 f = tr^2 g$. Let now $\text{tr}^2 f = tr^2 g$. By Lemma 9.2.1 both f and g are conjugate to some standard forms: $f \sim m_p$ and $g \sim m_q$. Thus

$$\text{tr}^2 m_p = \text{tr}^2 f = \text{tr}^2 g = \text{tr}^2 m_q.$$

Thus by (9.2.2) we have either $p = q$, or $p = \frac{1}{q}$. If $p = 1$, then always $p = q$. If $p \neq 1$ then for $h(z) = -\frac{1}{z}$ we have

$$h m_p h^{-1} = m_{\frac{1}{p}}.$$

Thus

$$f \sim m_p \sim m_q \sim g. \qquad \square$$

Definition 9.2.3. Let g be any non-identical Möbius transformation. Then

1. g is *parabolic* if $g \sim m_1$;

2. g is *elliptic* if $g \sim m_k$ with $|k| = 1$;

3. g is *hyperbolic* if $g \sim m_k$ with $k > 0$ and $k \neq 1$;

4. otherwise g is *loxodromic*, i.e., $g \sim m_k$ with non-positive k and $|k| \neq 1$.

Corollary 9.2.4. *Let g be any non-identical Möbius transformation, then*

1. *g is parabolic if and only if $\text{tr}^2 g = 4$;*

2. *g is elliptic if and only if $\text{tr}^2 g \in [0, 4)$;*

3. *g is hyperbolic if and only if $\text{tr}^2 g \in (4, +\infty)$;*

4. *g is loxodromic if and only if $\text{tr}^2 g \notin [0, +\infty)$.*

Proof. It is sufficient to calculate $\text{tr}^2 m_k$ for the first three cases, the remaining traces will correspond to the fourth case. Using (9.2.2) we have

1. let $k = 1$, then $\text{tr}^2 m_1 = 4$;

2. let $k = e^{i\theta}$ with $\theta \neq 2n\pi$, then $\text{tr}^2 = e^{i\theta} + e^{-i\theta} + 2 = 2 + 2\cos\theta \in [0, 4)$;

3. let $k > 0$ and $k \neq 1$, then $\text{tr}^2 m_k = k + \frac{1}{k} + 2 \in (4, +\infty)$. $\square$

Remark 9.2.5. The groups $\text{Möb}(\mathbb{D})$ and $\text{Möb}(\Pi)$ of the Möbius transformations preserving the unit disk and the upper half-plane, respectively, consist only of elliptic, parabolic, and hyperbolic transformations, and for elements of both of these groups the statement of Theorem 9.2.2 remains valid.

9.3   Fixed points and commuting Möbius transformations

Observe first that two Möbius transformations g and f, neither being the identity, commute if and only if for each $h \in \text{Möb}(\mathbb{C})$ the transformations hgh^{-1} and hfh^{-1} commute. That is the property to be a pair of commuting Möbius transformations is invariant under conjugation.

Theorem 9.3.1. *Two Möbius transformations g and f, neither being the identity, which have the same fixed points, commute.*

Proof. Suppose first that g and f are parabolic. Then applying if necessary conjugation we may assume the unique fixed point of both of them is ∞. Then necessarily for some $h_1, h_2 \in \mathbb{R}$ we have $g(z) = z + h_1$ and $f(z) = z + h_2$, and these transformations obviously commute.

Let now g and f have a common pair of fixed points. Again applying if necessary conjugation we may assume that those points are 0 and ∞. Then necessarily for some $k_1, k_2 \in \mathbb{C} \setminus \{0, 1\}$ we have $g(z) = k_1 z$ and $f(z) = k_2 z$, and again these transformations obviously commute. $\qquad\square$

Theorem 9.3.2. *If two Möbius transformations g and f, neither being the identity, commute then either they have the same set of fixed points, or $g^2 = I$ and $f^2 = I$.*

Proof. Assume first that g is parabolic with fixed point at ∞, that is $g(z) = z + h$, $h \in \mathbb{R} \setminus \{0\}$, and that

$$ h(z) = \frac{az + b}{cz + d}, \qquad ad - bc = 1. $$

Now, g and f commute if and only if the commutator $[g, f] = gfg^{-1}f^{-1} = I$. Calculate

$$ [g, f](z) = \frac{(1 + ach + c^2 h^2)z + h(1 - a^2 - ac)}{c^2 h z + 1 - ach}. $$

Thus $[g, f] = I$ if and only if $c = 0$ and $1 - a^2 - ac = 0$, which implies $a = \pm 1$ and, taking into account $ad - bc = 1$, $d = a = \pm 1$. Thus f is parabolic with fixed point at ∞ as well.

Let now g have two fixed points 0 and ∞, i.e., $g(z) = kz$, $k \in \mathbb{C} \setminus \{0, 1\}$. In this case

$$ [g, f](z) = k\frac{(ad - bck)z + ab(k - 1)}{cd(1 - k)z + (adk - bc)}. $$

Assuming that $[g, f] = I$ we have $ab = cd = 0$. If now $bc = 0$ then $ad - bc = 1$ implies $a \neq 0$, $d \neq 0$ and $b = c = 0$. That is f has the same fixed points 0 and ∞. If $bc \neq 0$ then by the same arguments we have $a = d = 0$ and $bc = -1$, thus

$$f(z) = \frac{b}{cz} = -\frac{b^2}{z},$$

and obviously $f^2 = I$. Moreover we have $z = I(z) = [g, f](z) = k^2 z$, thus $k = -1$ and $g^2 = I$. $\qquad\square$

Note that for Möbius transformations from $\text{Möb}(\mathbb{D})$ and $\text{Möb}(\Pi)$ the second option of Theorem 9.3.2 is excluded (because neither of the transformations from $\text{Möb}(\mathbb{D})$ or $\text{Möb}(\Pi)$ can be loxodromic, as f is in the second option). Thus for $\text{Möb}(\mathbb{D})$ and $\text{Möb}(\Pi)$ we have the following assertion.

Corollary 9.3.3. *Two non-identical Möbius transformations from* $\text{Möb}(\mathbb{D})$ *(or from* $\text{Möb}(\Pi)$*) commute if and only if they have the same set of fixed points.*

9.4 Elements of hyperbolic geometry

Denote by H the hyperbolic plane, which could be the unit disk $\mathbb{D}$ or the upper half-plane Π depending on the model under consideration.

A geodesic, or a hyperbolic straight line, on H is (a part of) a Euclidean circle or straight line orthogonal to the boundary of H. We collect here their important properties (for proofs and details see, for example, [21]).

There is a unique geodesic passing through any two distinct points of the hyperbolic plane.

Two different geodesics intersect in at most one point in the hyperbolic plane.

Given any two geodesics L_1 and L_2, there is a Möbius transformation $g \in \text{Möb}(H)$ such that $g(L_1) = L_2$.

Given any geodesics L and any point w there is a unique geodesic L_1 passing through w and orthogonal to L, and the distance $\rho(w, L) = \inf\{\rho(w, z) : z \in L\}$ from w to L is measured along L_1.

Given two distinct points z_1 and z_2, denote by $[z_1, z_2]$ the part of the unique geodesic passing through these points, whose length is the distance between them. Let w be a mid-point of $[z_1, z_2]$, then

$$L = \{z : \rho(z, z_1) = \rho(z, z_2)\}$$

is the unique geodesic through w and orthogonal to $[z_1, z_2]$, which is called the *perpendicular bisector* of $[z_1, z_2]$.

Given two disjoint geodesics L_1 and L_2, there is a unique geodesic which is orthogonal to both L_1 and L_2. The distance between them

$$\rho(L_1, L_2) = \inf\{\rho(z_1, z_2) : z_1 \in L_1, z_2 \in L_2\}$$

is measured along their common orthogonal.

Each pair of geodesics, say L_1 and L_2, lie in a geometrically defined object, one-parameter family $\mathcal{P}$ of geodesics, which is called the *pencil* determined by L_1 and L_2. Each pencil has an associated family $\mathcal{C}$ of lines, called *cycles*, which are orthogonal trajectories to the geodesics forming the pencil. In Figures 9.1, 9.2, and 9.3, illustrating in $\mathbb{D}$ possible pencils, the cycles are drawn in bold lines.

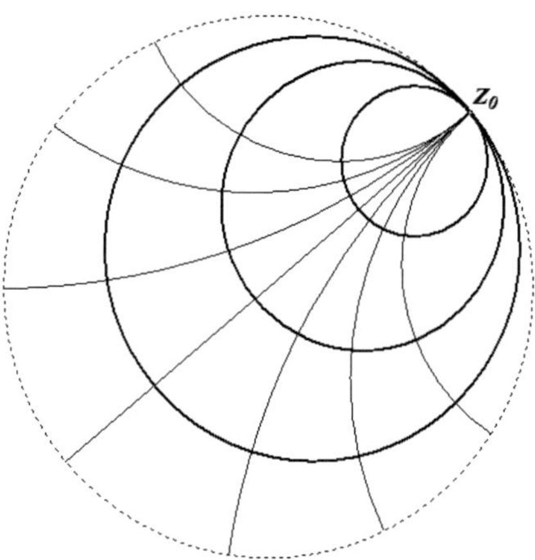

Figure 9.1: Parabolic pencil and corresponding horocycles.

The pencil $\mathcal{P}$ determined by L_1 and L_2 is called

1. *parabolic* if L_1 and L_2 are parallel; in this case $\mathcal{P}$ is the set of all geodesics parallel to both L_1 and L_2, and cycles are called *horocycles*;

2. *elliptic* if L_1 and L_2 are intersecting; in this case $\mathcal{P}$ is the set of all geodesics passing through the common point of L_1 and L_2;

3. *hyperbolic* if L_1 and L_2 are disjoint; in this case $\mathcal{P}$ is the set of all geodesics orthogonal to the common orthogonal of L_1 and L_2, and cycles are called *hypercycles*.

Let us mention the following joint properties of $\mathcal{P}$ and $\mathcal{C}$.

1. each point in the hyperbolic plane lies on exactly one cycle in $\mathcal{C}$;

2. with possibly one exception (common point of geodesics in an elliptic pencil), each point in the hyperbolic plane lies on exactly one geodesic in $\mathcal{P}$;

3. all geodesics in $\mathcal{P}$ are orthogonal to every cycle in $\mathcal{C}$;

4. every cycle in $\mathcal{C}$ is invariant under the reflection in any geodesic in $\mathcal{P}$;

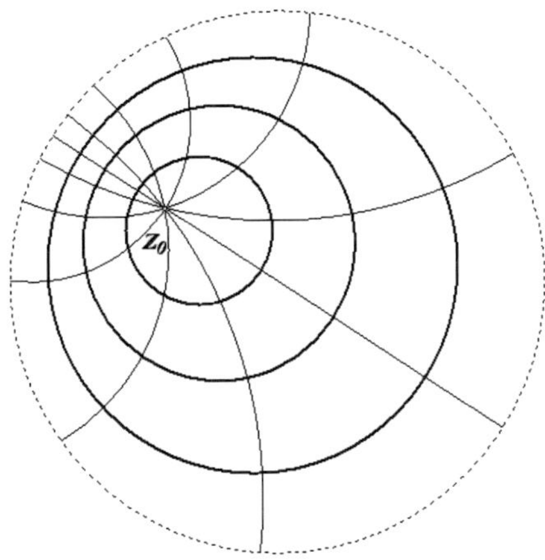

Figure 9.2: Elliptic pencil and corresponding cycles.

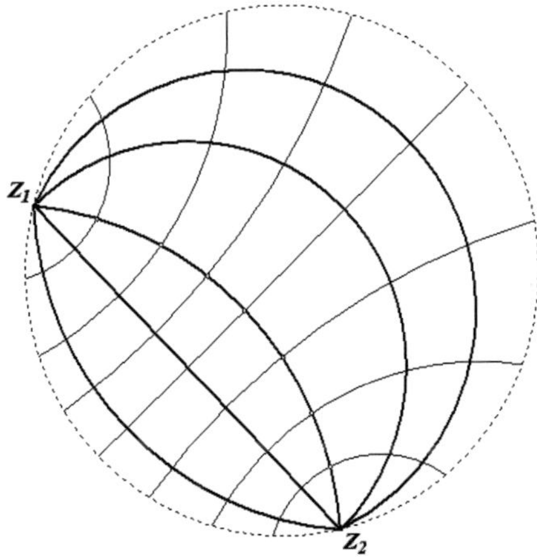

Figure 9.3: Hyperbolic pencil and corresponding hypercycles.

5. any two cycles, say C_1 and C_2, are *equidistant*, that is, for each $z_1 \in C_1$ there is $z_2 \in C_2$ such that

$$\rho(z_1, z_2) = \rho(C_1, C_2),$$

moreover z_1 and z_2 lie on the same geodesic in $\mathcal{P}$;

6. two points z_1 and z_2 lie on the same cycle in $\mathcal{C}$ if and only if the perpendicular bisector of $[z_1, z_2]$ is in $\mathcal{P}$;

7. the set $\mathcal{P}$ is precisely the set of geodesics of the form

$$L = \{z : a \sinh \rho(z, L_1) = b \sinh \rho(z, L_2), \ a, b > 0\}.$$

9.5 Action of Möbius transformations

Recall first that every conformal isometry (a movement) of the hyperbolic plane is given by a Möbius transformation from $\mathrm{M\ddot{o}b}(\mathbb{D})$.

A Möbius transformation $g \in \mathrm{M\ddot{o}b}(\mathbb{D})$ is either parabolic, or elliptic, or hyperbolic. Furthermore, every $g \in \mathrm{M\ddot{o}b}(\mathbb{D})$ can be represented as $g = \sigma_2\,\sigma_1$, where σ_k is a reflection in a certain geodesic L_k, $k = 1, 2$.

A transformation g is *parabolic* if and only if the above geodesics L_1 and L_2 determine a *parabolic* pencil. Given parabolic g, the associated parabolic pencil is the pencil containing all geodesics which end at the fixed point of g. Moreover, either L_1 or L_2 may be chosen arbitrarily from this pencil, and L_2 is the bisector of L_1 and $g(L_1)$.

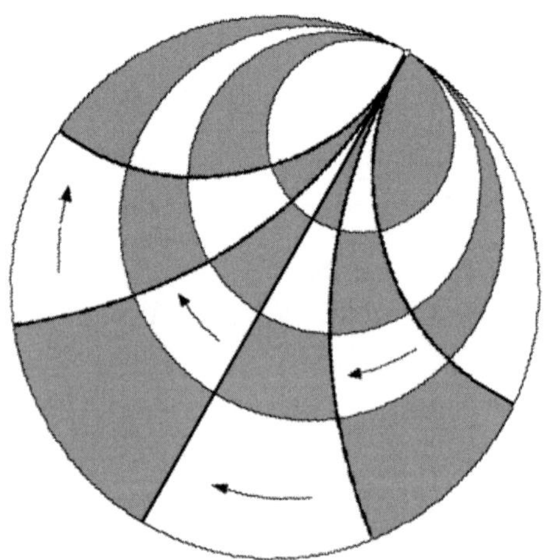

Figure 9.4: Parabolic transformation.

A transformation g is *elliptic* if and only if the above geodesics L_1 and L_2 determine an *elliptic* pencil. Given elliptic g, the associated elliptic pencil is the pencil containing all geodesics passing through the fixed point z_0 of g in the hyperbolic plane. Moreover, either L_1 or L_2 may be chosen arbitrarily from this pencil, and the other L_k is then uniquely determined by g.

An elliptic transformation g is completely determined by and completely determines its fixed point z_0 in the hyperbolic plane and a real number $\theta \in [0, 2\pi)$. Indeed, let z_1 be another fixed point of g, the reflection of z_0 in the circle at infinity.

Then

$$\frac{g(z) - z_0}{g(z) - z_1} = e^{i\theta}\left(\frac{z - z_0}{z - z_1}\right),$$

thus $g'(z_0) = e^{i\theta}$. The real number θ is called the *angle of rotation* of g. In this case g is conjugate to $m_\theta(z) = e^{i\theta}z$, and

$$\mathrm{tr}^2 g = 2(1 + \cos\theta).$$

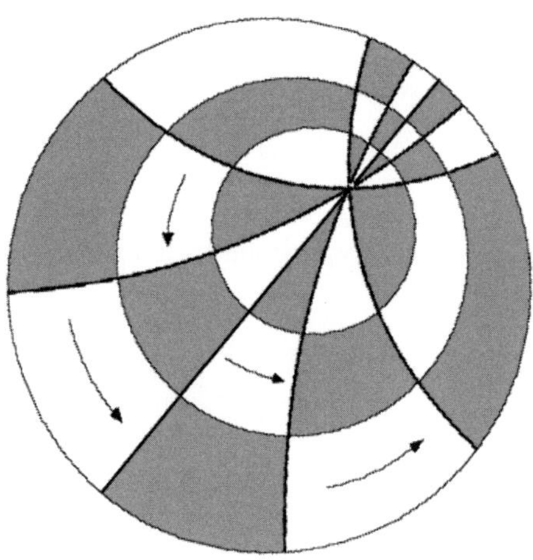

Figure 9.5: Elliptic transformation.

A transformation g is *hyperbolic* if and only if the above geodesics L_1 and L_2 determine a *hyperbolic* pencil. The *axis* of g (in the hyperbolic plane) is the axis of the pencil, that is the unique geodesic orthogonal to all lines in the pencil, and ends at the fixed points of g. We mention, that the axis of g is the unique g-invariant geodesic. Again either L_1 or L_2 may be chosen arbitrarily from this pencil, and the other L_k is then uniquely determined by g.

In all cases the action of $g \in \mathrm{M\ddot{o}b}(\mathbb{D})$ is as follows: each geodesic L from the pencil $\mathcal{P}$, determined by g, moves along the cycles in $\mathcal{C}$ to the geodesic $g(L) \in \mathcal{P}$, while each cycle in $\mathcal{C}$ is invariant under the action of g.

An elliptic transformation is called the *rotation* around its fixed point. A parabolic transformation is called the *parallel displacement*. A hyperbolic transformation is called the *translation* along its (unique translation) axis.

In the Euclidean geometry parallel displacements and translations coincide and they have infinitely many axes.

Given $g \in \mathrm{M\ddot{o}b}(\mathbb{D})$, consider corresponding pencil $\mathcal{P}$ and denote by $G_{\mathcal{P}}$ the set of all $h \in \mathrm{M\ddot{o}b}(\mathbb{D})$ having the same fixed points as g, and thus determining the same pencil $\mathcal{P}$. The set $G_{\mathcal{P}}$ is obviously a commutative group. Starting with any

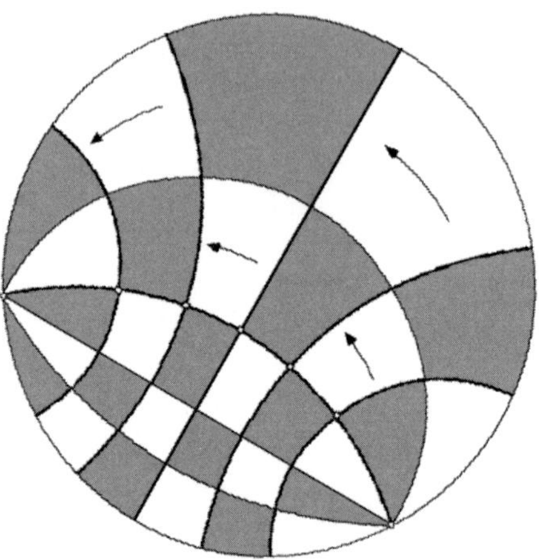

Figure 9.6: Hyperbolic transformation.

geodesic $L_0 \in \mathcal{P}$ the set of all geodesics in $\mathcal{P}$ has the form

$$\{h(L_0) : h \in G_{\mathcal{P}}\},$$

and all cycles in $\mathcal{C}$ are invariant under the action of each $h \in G_{\mathcal{P}}$.

Furthermore, $G_{\mathcal{P}}$ is a (commutative) one-parametric group generated by any non-identical element of $G_{\mathcal{P}}$, and each one-parametric group in $\mathrm{M\ddot{o}b}(\mathbb{D})$ has the above form.

Note finally, that each of the three following objects:

(i) pencil of hyperbolic straight lines,

(ii) one-parametric group of Möbius transformations,

(iii) fixed point set of a Möbius transformation

determines uniquely two others.

9.6　Classification theorem

It turns out that each of the above three objects may be chosen for the classification of known commutative C^*-algebras of Toeplitz operators. Our main result is as follows.

Theorem 9.6.1. *Given a pencil $\mathcal{P}$ of geodesics, consider the set of symbols which are constant on corresponding cycles. The C^*-algebra generated by Toeplitz operators with such defining symbols is commutative.*

In terms of the one-parameter groups the theorem reads as follows.

Theorem 9.6.2. *Given a non-identical Möbius transformation $g \in \mathrm{M\ddot{o}b}(\mathbb{D})$, consider the set of symbols which are invariant with respect to the one-parameter group generated by g. The C^*-algebra generated by Toeplitz operators with such defining symbols is commutative.*

That is, *each pencil of geodesics generates a commutative C^*-algebra of Toeplitz operators*, or equivalently *each maximal commutative subgroup in $\mathrm{M\ddot{o}b}(\mathbb{D})$ generates a commutative C^*-algebra of Toeplitz operators.*

We postpone the proof of Theorem 9.6.1 to the next section. Meanwhile we repeat the results from Chapters 5, 6, and 7, making them uniform and labeling γ's by the fixed points of the corresponding Möbius transformations.

We begin with the upper half-plane Π in $\mathbb{C}$. Introduce the C^*-algebra $\mathcal{A}(\infty)$ of all bounded measurable symbols which depend only on v (the imaginary part of a variable $w = u + iv$), and consider the Toeplitz operator algebra $\mathcal{T}(\mathcal{A}(\infty))$ generated by all operators of the form

$$T_a : \varphi \in \mathcal{A}^2(\Pi) \longmapsto B_\Pi a \varphi \in \mathcal{A}^2(\Pi),$$

where $a = a(v) \in \mathcal{A}(\infty)$.

Theorem 9.6.3. *Let $a = a(v) \in \mathcal{A}(\infty)$. Then the Toeplitz operator T_a acting on $\mathcal{A}^2(\Pi)$ is unitary equivalent to the multiplication operator $\gamma_a^{(\infty)} I$, acting on $L_2(\mathbb{R}_+)$. The function $\gamma_a^{(\infty)}(x)$ is given by*

$$\gamma_a^{(\infty)}(x) = \int_{\mathbb{R}_+} a\Big(\frac{y}{2x}\Big) e^{-y}\, dy, \quad x \in \mathbb{R}_+.$$

The algebra $\mathcal{T}(\mathcal{A}(\infty))$ is commutative. The isomorphic imbedding

$$\tau_\infty : \mathcal{T}(\mathcal{A}(\infty)) \longrightarrow C_b(\mathbb{R}_+)$$

is generated by the following mapping of generators of the algebra $\mathcal{T}(\mathcal{A}(\infty))$,

$$\tau_\infty : T_a \longmapsto \gamma_a^{(\infty)}(x),$$

where $a = a(v) \in \mathcal{A}(\infty)$.

Consider now the unit disk $\mathbb{D}$ in $\mathbb{C}$. Denote by $\mathcal{A}(0)$ ($\subset L_\infty(\mathbb{D})$) the C^*-algebra of bounded measurable symbols which depend only on $r = \sqrt{z\bar{z}}$, and consider the Toeplitz operator algebra $\mathcal{T}(\mathcal{A}(0))$ generated by all operators of the form

$$T_a : \varphi \in \mathcal{A}^2(\mathbb{D}) \longmapsto B_\mathbb{D} a \varphi \in \mathcal{A}^2(\mathbb{D}),$$

where $a = a(r) \in \mathcal{A}(0)$.

Theorem 9.6.4. *Let $a = a(r) \in \mathcal{A}(0)$. Then the Toeplitz operator T_a acting on $\mathcal{A}^2(\mathbb{D})$ is unitary equivalent to the multiplication operator $\gamma_a^{(0)} I$, acting on l_2^+. The sequence $\gamma_a^{(0)} = \{\gamma_a^{(0)}(n)\}_{n \in \mathbb{Z}_+}$ is given by*

$$\gamma_a^{(0)}(n) = \int_0^1 a\left(r^{\frac{1}{2(n+1)}}\right) dr = (n+1) \int_0^1 a(\sqrt{r}) \, r^n \, dr, \quad n \in \mathbb{Z}_+.$$

The algebra $\mathcal{T}(\mathcal{A}(0))$ is commutative. The isomorphic imbedding

$$\tau_0 : \mathcal{T}(\mathcal{A}(0)) \longrightarrow l_\infty$$

is generated by the following mapping of generators of the algebra $\mathcal{T}(\mathcal{A}(0))$,

$$\tau_0 : T_a \longmapsto \gamma_a^{(0)},$$

where $a = a(r) \in \mathcal{A}(0)$.

Return again to the upper half-plane Π. Denote by $\mathcal{A}(0, \infty)$ the C^*-algebra of bounded measurable homogeneous functions on Π of order zero, or functions depending only on the polar coordinate θ. Also consider the Toeplitz operator algebra $\mathcal{T}(\mathcal{A}(0, \infty))$ which is generated by all the operators of the form

$$T_a : \varphi \in \mathcal{A}^2(\Pi) \longmapsto B_\Pi a \varphi \in \mathcal{A}^2(\Pi),$$

where $a = a(\theta) \in \mathcal{A}(0, \infty)$.

Theorem 9.6.5. *Let $a = a(\theta) \in \mathcal{A}(0, \infty)$. Then the Toeplitz operator T_a acting on $\mathcal{A}^2(\Pi)$ is unitary equivalent to the multiplication operator $\gamma_a^{(0,\infty)} I$ acting on $L_2(\mathbb{R})$. The function $\gamma_a^{(0,\infty)}(\lambda)$ is given by*

$$\gamma_a^{(0,\infty)}(\lambda) = \frac{2\lambda}{1 - e^{-2\pi\lambda}} \int_0^\pi a(\theta) \, e^{-2\lambda\theta} \, d\theta, \qquad \lambda \in \mathbb{R}.$$

The algebra $\mathcal{T}(\mathcal{A}(0, \infty))$ is commutative. The isomorphic imbedding

$$\tau_{(0,\infty)} : \mathcal{T}(\mathcal{A}(0, \infty)) \longrightarrow C_b(\mathbb{R})$$

is generated by the following mapping of generators of the algebra $\mathcal{T}(\mathcal{A}(0, \infty))$,

$$\tau_{(0,\infty)} : T_a \longmapsto \gamma_a^{(0,\infty)}(\lambda),$$

where $a = a(\theta) \in \mathcal{A}(0, \infty)$.

9.7   Proof of the classification theorem

With each pencil of hyperbolic geodesics $\mathcal{P}$ we associate the class $\mathcal{A}_\mathcal{P}$ of measurable functions (defining symbols) $a = a(z)$ which are constant on cycles, and for which the corresponding Toeplitz operators T_a are bounded on the Bergman space $\mathcal{A}^2(\mathbb{D})$. In particular, the class $\mathcal{A}_{\mathcal{P}}$ contains all $L_\infty(\mathbb{D})$-functions which are constant on cycles. For simplicity we restrict ourselves here to the case of bounded symbols. That is, in what follows we mean by $\mathcal{A}_{\mathcal{P}}$ the algebra of all $L_\infty(\mathbb{D})$-functions which are constant on cycles of the pencil $\mathcal{P}$.

We will show that each Toeplitz-operator C^*-algebra $\mathcal{T}(\mathcal{A}_{\mathcal{P}})$ which is generated by all operators T_a with $a \in \mathcal{A}_{\mathcal{P}}$ is commutative, and we will describe its structure.

Start with an elliptic pencil $\mathcal{P} = \mathcal{P}(z_0)$, which consists of all hyperbolic straight lines passing through a point $z_0 \in \mathbb{D}$. The functions $a = a(z)$ from $\mathcal{A}(z_0) = \mathcal{A}_{\mathcal{P}(z_0)}$ now are constant on each hyperbolic circle $\mathcal{S}(z_0, r)$, $r > 0$.

The simplest case of the above situation is the case of $z_0 = 0$. The pencil $\mathcal{P}(0)$ consists of all (Euclidean) diameters of the unit circle $\mathbb{D}$, the set of cycles coincides with the set of all (Euclidean) circles having common center $z_0 = 0$, and the class of defining symbols $\mathcal{A}(0)$ is the class of radial symbols.

Passing to the general case of an arbitrary point $z_0 \in \mathbb{D}$, introduce the Möbius transformation

$$\alpha_{z_0}(z) = \frac{z_0 - z}{1 - z\overline{z_0}}$$

of $\mathbb{D}$ onto itself, which maps the point z_0 to the point 0. Note that this mapping is self-inverse: $\alpha_{z_0}^{-1}(z) = \alpha_{z_0}(z)$. The biholomorphic invariance of the hyperbolic metric implies that $\mathcal{P}(0) = \alpha_{z_0}(\mathcal{P}(z_0))$ and

$$\mathcal{A}(0) = \{a(\alpha_{z_0}(z)) : a(z) \in \mathcal{A}(z_0)\}.$$

Define the unitary operator $U_{z_0} = U_{z_0}^{-1} : L_2(\mathbb{D}) \longrightarrow L_2(\mathbb{D})$ by

$$(U_{z_0}\varphi)(z) = \alpha'_{z_0}(z)\varphi[\alpha_{z_0}(z)].$$

It is easy to see that for each $a(z) \in \mathcal{A}(z_0)$ we have

$$T_{a(z)} = U_{z_0} T_{a(\alpha_{z_0}(z))} U_{z_0}.$$

Thus by Theorem 9.6.4 we have

Theorem 9.7.1. *The C^*-algebra $\mathcal{T}(\mathcal{A}(z_0))$ is commutative and isomorphically embedded into the algebra l_∞. This embedding*

$$\nu_{z_0} : \mathcal{T}(\mathcal{A}(z_0)) \longrightarrow l_\infty$$

is generated by the mapping

$$\nu_{z_0} : T_a \longmapsto \gamma_a^{(z_0)},$$

where $a(z) \in \mathcal{A}(z_0)$, and the sequence $\gamma_a^{(z_0)}$ is given by formula

$$\gamma_a^{(z_0)}(n) = (n + 1) \int_0^1 a(\sqrt{\alpha_{z_0}(r)})\, r^n dr, \qquad n \in \mathbb{Z}_+.$$

Consider now a parabolic pencil $\mathcal{P} = \mathcal{P}(z_0)$, which consists of all (straight) lines parallel to a given line, and tending to a same point $z_0 \in \partial\mathbb{D}$. The functions $a = a(z)$ from $\mathcal{A}(z_0) = \mathcal{A}_{\mathcal{P}(z_0)}$ now are constant on each horocycle, i.e., on each Euclidean circle tangent to $\partial\mathbb{D}$ at the point z_0.

Introduce the Möbius transformation

$$w = \alpha_{z_0}(z) = i\frac{z_0 + z}{z_0 - z} \tag{9.7.1}$$

of the unit disk $\mathbb{D}$ onto the upper half-plane Π, which maps the point $z_0 \in \partial\mathbb{D}$ to the point $\infty \in \Pi$. Then the pencil $\mathcal{P}(\infty) = \alpha_{z_0}(\mathcal{P}(z_0))$ on the upper half-plane Π consists of all semi-lines which are parallel in Euclidean sense to positive semi-axis $\{w = 0 + iv : v \in \mathbb{R}_+\}$, the set of all horocycles coincides with the set of all Euclidean straight lines parallel to the real axis being the boundary $\mathbb{R} = \partial\Pi$ of the upper half-plane, and the set

$$\mathcal{A}(\infty) = \{a(\alpha_{z_0}^{-1}(w)) : a(z) \in \mathcal{A}(z_0)\}$$

consists of functions depending only on v, the imaginary part of $w = u + iv \in \Pi$.

The unitary operator $V_{z_0} : L_2(\Pi) \longrightarrow L_2(\mathbb{D})$, where

$$(V_{z_0}\varphi)(z) = i\frac{2z_0}{(z_0 - z)^2}\, \varphi\left(i\frac{z_0 + z}{z_0 - z}\right),$$

provides the following unitary equivalence of the Toeplitz operators,

$$T_{a(z)} = V_{z_0}^{-1} T_{a(\alpha_{z_0}^{-1}(w))} V_{z_0}$$

for each $a(z) \in \mathcal{A}(z_0)$, and where $\alpha_{z_0}^{-1}(w)$ is the mapping inverse to (9.7.1). Thus as a direct corollary of Theorem 9.6.3 we have

Theorem 9.7.2. *The algebra $\mathcal{T}(\mathcal{A}(z_0))$ is commutative. The isomorphic imbedding*

$$\tau_{z_0} : \mathcal{T}(\mathcal{A}(z_0)) \longrightarrow C_b(\mathbb{R}_+)$$

is generated by the following mapping of generators of the algebra $\mathcal{T}(\mathcal{A}(z_0))$,

$$\tau_{z_0} : T_a \longmapsto \gamma_a^{(z_0)}(x) = \int_{\mathbb{R}_+} a\left(\alpha_{z_0}^{-1}\left(\frac{y}{2x}\right)\right) e^{-y}\, dy, \quad x \in \mathbb{R}_+,$$

where $a = a(z) \in \mathcal{A}(z_0)$, and $\alpha_{z_0}^{-1}(w)$ is the mapping inverse to (9.7.1).

Consider finally a hyperbolic pencil $\mathcal{P} = \mathcal{P}(z_1, z_2)$, which consists of all (hyperbolic straight) lines orthogonal to a given line with endpoint $z_1, z_2 \in \partial\mathbb{D}$. The functions $a = a(z)$ from $\mathcal{A}(z_1, z_2) = \mathcal{A}_{\mathcal{P}(z_1,z_2)}$ now are constant on each hypercycle, i.e., on each Euclidean arc connecting the points z_1 and z_2.

Introduce the following Möbius transformation of the unit disk $\mathbb{D}$ onto the upper half-plane Π:

$$w = \alpha_{z_1,z_2}(z) = \sqrt{\frac{z_2}{z_1}}\frac{z_1 - z}{z - z_2}, \tag{9.7.2}$$

where the value of the square root is selected to be in the upper half-plane. This transformation maps the points z_1 and z_2 on $\partial\mathbb{D}$ to the points 0 and ∞ on Π, respectively. Then the pencil $\mathcal{P}(0,\infty) = \alpha_{z_1,z_2}(\mathcal{P}(z_1,z_2))$ on the upper half-plane Π consists of all Euclidean semi-circles centered at the origin, the set of all hypercycles coincides with the set of all Euclidean rays outgoing from the origin, and the set

$$\mathcal{A}(0,\infty) = \{a(\alpha^{-1}_{z_1,z_2}(w)) : a(z) \in \mathcal{A}(z_1, z_2)\},$$

where $\alpha^{-1}_{z_1,z_2}$ is the mapping inverse to (9.7.2), consists of homogeneous zero order functions.

Introduce the unitary operator $W_{z_1,z_2} : L_2(\Pi) \longrightarrow L_2(\mathbb{D})$ as

$$(W_{z_1,z_2}\varphi)(z) = \sqrt{\frac{z_2}{z_1}\frac{z_2 - z_1}{(z - z_2)^2}}\,\varphi\left(\sqrt{\frac{z_2}{z_1}}\frac{z_1 - z}{z - z_2}\right),$$

then the Toeplitz operator algebra $T(\mathcal{A}(z_1, z_2))$, we are interested in, is unitary equivalent to the algebra

$$T(\mathcal{A}(0,\infty)) = W^{-1}_{z_1,z_2}T(\mathcal{A}(z_1, z_2))W_{z_1,z_2}.$$

Thus by Theorem 9.6.5 we have

Theorem 9.7.3. *The algebra $T(\mathcal{A}(z_1, z_2))$ is commutative. The isomorphic imbedding*

$$\tau_{(z_1,z_2)} : T(\mathcal{A}(z_1, z_2)) \longrightarrow C_b(\mathbb{R})$$

is generated by the following mapping of generators of the algebra $T(\mathcal{A}(z_1, z_2))$,

$$\tau_{(z_1,z_2)} : T_a \longmapsto \gamma_a^{(z_1,z_2)}(\lambda) = \frac{2\lambda}{1 - e^{-2\lambda\pi}}\int_0^\pi a(\alpha^{-1}_{z_1,z_2}(e^{i\theta}))\,e^{-2\lambda\theta}\,d\theta, \quad \lambda \in \mathbb{R},$$

where $a(z) \in \mathcal{A}(z_1, z_2)$, and $\alpha^{-1}_{z_1,z_2}(w)$ is the mapping inverse to (9.7.2).

Remark 9.7.4. The pencils $\mathcal{P}$ and corresponding algebras of defining symbols $\mathcal{A}$ considered in this section are labelled by the fixed point(s) of the Möbius transformations defining the above pencils and generating the corresponding one-parameter groups of Möbius transformations. Thus in fact we use here the third possibility to classify the commutative C^*-algebras of Toeplitz operators mentioned at the end of Section 9.5.

Chapter 10

Weighted Bergman Spaces

10.1 Unit disk

We recall here some necessary facts on weighted Bergman spaces and the corresponding Bergman projections; for further details see, for example [102, 240].

Let $dv(z) = dx\,dy$, $z = x + iy$, be the standard Lebesgue plane measure, as previously. For each $\lambda \in (-1, \infty)$, introduce the measure

$$d\mu_\lambda(z) = \frac{\lambda + 1}{\pi}(1 - |z|^2)^\lambda \, dv(z), \tag{10.1.1}$$

which is normalized to be a probability measure in $\mathbb{D}$. Let $L_2(\mathbb{D}, d\mu_\lambda)$ be the Hilbert space with the scalar product

$$\langle f, g \rangle_\lambda = \int_{\mathbb{D}} f(z)\overline{g(z)} \, d\mu_\lambda(z).$$

The (weighted) Bergman space $\mathcal{A}_\lambda^2(\mathbb{D})$, where $\lambda \in (-1, \infty)$, is defined as a subspace of $L_2(\mathbb{D}, d\mu_\lambda)$ which consists of functions analytic in $\mathbb{D}$.

As in Section 2.2, the key result is given by the following statement.

Theorem 10.1.1. *For each compact subset K of the unit disk $\mathbb{D}$ there is a constant $C = C(K, \lambda)$ such that for all $f \in \mathcal{A}_\lambda^2(\mathbb{D})$,*

$$\sup_{z \in K} |f(z)| \le C\|f\|_\lambda.$$

Proof. Let $D_{z,r}$ be the open disk centered in z and having the radius r. Given $f \in \mathcal{A}_\lambda^2(\mathbb{D})$ and $D_{z,r} \subset \mathbb{D}$, consider the Taylor series

$$f(\zeta) = f(z) + \sum_{n=1}^\infty a_n(\zeta - z)^n, \qquad \zeta \in D_{z,r}.$$

This series converges uniformly on each closed disk $\overline{D}_0 = \overline{D}_{z,r_0}$, where $r_0 < r$. Thus

$$\int_{\overline{D}_0} f(\zeta)\,dv(\zeta) = f(z)\int_{\overline{D}_0} dv(\zeta) + \sum_{n=1}^{\infty} a_n \int_{\overline{D}_0} (\zeta - z)^n\,dv(\zeta) = \pi r_0^2 f(z).$$

Then

$$\begin{aligned}
|f(z)| \;&=\; \frac{1}{\pi r_0^2}\left|\int_{\mathbb{D}} f(\zeta)\chi_{\overline{D}_0}(\zeta)\,dv(\zeta)\right| \\[2mm]
&\leq\; \frac{1}{\pi r_0^2}\left(\int_{\mathbb{D}} |f(\zeta)|^2\,d\mu_\lambda(\zeta)\right)^{1/2}\left(\frac{\pi}{\lambda+1}\int_{\overline{D}_0}\frac{dv(\zeta)}{(1-|\zeta|^2)^\lambda}\right)^{1/2} \\[2mm]
&\leq\; \mathrm{const}\,\|f\|_{\mathcal{A}^2(D)},
\end{aligned}$$

where $\chi_{\overline{D}_0}$ is the characteristic function of $\overline{D}_0$, and the constant depends on r_0 and λ only. To finish the proof we cover the compact set K by a finite number of such disks.　　$\square$

Corollary 10.1.2. *The Bergman space $\mathcal{A}^2_\lambda(\mathbb{D})$ is a closed subspace of $L_2(\mathbb{D}, d\mu_\lambda)$.*

Proof. Let $\{f_n\}$ be a fundamental sequence of analytic functions from $\mathcal{A}^2_\lambda(\mathbb{D})$ converging on $L_2(D)$ to a certain function $f \in L_2(\mathbb{D}, d\mu_\lambda)$. For any compact $K \subset \mathbb{D}$ we have

$$|f_n(z) - f_m(z)| \leq C\|f_n - f_m\|_{\mathcal{A}^2_\lambda(\mathbb{D})}.$$

Thus the sequence $\{f_n\}$ converges uniformly on every compact subset of $\mathbb{D}$ to the function f. Hence f is analytic in $\mathbb{D}$ and belongs to $\mathcal{A}^2_\lambda(\mathbb{D})$.　　$\square$

As in Section 2.2 we have that for any fixed point $z \in \mathbb{D}$ the evaluation functional $\varphi_z : f \longmapsto f(z)$ is linear and bounded. Thus by the Riesz representation theorem there exists a unique element $k_z \in \mathcal{A}^2_\lambda(\mathbb{D})$ such that $\varphi_z = \langle \cdot, k_z\rangle_\lambda$; that is,

$$f(z) = \int_{\mathbb{D}} f(\zeta)\,\overline{k_z(\zeta)}\,d\mu_\lambda(\zeta).$$

The function $K_{\mathbb{D},\lambda}(z,\zeta) = \overline{k_z(\zeta)}$ is called the (weighted) Bergman kernel function of the unit disk $\mathbb{D}$, and it has the reproducing property

$$f(z) = \int_{\mathbb{D}} K_{\mathbb{D},\lambda}(z,\zeta)\,f(\zeta)\,d\mu_\lambda(\zeta),$$

for all $f(z) \in \mathcal{A}^2_\lambda(\mathbb{D})$.

It is well known that the polynomials are dense in $\mathcal{A}^2_\lambda(\mathbb{D})$ and is a matter of a simple calculation to check that the system of functions

$$e_n(z) = \sqrt{\frac{\Gamma(n+2+\lambda)}{n!\,\Gamma(2+\lambda)}}\, z^n, \qquad n \in \mathbb{Z}_+, \tag{10.1.2}$$

forms an orthonormal basis in $\mathcal{A}_\lambda^2(\mathbb{D})$.

Not repeating all the steps of Section 2.2, we give only a description of the (weighted) Bergman projection $B_{\mathbb{D},\lambda}$ of $L_2(\mathbb{D}, d\mu_\lambda)$ onto $\mathcal{A}_\lambda^2(\mathbb{D})$.

Theorem 10.1.3. *For any $\lambda \in (-1, \infty)$, the (weighted) Bergman projection $B_{\mathbb{D},\lambda}$ of $L_2(\mathbb{D}, d\mu_\lambda)$ onto $\mathcal{A}_\lambda^2(\mathbb{D})$ is given by*

$$(B_{\mathbb{D},\lambda} f)(z) = \int_{\mathbb{D}} \frac{f(\zeta)}{(1 - z\bar{\zeta})^{2+\lambda}} \, d\mu_\lambda(\zeta). \tag{10.1.3}$$

Proof. For any $f \in L_2(\mathbb{D}, d\mu_\lambda)$, we have

$$(B_{\mathbb{D},\lambda} f)(z) = \sum_{n=0}^{\infty} \langle B_{\mathbb{D},\lambda} f, e_n \rangle_\lambda e_n(z),$$

where $\{e_n\}_{n \in \mathbb{Z}_+}$ is the orthonormal basis (10.1.2) in $\mathcal{A}_\lambda^2(\mathbb{D})$, and the series converges uniformly on each compact subset in $\mathbb{D}$. Further,

$$\langle B_{\mathbb{D},\lambda} f, e_n \rangle_\lambda = \langle f, B_{\mathbb{D},\lambda} e_n \rangle_\lambda = \langle f, e_n \rangle_\lambda.$$

For each fixed $z \in \mathbb{D}$, the series

$$\sum_{n=0}^{\infty} \frac{\Gamma(n + 2 + \lambda)}{n!\, \Gamma(2 + \lambda)} (z\bar{\zeta})^n = \frac{1}{(1 - z\bar{\zeta})^{2+\lambda}} \tag{10.1.4}$$

converges uniformly in $\zeta \in \mathbb{D}$, thus, interchanging the summation and integration, we have

$$
\begin{aligned}
(B_{\mathbb{D},\lambda} f)(z) &= \sum_{n=0}^{\infty} \langle f, e_n \rangle_\lambda e_n(z) = \sum_{n=0}^{\infty} \frac{\Gamma(n + 2 + \lambda)}{n!\, \Gamma(2 + \lambda)} \int_{\mathbb{D}} f(\zeta)\,(z\bar{\zeta})^n \, d\mu_\lambda(\zeta) \\
&= \int_{\mathbb{D}} f(\zeta) \left[\sum_{n=0}^{\infty} \frac{\Gamma(n + 2 + \lambda)}{n!\, \Gamma(2 + \lambda)} (z\bar{\zeta})^n \right] d\mu_\lambda(\zeta) \\
&= \int_{\mathbb{D}} \frac{f(\zeta)}{(1 - z\bar{\zeta})^{2+\lambda}} \, d\mu_\lambda(\zeta).
\end{aligned}
$$

$\square$

An alternative definition of the weighted Bergman space (see, for example, [25]) can be given in terms of another parameter $h \in (0, 1)$. The weighted Bergman space $\mathcal{A}_h^2(\mathbb{D})$ on the unit disk is the space of analytic functions in $L_2(\mathbb{D}, d\mu_h)$, with

$$d\mu_h(z) = (\frac{1}{h} - 1)(1 - |z|^2)^{\frac{1}{h} - 2} \frac{1}{\pi}\, dv(z) = (\frac{1}{h} - 1)(1 - |z|^2)^{\frac{1}{h}}\, d\mu(z), \tag{10.1.5}$$

and where the measure

$$d\mu(z) = \frac{1}{\pi} \frac{dv(z)}{(1 - |z|^2)^2}$$

is Möbius invariant.

For $\lambda + 2 = \frac{1}{h}$ we have the same space, and for $\lambda = 0$ or $h = \frac{1}{2}$ we have the classical weightless case (with normalized measure).

In terms of h, the weighted Bergman kernel function has the form

$$K_{\mathbb{D},h}(z,\zeta) = \frac{1}{(1 - z\overline{\zeta})^{\frac{1}{h}}},$$

and the orthogonal Bergman projection is

$$(B_{\mathbb{D},h}f)(z) = \int_{\mathbb{D}} \frac{f(\zeta)}{(1 - z\overline{\zeta})^{\frac{1}{h}}} \, d\mu_h(\zeta) = (\frac{1}{h} - 1) \int_{\mathbb{D}} f(\zeta) \left(\frac{1 - \zeta\overline{\zeta}}{1 - z\overline{\zeta}}\right)^{\frac{1}{h}} d\mu(\zeta). \quad (10.1.6)$$

Recall that all Möbius transformations of the unit disk are of the form

$$g(z) = \frac{az + \overline{b}}{bz + \overline{a}},$$

where a, $b \in \mathbb{C}$ and $|a|^2 - |b|^2 = 1$. We have

$$g'(z) = \frac{1}{(bz + \overline{a})^2}, \qquad |g'(z)| = \frac{1 - |g(z)|^2}{1 - |z|^2},$$

and

$$g^{-1}(z) = \frac{\overline{a}z - \overline{b}}{-bz + a}.$$

It is easy to see that the operator

$$\begin{aligned}
(U_{g,\lambda}f)(z) &= g'(z)^{\frac{\lambda}{2}+1} f(g(z)) \\
&= \frac{1}{(bz + \overline{a})^{\lambda+2}} g(\frac{az + \overline{b}}{bz + \overline{a}})
\end{aligned}$$

is unitary on both $L_2(\mathbb{D}, d\mu_\lambda)$ and $\mathcal{A}_\lambda^2(\mathbb{D})$, and that $U_{g,\lambda}^{-1} = U_{g^{-1},\lambda}$.

Lemma 10.1.4. *The Bergman projection $B_{\mathbb{D},\lambda}$ is Möbius invariant, that is* $U_{g,\lambda}B_{\mathbb{D},\lambda}U_{g,\lambda}^{-1} = B_{\mathbb{D},\lambda}$.

Proof. Calculate

$$(U_{g,\lambda}B_{\mathbb{D},\lambda}U_{g,\lambda}^{-1}f)(z)$$

$$= (\lambda + 1) g'(z)^{\frac{\lambda}{2}+1} \int_{\mathbb{D}} \left(\frac{1 - \zeta\overline{\zeta}}{1 - g(z)\overline{\zeta}}\right)^{\lambda+2} (g^{-1})'(\zeta)^{\frac{\lambda}{2}+1} f(g^{-1}(\zeta)) \, d\mu(\zeta)$$

$$= (\lambda + 1) g'(z)^{\frac{\lambda}{2}+1} \int_{\mathbb{D}} \left(\frac{1 - g(\zeta)\overline{g(\zeta)}}{1 - g(z)\overline{g(\zeta)}}\right)^{\lambda+2} (g^{-1})'(g(\zeta))^{\frac{\lambda}{2}+1} f(\zeta) \, d\mu(\zeta)$$

$$= (\lambda + 1) g'(z)^{\frac{\lambda}{2}+1} \int_{\mathbb{D}} \left(\frac{1 - \zeta\overline{\zeta}}{1 - g(z)\overline{g(\zeta)}}\right)^{\lambda+2} \overline{g'(\zeta)}^{\frac{\lambda}{2}+1} f(\zeta) \, d\mu(\zeta)$$

$$= (B_{\mathbb{D},\lambda}f)(z).$$

We have used the formula

$$\left(g'(z)\,\frac{1}{(1-g(z)\overline{g(\zeta)})^2}\,\overline{g'(\zeta)}\right)^{\frac{\lambda+2}{2}} = \left(\frac{1}{(1-z\overline{\zeta})^2}\right)^{\frac{\lambda+2}{2}},$$

which follows from (2.2.5). $\qquad\qquad\square$

The reproducing property Bergman kernel function shows that the system of functions $k_{h,\zeta}(z) = K_h(z,\overline{\zeta})$, $\zeta \in \mathbb{D}$, forms a system of coherent states (see Section A.1) in the space $\mathcal{A}_h^2(\mathbb{D})$. Thus, given a Toeplitz operator T_a with the defining symbol $a = a(z)$ on the weighted Bergman space $\mathcal{A}_h^2(\mathbb{D})$, the corresponding Wick function, see (A.1.3), is calculated by the formula (following tradition we put the bar over the second argument)

$$\begin{aligned}
\widetilde{a}_h(z,\overline{\zeta}) &= f_\zeta^{-1}(z)\langle a(w)f_\zeta(w), f_z(w)\rangle \\
&= (1-z\overline{\zeta})^{\frac{1}{h}} \int_{\mathbb{D}} \frac{a(w)}{(1-z\overline{w})^{\frac{1}{h}}(1-w\overline{\zeta})^{\frac{1}{h}}} \, d\mu_h(w) \\
&= \left(\frac{1}{h}-1\right)\int_{\mathbb{D}} a(w) \left(\frac{(1-z\overline{\zeta})(1-w\overline{w})}{(1-z\overline{w})(1-w\overline{\zeta})}\right)^{\frac{1}{h}} d\mu(w).
\end{aligned}$$

Thus the Wick form of the Toeplitz operator T_a is

$$\begin{aligned}
(T_a)(z) &= \int_{\mathbb{D}} \frac{\widetilde{a}_h(z,\overline{\zeta})f(\zeta)}{(1-z\overline{\zeta})^{\frac{1}{h}}} \, d\mu_h(\zeta) \\
&= \left(\frac{1}{h}-1\right)\int_{\mathbb{D}} \widetilde{a}_h(z,\overline{\zeta})f(\zeta) \left(\frac{1-|\zeta|^2}{1-z\overline{\zeta}}\right)^{\frac{1}{h}} d\mu(\zeta),
\end{aligned}$$

the Wick symbol, or the Berezin transform, has the form

$$\widetilde{a}_h(z,\overline{z}) = \left(\frac{1}{h}-1\right)\int_{\mathbb{D}} a(\zeta) \left(\frac{(1-|z|^2)(1-|\zeta|^2)}{(1-z\overline{\zeta})(1-\zeta\overline{z})}\right)^{\frac{1}{h}} d\mu(\zeta), \qquad (10.1.7)$$

and the composition formula for Wick symbols by (A.1.5) is given by

$$(\widetilde{a}_h \star \widetilde{b}_h)(z,\overline{z}) = \left(\frac{1}{h}-1\right)\int_{\mathbb{D}} \widetilde{a}_h(z,\overline{\zeta})\,\widetilde{b}_h(\zeta,\overline{z}) \left(\frac{(1-|z|^2)(1-|\zeta|^2)}{(1-z\overline{\zeta})(1-\zeta\overline{z})}\right)^{\frac{1}{h}} d\mu(\zeta). \qquad (10.1.8)$$

10.2 Upper half-plane

We will use the following normalized invariant measure on the upper plane Π,

$$d\nu(z) = \frac{1}{\pi}\frac{dx\,dy}{(2\,\mathrm{Im}\,z)^2} = \frac{1}{2\pi i}\frac{d\overline{z}\,dz}{(2\,\mathrm{Im}\,z)^2}. \qquad (10.2.1)$$

For $\lambda \in (-1, \infty)$, the weighted Bergman space $\mathcal{A}_\lambda^2(\Pi)$ on the upper half-plane is the space of functions analytic in $L_2(\Pi, d\nu_\lambda)$, where

$$d\nu_\lambda(z) = (\lambda + 1)(2\operatorname{Im} z)^{\lambda+2} d\nu(z) = (\lambda + 1)(2\operatorname{Im} z)^{\lambda} \frac{1}{2\pi i} d\bar{z}\, dz$$

and

$$\|f\|_\lambda = \left(\int_\Pi |f(z)|^2\, d\nu_\lambda(z) \right)^{\frac{1}{2}}.$$

The Möbius transformation (2.2.6)

$$z = \frac{w - i}{1 - iw}$$

maps the upper half-plane Π onto the unit disk $\mathbb{D}$, and its inverse transformation is given by

$$w = \frac{z + i}{1 + iz}.$$

We have

$$1 - iw = 1 - \frac{iz - 1}{1 + iz} = \frac{2}{1 + iz},$$

thus

$$\frac{1 - iw}{\sqrt{2}} = \frac{\sqrt{2}}{1 + iz} \qquad \text{and} \qquad \frac{1 + i\overline{w}}{\sqrt{2}} = \frac{\sqrt{2}}{1 - i\overline{z}}; \tag{10.2.2}$$

and now

$$
\begin{aligned}
1 - z\bar{z} &= 1 - \frac{w - i}{1 - iw} \cdot \frac{\overline{w} + i}{1 + i\overline{w}} \\
&= \frac{1 - iw + i\overline{w} + w\overline{w} - w\overline{w} + i\overline{w} - iw + 1}{(1 - iw)(1 + i\overline{w})} \\
&= -2i \frac{w - \overline{w}}{(1 - iw)(1 + i\overline{w})} \\
&= 2\operatorname{Im} w \cdot \frac{\sqrt{2}}{1 - iw} \cdot \frac{\sqrt{2}}{1 + i\overline{w}} = 2\operatorname{Im} w \cdot \frac{1 + iz}{\sqrt{2}} \cdot \frac{1 - i\overline{z}}{\sqrt{2}};
\end{aligned}
\tag{10.2.3}
$$

and

$$d\bar{z}\, dz = \left(\frac{\sqrt{2}}{1 - iw} \right)^2 \left(\frac{\sqrt{2}}{1 + i\overline{w}} \right)^2 d\overline{w}\, dw. \tag{10.2.4}$$

Introduce the operator $U_\lambda : \mathcal{A}_\lambda^2(\mathbb{D}) \longrightarrow \mathcal{A}_\lambda^2(\Pi)$ by the rule

$$(U_\lambda f)(w) = \left(\frac{\sqrt{2}}{1 - iw} \right)^{\lambda+2} f\left(\frac{w - i}{1 - iw} \right),$$

and its inverse $U_\lambda^{-1} : \mathcal{A}_\lambda^2(\Pi) \longrightarrow \mathcal{A}_\lambda^2(\mathbb{D})$ which is given by

$$(U_\lambda^{-1} f)(z) = \left(\frac{\sqrt{2}}{1+iz}\right)^{\lambda+2} f\left(\frac{z+i}{1+iz}\right).$$

We check now that the operator U_λ is unitary,

$$
\begin{aligned}
&\overline{\langle \varphi(z), (U_\lambda^{-1} f)(z)\rangle}_{\mathcal{A}_\lambda^2(\mathbb{D})} \\
={}& (\lambda+1) \int_{\mathbb{D}} \varphi(z) \overline{\left(\frac{\sqrt{2}}{1+iz}\right)^{\lambda+2} f(\frac{z+i}{1+iz})} \, (1 - z\overline{z})^\lambda \frac{d\overline{z}dz}{2\pi i} \\
={}& (\lambda+1) \int_{\Pi} \varphi\left(\frac{w-i}{1-iw}\right) \left(\frac{1+i\overline{w}}{\sqrt{2}}\right)^{\lambda+2} \overline{f(w)}\, (2\,\mathrm{Im}\,w)^\lambda \\
& \cdot \left(\frac{1-iw}{\sqrt{2}}\right)^{-\lambda} \left(\frac{1+i\overline{w}}{\sqrt{2}}\right)^{-\lambda} \cdot \left(\frac{\sqrt{2}}{1-iw}\right)^2 \left(\frac{\sqrt{2}}{1+i\overline{w}}\right)^2 \frac{d\overline{w}dw}{2\pi i} \\
={}& (\lambda+1) \int_{\Pi} \left(\frac{\sqrt{2}}{1-iw}\right)^{\lambda+2} \varphi\left(\frac{w-i}{1-iw}\right) \overline{f(w)}\, (2\,\mathrm{Im}\,w)^\lambda \frac{d\overline{w}dw}{2\pi i} \\
={}& \langle (U_\lambda \varphi)(w), f(w)\rangle_{\mathcal{A}_\lambda^2(\Pi)}.
\end{aligned}
$$

Note that the operator

$$U_\lambda : \ L_2(\mathbb{D}, d\mu_\lambda) \longrightarrow L_2(\Pi, d\nu_\lambda),$$

given by the above formula is unitary as well.

The weighted Bergman projection $B_{\Pi,\lambda} : L_2(\Pi, d\nu_\lambda) \longrightarrow \mathcal{A}_\lambda^2(\Pi)$ has obviously the form $B_{\Pi,\lambda} = U_\lambda B_{\mathbb{D},\lambda} U_\lambda^{-1}$, where the weighted Bergman projection $B_{\mathbb{D},\lambda} : L_2(\mathbb{D}, d\mu_\lambda) \longrightarrow \mathcal{A}_\lambda^2(\mathbb{D})$ is given by (10.1.3).

Calculate

$$
\begin{aligned}
(U_\lambda B_{\mathbb{D},\lambda} U_\lambda^{-1} f)(w) ={}& \left(\frac{\sqrt{2}}{1-iw}\right)^{\lambda+2} (\lambda+1) \int_{\mathbb{D}} \left(\frac{\sqrt{2}}{1+i\zeta}\right)^{\lambda+2} f\left(\frac{\zeta+i}{1+i\zeta}\right) \\
& \cdot \left(\frac{1 - \zeta\overline{\zeta}}{1 - \frac{w-i}{1-iw}\overline{\zeta}}\right)^{\lambda+2} \frac{1}{2\pi i} \frac{d\overline{\zeta}d\zeta}{(1 - \zeta\overline{\zeta})^2}.
\end{aligned}
$$

Substituting $\tau = \frac{\zeta + i}{1 + i\zeta}$ and using formulas (10.2.2–10.2.4), we have

$$(U_\lambda B_{\mathbb{D},\lambda} U_\lambda^{-1} f)(w)$$

$$= \left(\frac{\sqrt{2}}{1 - iw}\right)^{\lambda+2} (\lambda + 1) \int_\Pi \left(\frac{1 - i\tau}{\sqrt{2}}\right)^{\lambda+2} f(\tau) \left(\frac{-2i\frac{\tau - \bar{\tau}}{(1 - i\tau)(1 + i\bar{\tau})}}{-2i\frac{w - \bar{\tau}}{(1 - iw)(1 + i\bar{\tau})}}\right)^{\lambda+2}$$

$$\cdot \frac{1}{2\pi i} \left(\frac{\sqrt{2}}{1 - i\tau}\right)^2 \left(\frac{\sqrt{2}}{1 + i\bar{\tau}}\right)^2 d\bar{\tau}\, d\tau \left(2\,\mathrm{Im}\,\tau \cdot \frac{\sqrt{2}}{1 - i\tau} \cdot \frac{\sqrt{2}}{1 + i\bar{\tau}}\right)^{-2}$$

$$= (\lambda + 1) \int_\Pi f(\tau) \left(\frac{\tau - \bar{\tau}}{w - \bar{\tau}}\right)^{\lambda+2} \frac{1}{2\pi i} \frac{d\bar{\tau}\, d\tau}{(2\,\mathrm{Im}\,\tau)^2}.$$

Thus, finally, the weighted Bergman projection $B_{\Pi,\lambda} : L_2(\Pi, d\nu_\lambda) \longrightarrow \mathcal{A}_\lambda^2(\Pi)$ has the form

$$(B_{\Pi,\lambda} f)(z) = (\lambda + 1) \int_\Pi f(\zeta) \left(\frac{\zeta - \bar{\zeta}}{z - \bar{\zeta}}\right)^{\lambda+2} d\nu(\zeta),$$

where the invariant measure $d\nu(\zeta)$ is given by (10.2.1).

10.3　Representations of the weighted Bergman space

10.3.1　Unit disk, polar coordinates

The Bergman space can be characterized as the set of $L_2(\mathbb{D}, d\mu_\lambda)$ functions which satisfy the Cauchy-Riemann equation

$$\frac{\partial}{\partial \bar{z}} f(z) = 0.$$

Passing to polar coordinates ($z = rt$, $r \in [0, 1)$, $t \in S^1$), we have

$$L_2(\mathbb{D}, d\mu_\lambda) = L_2\left([0, 1), \frac{\lambda + 1}{\pi}(1 - r^2)^\lambda\, rdr\right) \otimes L_2\left(S^1, \frac{dt}{it}\right)$$

$$= L_2\left([0, 1), \frac{\lambda + 1}{\pi}(1 - r^2)^\lambda\, rdr\right) \otimes L_2(S^1).$$

Recall that in polar coordinates

$$\frac{\partial}{\partial \bar{z}} = \frac{t}{2}\left(\frac{\partial}{\partial r} - \frac{t}{r}\frac{\partial}{\partial t}\right).$$

Introduce the unitary operator

$$U_1 = I \otimes \mathcal{F} : \quad L_2(\mathbb{D}, d\mu_\lambda) \longrightarrow L_2\left([0, 1), \frac{\lambda + 1}{\pi}(1 - r^2)^\lambda\, rdr\right) \otimes l_2,$$

where the discrete Fourier transform $\mathcal{F} : L_2(S^1) \to l_2$ is given by (see (4.1.1))

$$\mathcal{F} : f \longmapsto c_n = \frac{1}{\sqrt{2\pi}} \int_{S^1} f(t)\, t^{-n} \frac{dt}{it}, \quad n \in \mathbb{Z}.$$

As in Section 4.1, we have

$$(I \otimes \mathcal{F}) \frac{t}{2} \left(\frac{\partial}{\partial r} - \frac{t}{r} \frac{\partial}{\partial t} \right) (I \otimes \mathcal{F}^{-1}) \{ c_n(r) \}_{n \in \mathbb{Z}} = \left\{ \frac{1}{2} \left(\frac{\partial}{\partial r} - \frac{n-1}{r} \right) c_{n-1}(r) \right\}_{n \in \mathbb{Z}}.$$

Thus the image of the Bergman space $\mathcal{A}^2_{1,\lambda} = U_1(\mathcal{A}^2_\lambda(\mathbb{D}))$ can be described as the (closed) subspace of

$$L_2\!\left([0,1), \frac{\lambda+1}{\pi}(1-r^2)^\lambda\, rdr\right) \otimes l_2 = l_2\!\left(L_2\!\left([0,1), \frac{\lambda+1}{\pi}(1-r^2)^\lambda\, rdr\right)\right),$$

which consists of all sequences $\{ c_n(r) \}_{n \in \mathbb{Z}}$ satisfying the equations

$$\frac{1}{2} \left(\frac{\partial}{\partial r} - \frac{n}{r} \right) c_n(r) = 0, \quad n \in \mathbb{Z}.$$

Their general solutions have obviously the form

$$c_n(r) = c'_n\, r^n, \quad n \in \mathbb{Z}.$$

Each function $c_n(r) = c'_n\, r^n$ has to be in $L_2([0,1), \frac{\lambda+1}{\pi}(1-r^2)^\lambda\, rdr)$, which implies that $c_n(r) \equiv 0$, for all $n < 0$. That is, the space $\mathcal{A}^2_1$ coincides with the space of all two-sided sequences $\{ c_n(r) \}_{n \in \mathbb{Z}}$ with

$$c_n(r) = \begin{cases} \sqrt{2}\, \alpha_{n,\lambda} c_n r^n, & \text{if } n \in \mathbb{Z}_+ \\ 0, & \text{if } n \in \mathbb{Z} \setminus \mathbb{Z}_+ \end{cases},$$

where the normalizing constants

$$\alpha_{n,\lambda} = \left(\frac{\lambda+1}{\pi} B(n+1, \lambda+1) \right)^{-\frac{1}{2}} = \left(\frac{\pi\, \Gamma(n+2+\lambda)}{n!\, \Gamma(2+\lambda)} \right)^{\frac{1}{2}}, \quad n \in \mathbb{Z}_+ \quad (10.3.1)$$

are selected so that

$$\| \{ c_n(r) \}_{n \in \mathbb{Z}} \| = \left(\sum_{n \in \mathbb{Z}_+} |c_n|^2 \right)^{1/2} = \| \{ c_n \}_{n \in \mathbb{Z}_+} \|_{l_2}.$$

For each $n \in \mathbb{Z}_+$ introduce the unitary operator

$$u_{n,\lambda} : L_2\!\left([0,1), \frac{\lambda+1}{\pi}(1-r^2)^\lambda\, rdr\right) \longrightarrow L_2([0,1), rdr)$$

by the rule

$$(u_{n,\lambda} f)(r) = \alpha_{n,\lambda}^{-1}\, \omega_{n,\lambda}^{-n}(r)\, f(\omega_{n,\lambda}(r)),$$

where $r = \omega_{n,\lambda}(s)$ is the inverse function to the function

$$
\begin{aligned}
\sigma_{n,\lambda}(r) &= \alpha_{n,\lambda}\left(\frac{\lambda+1}{\pi}\int_0^r u^{2n+1}(1-u^2)^\lambda du\right)^{\frac{1}{2}} \\
&= \alpha_{n,\lambda}\left(\frac{\lambda+1}{\pi}B_{r^2}(n+1,\lambda+1)\right)^{\frac{1}{2}} \\
&= I_{r^2}(n+1,\lambda+1)^{\frac{1}{2}}.
\end{aligned}
$$

Here the incomplete B-function B_x and the function I_x are given by formulas 8.391 and 8.392 in [86].

The inverse operator

$$u_{n,\lambda}^{-1} = u_{n,\lambda}^* : \ L_2([0,1), r\,dr) \longrightarrow L_2\left([0,1), \frac{\lambda+1}{\pi}(1-r^2)^\lambda\, r\,dr\right)$$

is given by

$$(u_{n,\lambda}^{-1} f)(r) = \alpha_{n,\lambda}\, r^n\, f(\sigma_{n,\lambda}(r)).$$

Finally, define the unitary operator

$$U_{2,\lambda} : l_2\left(L_2([0,1), \frac{\lambda+1}{\pi}(1-r^2)^\lambda\, r\,dr)\right) \longrightarrow l_2(L_2([0,1), r\,dr)) = L_2([0,1), r\,dr)\otimes l_2$$

as follows:

$$U_{2,\lambda} : \ \{c_n(r)\}_{n\in\mathbb{Z}} \longmapsto \{(u_{|n|,\lambda}c_n)(r)\}_{n\in\mathbb{Z}}.$$

Then the space $\mathcal{A}_{2,\lambda}^2 = U_{2,\lambda}(\mathcal{A}_{1,\lambda}^2)$ coincides with the space of all sequences $\{d_n(r)\}_{n\in\mathbb{Z}}$, where

$$d_n = u_{n,\lambda}(\sqrt{2}\,\alpha_{n,\lambda}c_n r^n) = \sqrt{2}\,c_n,$$

for $n \in \mathbb{Z}_+$, and $d_n(r) \equiv 0$, for $n \in \mathbb{Z}\setminus\mathbb{Z}_+$.

As in Section 4.1, let $\ell_0(r) = \sqrt{2}$, and let L_0 be the one-dimensional subspace of $L_2([0,1), r\,dr)$ generated by $\ell_0(r)$. The one-dimensional projection P_0 of $L_2([0,1), r\,dr)$ onto L_0 has the form (4.1.3):

$$(P_0 f)(r) = \langle f, \ell_0\rangle \cdot \ell_0 = \sqrt{2}\int_0^1 f(\rho)\,\sqrt{2}\rho\,d\rho.$$

As previously, let l_2^+ be the subspace of (two-sided) l_2, consisting of all sequences $\{c_n\}_{n\in\mathbb{Z}}$, such that $c_n = 0$ for all $n \in \mathbb{Z}\setminus\mathbb{Z}_+$, and let p^+ be the orthogonal projection of l_2 onto l_2^+.

Now $\mathcal{A}_{2,\lambda}^2 = L_0 \otimes l_2^+$, and the orthogonal projection

$$B_{2,\lambda} : \ l_2(L_2([0,1), r\,dr)) = L_2([0,1), r\,dr)\otimes l_2 \longrightarrow \mathcal{A}_{2,\lambda}^2$$

obviously has the form

$$B_2 = P_0 \otimes p^+.$$

Thus we arrive at the following theorem.

Theorem 10.3.1. *The unitary operator $U_\lambda = U_{2,\lambda} U_1$ gives an isometric isomorphism of the space $L_2(\mathbb{D}, d\mu_\lambda)$ onto $L_2([0,1), r dr) \otimes l_2$ under which*

   1. the weighted Bergman space $\mathcal{A}_\lambda^2(\mathbb{D})$ is mapped onto $L_0 \otimes l_2^+$,

$$U_\lambda : \mathcal{A}_\lambda^2(\mathbb{D}) \longrightarrow L_0 \otimes l_2^+,$$

   where L_0 is the one-dimensional subspace of $L_2([0,1), r dr)$, generated by $\ell_0(r) = \sqrt{2}$,

   2. the weighted Bergman projection $B_{\mathbb{D},\lambda}$ is unitary equivalent to

$$U_\lambda B_{\mathbb{D},\lambda} U_\lambda^{-1} = P_0 \otimes p^+,$$

   where P_0 is the one-dimensional projection of $L_2([0,1), r dr)$ onto L_0.

As in Section 4.1, we use the isometric imbedding

$$R_0 : l_2^+ \longrightarrow L_2([0,1), r dr) \otimes l_2$$

given by

$$R_0 : \{c_n\}_{n \in \mathbb{Z}_+} \longmapsto \ell_0(r)\{\chi_+(n) c_n\}_{n \in \mathbb{Z}},$$

where we extend the sequence $\{c_n\}_{n \in \mathbb{Z}_+}$ to an element of l_2 setting $c_n = 0$ for negative indices $n < 0$. The image of R_0 coincides with the space $\mathcal{A}_{2,\lambda}^2$. The adjoint operator $R_0^* : L_2([0,1), r dr) \otimes l_2 \to l_2^+$ has the form

$$R_0^* : \{c_n(r)\}_{n \in \mathbb{Z}} \longmapsto \left\{ \int_0^1 c_n(\rho)\sqrt{2}\, \rho\, d\rho \right\}_{n \in \mathbb{Z}_+},$$

and, as previously,

$$R_0^* R_0 = I \quad : \quad l_2^+ \longrightarrow l_2^+,$$
$$R_0 R_0^* = B_2 \quad : \quad L_2([0,1), r dr) \otimes l_2 \longrightarrow \mathcal{A}_{2,\lambda}^2 = L_0 \otimes l_2^+.$$

The operator $R_\lambda = R_0^* U_\lambda$ maps the space $L_2(\mathbb{D}, d\mu_\lambda)$ onto l_2^+, and the restriction

$$R_\lambda|_{\mathcal{A}_\lambda^2(\mathbb{D})} : \mathcal{A}_\lambda^2(\mathbb{D}) \longrightarrow l_2^+$$

is an isometric isomorphism. The adjoint operator

$$R_\lambda^* = U_\lambda^* R_0 : l_2^+ \longrightarrow \mathcal{A}_\lambda^2(\mathbb{D}) \subset L_2(\mathbb{D}, d\mu_\lambda)$$

is an isometric isomorphism of l_2^+ onto the subspace $\mathcal{A}_\lambda^2(\mathbb{D})$ of the space $L_2(\mathbb{D}, d\mu_\lambda)$.

Remark 10.3.2. We have

$$R_\lambda R_\lambda^* = I \quad : \quad l_2^+ \longrightarrow l_2^+,$$
$$R_\lambda^* R_\lambda = B_{\mathbb{D},\lambda} \quad : \quad L_2(\mathbb{D}, d\mu_\lambda) \longrightarrow \mathcal{A}_\lambda^2(\mathbb{D}).$$

Theorem 10.3.3. *The isometric isomorphism*

$$R_\lambda^* = U_\lambda^* R_0 : l_2^+ \longrightarrow \mathcal{A}_\lambda^2(\mathbb{D})$$

is given by

$$R_\lambda^* : \{c_n\}_{n\in\mathbb{Z}_+} \longmapsto \frac{1}{\sqrt{\pi}} \sum_{n\in\mathbb{Z}_+} \alpha_{n,\lambda}\, c_n z^n, \tag{10.3.2}$$

where the constants $\alpha_{n,\lambda}$ are given by (10.3.1).

Proof. Calculate

$$\begin{aligned}
R_\lambda^* = U_1^* U_{2,\lambda}^* R_0 \quad : \quad & \{c_n\}_{n\in\mathbb{Z}_+} \longmapsto U_1^* U_{2,\lambda}^* (\{\sqrt{2}\, c_n\}_{n\in\mathbb{Z}_+}) \\
= \quad & U_1^* (\{\sqrt{2}\, \alpha_{n,\lambda}\, c_n r^n\}_{n\in\mathbb{Z}_+}) \\
= \quad & \frac{1}{\sqrt{2\pi}} \sum_{n\in\mathbb{Z}_+} \sqrt{2}\, \alpha_{n,\lambda}\, c_n (rt)^n = \frac{1}{\sqrt{\pi}} \sum_{n\in\mathbb{Z}_+} \alpha_{n,\lambda}\, c_n z^n.
\end{aligned}$$

$\square$

Corollary 10.3.4. *The inverse isomorphism*

$$R_\lambda : \mathcal{A}_\lambda^2(\mathbb{D}) \longrightarrow l_2^+$$

is given by

$$R_\lambda : \quad \varphi(z) \longmapsto \left\{ \frac{\alpha_{n,\lambda}}{\sqrt{\pi}} \int_{\mathbb{D}} \varphi(z)\, \overline{z}^n\, d\mu_\lambda(z) \right\}_{n\in\mathbb{Z}_+}. \tag{10.3.3}$$

10.3.2   Upper half-plane, Cartesian coordinates

Again, the Bergman space can be characterized as the set of $L_2(\Pi, d\nu_\lambda)$ functions which satisfy the Cauchy-Riemann equation

$$\frac{\partial}{\partial \overline{z}}\, f(z) = 0. \tag{10.3.4}$$

Introduce the unitary operator

$$U_1 = 1/\sqrt{\pi}\, (F \otimes I) : L_2(\Pi, d\nu_\lambda) \longrightarrow L_2(\mathbb{R}, dx) \otimes L_2(\mathbb{R}_+, (\lambda+1)(2y)^\lambda dy),$$

where the Fourier integral transform $F : L_2(\mathbb{R}) \to L_2(\mathbb{R})$ is given by

$$(Ff)(u) = \frac{1}{\sqrt{2\pi}} \int_{\mathbb{R}} e^{-iux} f(x)\, dx.$$

The image $\mathcal{A}_{1,\lambda}^2(\Pi) = U_1(\mathcal{A}_\lambda^2(\Pi))$ consists of all functions $\varphi = \varphi(x,y)$ satisfying the equation

$$U_1 \frac{\partial}{\partial \bar{z}} U_1^{-1} \varphi = \frac{i}{2}\left(x + \frac{\partial}{\partial y}\right)\varphi = 0,$$

whose general solution has obviously the form

$$\varphi(x,y) = \psi(x)e^{-xy}.$$

The function φ has to be in $L_2(\mathbb{R}, dx) \otimes L_2(\mathbb{R}_+, (\lambda+1)(2y)^\lambda dy)$, thus $\mathcal{A}_{1,\lambda}^2(\Pi)$ is the space of all functions

$$\varphi(x,y) = \chi_+(x)\theta_\lambda(x)f(x)e^{-xy}, \quad f \in L_2(\mathbb{R}),$$

where

$$\theta_\lambda(x) = \left((\lambda+1)\int_{\mathbb{R}_+} e^{-2xv}(2v)^\lambda dv\right)^{-1/2} = \left(\frac{2x^{\lambda+1}}{(\lambda+1)\Gamma(\lambda+1)}\right)^{1/2}, \quad x \geq 0.$$

$$(10.3.5)$$

Moreover,

$$\|\varphi\|_{\mathcal{A}_{1,\lambda}^2(\Pi)} = \|f\|_{L_2(\mathbb{R}_+)}.$$

Introduce the unitary operator

$$U_{2,\lambda}: \ L_2(\mathbb{R}, dx) \otimes L_2(\mathbb{R}_+, (\lambda+1)(2y)^\lambda dy) \longrightarrow L_2(\mathbb{R}, dx) \otimes L_2(\mathbb{R}_+, dy)$$

as follows,

$$(U_{2,\lambda}\varphi)(x,y) = \frac{1}{\theta_\lambda(|x|)} e^{-y/2+|x|\beta(|x|,y)}\varphi(x, \beta(|x|, y)),$$

where for each fixed $x > 0$ the function $\beta(x,y)$ is the inverse function to

$$\gamma(x,t) = -\ln\left\{\theta_\lambda^2(x)(\lambda+1)\int_t^\infty (2\eta)^\lambda e^{-2x\eta}d\eta\right\}, \quad (10.3.6)$$

i.e., $\beta(x, \gamma(x,t)) = t, \ x > 0$.

The inverse operator

$$U_{2,\lambda}^{-1}: \ L_2(\mathbb{R}, dx) \otimes L_2(\mathbb{R}_+, dy) \longrightarrow L_2(\mathbb{R}, dx) \otimes L_2(\mathbb{R}_+, (\lambda+1)(2y)^\lambda dy)$$

has the form

$$(U_{2,\lambda}^{-1}\varphi)(x,y) = \theta_\lambda(|x|)\, e^{\gamma(|x|,y)/2-|x|y}\varphi(x, \gamma(|x|, y)).$$

For each $f \in L_2(\mathbb{R})$, one has

$$U_{2,\lambda}: \chi_+(x)\theta_\lambda(x)f(x)e^{-xy} \longmapsto \chi_+(x)f(x)e^{-y/2}.$$

Thus, the image $\mathcal{A}_{2,\lambda}^2 = U_{2,\lambda}(\mathcal{A}_{1,\lambda}^2(\Pi))$ is the set of all functions of the form

$$\psi(x,y) = \chi_+(x) f(x) e^{-y/2}, \qquad f \in L_2(\mathbb{R}),$$

where $\chi_+(x)$ is the characteristic function of $\mathbb{R}_+$.

We summarize the above in the following theorem.

Theorem 10.3.5. *The unitary operator $U_\lambda = U_{2,\lambda}U_1$ gives an isometric isomorphism of $L_2(\Pi, d\nu_\lambda)$ onto $L_2(\mathbb{R}, dx) \otimes L_2(\mathbb{R}_+, dy)$ under which*

1. *The Bergman space $\mathcal{A}_\lambda^2(\Pi)$ is mapped onto $L_2(\mathbb{R}_+) \otimes L_0$, where L_0 is the one-dimensional subspace of $L_2(\mathbb{R}_+, dy)$ generated by $l_0(y) = e^{-y/2}$.*

2. *The Bergman projection $B_{\Pi,\lambda}$ is unitary equivalent to*

$$U_\lambda B_{\Pi,\lambda} U_\lambda^{-1} = \chi_+ I \otimes P_0,$$

where P_0 is the one-dimensional projection of $L_2(\mathbb{R}_+, dy)$ onto L_0:

$$(P_0 \psi)(y) = e^{-y/2} \int_0^\infty \psi(v) e^{-v/2} dv.$$

As in Section 3.1, introduce the isometric imbedding

$$R_0 : L_2(\mathbb{R}_+) \longrightarrow L_2(\mathbb{R}) \otimes L_2(\mathbb{R}_+)$$

by the rule

$$(R_0 f)(x, y) = \chi_+(x)\, f(x)\, \ell_0(y).$$

Here the function $f(x)$ is extended to an element of $L_2(\mathbb{R})$ by setting $f(x) \equiv 0$, for $x < 0$. The image of R_0 obviously coincides with the space $\mathcal{A}_2^2$. The adjoint operator $R_0^* : L_2(\Pi) \to L_2(\mathbb{R}_+)$ is given by

$$(R_0^* \varphi)(x) = \chi_+(x) \int_{\mathbb{R}_+} \varphi(x, \eta)\, \ell_0(\eta)\, d\eta,$$

and

$$R_0^* R_0 = I \quad : \quad L_2(\mathbb{R}_+) \longrightarrow L_2(\mathbb{R}_+),$$
$$R_0 R_0^* = B_2 \quad : \quad L_2(\Pi) \longrightarrow \mathcal{A}_2^2 = L_2(\mathbb{R}_+) \otimes L_0.$$

Now the operator $R_\lambda = R_0^* U_\lambda$ maps the space $L_2(\Pi, d\nu_\lambda)$ onto $L_2(\mathbb{R}_+)$, and the restriction

$$R_\lambda|_{\mathcal{A}_\lambda^2(\Pi)} : \mathcal{A}_\lambda^2(\Pi) \longrightarrow L_2(\mathbb{R}_+)$$

is an isometric isomorphism. The adjoint operator

$$R_\lambda^* = U_\lambda^* R_0 : L_2(\mathbb{R}_+) \longrightarrow \mathcal{A}_\lambda^2(\Pi) \subset L_2(\Pi, d\nu_\lambda)$$

is an isometric isomorphism of $L_2(\mathbb{R}_+)$ onto the subspace $\mathcal{A}_\lambda^2(\Pi)$ of the space $L_2(\Pi, d\nu_\lambda)$.

Remark 10.3.6. We have

$$R_\lambda R_\lambda^* = I \quad : \quad L_2(\mathbb{R}_+) \longrightarrow L_2(\mathbb{R}_+),$$
$$R_\lambda^* R_\lambda = B_{\Pi,\lambda} \quad : \quad L_2(\Pi, d\nu_\lambda) \longrightarrow \mathcal{A}_\lambda^2(\Pi).$$

Theorem 10.3.7. *The isometric isomorphism*

$$R_\lambda^* = U_\lambda^* R_0 : L_2(\mathbb{R}_+) \longrightarrow \mathcal{A}_\lambda^2(\Pi)$$

is given by

$$(R_\lambda^* f)(z) = \frac{1}{\sqrt{\Gamma(\lambda + 2)}} \int_{\mathbb{R}_+} f(\xi)\, \xi^{\frac{\lambda+1}{2}}\, e^{iz\cdot\xi}\, d\xi. \tag{10.3.7}$$

Proof. Calculate

$$
\begin{aligned}
(R_\lambda^* f)(z) &= (U_1^* U_{2,\lambda}^* R_0 f)(z) \\
&= \sqrt{\pi}(F^{-1} \otimes I)(\chi_+(\xi)\, f(\xi)\, \theta_\lambda(\xi)\, e^{\gamma(\xi,y)/2 - \xi y}\, e^{-\gamma(\xi,y)/2}) \\
&= \frac{1}{\sqrt{2}} \int_{\mathbb{R}} \chi_+(\xi)\, f(\xi)\, \frac{\sqrt{2}\,\xi^{\frac{\lambda+1}{2}}}{\sqrt{(\lambda+1)\Gamma(\lambda+1)}}\, e^{-\xi y}\, e^{ix\xi}\, d\xi \\
&= \frac{1}{\sqrt{\Gamma(\lambda+2)}} \int_{\mathbb{R}_+} f(\xi)\, \xi^{\frac{\lambda+1}{2}}\, e^{i(x+iy)\cdot\xi}\, d\xi.
\end{aligned}
$$

$\square$

Corollary 10.3.8. *The inverse isomorphism*

$$R_\lambda : \mathcal{A}_\lambda^2(\Pi) \longrightarrow L_2(\mathbb{R}_+)$$

is given by

$$
\begin{aligned}
(R_\lambda \varphi)(x) &= \frac{x^{\frac{\lambda+1}{2}}}{\sqrt{\Gamma(\lambda+2)}} \int_\Pi \varphi(w)\, e^{-i\,\overline{w}\cdot x}\, \mu_\lambda(w)\, dv(w) \tag{10.3.8} \\
&= \frac{(\lambda+1)\, x^{\frac{\lambda+1}{2}}}{\sqrt{\Gamma(\lambda+2)}} \int_\Pi \varphi(\xi + i\eta)\, e^{-i\,(\xi-i\eta)\cdot x}\, (2\eta)^\lambda\, \frac{1}{\pi}\, d\xi d\eta.
\end{aligned}
$$

10.3.3 Upper half-plane, polar coordinates

Passing to polar coordinates we have the tensor decomposition

$$L_2(\Pi, d\nu_\lambda) = L_2(\mathbb{R}_+, r^{\lambda+1} dr) \otimes L_2([0, \pi], 1/\pi 2^\lambda(\lambda+1)\sin^\lambda \theta d\theta).$$

Rewriting the equation (10.3.4) in polar coordinates, we have that the Bergman space $\mathcal{A}_\lambda^2(\Pi)$ is the set of all functions satisfying the equation

$$\left(r\frac{\partial}{\partial r} + i\frac{\partial}{\partial \theta}\right)\varphi(r, \theta) = 0.$$

Introduce the unitary operator

$$U_1 = 1/\sqrt{\pi}(M \otimes I) \quad : \quad L_2(\mathbb{R}_+, r^{\lambda+1}dr) \otimes L_2([0,\pi], 1/\pi 2^\lambda(\lambda+1)\sin^\lambda\theta d\theta)$$
$$\longrightarrow \quad L_2(\mathbb{R}) \otimes L_2([0,\pi], 2^\lambda(\lambda+1)\sin^\lambda\theta d\theta),$$

where the Mellin transform $M : L_2(\mathbb{R}_+, r^{\lambda+1}dr) \longrightarrow L_2(\mathbb{R})$ is given by

$$(M\psi)(\xi) = \frac{1}{\sqrt{2\pi}} \int_{\mathbb{R}_+} r^{-i\xi+\lambda/2}\,\psi(r)\,dr.$$

The inverse Mellin transform $M^{-1} : L_2(\mathbb{R}) \longrightarrow L_2(\mathbb{R}_+, r^{\lambda+1}dr)$ has the form

$$(M^{-1}\psi)(r) = \frac{1}{\sqrt{2\pi}} \int_{\mathbb{R}} r^{i\xi-\lambda/2-1}\,\psi(\xi)\,d\xi.$$

It is easy to see that

$$U_1\left(r\frac{\partial}{\partial r} + i\frac{\partial}{\partial\theta}\right)U_1^{-1} = i(\xi + (\lambda/2+1)i)I + i\frac{\partial}{\partial\theta}.$$

Thus, the image of the Bergman space $\mathcal{A}^2_{1,\lambda} = U_1(\mathcal{A}^2_\lambda(\Pi))$ can be described as the (closed) subspace of $L_2(\mathbb{R}) \otimes L_2([0,\pi], 2^\lambda(\lambda+1)\sin^\lambda\theta d\theta)$ which consists of all functions $\varphi(\lambda,\theta)$ satisfying the equation

$$\left((\xi + (\lambda/2+1)i)I + \frac{\partial}{\partial\theta}\right)\varphi(\xi,\theta) = 0.$$

The general $L_2(\mathbb{R}) \otimes L_2([0,\pi], 2^\lambda(\lambda+1)\sin^\lambda\theta d\theta)$ solution of this equation has the form

$$\varphi(\xi,\theta) = f(\xi)\,\vartheta_\lambda(\xi)\,e^{-(\xi+(1+\lambda/2)i)\theta}, \qquad f(\xi) \in L_2(\mathbb{R}), \tag{10.3.9}$$

where

$$\vartheta_\lambda(\xi) = \left(2^\lambda(\lambda+1)\int_0^\pi e^{-2\xi\theta}\sin^\lambda\theta d\theta\right)^{-1/2}$$
$$= \frac{B(\frac{\lambda+2}{2} + i\xi, \frac{\lambda+2}{2} - i\xi)^{1/2}}{\sqrt{\pi}}\,e^{\pi\xi/2} = \frac{|\Gamma(\frac{\lambda+2}{2} + i\xi)|}{\sqrt{\pi\Gamma(\lambda+2)}}\,e^{\pi\xi/2}, \tag{10.3.10}$$

and

$$\|\varphi(\xi,\theta)\|_{L_2(\mathbb{R})\otimes L_2([0,\pi], 2^\lambda(\lambda+1)\sin^\lambda\theta d\theta)} = \|f(\xi)\|_{L_2(\mathbb{R})}.$$

Lemma 10.3.9. *The unitary operator $U_1 = 1/\sqrt{\pi}(M \otimes I)$ is an isometric isomorphism of the space $L_2(\Pi, dv_\lambda)$ onto $L_2(\mathbb{R}) \otimes L_2([0,\pi], 2^\lambda(\lambda+1)\sin^\lambda\theta d\theta)$ under which the Bergman space $\mathcal{A}^2_\lambda(\Pi)$ is mapped onto*

$$\mathcal{A}^2_{1,\lambda} = \left\{\varphi(\xi,\theta) = f(\xi)\,\vartheta_\lambda(\xi)\,e^{-(\xi+(1+\lambda/2)i)\theta} : \quad f(\xi) \in L_2(\mathbb{R})\right\}.$$

Let $R_0 : L_2(\mathbb{R}) \longrightarrow \mathcal{A}^2_{1,\lambda}(\Pi) \subset L_2(\mathbb{R}) \otimes L_2([0,\pi], 2^\lambda(\lambda+1)\sin^\lambda \theta d\theta)$ be the isometric imbedding given by

$$(R_0 f)(\xi, \theta) = f(\xi)\, \vartheta_\lambda(\xi)\, e^{-(\xi+(1+\lambda/2)i)\theta}.$$

The adjoint operator $R_0^* : L_2(\mathbb{R}) \otimes L_2([0,\pi], 2^\lambda(\lambda+1)\sin^\lambda \theta d\theta) \longrightarrow L_2(\mathbb{R})$ has the form

$$(R_0^*\psi)(\xi) = 2^\lambda(\lambda+1)\vartheta_\lambda(\xi) \int_0^\pi \psi(\xi,\theta)\, e^{-(\xi-(1+\lambda/2)i)\theta}\, \sin^\lambda \theta\, d\theta,$$

and

$$\begin{aligned}
R_0^* R_0 &= I &:&\quad L_2(\mathbb{R}) \longrightarrow L_2(\mathbb{R}),\\
R_0 R_0^* &= B_1 &:&\quad L_2(\mathbb{R}) \otimes L_2([0,\pi], 2^\lambda(\lambda+1)\sin^\lambda \theta d\theta) \longrightarrow \mathcal{A}^2_{1,\lambda},
\end{aligned}$$

where $B_1 = U_1 B_{\Pi,\lambda} U_1^{-1}$ is the orthogonal projection of $L_2(\mathbb{R}) \otimes L_2([0,\pi], 2^\lambda(\lambda+1)\sin^\lambda \theta d\theta)$ onto $\mathcal{A}^2_{1,\lambda}$.

Now the operator $R_\lambda = R_0^* U_1$ maps the space $L_2(\Pi, d\nu_\lambda)$ onto $L_2(\mathbb{R})$, and its restriction

$$R_\lambda|_{\mathcal{A}^2_\lambda(\Pi)} : \mathcal{A}^2_\lambda(\Pi) \longrightarrow L_2(\mathbb{R})$$

is an isometric isomorphism. The adjoint operator

$$R_\lambda^* = U_1^* R_0 : L_2(\mathbb{R}) \longrightarrow \mathcal{A}^2_\lambda(\Pi) \subset L_2(\Pi, d\nu_\lambda)$$

is an isometric isomorphism of $L_2(\mathbb{R})$ onto $\mathcal{A}^2_\lambda(\Pi)$.

Remark 10.3.10. We have

$$\begin{aligned}
R_\lambda R_\lambda^* &= I &:&\quad L_2(\mathbb{R}) \longrightarrow L_2(\mathbb{R}),\\
R_\lambda^* R_\lambda &= B_{\Pi,\lambda} &:&\quad L_{2,\lambda}(\Pi) \longrightarrow \mathcal{A}^2_\lambda(\Pi).
\end{aligned}$$

Theorem 10.3.11. *The isometric isomorphism*

$$R_\lambda^* = U_1^* R_0 : L_2(\mathbb{R}) \longrightarrow \mathcal{A}^2_\lambda(\Pi)$$

is given by

$$(R_\lambda^* f)(z) = \frac{1}{\sqrt{2}} \int_{\mathbb{R}} z^{i\xi-(1+\lambda/2)}\, \vartheta_\lambda(\xi)\, f(\xi)\, d\xi. \tag{10.3.11}$$

Proof. Calculate

$$\begin{aligned}
(R_\lambda^* f)(z) &= (U_1^* R_0 f)(z)\\
&= \sqrt{\pi}(M^{-1} \otimes I) f(\xi)\, \vartheta_\lambda(\xi)\, e^{-(\xi+(1+\lambda/2)i)\theta}\\
&= \frac{1}{\sqrt{2}} \int_{\mathbb{R}} r^{i\xi-(1+\lambda/2)} f(\xi)\, \vartheta_\lambda(\xi)\, e^{-(\xi+(1+\lambda/2)i)\theta}\, d\xi\\
&= \frac{1}{\sqrt{2}} \int_{\mathbb{R}} z^{i\xi-(1+\lambda/2)}\, \vartheta_\lambda(\xi)\, f(\xi)\, d\xi.
\end{aligned}$$

$\square$

Corollary 10.3.12. *The inverse isomorphism*

$$R_\lambda : \mathcal{A}^2_\lambda(\Pi) \longrightarrow L_2(\mathbb{R}_+)$$

is given by

$$(R_\lambda \varphi)(\xi) = \frac{\vartheta_\lambda(\xi)}{\sqrt{2}} \int_\Pi (\bar{z})^{-i\xi - (1+\lambda/2)} \, \varphi(z) \, \mu_\lambda(z) \, dv(z). \qquad (10.3.12)$$

10.4   Model classes of Toeplitz operators

In the next three subsections we will consider three classes of Toeplitz operators which correspond to the three model cases for pencils. That is, we will consequently study Toeplitz operators on the unit disk with radial defining symbols (model elliptic case), Toeplitz operators on the upper half-plane with defining symbol depending on y (model parabolic case), and Toeplitz operators on the upper half-plane with defining symbols depending in θ (model hyperbolic case). We will show, in particular, that in all these three cases the corresponding Toeplitz operator C^*-algebras are commutative.

Thus, extending this result by means of Möbius transformations to symbols constant on cycles of an arbitrary pencil of hyperbolic geodesics (as in Chapter 9), we come to the following generalization of Theorem 9.6.1.

Theorem 10.4.1. *Given a pencil $\mathcal{P}$ of geodesics, consider the set of L_∞-symbols which are constant on corresponding cycles. Then the C^*-algebra generated by Toeplitz operators with such defining symbols is commutative on each weighted Bergman space $A^2_\lambda(\mathbb{D})$.*

10.4.1   Toeplitz operators with radial symbols

Given a radial function $a = a(r) \in L_\infty(0,1)$, consider the Toeplitz operator

$$T_a : \varphi \in \mathcal{A}^2_\lambda(\mathbb{D}) \longmapsto B_{\mathbb{D},\lambda} a\varphi \in \mathcal{A}^2_\lambda(\mathbb{D}).$$

Theorem 10.4.2. *For any $a = a(r) \in L_\infty[0,1)$, the Toeplitz operator T_a acting on $\mathcal{A}^2_\lambda(\mathbb{D})$ is unitary equivalent to the multiplication operator $\gamma_{a,\lambda} I = R_\lambda T_a R^*_\lambda$, acting on l_2. The sequence $\gamma_{a,\lambda} = \{\gamma_{a,\lambda}(n)\}_{n\in\mathbb{Z}_+}$ is*

$$\gamma_{a,\lambda}(n) = \frac{1}{B(n+1, \lambda+1)} \int_0^1 a(\sqrt{r}) r^n (1-r)^\lambda dr, \qquad n \in \mathbb{Z}_+, \qquad (10.4.1)$$

and the operators R^*_λ and R_λ are given by (10.3.2) and (10.3.3), respectively.

Proof. Calculate

$$
\begin{aligned}
R_\lambda T_a R_\lambda^* &= R_\lambda B_{\mathbb{D},\lambda} a B_{\Pi,\lambda} R_\lambda^* = R_\lambda (R_\lambda^* R_\lambda) a (R_\lambda^* R_\lambda) R_\lambda^* \\
&= (R_\lambda R_\lambda^*) R_\lambda a R_\lambda^* (R_\lambda R_\lambda^*) = R_\lambda a R_\lambda^* \\
&= R_0^* U_{2,\lambda} U_1 a(r) U_1^{-1} U_{2,\lambda}^{-1} R_0 \\
&= R_0^* U_{2,\lambda} a(r) U_{2,\lambda}^{-1} R_0 \\
&= R_0^* a(\omega_{n,\lambda}(r)) R_0 .
\end{aligned}
$$

Now

$$
R_0^* a(\omega_{n,\lambda}(r)) R_0 \, \{c_n\}_{n\in\mathbb{Z}_+} = \left\{ 2 \int_0^1 a(\omega_{n,\lambda}(r)) \, c_n r \, dr \right\}_{n\in\mathbb{Z}_+} = \{\gamma_{a,\lambda}(n) \, c_n\}_{n\in\mathbb{Z}_+},
$$

where the sequence $\gamma_{a,\lambda} = \{\gamma_{a,\lambda}(n)\}_{n\in\mathbb{Z}_+}$ has the form

$$
\begin{aligned}
\gamma_{a,\lambda}(n) &= 2 \int_0^1 a(\omega_{n,\lambda}(r)) \, r \, dr \\
&= \alpha_{n,\lambda}^2 \frac{\lambda+1}{\pi} \int_0^1 a(r) \, r^{2n+1} (1-r^2)^\lambda dr \\
&= \frac{1}{B(n+1,\lambda+1)} \int_0^1 a(\sqrt{r}) r^n (1-r)^\lambda dr .
\end{aligned}
$$

Here the constant $\alpha_{n,\lambda}$ is given by (10.3.1). $\qquad\square$

Corollary 10.4.3. *The C^*-algebra $\mathcal{T}_\lambda$ generated by all Toeplitz operators T_a with symbols $a = a(r) \in L_\infty[0,1)$ is commutative and is isometrically imbedded in l_∞. The isometric imbedding τ_λ is generated by the mapping*

$$
\tau_\lambda : \; T_a \longmapsto \gamma_{a,\lambda} .
$$

Rewriting the statement of Theorem 10.4.2 we come to the following spectral-type representation of a Toeplitz operator.

Theorem 10.4.4. *Let $a = a(r)$. Then the Toeplitz operator T_a acting on $\mathcal{A}_\lambda^2(\mathbb{D})$ admits the representation*

$$
(T_a \varphi)(z) = \sum_{n=0}^\infty \gamma_{a,\lambda}(n) \frac{\alpha_{n,\lambda}^2}{\pi} \int_{\mathbb{D}} \varphi(\zeta)(z\bar{\zeta})^n d\mu_\lambda(\zeta),
$$

where the constants $\alpha_{n,\lambda}$ are given by (10.3.1).

Proof. Follows directly from Theorems 10.4.2, and 10.3.3, and Corollary 10.3.4. $\qquad\square$

The standard orthonormal monomial basis in $\mathcal{A}_\lambda^2(\mathbb{D})$ is obviously given by

$$e_{n,\lambda}(z) = \frac{\alpha_{n,\lambda}}{\sqrt{\pi}}\, z^n, \qquad n \in \mathbb{Z}_+,$$

and the orthonormal projection $P_{n,\lambda}$ of $\mathcal{A}_\lambda^2(\mathbb{D})$ onto the one-dimensional space $L_{n,\lambda}$, generated by $e_{n,\lambda}(z)$, has the form

$$(P_{n,\lambda}\varphi)(z) = \langle \varphi, e_{n,\lambda}\rangle\, e_{n,\lambda} = \frac{\alpha_{n,\lambda}^2}{\pi} \int_{\mathbb{D}} \varphi(\zeta)(z\overline{\zeta})^n d\mu_\lambda(\zeta). \tag{10.4.2}$$

Thus the above theorem states that for a radial symbol $a = a(r)$, we have

$$T_a\varphi = \sum_{n=0}^{\infty} \gamma_{a,\lambda}(n)\, P_{n,\lambda}\varphi. \tag{10.4.3}$$

We give now an independent direct proof of the equality

$$T_a e_{n,\lambda} = \gamma_{n,\lambda}\, e_{n,\lambda}.$$

Note that the series (10.1.4)

$$\frac{1}{(1 - z\overline{\zeta})^{2+\lambda}} = \sum_{n=0}^{\infty} \frac{\Gamma(n+2+\lambda)}{n!\,\Gamma(2+\lambda)}\, (z\overline{\zeta})^n$$

converges uniformly for any fixed $z \in \mathbb{D}$, which permits us to interchange the summation and integration on the next formula.

For $a = a(r)$, we have

$$
\begin{aligned}
T_a e_{n,\lambda} &= \frac{\alpha_{n,\lambda}}{\sqrt{\pi}} \frac{\lambda+1}{\pi} \int_{\mathbb{D}} \frac{a(r)\,\zeta^n}{(1-z\overline{\zeta})^{\lambda+2}} (1-r^2)^\lambda dv(\zeta) \\
&= \frac{\alpha_{n,\lambda}}{\sqrt{\pi}} \frac{\lambda+1}{\pi} \sum_{k=0}^{\infty} \frac{\Gamma(k+2+\lambda)}{k!\,\Gamma(\lambda+2)}\, z^n \int_0^1 a(r) r^{n+k}(1-r^2)^\lambda r dr \int_{S^1} t^{n-k} \frac{dt}{it} \\
&= \frac{\alpha_{n,\lambda}}{\sqrt{\pi}} 2(\lambda+1) \frac{\Gamma(n+2+\lambda)}{n!\,\Gamma(\lambda+2)}\, z^n \int_0^1 a(r) r^{2n}(1-r^2)^\lambda r dr \\
&= \frac{1}{B(n+1,\lambda+1)} \int_0^1 a(\sqrt{r}) r^n (1-r)^\lambda dr\, \frac{\alpha_{n,\lambda}}{\sqrt{\pi}}\, z^n \\
&= \gamma_{a,\lambda}\, e_{n,\lambda}(z).
\end{aligned}
$$

Theorem 10.4.5. *Let T_a be the Toeplitz operator with radial symbol $a(r) \in L_\infty[0,1)$. Then the corresponding Wick function has the form*

$$
\begin{aligned}
\widetilde{a}(z,\overline{\zeta}) &= (1 - z\overline{\zeta})^{\lambda+2} \sum_{n=0}^{\infty} \frac{\Gamma(n+2+\lambda)}{n!\,\Gamma(2+\lambda)} (z\overline{\zeta})^n\, \gamma_{a,\lambda}(n) \\
&= (1 - z\overline{\zeta})^{\lambda+2} \sum_{n=0}^{\infty} \frac{\alpha_{n,\lambda}^2}{\pi} (z\overline{\zeta})^n\, \gamma_{a,\lambda}(n).
\end{aligned}
$$

Proof. The reproducing property of the Bergman kernel shows that the system of functions $k_\zeta(z) = \overline{k_z(\zeta)} = K_{\mathbb{D}}(z,\zeta)$, $\zeta \in \mathbb{D}$, forms the system of coherent states (see Appendix A, Section A.1) in the Bergman space $\mathcal{A}_\lambda^2(\mathbb{D})$. Thus

$$
\begin{aligned}
\widetilde{a}(z,\overline{\zeta}) \;=\;& \frac{\langle T_a k_\zeta, k_z \rangle}{\langle k_\zeta, k_z \rangle} = k_\zeta(z)^{-1} \langle a k_\zeta, k_z \rangle \\[2mm]
=\;& (1 - z\overline{\zeta})^{\lambda+2} \int_{\mathbb{D}} \frac{a(|w|)}{(1 - w\overline{\zeta})^{\lambda+2}(1 - z\overline{w})^{\lambda+2}} \frac{\lambda+1}{\pi} (1 - |w|^2)^\lambda dv(w) \\[2mm]
=\;& (1 - z\overline{\zeta})^{\lambda+2} \int_{\mathbb{D}} a(|w|) \sum_{k=0}^{\infty} \frac{\Gamma(k+2+\lambda)}{k!\,\Gamma(2+\lambda)} (w\overline{\zeta})^k \\[2mm]
& \cdot \sum_{n=0}^{\infty} \frac{\Gamma(n+2+\lambda)}{n!\,\Gamma(2+\lambda)} (z\overline{w})^n \frac{\lambda+1}{\pi} (1 - |w|^2)^\lambda dv(w) \\[2mm]
=\;& (1 - z\overline{\zeta})^{\lambda+2} \frac{\lambda+1}{\pi} \sum_{k=0}^{\infty} \sum_{n=0}^{\infty} \frac{\Gamma(k+2+\lambda)}{k!\,\Gamma(2+\lambda)} \frac{\Gamma(k+2+\lambda)}{k!\,\Gamma(2+\lambda)} \overline{\zeta}^k z^n \\[2mm]
& \cdot \int_0^1 a(r)\, r^{k+n}(1-r^2)^\lambda r\,dr \int_{S^1} t^{k-n} \frac{dt}{it} \\[2mm]
=\;& (1 - z\overline{\zeta})^{\lambda+2}\, 2(\lambda+1) \sum_{n=0}^{\infty} \left(\frac{\Gamma(n+2+\lambda)}{n!\,\Gamma(2+\lambda)} \right)^2 (z\overline{\zeta})^n \\[2mm]
& \cdot \int_0^1 a(r)\, r^{2n_1}(1-r^2)^\lambda dr \\[2mm]
=\;& (1 - z\overline{\zeta})^{\lambda+2} \sum_{n=0}^{\infty} \frac{\Gamma(n+2+\lambda)}{n!\,\Gamma(2+\lambda)} (z\overline{\zeta})^n\, \gamma_{a,\lambda}(n) \\[2mm]
=\;& (1 - z\overline{\zeta})^{\lambda+2} \sum_{n=0}^{\infty} \frac{\alpha_{n,\lambda}^2}{\pi} (z\overline{\zeta})^n\, \gamma_{a,\lambda}(n).
\end{aligned}
$$

$\square$

Corollary 10.4.6. *Let T_a be the Toeplitz operator with radial symbol $a(r) \in L_\infty[0,1)$. Then the Wick symbol of T_a is radial as well and is given by the formula*

$$
\widetilde{a}(r) = \widetilde{a}(z,\overline{z}) = (1 - r^2)^{\lambda+2} \sum_{n=0}^{\infty} \frac{\Gamma(n+2+\lambda)}{n!\,\Gamma(2+\lambda)} r^{2n}\, \gamma_{a,\lambda}(n).
$$

Remark 10.4.7. Given a radial symbol $a = a(r)$, writing the Toeplitz operator T_a in terms of its Wick symbol we get formula (10.4.3). Indeed,

$$
\begin{aligned}
(T_a\varphi)(z) &= \int_{\mathbb{D}} \widetilde{a}(z,\overline{\zeta})\varphi(\zeta)k_\zeta(z)d\mu_\lambda(\zeta) \\
&= \int_{\mathbb{D}} \sum_{n=0}^{\infty} \frac{\alpha_{n,\lambda}^2}{\pi}\,(z\overline{\zeta})^n\,\gamma_{a,\lambda}(n)\,\varphi(\zeta)\,d\mu_\lambda(\zeta) \\
&= \sum_{n=0}^{\infty} \gamma_{a,\lambda}(n)\,\frac{\alpha_{n,\lambda}^2}{\pi}\int_{\mathbb{D}}\varphi(\zeta)\,(z\overline{\zeta})^n\,d\mu_\lambda(\zeta) \\
&= \sum_{n=0}^{\infty} \gamma_{a,\lambda}(n)\,(P_{n,\lambda}\varphi)(z),
\end{aligned}
$$

where the one-dimensional projections $P_{n,\lambda}$, $n \in \mathbb{Z}_+$ are given by (10.4.2).

Corollary 10.4.8. *Let T_{a_1} and T_{a_2} be two Toeplitz operators with symbols $a_1(r)$ and $a_2(r)$ respectively, and let $\widetilde{a}_1(r)$ and $\widetilde{a}_2(r)$ be their Wick symbols. Then the Wick symbol $\widetilde{a}(r)$ of the composition $T_{a_1}T_{a_2}$ is given by*

$$
\widetilde{a}(r) = (\widetilde{a}_1 \star \widetilde{a}_2)(r) = (1-r^2)^{\lambda+2}\sum_{n=0}^{\infty} \frac{\Gamma(n+2+\lambda)}{n!\,\Gamma(2+\lambda)}\,r^{2n}\,\gamma_{a_1,\lambda}(n)\,\gamma_{a_2,\lambda}(n).
$$

10.4.2   Toeplitz operators with symbols depending on y

Given a function $a = a(y) \in L_\infty(\mathbb{R}_+)$, which depends only on vertical variable $y = \mathrm{Im}\,z$, consider the Toeplitz operator

$$
T_a : \varphi \in \mathcal{A}_\lambda^2(\Pi) \longmapsto B_{\Pi,\lambda}a\varphi \in \mathcal{A}_\lambda^2(\Pi).
$$

Theorem 10.4.9. *For any $a = a(y) \in L_\infty(\mathbb{R}_+)$, the Toeplitz operator T_a acting on $\mathcal{A}_\lambda^2(\Pi)$ is unitary equivalent to the multiplication operator $\gamma_{a,\lambda}I = R_\lambda T_a R_\lambda^*$, acting on $L_2(\mathbb{R}_+)$. The function $\gamma_{a,\lambda}(x)$ is*

$$
\begin{aligned}
\gamma_{a,\lambda}(x) &= \frac{x^{\lambda+1}}{\Gamma(\lambda+1)}\int_0^\infty a(t/2)t^\lambda e^{-xt}dt \\
&= \frac{1}{\Gamma(\lambda+1)}\int_0^\infty a(\frac{t}{2x})t^\lambda e^{-t}dt, \quad x \in \mathbb{R}_+, \quad\quad (10.4.4)
\end{aligned}
$$

and the operators R_λ^* and R_λ are given by (10.3.7) and (10.3.8), respectively.

Proof. Calculate

$$
\begin{aligned}
R_\lambda T_a R_\lambda^* &= R_\lambda B_{\Pi,\lambda}aB_{\Pi,\lambda}R_\lambda^* = R_\lambda(R_\lambda^* R_\lambda)a(R_\lambda^* R_\lambda)R_\lambda^* \\
&= (R_\lambda R_\lambda^*)R_\lambda a R_\lambda^*(R_\lambda R_\lambda^*) = R_\lambda a R_\lambda^* \\
&= R_0^* U_{2,\lambda}U_1 a(y)U_1^{-1}U_{2,\lambda}^{-1}R_0 \\
&= R_0^* U_{2,\lambda}a(y)U_{2,\lambda}^{-1}R_0 \\
&= R_0^* a(\beta(|x|,y))R_0.
\end{aligned}
$$

Now

$$(R_0^* a(\beta(|x|, y)) R_0 f)(x) = \int_{\mathbb{R}_+} a(\beta(|x|, \eta)) f(x) e^{-\eta} d\eta = \gamma_{a,\lambda}(x) \cdot f(x),$$

where

$$
\begin{aligned}
\gamma_{a,\lambda}(x) &= \int_{\mathbb{R}_+} a(\beta(|x|, \eta)) e^{-\eta} d\eta = \int_{\mathbb{R}_+} a(t) e^{-\gamma(x,t)} d\gamma(x,t) \\
&= \int_{\mathbb{R}_+} a(t) \theta_\lambda^2(x)(\lambda + 1)(2t)^\lambda e^{-2tx} dt \\
&= \frac{x^{\lambda+1}}{\Gamma(\lambda+1)} \int_0^\infty a(t/2) t^\lambda e^{-xt} dt \\
&= \frac{1}{\Gamma(\lambda+1)} \int_0^\infty a(\frac{t}{2x}) t^\lambda e^{-t} dt, \quad x \in \mathbb{R}_+.
\end{aligned}
$$

Here the functions $\gamma(x,t)$ and $\theta_\lambda(x)$ are given by (10.3.6) and (10.3.5), respectively.

$\square$

Corollary 10.4.10. *The C^*-algebra $\mathcal{T}_\lambda$ generated by all Toeplitz operators T_a with symbols $a = a(y) \in L_\infty(\mathbb{R}_+)$ is commutative and is isometrically imbedded in $C_b(\mathbb{R}_+)$. The isometric imbedding τ_λ is generated by the mapping*

$$\tau_\lambda : T_a \longmapsto \gamma_{a,\lambda}.$$

Rewriting the statement of Theorem 10.4.9 we come to the following spectral-type representation of a Toeplitz operator.

Theorem 10.4.11. *Let $a = a(y)$. Then the Toeplitz operator T_a acting on $\mathcal{A}_\lambda^2(\Pi)$ admits the representation*

$$(T_a \varphi)(z) = \frac{1}{\sqrt{\Gamma(\lambda+2)}} \int_{\mathbb{R}_+} t^{\frac{\lambda+1}{2}} \gamma_{a,\lambda}(t) f(t) e^{iz \cdot t} dt, \qquad (10.4.5)$$

where $f(x) = (R_\lambda \varphi)(x)$.

Proof. Follows directly from Theorems 10.4.9, and 10.3.7, and Corollary 10.3.8.

$\square$

At the same time it is instructive to give a direct proof of the theorem which does not use the results of the previous section. Indeed, for a symbol $a = a(y)$ depending only on y consider the Toeplitz operator

$$(T_a \varphi)(z) = (\lambda + 1) \int_\Pi a(\eta) \varphi(\zeta) \left(\frac{\zeta - \bar{\zeta}}{z - \bar{\zeta}} \right)^{\lambda+2} d\nu(\zeta),$$

where $\zeta = \xi + i\eta$. Represent the function $\varphi(\zeta)$ in the form of the Fourier integral

$$\varphi(\xi + i\eta) = \frac{1}{\sqrt{\Gamma(\lambda + 2)}} \int_{\mathbb{R}_+} t^{\frac{\lambda+1}{2}} f(t)\, e^{it(\xi+i\eta)}\, dt, \quad \eta > 0,$$

where $f \in L_2(\mathbb{R}_+)$. Now

$$(T_a\varphi)(z) = \frac{i^{\lambda+2}}{\pi\sqrt{\Gamma(\lambda+2)}} \int_{\mathbb{R}_+} a(\eta)(2\eta)^\lambda d\eta \int_{\mathbb{R}_+} t^{\frac{\lambda+1}{2}} f(t)\, e^{-t\eta} dt \int_{\mathbb{R}} \frac{e^{it\xi}\, d\xi}{(z + i\eta - \xi)^{\lambda+2}}.$$

Using the formula, see [86] 3.382.6,

$$\int_{\mathbb{R}} (i\beta - \xi)^{-(\lambda+2)} e^{it\xi} d\xi = \frac{2\pi}{i^{\lambda+2}} \frac{t^{\lambda+1} e^{-\beta t}}{\Gamma(\lambda+2)}, \quad t > 0,$$

we have

$$\begin{aligned}
(T_a\varphi)(z) &= \frac{2(\lambda+1)}{\Gamma(\lambda+2)^{3/2}} \int_{\mathbb{R}_+} a(\eta)(2\eta)^\lambda d\eta \int_{\mathbb{R}_+} t^{\frac{\lambda+1}{2}+(\lambda+1)} f(t)\, e^{-2t\eta+izt} dt \\
&= \frac{2}{\Gamma(\lambda+2)^{1/2}} \int_{\mathbb{R}_+} t^{\frac{\lambda+1}{2}} f(t)\, e^{izt} dt \frac{t^{\lambda+1}}{\Gamma(\lambda+1)} \int_{\mathbb{R}_+} a(\eta)(2\eta)^\lambda e^{-2t\eta} d\eta \\
&= \frac{1}{\sqrt{\Gamma(\lambda+2)}} \int_{\mathbb{R}_+} t^{\frac{\lambda+1}{2}} f(t)\, \gamma_{a,\lambda}(t)\, e^{izt} dt,
\end{aligned}$$

where

$$\gamma_{a,\lambda}(t) = \frac{t^{\lambda+1}}{\Gamma(\lambda+1)} \int_0^\infty a\left(\frac{\eta}{2}\right) \eta^\lambda e^{-t\eta}\, d\eta.$$

Theorem 10.4.12. *Given $a = a(y)$, the Wick symbol $\widetilde{a}(z, \overline{z})$ of the Toeplitz operator T_a depends only on y as well, and has the form*

$$\widetilde{a}(y) = \widetilde{a}(z, \overline{z}) = \frac{\langle T_a k_z, k_z \rangle}{\langle k_z, k_z \rangle} = \frac{(2y)^{\lambda+2}}{\Gamma(\lambda+2)} \int_{\mathbb{R}_+} u^{\lambda+1} \gamma_{a,\lambda}(u)\, e^{-2yu}\, du, \qquad (10.4.6)$$

and the corresponding Wick function is given by formula

$$\widetilde{a}(z, \overline{w}) = \frac{\langle T_a k_w, k_z \rangle}{\langle k_w, k_z \rangle} = \frac{[-i(z - \overline{w})]^{\lambda+2}}{\Gamma(\lambda+2)} \int_{\mathbb{R}_+} u^{\lambda+1} \gamma_{a,\lambda}(u)\, e^{i(z-\overline{w})u}\, du. \qquad (10.4.7)$$

Proof. Consider $k_z(w) = i^{2+\lambda}(w - \overline{z})^{-(\lambda+2)} = i^{2+\lambda}(u + iv - x + iy)^{-(\lambda+2)}$ and calculate

$$\begin{aligned}
(U_1 k_z)(u, v) &= \frac{i^{2+\lambda}}{\pi\sqrt{2}} \int_{\mathbb{R}} (\xi + iv - x + iy)^{-(\lambda+2)} e^{-i\xi u} d\xi \\
&= \frac{1}{i^{2+\lambda}\pi\sqrt{2}} \int_{\mathbb{R}} (\xi + i(y + v + ix))^{-(\lambda+2)} e^{-i\xi u} d\xi.
\end{aligned}$$

Using formula 3.382.7 from [86], we have

$$(U_1 k_z)(u, v) = (-1)^{\lambda+2} \chi_+(u) \frac{\sqrt{2}\, u^{\lambda+1}}{\Gamma(\lambda+2)}\, e^{-u(y+\eta)-iux}.$$

Thus

$$
\begin{aligned}
\langle T_a k_z, k_z \rangle &= \langle a k_z, k_z \rangle = \langle U_1 a k_z, U_1 k_z \rangle = \langle a U_1 k_z, U_1 k_z \rangle \\
&= \frac{2}{[\Gamma(\lambda+2)]^2} \int_0^\infty \int_0^\infty a(v)\, u^{2(\lambda+1)}\, e^{-2u(y+v)} (\lambda+1)\, (2v)^\lambda du\, dv \\
&= \frac{1}{\Gamma(\lambda+2)} \int_0^\infty u^{\lambda+1} e^{-2yu} du\, \frac{2u^{\lambda+1}}{\Gamma(\lambda+1)} \int_0^\infty a(v)(2v)^\lambda e^{-2uv}\, dv \\
&= \frac{1}{\Gamma(\lambda+2)} \int_{\mathbb{R}_+} u^{\lambda+1}\, \gamma_{a,\lambda}(u)\, e^{-2yu}\, du.
\end{aligned}
$$

Thus we have (10.4.6). The equality (10.4.7) follows from (10.4.6) by the analytic continuation principle, or can be verified by direct calculations. $\qquad\square$

Remark 10.4.13. Formula (10.4.6) admits an interesting interpretation. Start with a symbol $a = a(y)$ and the Toeplitz operator T_a^λ acting on $\mathcal{A}_\lambda^2(\Pi)$, calculate the corresponding function $\gamma_{a,\lambda}(x)$, $x > 0$, and consider now the Toeplitz operator $T_{\gamma_{a,\lambda}}^{\lambda+1}$ with symbol $\gamma_{a,\lambda}(y)$ acting on $\mathcal{A}_{\lambda+1}^2(\Pi)$. Then the corresponding function $\gamma_{\gamma_{a,\lambda},\lambda+1}$ coincides with the Wick symbol of the initial Toeplitz operator T_a^λ, i.e.,

$$\widetilde{a}(y) = \widetilde{a}(z, \overline{z}) = \gamma_{\gamma_{a,\lambda},\lambda+1}(y).$$

Remark 10.4.14. Given a symbol $a = a(y)$, writing the Toeplitz operator T_a in terms of its Wick symbol we get the formula (10.4.5). Indeed

$$
\begin{aligned}
(T_a \varphi)(z) &= \int_\Pi \widetilde{a}(z, \overline{w}) \frac{\varphi(w)\, i^{\lambda+2}}{(z-\overline{w})^{\lambda+2}}\, \mu_\lambda(w)\, dv(w) \\
&= \int_\Pi \frac{[-i(z-\overline{w})]^{\lambda+2}}{\Gamma(\lambda+2)} \int_{\mathbb{R}_+} u^{\lambda+1}\, \gamma_{a,\lambda}(u)\, e^{i(z-\overline{w})u}\, du \\
&\qquad \cdot \frac{\varphi(w)\, i^{\lambda+2}}{(z-\overline{w})^{\lambda+2}}\, \mu_\lambda(w)\, dv(w) \\
&= \frac{1}{\sqrt{\Gamma(\lambda+2)}} \int_{\mathbb{R}_+} u^{\frac{\lambda+1}{2}}\, \gamma_{a,\lambda}(u)\, e^{izu}\, du \\
&\qquad \cdot \frac{u^{\frac{\lambda+1}{2}}}{\sqrt{\Gamma(\lambda+2)}} \int_\Pi \varphi(w)\, e^{-i\,\overline{w}\cdot u}\, \mu_\lambda(w)\, dv(w) \\
&= \frac{1}{\sqrt{\Gamma(\lambda+2)}} \int_{\mathbb{R}_+} u^{\frac{\lambda+1}{2}}\, \gamma_{a,\lambda}(u)\, (R_\lambda \varphi)(u)\, e^{izu}\, du.
\end{aligned}
$$

Corollary 10.4.15. *Let T_{a_1} and T_{a_2} be two Toeplitz operators with symbols $a_1(y)$ and $a_2(y)$ respectively, and let $\widetilde{a}_1(y)$ and $\widetilde{a}_2(y)$ be their Wick symbols. Then the Wick symbol $\widetilde{a}(y)$ of the composition $T_{a_1} T_{a_2}$ is given by*

$$\widetilde{a}(y) = (\widetilde{a}_1 \star \widetilde{a}_2)(y) = \frac{(2y)^{\lambda+2}}{\Gamma(\lambda+2)} \int_{\mathbb{R}_+} u^{\lambda+1}\, \gamma_{a_1,\lambda}(u)\, \gamma_{a_2,\lambda}(u)\, e^{-2yu}\, du.$$

10.4.3   Toeplitz operators with symbols depending on θ

Let now $a = a(\theta)$ depend only on the angular variable θ.

Theorem 10.4.16. *Given $a = a(\theta) \in L_\infty(0, \pi)$, the Toeplitz operator T_a^λ acting on $\mathcal{A}_\lambda^2(\Pi)$ is unitary equivalent to the multiplication operator $\gamma_{a,\lambda} I = R_\lambda T_a^\lambda R_\lambda^*$, acting on $L_2(\mathbb{R})$. The function $\gamma_{a,\lambda}(\xi)$ is*

$$
\begin{aligned}
\gamma_{a,\lambda}(\xi) &= 2^\lambda (\lambda+1)\vartheta_\lambda^2(\xi) \int_0^\pi a(\theta)\, e^{-2\xi\theta}\, \sin^\lambda \theta\, d\theta, &&(10.4.8)\\
&= \left(\int_0^\pi e^{-2\xi\theta}\, \sin^\lambda \theta\, d\theta \right)^{-1} \int_0^\pi a(\theta)\, e^{-2\xi\theta}\, \sin^\lambda \theta\, d\theta, && \xi \in \mathbb{R},
\end{aligned}
$$

and the operators R_λ^* and R_λ are given by (10.3.11) and (10.3.12), respectively.

Proof. Calculate

$$
\begin{aligned}
R_\lambda T_a\, R_\lambda^* &= R_\lambda B_{\Pi,\lambda} a B_{\Pi,\lambda} R_\lambda^* = R_\lambda (R_\lambda^* R_\lambda) a (R_\lambda^* R_\lambda) R_\lambda^* \\
&= (R_\lambda R_\lambda^*) R_\lambda a R_\lambda^* (R_\lambda R_\lambda^*) = R_\lambda a R_\lambda^* \\
&= R_0^* U_1 a(\theta) U_1^{-1} R_0 \\
&= R_0^* a(\theta) R_0.
\end{aligned}
$$

Thus

$$
\begin{aligned}
(R_0^* a(\theta) R_0 f)(\xi) &= 2^\lambda(\lambda+1)\vartheta_\lambda(\xi) \int_0^\pi a(\theta)\, e^{-(\xi-(1+\lambda/2)i)\theta}\, f(x) \\
&\qquad \cdot\ \vartheta_\lambda(\xi)\, e^{-(\xi+(1+\lambda/2)i)\theta}\, \sin^\lambda \theta\, d\theta \\
&= \gamma_{a,\lambda}(\xi) \cdot f(\xi),
\end{aligned}
$$

where

$$\gamma_{a,\lambda}(\xi) = 2^\lambda (\lambda+1)\vartheta_\lambda^2(\xi) \int_0^\pi a(\theta)\, e^{-2\xi\theta}\, \sin^\lambda \theta\, d\theta, \qquad \xi \in \mathbb{R}.$$

Here the function $\vartheta_\lambda(\xi)$ is given by (10.3.10). $\qquad\square$

Corollary 10.4.17. *The C^*-algebra $\mathcal{T}_\lambda$ generated by all Toeplitz operators T_a with symbols $a = a(\theta) \in L_\infty(0, \pi)$ is commutative and is isometrically imbedded in $C_b(\mathbb{R})$. The isometric imbedding τ_λ is generated by the mapping*

$$\tau_\lambda : \ T_a \longmapsto \gamma_{a,\lambda}.$$

Rewriting the statement of Theorem 10.4.16 we come to the following spectral-type representation of a Toeplitz operator.

Theorem 10.4.18. *Let* $a = a(\theta)$. *Then the Toeplitz operator* T_a *acting on* $\mathcal{A}_\lambda^2(\Pi)$ *admits the representation*

$$(T_a\varphi)(z) = \frac{1}{\sqrt{2}} \int_{\mathbb{R}} z^{i\xi-(1+\lambda/2)}\, \vartheta_\lambda(\xi)\, \gamma_{a,\lambda}(\xi)\, f(\xi)\, d\xi, \qquad (10.4.9)$$

where $f(\xi) = (R_\lambda\varphi)(\xi)$.

Proof. Follows directly from Theorems 10.4.16, and 10.3.11, and Corollary 10.3.12. $\qquad\square$

Theorem 10.4.19. *Given* $a = a(\theta)$, *the Wick symbol* $\widetilde{a}(z, \overline{z})$ *of the Toeplitz operator* T_a *depends only on* θ ($= \arg z$) *as well, and has the form*

$$\widetilde{a}(\theta) = \widetilde{a}(z, \overline{z}) = 2^{\lambda+1} \sin^{\lambda+2}\theta \int_{\mathbb{R}} e^{-2\xi\theta}\, \vartheta_\lambda^2(\xi)\, \gamma_{a,\lambda}(\xi)\, d\xi \qquad (10.4.10)$$

and the corresponding Wick function is given by formula

$$\widetilde{a}(z, \overline{w}) = \frac{\langle T_a k_w, k_z \rangle}{\langle k_w, k_z \rangle} = (z - \overline{w})^{\lambda+2}\, (z\overline{w})^{-(\lambda+2)/2}\, \frac{i^{-(\lambda+2)}}{2} \int_{\mathbb{R}} \left(\frac{z}{\overline{w}}\right)^{i\xi} \vartheta_\lambda^2(\xi)\, \gamma_{a,\lambda}(\xi)\, d\xi. \tag{10.4.11}$$

Proof. Consider $k_z(w) = i^{2+\lambda}(w - \overline{z})^{-(\lambda+2)} = i^{2+\lambda}(\rho e^{i\lambda} - r e^{-i\theta})^{-(\lambda+2)}$ and calculate

$$(U_1 k_z)(\xi, \lambda) = \frac{i^{2+\lambda}}{\pi\sqrt{2}} \int_{\mathbb{R}_+} \rho^{-i\xi+\lambda/2}(\rho e^{i\lambda} - \overline{z})^{-(\lambda+2)}\, d\rho.$$

Using formula 3.194.3 from [86] and (10.3.10), we have

$$(U_1 k_z)(\xi, \lambda) = \frac{B(\frac{\lambda+2}{2} - i\xi, \frac{\lambda+2}{2} + i\xi)}{\sqrt{2\pi}}\, e^{\xi\lambda}\, e^{-\xi\lambda - i\frac{\lambda+2}{2}\lambda}\, (\overline{z})^{-i\xi - \frac{\lambda+2}{2}}$$

$$= \frac{\vartheta_\lambda^2(\xi)}{\sqrt{2}}\, e^{-\xi\lambda - i\frac{\lambda+2}{2}\lambda}\, (\overline{z})^{-i\xi - \frac{\lambda+2}{2}}.$$

Thus

$$\begin{aligned}
\langle T_a k_z, k_z \rangle &= \langle a k_z, k_z \rangle = \langle U_1 a k_z, U_1 k_z \rangle = \langle a U_1 k_z, U_1 k_z \rangle \\
&= \frac{1}{2} \int_{\mathbb{R}} \int_0^\pi a(\lambda)\, \vartheta_\lambda^4(\xi) e^{-2\xi\lambda} (\overline{z})^{-i\xi - \frac{\lambda+2}{2}}\, z^{i\xi - \frac{\lambda+2}{2}}\, 2^\lambda(\lambda+1) \sin^\lambda\theta\, d\xi d\theta \\
&= \frac{r^{-(\lambda+2)}}{2} \int_{\mathbb{R}} \vartheta_\lambda^2(\xi)\, e^{-2\xi\theta}\, d\xi\, 2^\lambda(\lambda+1)\vartheta_\lambda^2(\xi) \int_0^\pi a(\lambda)\, e^{-2\xi\lambda}\, \sin^\lambda\lambda\, d\lambda \\
&= \frac{r^{-(\lambda+2)}}{2} \int_{\mathbb{R}} \vartheta_\lambda^2(\xi)\, e^{-2\xi\theta}\, \gamma_{a,\lambda}(\xi)\, d\xi.
\end{aligned}$$

Similarly

$$\langle T_a k_w, k_z \rangle = \frac{(z\overline{w})^{-(\lambda+2)/2}}{2} \int_{\mathbb{R}} \left(\frac{z}{\overline{w}}\right)^{i\xi} \vartheta_\lambda^2(\xi)\, \gamma_{a,\lambda}(\xi)\, d\xi.$$

Furthermore $\langle k_w, k_z \rangle = k_w(z) = i^{\lambda+2}(z - \overline{w})^{-(\lambda+2)}$, and $\langle k_z, k_z \rangle = k_z(z) = (2\operatorname{Im} z)^{-(\lambda+2)}$. Thus we have both (10.4.10) and (10.4.11). $\qquad\square$

Remark 10.4.20. Given a symbol $a = a(\theta)$, writing the Toeplitz operator T_a in terms of its Wick symbol we get the formula (10.4.9). Indeed,

$$
\begin{aligned}
(T_a\varphi)(z) &= \int_\Pi \widetilde{a}(z,\overline{w})\, \frac{\varphi(w)\, i^{\lambda+2}}{(z - \overline{w})^{\lambda+2}}\, \mu_\lambda(w)\, dv(w) \\
&= \frac{1}{2} \int_\Pi (z\overline{w})^{-(\lambda+2)/2} \varphi(w)\, \mu_\lambda(w)\, dv(w) \int_{\mathbb{R}} \left(\frac{z}{\overline{w}}\right)^{i\xi} \vartheta_\lambda^2(\xi)\, \gamma_{a,\lambda}(\xi)\, d\xi \\
&= \frac{1}{\sqrt{2}} \int_{\mathbb{R}} z^{i\xi - \frac{\lambda+2}{2}}\, \vartheta_\lambda(\xi)\, \gamma_{a,\lambda}(\xi)\, d\xi \\
&\quad\cdot \frac{\vartheta_\lambda(\xi)}{\sqrt{2}} \int_\Pi (\overline{w})^{-i\xi - \frac{\lambda+2}{2}}\, \varphi(w)\, \mu_\lambda(w)\, dv(w) \\
&= \frac{1}{\sqrt{2}} \int_{\mathbb{R}} z^{i\xi - \frac{\lambda+2}{2}}\, \vartheta_\lambda(\xi)\, \gamma_{a,\lambda}(\xi)\, (R_\lambda\varphi)(\xi)\, d\xi.
\end{aligned}
$$

Corollary 10.4.21. *Let T_{a_1} and T_{a_2} be two Toeplitz operators with symbols $a_1(\theta)$ and $a_2(\theta)$ respectively, and let $\widetilde{a}_1(\theta)$ and $\widetilde{a}_2(\theta)$ be their Wick symbols. Then the Wick symbol $\widetilde{a}(\theta)$ of the composition $T_{a_1}T_{a_2}$ is given by*

$$\widetilde{a}(\theta) = (\widetilde{a}_1 \star \widetilde{a}_2)(\theta) = 2^{\lambda+1} \sin^{\lambda+2}\theta \int_{\mathbb{R}} e^{-2\xi\theta}\, \vartheta_\lambda^2(\xi)\, \gamma_{a_1,\lambda}(\xi)\, \gamma_{a_2,\lambda}(\xi)\, d\xi.$$

10.5   Boundedness, spectra, and invariant subspaces

As was shown in Theorems 10.4.2, 10.4.9, and 10.4.16 for each type of model pencil (elliptic, parabolic, and hyperbolic) and with the corresponding set of L_∞ defining symbols constant on cycles, the corresponding Toeplitz operators are unitary equivalent to the multiplication operators $\gamma_{a,\lambda} I$, where $\gamma_{a,\lambda}$ are given by (10.4.1), (10.4.4), and (10.4.8), respectively. These theorems can be easily extended for the case of unbounded (measurable) defining symbols constant on cycles. Indeed, having a symbol a such that the corresponding $\gamma_{a,\lambda}$ exists (the integral defining $\gamma_{a,\lambda}$ converges) we can define first the operator T_a on an appropriate dense set in the weighted Bergman space $\mathcal{A}_\lambda^2$ and then in case of boundedness of $\gamma_{a,\lambda}$ extend T_a to a bounded operator on all of $\mathcal{A}_\lambda^2$. That is, we have

Theorem 10.5.1. *Given an unbounded (measurable) defining symbol a constant on cycles, the corresponding Toeplitz operator T_a is bounded if and only if $\gamma_{a,\lambda}$ is bounded, and in case of boundedness the assertions of Theorems 10.4.2, 10.4.9, and 10.4.16 remain true.*

Further in the elliptic case, the Toeplitz operator with (unbounded measurable) radial defining symbol $a(r)$ is compact on $\mathcal{A}^2_\lambda(\mathbb{D})$ if and only if

$$\lim_{n\to\infty} \gamma_{a,\lambda}(n) = 0.$$

We note that the consideration of unbounded defining symbols is quite natural. In particular, as will be shown in the next chapter, even if one starts with bounded defining symbols only the Toeplitz operators with unbounded defining symbols will appear both under the uniform limit of Toeplitz operators with bounded symbols (see Theorem 11.1.1) and as a result of algebraic operations with Toeplitz operators having bounded symbols (see Lemma 11.2.1).

The next assertion follows directly from Theorem 10.5.1 and describes the spectral properties.

Corollary 10.5.2. *Given a model pencil, let a be an unbounded measurable defining symbol constant on cycles such that the Toeplitz operator T_a is bounded on $\mathcal{A}^2_\lambda$. Then the spectrum of T_a in $\mathcal{A}^2_\lambda$ coincides with the closure of the image of the corresponding $\gamma_{a,\lambda}$,*

$$\operatorname{sp} T_a = \operatorname{clos}(\operatorname{Range} \gamma_{a,\lambda})$$

Further in the elliptic case, the essential spectrum of a bounded Toeplitz operator with (unbounded measurable) radial defining symbol $a(r)$ coincides with the set of the limit points of the sequence $\gamma_{a,\lambda} = \{\gamma_{a,\lambda}(n)\}_{n\in\mathbb{Z}_+}$.

The property of being unitary equivalent to a multiplication operator permits us to describe easily the invariant subspaces of each C^*-algebra generated by Toeplitz operators with defining symbols constant on cycles.

Denote by X the domain of $\gamma_{a,\lambda}$, that is $\mathbb{Z}_+$, $\mathbb{R}_+$, or $\mathbb{R}$, depending on the type of the model pencil, elliptic, parabolic, or hyperbolic, respectively. Now as a direct corollary of Theorems 10.4.2, 10.4.9, and 10.4.16 we have

Theorem 10.5.3. *Given a model pencil $\mathcal{P}$, consider the C^*-algebra $\mathcal{T}$ generated by bounded Toeplitz operators T_a with (unbounded measurable) defining symbols constant on cycles. The commutative algebra $\mathcal{T}$ is reducible and every invariant subspace of $\mathcal{T}$ is defined by a measurable subset M of X and has the form*

$$X_M = (R^*_\lambda \chi_M I) L_2(X),$$

where χ_M is the characteristic function of M and the operator R^*_λ is given by (10.3.2), (10.3.7), or (10.3.11), depending on a type of the pencil $\mathcal{P}$.

Chapter 11

Commutative Algebras of Toeplitz Operators

The commutative C^*-algebras of Toeplitz operators on the classical (weightless) Bergman space were classified in Chapter 9 by pencils of geodesics on the unit disk, considered as the hyperbolic plane. Theorem 10.4.1 shows that the same classes of defining symbols generate commutative C^*-algebras of Toeplitz operators on *each weighted* Bergman space. At the same time the principal question, *whether the above cases are the only possible sets of defining symbols which might generate the commutative C^*-algebras of Toeplitz operators on each weighted Bergman space,* has remained open.

In this chapter we give the affirmative answer to the above question.

We note that there is a trivial case having in fact no connection with specific properties of Toeplitz operators. Every C^*-algebra with identity (Toeplitz operators with the defining symbol $e(z) \equiv 1$) generated by a self-adjoint element (Toeplitz operator with a real-valued defining symbol $a = a(z)$) is obviously commutative. We exclude this obvious case from further considerations.

The commutativity of the Toeplitz operator algebras on *each* weighted Bergman space is of great importance and permits us to make use of the Berezin quantization procedure (see Appendix B). At the same time to obtain the necessary information about potential defining symbols we need to calculate additionally the *second and third term* in the asymptotic expansion of a commutator. It turns out that the first, second, and third terms of this expansion together provide us with the exact geometric information: *in order to generate a commutative C^*-algebra of Toeplitz operators on each weighted Bergman space their defining symbols must be constant on the orthogonal trajectories to the geodesics of a certain pencil.*

We show as well that there exist non-typical, in a sense, C^*-algebras of Toeplitz operators which are commutative *only on a single* weighted Bergman space.

11.1   On symbol classes

Although the bounded Toeplitz operators are normally considered with defining symbols belonging to a certain (Banach or C^*-) algebra, the only a priori natural structure on a set of symbols is a linear space (without any topology). This is quite fair. Indeed, considering bounded Toeplitz operators with unbounded defining symbols, we gave an example (see Example 6.2.1) of two bounded Toeplitz operators with symbols a_1 and a_2 such that the Toeplitz operator with symbol $a_1 \cdot a_2$ is unbounded. We note that Toeplitz operators with unbounded defining symbols can easily appear both as uniform limits of Toeplitz operators with bounded defining symbols (see Theorem 11.1.1) and as results of algebraic operations with Toeplitz operators having bounded defining symbols (see Lemma 11.2.1).

Consequently, discussing commutative C^*-algebras of Toeplitz operators we will always assume that the corresponding generating class of symbols is a linear space. As was already mentioned, there is a trivial case having in fact no connection with specific properties of Toeplitz operators. Each C^*-algebra with identity (Toeplitz operators with the defining symbol $e(z) \equiv 1$) generated by a self-adjoint element (Toeplitz operator with a real-valued defining symbol $a = a(z)$) is obviously commutative. The set of generating symbols here is quite restricted, and coincides with the two-dimensional linear space generated by $e(z)$ and $a(z)$. We exclude this obvious case from further consideration.

To underline the geometric nature of symbol classes which generate the commutative C^*-algebras of Toeplitz operators we considered bounded measurable defining symbols in the previous section. This also agrees with the desire for such (commutative) algebras to be, in a sense, maximal. We note that the arguments used in the proof do not require any assumption on smoothness properties of the symbols. The same result (commutativity of Toeplitz operator C^*-algebra) remains valid for each linear subspace of L_∞-symbols (constant on cycles). Moreover, we can start as well with a much more restricted set of defining symbols (say, smooth symbols only) and extend them to all L_∞-symbols by means of uniform and strong operator limits of sequences of Toeplitz operators. Thus it is irrelevant which class of symbols, smooth or L_∞, one starts with.

In Section 11.4 we assume that a commutative C^*-algebra of Toeplitz operators is generated by symbols from a certain linear space of smooth functions, and then prove in Section 11.8 that the symbols are constant on the cycles of a pencil. Then the initial class of defining symbols can be extended, knowing the result of Theorem 10.4.1, to all L_∞-functions constant on the cycles of the corresponding pencil.

As a result, in what follows it is sufficient to deal with the smooth symbols only. As we consider the C^*-algebra generated by Toeplitz operators, we can always assume, without loss of generality, that our set of defining symbols is closed under complex conjugation and contains the function $e(z) \equiv 1$.

We show now how Toeplitz operators with unbounded defining symbols can appear as uniform limits of Toeplitz operators with bounded symbols. In the next theorem we use the weighted Bergman spaces $\mathcal{A}_\lambda^2(\Pi)$ labeled by $\lambda = \frac{1}{h} - 2$.

Theorem 11.1.1. *The Toeplitz operator $T_a^{(\lambda)}$ with unbounded defining symbol*

$$a(v) = (2v)^{-\beta}\sin(2v)^{-\alpha}, \qquad v = \operatorname{Im} w \in \mathbb{R}_+, \tag{11.1.1}$$

where $0 < \beta < 1$ and $\alpha > \beta$, is bounded and belongs to the C^*-algebra generated by Toeplitz operators with smooth bounded defining symbols on each weighted Bergman space $\mathcal{A}_\lambda^2(\Pi)$.

Proof. First of all by Example 13.1.4, the Toeplitz operator $T_a^{(\lambda)}$ with defining symbol (11.1.1) is bounded on each $\mathcal{A}_\lambda^2(\Pi)$, where $\lambda \geq 0$.

Consider now the sequence $\{a_n\}$, where

$$a_n(v) = \begin{cases} a(v), & v \in [v_n, \infty) \\ 0, & v \in [0, v_n) \end{cases},$$

and $v_n = \frac{1}{2}(\pi n)^{-1/\alpha}$ are zeros of the function (11.1.1). Note that each symbol $a_n(v)$ is bounded and continuous. Further each $a_n(v)$ can be uniformly approximated by smooth symbols and thus belongs to the C^*-algebra generated by the Toeplitz operators with smooth bounded defining symbols. By Theorem 10.4.9 the Toeplitz operator $T_a^{(\lambda)}$, acting on $\mathcal{A}_\lambda^2(\Pi)$, is unitary equivalent to the multiplication operator $\gamma_{a,\lambda} I$ acting on $L_2(\mathbb{R}_+)$, where the function $\gamma_{a,\lambda}(x)$ is given by (10.4.4). Thus

$$\begin{aligned}
\|T_a^{(\lambda)} - T_{a_n}^{(\lambda)}\| &= \|T_{(a-a_n)}^{(\lambda)}\| = \sup_{x \in \mathbb{R}_+} |\gamma_{(a-a_n),\lambda}(x)| \\
&= \sup_{x \in \mathbb{R}_+} \left| \frac{x^{\lambda+1}}{\Gamma(\lambda+1)} \int_0^{v_n} a(t/2)t^\lambda e^{-xt}\,dt \right| \\
&= \sup_{x \in \mathbb{R}_+} \left| \frac{x^{\lambda+1}}{\Gamma(\lambda+1)} \left(B_a(v_n)v_n^\lambda e^{-xv_n} \right. \right. \\
&\qquad \left. \left. - \int_0^{v_n} B_a(t)(\lambda - xt)t^{\lambda-1}e^{-xt}\,dt \right) \right|,
\end{aligned}$$

where

$$\begin{aligned}
B_a(t) &= \int_0^t a(v/2)\,dv \\
&= \int_0^t v^{-\beta}\sin v^{-\alpha}\,dv = \frac{1}{\alpha}\int_{t^{-\alpha}}^\infty y^{\frac{\beta-1}{\alpha}-1}\sin y\,dy.
\end{aligned}$$

Integrating by parts twice we get

$$\begin{aligned}
B_a(t) &= \frac{t^{\alpha-\beta+1}}{\alpha}\cos t^{-\alpha} - \frac{(\beta-\alpha-1)}{\alpha^2}t^{2\alpha-\beta+1}\sin t^{-\alpha} \\
&\quad - \frac{(\beta-\alpha-1)(\beta-2\alpha-1)}{\alpha^3}\int_{t^{-\alpha}}^\infty y^{\frac{\beta-1}{\alpha}-3}\sin y\,dy.
\end{aligned}$$

Thus we have

$$B_a(t) = \frac{t^{\alpha-\beta+1}}{\alpha} \cos t^{-\alpha} + O(t^{2\alpha-\beta+1}), \quad t \to 0,$$

or

$$|B_a(t)| \leq \text{const } t^{\alpha-\beta+1},$$

where the constant does not depend on $t \in (0,1)$. Thus

$$
\begin{aligned}
\|T^{(\lambda)}_{(a-a_n)}\| \;\leq\;& \text{const} \sup_{x\in\mathbb{R}_+} \left| \frac{x^{\lambda+1}}{\Gamma(\lambda+1)} \left(v_n^{\alpha-\beta+1+\lambda} e^{-xv_n} + \lambda \int_0^{v_n} t^{\alpha-\beta+\lambda} e^{-xt} dt \right.\right.\\
& \left.\left. + x \int_0^{v_n} t^{\alpha-\beta+\lambda+1} e^{-xt} dt \right) \right| \\
\leq\;& \text{const} \frac{1}{\Gamma(\lambda+1)} \sup_{x\in\mathbb{R}_+} \left(v_n^{\alpha-\beta} (v_n x)^{\lambda+1} e^{-v_n x} \right) \\
& +\text{const} \frac{1}{\Gamma(\lambda+1)} \sup_{x\in\mathbb{R}_+} \left(\frac{\lambda}{x^{\alpha-\beta}} \int_0^{v_n x} v^{\alpha-\beta+\lambda} e^{-v} dv \right) \\
& +\text{const} \frac{1}{\Gamma(\lambda+1)} \sup_{x\in\mathbb{R}_+} \left(\frac{1}{x^{\alpha-\beta}} \int_0^{v_n x} v^{\alpha-\beta+\lambda+1} e^{-v} dv \right) \\
:=\;& I_1 + I_2 + I_3.
\end{aligned}
$$

To evaluate I_1 note that

$$\sup_{x\in\mathbb{R}_+} \left((v_n x)^{\lambda+1} e^{-v_n x} \right) < \infty,$$

thus

$$I_1 \leq c_1(\lambda)\, v_n^{\alpha-\beta}.$$

Evaluating I_2 we assume first that $v_n x \leq 1$, then

$$
\begin{aligned}
I_2 \;\leq\;& \text{const} \frac{1}{\Gamma(\lambda+1)} \sup_{x\in\mathbb{R}_+} \frac{\lambda}{x^{\alpha-\beta}} \cdot (v_n x)^{\alpha-\beta+\lambda+1} \\
\leq\;& c_2(\lambda) \sup_{x\in\mathbb{R}_+} v_n^{\alpha-\beta} \cdot (v_n x)^{\lambda+1} \leq c_2(\lambda) v_n^{\alpha-\beta}.
\end{aligned}
$$

If $v_n x > 1$, that is $x > v_n^{-1}$, we have

$$I_2 \leq \text{const} \frac{\lambda}{\Gamma(\lambda+1)} v_n^{\alpha-\beta} \int_0^{\infty} v^{\alpha-\beta+\lambda} e^{-v} dv \leq c_3(\lambda) v_n^{\alpha-\beta}.$$

The evaluation of I_3 is quite analogous. Thus we have

$$\|T_a^{(\lambda)} - T_{a_n}^{(\lambda)}\| \leq c(\lambda) \cdot v_n^{\alpha-\beta},$$

where the constant $c(\lambda)$ depends on λ but does not depend on n, and where v_n tends to 0 as n tends to infinity. $\qquad\square$

11.2 Commutativity on a single Bergman space

Before passing to Toeplitz operator algebras commutative on *each* weighted Bergman space, we present results on the problem: *whether there exist C^*-algebras of Toeplitz operators commutative on a* single *Bergman space.*

We start with the Toeplitz operator C^*-algebra with identity generated by the (single) Toeplitz operator with defining symbol $a(z) = \operatorname{Re} z = x$. This algebra is commutative on *each* weighted Bergman space $\mathcal{A}^2_\lambda(\mathbb{D})$. Observe now that it contains at least one Toeplitz operator which is different from a linear combination of the initial generators.

Lemma 11.2.1. *Given a weighted Bergman space $\mathcal{A}^2_\lambda(\mathbb{D})$, the equality*

$$(T_x^{(\lambda)})^2 = T_{x^2}^{(\lambda)} + K^{(\lambda)}$$

holds, where the compact operator $K^{(\lambda)} = T_{k_\lambda(r)}^{(\lambda)}$ is the Toeplitz operator with a certain radial symbol $k_\lambda(r)$, $r = |z|$. In particular,

$$
\begin{aligned}
k_0(r) &= \frac{1}{4}\left(1 - r^2 + \ln r^2\right), \\
k_1(r) &= -\frac{1}{4}\left(1 + r^2 + \frac{2r^2}{1 - r^2}\ln r^2\right).
\end{aligned}
$$

Proof. Consider the following orthogonal (not orthonormal) basis, common for all weighted Bergman spaces $\mathcal{A}^2_\lambda(\mathbb{D})$,

$$e_n(z) = z^n, \qquad n = 0, 1, 2, \dots .$$

We prove first that the operator $T_z^{(\lambda)} T_{\bar{z}}^{(\lambda)}$ is diagonal with respect to this basis, and moreover

$$\left(T_z^{(\lambda)} T_{\bar{z}}^{(\lambda)} e_n\right)(z) = \frac{n}{n + \lambda + 1}\, e_n(z). \tag{11.2.1}$$

We have obviously

$$\left(T_z^{(\lambda)} e_n\right)(z) = e_{n+1}(z). \tag{11.2.2}$$

Calculate now

$$\left(T_{\bar{z}}^{(\lambda)} e_n\right)(z) = \frac{\lambda + 1}{\pi} \int_{\mathbb{D}} \frac{(1 - |\zeta|^2)^\lambda\, \overline{\zeta}\, \zeta^n}{(1 - z\overline{\zeta})^{\lambda+2}}\, d\nu(\zeta),$$

where $d\nu$ is the usual Lebesgue plane measure. Changing the variable $\zeta = r t$, where $r \in [0, 1]$ and t belongs to the unit circle $\mathbb{T}$, we have

$$\left(T_{\bar{z}}^{(\lambda)} e_n\right)(z) = \frac{\lambda + 1}{\pi i} \int_0^1 \int_{\mathbb{T}} \frac{(1 - r^2)^\lambda\, r^{n+2}\, t^{n-2}}{(1 - zrt^{-1})^{\lambda+2}}\, dt dr.$$

Substitute

$$(1 - zrt^{-1})^{-(\lambda+2)} = \sum_{k=0}^{\infty} (-1)^k C_k^{-(\lambda+2)} (zrt^{-1})^k,$$

where

$$(-1)^k C_k^{-(\lambda+2)} = \frac{(\lambda+2)(\lambda+3)\ldots(\lambda+k+1)}{k!}.$$

Then according to the residue theory we have

$$
\begin{aligned}
\left(T_{\bar{z}}^{(\lambda)} e_n\right)(z) &= 2(\lambda+1)(-1)^{n-1} C_{n-1}^{-(\lambda+2)} z^{n-1} \int_0^1 (1-r^2)^\lambda r^{2n+1} dr \\
&= (\lambda+1)(-1)^{n-1} C_{n-1}^{-(\lambda+2)} z^{n-1} \int_0^1 (1-r)^\lambda r^n dr \\
&= (\lambda+1)(-1)^{n-1} C_{n-1}^{-(\lambda+2)} B(\lambda+1, n+1) z^{n-1} \\
&= (\lambda+1) \frac{(\lambda+2)(\lambda+3)\ldots(\lambda+n)}{(n-1)!} \frac{\Gamma(\lambda+1)\Gamma(n+1)}{\Gamma(\lambda+n+2)} z^{n-1} \\
&= \frac{n}{n+\lambda+1} e_{n-1}(z), \qquad\qquad\qquad\qquad (11.2.3)
\end{aligned}
$$

where B and Γ are the classical Gauss functions. Now from (11.2.2) and (11.2.3) we have (11.2.1).

Recall that each Toeplitz operator with radial defining symbol, as well as the operator $T_z^{(\lambda)} T_{\bar{z}}^{(\lambda)}$, is diagonal in the basis $\{e_n(z)\}$. Moreover (see, for example, [44]) we have

$$T_z^{(0)} T_{\bar{z}}^{(0)} = T_{\tilde{k}_0(r)}^{(0)}, \qquad \text{where} \quad \tilde{k}_0(r) = 1 + \ln r^2. \qquad (11.2.4)$$

In fact, for each λ, the operator $T_z^{(\lambda)} T_{\bar{z}}^{(\lambda)}$ is the Toeplitz operator with a certain radial defining symbol $\tilde{k}_\lambda(r)$. We do not present the exact formula for $\tilde{k}_\lambda(r)$ here, but mention, for example, that

$$\tilde{k}_1(r) = -\left(1 + \frac{2r^2}{1-r^2} \ln r^2\right). \qquad (11.2.5)$$

Formulas (11.2.4) and (11.2.5) can be easily checked by comparing the spectral sequences given by (11.2.1), on the one hand, and by (10.4.1) on the other.

Calculate now

$$
\begin{aligned}
\left(T_x^{(\lambda)}\right)^2 &= \frac{1}{4}\left(T_z^{(\lambda)} + T_{\bar{z}}^{(\lambda)}\right)^2 \\
&= \frac{1}{4}\left(T_z^{(\lambda)} T_z^{(\lambda)} + T_z^{(\lambda)} T_{\bar{z}}^{(\lambda)} + T_{\bar{z}}^{(\lambda)} T_z^{(\lambda)} + T_{\bar{z}}^{(\lambda)} T_{\bar{z}}^{(\lambda)}\right) \\
&= \frac{1}{4}\left(T_{z^2}^{(\lambda)} + T_z^{(\lambda)} T_{\bar{z}}^{(\lambda)} + T_{|z|^2}^{(\lambda)} + T_{\bar{z}^2}^{(\lambda)}\right) \\
&= T_{x^2}^{(\lambda)} + \frac{1}{4}\left(T_z^{(\lambda)} T_{\bar{z}}^{(\lambda)} - T_{r^2}^{(\lambda)}\right) = T_{x^2}^{(\lambda)} + K^{(\lambda)}. \qquad (11.2.6)
\end{aligned}
$$

The operator $K^{(\lambda)}$ is obviously compact, and by the above is a Toeplitz operator with a certain radial defining symbol $k_\lambda(r)$. Moreover, by (11.2.4) and (11.2.5) we have

$$
\begin{aligned}
k_0(r) &= \frac{1}{4}\left(1 - r^2 + \ln r^2\right), \\
k_1(r) &= -\frac{1}{4}\left(1 + r^2 + \frac{2r^2}{1 - r^2}\ln r^2\right).
\end{aligned}
$$

$\square$

Let $\mathcal{A}(\mathbb{D})$ be a set (linear space) of defining symbols. By $\mathcal{T}_\lambda(\mathcal{A}(\mathbb{D}))$ we denote the C^*-algebra generated by Toeplitz operators with symbols from $\mathcal{A}(\mathbb{D})$, acting on the weighted Bergman space $\mathcal{A}_\lambda^2(\mathbb{D})$.

Theorem 11.2.2. *Given any weighted Bergman space $\mathcal{A}_{\lambda_0}^2(\mathbb{D})$, with $\lambda_0 \in (-1, +\infty)$, there exists a set of defining symbols $\mathcal{A}_{\lambda_0}(\mathbb{D})$ such that the Toeplitz operator C^*-algebra $\mathcal{T}_{\lambda_0}(\mathcal{A}_{\lambda_0}(\mathbb{D}))$ is commutative, while all other algebras $\mathcal{T}_\lambda(\mathcal{A}_{\lambda_0}(\mathbb{D}))$, $\lambda \neq \lambda_0$, are non-commutative.*

Proof. We prove the theorem for $\lambda_0 = 0$; all other λ are treated quite analogously. Introduce the set $\mathcal{A}_0(\mathbb{D})$ as the linear space generated by the following three functions:

$$
e(z) \equiv 1, \quad a(z) = \operatorname{Re} z = x, \quad a_0(z) = x^2 + k_0(r) = x^2 + \frac{1}{4}(1 - r^2 + \ln r^2). \quad (11.2.7)
$$

The algebra $\mathcal{T}_0(\mathcal{A}_0(\mathbb{D}))$ is obviously commutative, since by Lemma 11.2.1 $T_{a_0}^{(0)} = (T_x^{(0)})^2$.

To finish the proof we show that

$$
T_x^{(\lambda)} T_{a_0}^{(\lambda)} \neq T_{a_0}^{(\lambda)} T_x^{(\lambda)}, \tag{11.2.8}
$$

for each $\lambda \neq 0$. Since the operators $T_x^{(\lambda)}$ and $(T_x^{(\lambda)})^2$ obviously commute, we have that (11.2.8) is equivalent to

$$
T_x^{(\lambda)}(T_{a_0}^{(\lambda)} - (T_x^{(\lambda)})^2) \neq (T_{a_0}^{(\lambda)} - (T_x^{(\lambda)})^2)T_x^{(\lambda)},
$$

or by (11.2.6) to

$$
T_x^{(\lambda)}(T_{\tilde{k}_0}^{(\lambda)} - T_z^{(\lambda)} T_{\bar{z}}^{(\lambda)}) \neq (T_{\tilde{k}_0}^{(\lambda)} - T_z^{(\lambda)} T_{\bar{z}}^{(\lambda)})T_x^{(\lambda)}. \tag{11.2.9}
$$

By (11.2.2) and (11.2.3) we have

$$
(T_x^{(\lambda)} e_n)(z) = \frac{1}{2}\left((T_z^{(\lambda)} + T_{\bar{z}}^{(\lambda)})e_n\right)(z) = \frac{1}{2} e_{n+1}(z) + \frac{n}{2(n + \lambda + 1)} e_{n-1}(z).
$$

The Toeplitz operator $T^{(\lambda)}_{\widetilde{k}_0(r)}$ is diagonal in the basis $\{e_n(z)\}$, and moreover by (10.4.1) we have

$$(T^{(\lambda)}_{\widetilde{k}_0(r)} e_n)(z) = \gamma_{\widetilde{k}_0,\lambda}(n)\, e_n(z), \tag{11.2.10}$$

where

$$
\begin{aligned}
\gamma_{\widetilde{k}_0,\lambda}(n) &= \frac{1}{B(n+1,\lambda+1)} \int_0^1 \widetilde{k}_0(\sqrt{r})\,(1-r)^\lambda\, r^n dr \\
&= 1 + \frac{1}{B(n+1,\lambda+1)} \int_0^1 (1-r)^\lambda\, r^n \ln r\, dr \\
&= 1 + [\psi(n+1) - \psi(n+\lambda+2)].
\end{aligned}
$$

Passing to the last equality we use formula 4.253 from [86], where $\psi(x) = \frac{\Gamma'(x)}{\Gamma(x)}$ is the so called psi-function.

We prove now that the two sides of (11.2.9) are different, for example, on the first basis element $e_0(z)$. Indeed, by the above

$$I_1 := \left(T^{(\lambda)}_x (T^{(\lambda)}_{\widetilde{k}_0} - T^{(\lambda)}_z T^{(\lambda)}_{\overline{z}}) e_0 \right)(z) = \frac{1}{2}\left(1 + [\psi(1) - \psi(\lambda+2)]\right) e_1(z),$$

while

$$I_2 := \left((T^{(\lambda)}_{\widetilde{k}_0} - T^{(\lambda)}_z T^{(\lambda)}_{\overline{z}}) T^{(\lambda)}_x e_0 \right)(z) = \frac{1}{2}\left(1 + [\psi(2) - \psi(\lambda+3)] - \frac{1}{\lambda+2}\right) e_1(z).$$

Note that $\psi(x+1) = \psi(x) + \frac{1}{x}$. Thus

$$
\begin{aligned}
I_1 - I_2 &= \frac{1}{2}\left(\psi(1) - \psi(\lambda+2) - \psi(2) + \psi(\lambda+3) + \frac{1}{\lambda+2}\right) e_1(z) \\
&= -\frac{\lambda}{2(\lambda+2)}\, e_1(z),
\end{aligned}
$$

which is non-zero if and only if $\lambda \neq 0$. $\qquad\square$

11.3 Commutativity on each weighted Bergman space

We use the Berezin quantization procedure on the unit disk (see Appendix B for details). Consider the unit disk equipped with the symplectic form

$$\omega = d\mu(z) = \frac{1}{\pi}\frac{dx \wedge dy}{(1-(x^2+y^2))^2} = \frac{1}{2\pi i}\frac{d\overline{z} \wedge dz}{(1-|z|^2)^2}.$$

Given two functions $a, b \in C^\infty(\mathbb{D})$, their Poisson bracket is given by (B.2.2)

$$
\begin{aligned}
\{a, b\} &= \pi(1-(x^2+y^2))^2 \left(\frac{\partial a}{\partial y}\frac{\partial b}{\partial x} - \frac{\partial a}{\partial x}\frac{\partial b}{\partial y}\right) \\
&= 2\pi i(1 - z\overline{z})^2 \left(\frac{\partial a}{\partial z}\frac{\partial b}{\partial \overline{z}} - \frac{\partial a}{\partial \overline{z}}\frac{\partial b}{\partial z}\right). \tag{11.3.1}
\end{aligned}
$$

Recall that the Laplace-Beltrami operator has the form (B.2.3)

$$\Delta \;=\; \pi(1-(x^2+y^2))^2\left(\frac{\partial^2}{\partial x^2}+\frac{\partial^2}{\partial y^2}\right)$$

$$=\; 4\pi(1-z\bar z)^2\frac{\partial^2}{\partial z\partial\bar z}. \tag{11.3.2}$$

For each $h\in(0,1)$, introduce the weighted Bergman space $\mathcal{A}^2_h(\mathbb{D})$. For each function $a=a(z)\in C^\infty(\mathbb{D})$ consider the family of Toeplitz operators $T_a^{(h)}$ with (anti-Wick) symbol a acting on $\mathcal{A}^2_h(\mathbb{D})$, for $h\in(0,1)$, and denote by $\mathcal{T}_h$ the *-algebra generated by Toeplitz operators $T_a^{(h)}$ with defining symbols $a\in C^\infty(\mathbb{D})$.

The Wick symbols of the Toeplitz operator $T_a^{(h)}$ has the form (10.1.7)

$$\widetilde{a}_h(z,\bar z)=(\frac{1}{h}-1)\int_{\mathbb{D}}a(\zeta)\left(\frac{(1-|z|^2)(1-|\zeta|^2)}{(1-z\bar\zeta)(1-\zeta\bar z)}\right)^{\frac{1}{h}}d\mu(\zeta).$$

Let $T_a^{(h)}$ and $T_b^{(h)}$ be two Toeplitz operators with the Wick symbols $\widetilde{a}_h$ and $\widetilde{b}_h$, respectively. Then the star product $\widetilde{a}_h\star\widetilde{b}_h$ of the Wick symbols is defined as the Wick symbol of the composition $T_a^{(h)}T_b^{(h)}$ and is given by (10.1.8)

$$(\widetilde{a}_h\star\widetilde{b}_h)(z,\bar z)=(\frac{1}{h}-1)\int_{\mathbb{D}}\widetilde{a}_h(z,\bar\zeta)\,\widetilde{b}_h(\zeta,\bar z)\left(\frac{(1-|z|^2)(1-|\zeta|^2)}{(1-z\bar\zeta)(1-\zeta\bar z)}\right)^{\frac{1}{h}}d\mu(\zeta).$$

The correspondence principle says that under $\hbar=\frac{h}{2\pi}\to 0$ one has

$$\widetilde{a}_h(z,\bar z)\;=\; a(z,\bar z)+O(\hbar),$$

$$(\widetilde{a}_h\star\widetilde{b}_h-\widetilde{b}_h\star\widetilde{a}_h)(z,\bar z)\;=\; i\hbar\{a,b\}+O(\hbar^2). \tag{11.3.3}$$

The last formula immediately leads to certain information about the defining symbols which might generate a commutative Toeplitz operator algebra.

Theorem 11.3.1. *Let $\mathcal{A}(\mathbb{D})$ be a subalgebra of $C^\infty(\mathbb{D})$ such that for each $h\in(0,1)$ the Toeplitz operator algebra $\mathcal{T}_h(\mathcal{A}(\mathbb{D}))$, i.e., the C^*-algebra generated by the operators $T_a^{(h)}$, with $a\in\mathcal{A}(\mathbb{D})$, acting on the weighted Bergman space $\mathcal{A}^2_h(\mathbb{D})$, is commutative. Then $\mathcal{A}(\mathbb{D})$ is a commutative Lie algebra, i.e., $\{a,b\}=0$ for all $a,b\in\mathcal{A}(\mathbb{D})$.*

Proof. The commutativity of the Toeplitz operator algebra $\mathcal{T}_h(\mathcal{A}(\mathbb{D}))$ implies that the subalgebra $\widetilde{\mathcal{A}}'_h=\{\widetilde{a}_h:a\in\mathcal{A}(\mathbb{D})\}$ of the algebra $\widetilde{\mathcal{A}}_h$ is commutative as well. By (11.3.3) for all $a,b\in\mathcal{A}(\mathbb{D})$ we have

$$\frac{i}{2\pi}\{a,b\}=\lim_{h\to 0}\frac{1}{h}(\widetilde{a}_h\star\widetilde{b}_h-\widetilde{b}_h\star\widetilde{a}_h)=0. \qquad\square$$

The geometric information which follows from (11.3.3) and Theorem 11.3.1 is insufficient for our purposes. Our main results will follow from the second and third terms of the asymptotic expansion of the commutator of two Wick symbols, which are given by the following theorem.

Theorem 11.3.2. *For any pair $a = a(z, \overline{z})$ and $b = b(z, \overline{z})$ of six times continuously differentiable functions the following three-term asymptotic expansion formula holds,*

$$
\widetilde{a}_h \star \widetilde{b}_h - \widetilde{b}_h \star \widetilde{a}_h = \frac{ih}{2\pi} \{a, b\}
$$
$$
+ \frac{h^2}{2} \left[\frac{i}{8\pi^2} \left(\Delta\{a, b\} + \{a, \Delta b\} + \{\Delta a, b\} \right) + \frac{i}{\pi} \{a, b\} \right]
$$
$$
+ h^3 \left[\frac{i}{192\pi^3} \left(\{\Delta a, \Delta b\} + \{a, \Delta^2 b\} + \{\Delta^2 a, b\} \right. \right.
$$
$$
+ \Delta^2\{a, b\} + \Delta\{a, \Delta b\} + \Delta\{\Delta a, b\} \left. \right)
$$
$$
+ \frac{7i}{48\pi^2} \left(\Delta\{a, b\} + \{a, \Delta b\} + \{\Delta a, b\} \right) + \frac{i}{2\pi} \{a, b\} \left. \right] + o(h^3)
$$
$$
= i\hbar\,\{a, b\} + i\frac{\hbar^2}{4} \left(\Delta\{a, b\} + \{a, \Delta b\} + \{\Delta a, b\} + 8\pi\{a, b\} \right)
$$
$$
+ i\frac{\hbar^3}{24} \left[\{\Delta a, \Delta b\} + \{a, \Delta^2 b\} + \{\Delta^2 a, b\} + \Delta^2\{a, b\} \right.
$$
$$
+ \Delta\{a, \Delta b\} + \Delta\{\Delta a, b\} + 28\pi \left(\Delta\{a, b\} + \{a, \Delta b\} + \{\Delta a, b\} \right)
$$
$$
+ 96\pi^2\{a, b\} \left. \right] + o(\hbar^3),
$$

where the Poisson bracket $\{\,,\,\}$ and the Laplace-Beltrami operator Δ in coordinates $(z, \overline{z})$ are given by (11.3.1) and by (11.3.2), respectively.

The proof of the theorem is given in Section B.4.

Corollary 11.3.3. *Let $\mathcal{A}(\mathbb{D})$ be a subalgebra of $C^\infty(\mathbb{D})$ such that for each $h \in (0, 1)$ the Toeplitz operator algebra $\mathcal{T}_h(\mathcal{A}(\mathbb{D}))$ is commutative. Then for all $a, b \in \mathcal{A}(\mathbb{D})$ we have*

$$
\{a, b\} = 0, \tag{11.3.4}
$$
$$
\{a, \Delta b\} + \{\Delta a, b\} = 0, \tag{11.3.5}
$$
$$
\{\Delta a, \Delta b\} + \{a, \Delta^2 b\} + \{\Delta^2 a, b\} = 0. \tag{11.3.6}
$$

11.4　First term: common gradient and level lines

We start by introducing the symbol classes to be used. Let $\mathcal{A}(\mathbb{D})$ be a linear space of (smooth) functions. Denote by $\mathcal{T}(\mathcal{A}(\mathbb{D})) = \{\mathcal{T}_h(\mathcal{A}(\mathbb{D}))\}_h$ the family of C^*-algebras

$\mathcal{T}_h(\mathcal{A}(\mathbb{D}))$ generated by Toeplitz operators with defining symbols from $\mathcal{A}(\mathbb{D})$ and acting on the weighted Bergman spaces $\mathcal{A}_h^2(\mathbb{D})$.

We will call $\mathcal{A}(\mathbb{D})$ a generating space of symbols. In principle the same family of Toeplitz operator algebras can be generated by different generating spaces, i.e., there may exist different $\mathcal{A}_1(\mathbb{D})$ and $\mathcal{A}_2(\mathbb{D})$ such that $\mathcal{T}(\mathcal{A}_1(\mathbb{D})) = \mathcal{T}(\mathcal{A}_2(\mathbb{D}))$. The important thing here is that the generating set must be the same for all values of the parameter h.

We will call the family $\mathcal{T} = \{\mathcal{T}_h\}_h$ a single generated family if among its generating spaces there is a two-dimensional space $\mathcal{A}_a(\mathbb{D})$ which is generated by $e(z) \equiv 1$ and a real-valued function $a(z)$.

Consider in this connection the following two examples.

Example 11.4.1. Let $a(z) = \operatorname{Re} z = x$. Consider the corresponding single generated family of Toeplitz operator algebras $\mathcal{T}(\mathcal{A}_x(\mathbb{D})) = \{\mathcal{T}_h(\mathcal{A}_x(\mathbb{D}))\}_h$. As it was shown in the proof of Theorem 11.2.2, the algebra $\mathcal{T}_{h_0}(\mathcal{A}_x(\mathbb{D}))$ with $h_0 = 1/2$ is generated as well by $\mathcal{A}(\mathbb{D})$ which is the linear span of the three functions of (11.2.7). At the same time none of the other algebras $\mathcal{T}_h(\mathcal{A}_x(\mathbb{D}))$ is generated by this $\mathcal{A}(\mathbb{D})$.

Example 11.4.2. Let $a(z) = |z\bar{z}|^{1/2} = r$. Consider the corresponding single generated family of Toeplitz operator algebras $\mathcal{T}(\mathcal{A}_r(\mathbb{D})) = \{\mathcal{T}_h(\mathcal{A}_r(\mathbb{D}))\}_h$. It can be proved that the family $\mathcal{T}(\mathcal{A}_r(\mathbb{D}))$ can be generated as well by many other linear spaces of radial functions $a(r) \in C([0,1])$ which contain $e(z) \equiv 1$ and are closed under complex conjugation, and in particular it can be generated by the whole $C([0,1])$. At the same time, it can be proved that the family $\mathcal{T} = \{\mathcal{T}_h\}_h$, which is generated by the linear space of all smooth radial functions (not necessarily continuous at 1), is wider and can not be generated by $\mathcal{A}_r(\mathbb{D})$.

In what follows we will consider families of commutative Toeplitz operator algebras which contain among their generating spaces the ones given by Definition 11.4.3. To introduce them we need to consider the notion of the jet of a function, see, for example, [119, 174]. Given two complex-valued smooth functions f and g defined in a neighborhood of a point $z \in \mathbb{D}$, we say that they have the same jet of order k at z if their real partial derivatives at z up to order k are equal. It is easy to see that such a relation does not depend on the coordinate system and that it defines an equivalence relation. The corresponding equivalence class of a function f at z is denoted by $j_z^k(f)$ and is called the k-th order jet of f at z. Furthermore, given a complex vector space $\mathcal{A}(\mathbb{D})$ of smooth functions, we denote with $J_z^k(\mathcal{A}(\mathbb{D}))$ the space of k-jets at z of the elements in $\mathcal{A}(\mathbb{D})$. We observe that $J_z^k(\mathcal{A}(\mathbb{D}))$ is a finite dimensional complex vector space.

In what follows, for a differentiable function $f : \mathbb{D} \to \mathbb{C}$ we will say that $z \in \mathbb{D}$ is a non-singular point of f if $df_z \neq 0$.

The symbol classes that we will consider are given in the next definition.

Definition 11.4.3. Let $\mathcal{A}(\mathbb{D})$ be a complex vector space of smooth functions. We will say that $\mathcal{A}(\mathbb{D})$ is k-rich if it is closed under complex conjugation and the following conditions are satisfied:

(i) there is a finite set S such that for every $z \in \mathbb{D} \setminus S$ at least one element of $\mathcal{A}(\mathbb{D})$ is non-singular at z,

(ii) for every point $z \in \mathbb{D} \setminus S$ and $l = 0, \ldots, k$, the space of jets $J_z^l(\mathcal{A}(\mathbb{D}))$ has complex dimension at least $l + 1$.

Observe that k-richness implies l-richness for $l \leq k$. The following result ensures that k-richness, for each $k \geq 2$, excludes from consideration commutative Toeplitz C^*-algebras with identity generated by a single self-adjoint Toeplitz operator.

Lemma 11.4.4. *Let $\mathcal{A}(\mathbb{D})$ be a 2-rich space of smooth functions. Then, there is no open set V in $\mathbb{D}$ such that the restriction $\mathcal{A}(\mathbb{D})|_V$ is generated by a single real-valued function $a \in \mathcal{A}(\mathbb{D})$ and $e(z) \equiv 1$.*

Proof. Suppose such real-valued function a exists for some open set V. Then for every $b \in \mathcal{A}(\mathbb{D})|_V$ there exist $c_1, c_2 \in \mathbb{C}$ such that $b = c_1 a + c_2 e$. Thus $j_z^2(b) = c_1 j_z^2(a) + c_2 j_z^2(e)$, for each $z \in V$, or the complex dimension of $J_z^2(\mathcal{A}(\mathbb{D}))$ is at most 2. $\qquad\square$

The previous lemma shows the importance of considering a symbol set $\mathcal{A}(\mathbb{D})$ which is at least 2-rich. Having such a set $\mathcal{A}(\mathbb{D})$, assume now that the Toeplitz operator algebra $\mathcal{T}_h(\mathcal{A}(\mathbb{D}))$ is commutative for each $h \in (0, 1)$. Then by Corollary 11.3.3 we have that for all $a, b \in \mathcal{A}(\mathbb{D})$ the equalities (11.3.4), (11.3.5), and (11.3.6) must be satisfied.

Note that as the set $\mathcal{A}(\mathbb{D})$ is closed under complex conjugation, it is sufficient to consider the conditions (11.3.4), (11.3.5), and (11.3.6) for the real-valued functions only.

Each real-valued function $a \in \mathcal{A}(\mathbb{D})$ with a non-vanishing gradient in an open set $U \in \mathbb{D}$, has in U two systems of mutually orthogonal smooth lines, the system of level lines and the system of gradient lines. Given any such a pair, a function a and an open set U, it is easy to see that the above two systems of lines can be parameterized to be a new orthogonal coordinate system (u, v) in U. The level lines and the gradient lines of the function a in the coordinates (u, v) are given respectively as

$$u = u_0 = \text{const} \qquad \text{and} \qquad v = v_0 = \text{const}.$$

Thus, in particular, we have $a = a(u) = a(u(x, y))$.

The coordinate systems (u, v) and (x, y) are connected by

$$u = u(x, y), \quad v = v(x, y), \qquad \text{or} \qquad x = x(u, v), \quad y = y(u, v),$$

with

$$D = \frac{\partial x}{\partial u}\frac{\partial y}{\partial v} - \frac{\partial x}{\partial v}\frac{\partial y}{\partial u} \neq 0,$$

and the orthogonality of the coordinate system (u, v) is equivalent to

$$\frac{\partial x}{\partial u}\frac{\partial x}{\partial v} + \frac{\partial y}{\partial u}\frac{\partial y}{\partial v} \equiv 0.$$

In the coordinates (u, v) the metric, the symplectic form, and the Poisson brackets have respectively the form

$$ds^2 = \widetilde{g}_{11}(u, v)du^2 + \widetilde{g}_{22}(u, v)dv^2,$$

where

$$\widetilde{g}_{11} = g(x, y)\left[\left(\frac{\partial x}{\partial u}\right)^2 + \left(\frac{\partial y}{\partial u}\right)^2\right], \qquad \widetilde{g}_{22} = g(x, y)\left[\left(\frac{\partial x}{\partial v}\right)^2 + \left(\frac{\partial y}{\partial v}\right)^2\right],$$

with $g = g(x, y) = \pi^{-1}(1 - (x^2 + y^2))^{-2}$, and

$$\begin{aligned}
\omega &= g(x, y)\, D\, du \wedge dv, \\
\{f_1, f_2\} &= g^{-1}(x, y)\, D\left(\frac{\partial f_1}{\partial v}\frac{\partial f_2}{\partial u} - \frac{\partial f_1}{\partial u}\frac{\partial f_2}{\partial v}\right). \qquad (11.4.1)
\end{aligned}$$

The geometric information contained in the first term of the asymptotic expansion of a commutator, or equivalently in the condition (11.3.4), is given by the next lemma.

Lemma 11.4.5. *Let $\mathcal{A}(\mathbb{D})$ be a 2-rich space of smooth functions which generates for each $h \in (0, 1)$ the commutative C^*-algebra $\mathcal{T}_h(\mathcal{A}(\mathbb{D}))$ of Toeplitz operators. Then all real-valued functions in $\mathcal{A}(\mathbb{D})$ have (globally) the same set of level lines and the same set of gradient lines.*

Proof. Fix a real-valued function $a \in \mathcal{A}(\mathbb{D})$ and the local orthogonal coordinate system as above. For any other (real-valued) function $b \in \mathcal{A}(\mathbb{D})$ condition (11.3.4) implies

$$\{a, b\} = -g^{-1}D\, a'(u)\frac{\partial b}{\partial v} \equiv 0,$$

or

$$\frac{\partial b}{\partial v} \equiv 0;$$

that is, the function b has (in U_0) the *same* level lines, and thus has the *same* gradient lines as the function a. $\qquad \square$

11.5 Second term: gradient lines are geodesics

We continue to consider the above (local) orthogonal coordinate system (u, v). All real-valued functions from $\mathcal{A}(\mathbb{D})$ have the same set of level lines:

$$u = u_0 = \mathrm{const}, \quad \text{or} \quad \begin{cases} x = x(u_0, v) \\ y = y(u_0, v) \end{cases},$$

and the same set of gradient lines:

$$v = v_0 = \text{const}, \quad \text{or} \quad \begin{cases} x = x(u, v_0) \\ y = y(u, v_0) \end{cases}. \tag{11.5.1}$$

The Laplace-Beltrami operator in the coordinates (u, v) has the form (see, for example, [151], p. 87)

$$\Delta = \widetilde{g}^{11} \left(\frac{\partial^2}{\partial u^2} - \widetilde{\Gamma}^1_{11} \frac{\partial}{\partial u} - \widetilde{\Gamma}^2_{11} \frac{\partial}{\partial v} \right) + \widetilde{g}^{22} \left(\frac{\partial^2}{\partial v^2} - \widetilde{\Gamma}^1_{22} \frac{\partial}{\partial u} - \widetilde{\Gamma}^2_{22} \frac{\partial}{\partial v} \right),$$

where the matrix $(\widetilde{g}^{ij})$ is inverse to $(\widetilde{g}_{ij})$ and $\widetilde{\Gamma}^k_{ij}$ are the Schwarz-Christoffel symbols on (u, v).

For any function $c = c(u) \in \mathcal{A}(\mathbb{D})$ we have

$$\Delta c = c'' \widetilde{g}^{11} - c' \left(\widetilde{g}^{11} \widetilde{\Gamma}^1_{11} + \widetilde{g}^{22} \widetilde{\Gamma}^1_{22} \right). \tag{11.5.2}$$

The vanishing of the second term of the asymptotic in a commutator, or equivalently the condition (11.3.5), leads to the following theorem.

Theorem 11.5.1. *Let $\mathcal{A}(\mathbb{D})$ be a 2-rich space of smooth functions which generates for each $h \in (0,1)$ the commutative C^*-algebra $\mathcal{T}_h(\mathcal{A}(\mathbb{D}))$ of Toeplitz operators. Then the common gradient lines of all real-valued functions in $\mathcal{A}(\mathbb{D})$ are geodesics in the hyperbolic geometry of the unit disk $\mathbb{D}$.*

Proof. Given two real-valued functions $a = a(u)$, $b = b(u) \in \mathcal{A}(\mathbb{D})$, the condition (11.3.5) is equivalent to

$$\begin{aligned} 0 &\equiv a' \cdot \frac{\partial \Delta b}{\partial v} - b' \cdot \frac{\partial \Delta a}{\partial v} \\ &= (a'b'' - b'a'') \frac{\partial \widetilde{g}^{11}}{\partial v} - (a'b' - b'a') \frac{\partial}{\partial v} \left(\widetilde{g}^{11} \widetilde{\Gamma}^1_{11} + \widetilde{g}^{22} \widetilde{\Gamma}^1_{22} \right) \\ &= (a'b'' - b'a'') \frac{\partial \widetilde{g}^{11}}{\partial v}. \end{aligned}$$

Note that vanishing of $a'b'' - b'a''$ in an open subset of U is equivalent to the property that in this subset one of the functions, a or b, is a linear combination of the other and $e(z) \equiv 1$, which is impossible by Lemma 11.4.4. By Lemma 11.4.4 we can change, if necessary, in different parts of U the functions a and b from $\mathcal{A}(\mathbb{D})$ in order to have $a'b'' - b'a'' \neq 0$; then

$$\frac{\partial \widetilde{g}^{11}}{\partial v} = -\widetilde{g}^2_{11} \frac{\partial \widetilde{g}_{11}}{\partial v} \equiv 0, \tag{11.5.3}$$

or

$$\frac{\partial}{\partial v} \left\langle \frac{\partial}{\partial u}, \frac{\partial}{\partial u} \right\rangle \equiv 0, \tag{11.5.4}$$

where $\langle X_1, X_2 \rangle = ds^2(X_1, X_2)$ is the inner product of the vector fields X_1 and X_2.

Consider now any gradient line γ given by (11.5.1). The Frenet frame (e_1, e_2) of γ is given by

$$e_1 = \left\| \frac{\partial}{\partial u} \right\|^{-1} \frac{\partial}{\partial u}, \qquad e_2 = \left\| \frac{\partial}{\partial v} \right\|^{-1} \frac{\partial}{\partial v}.$$

By [118], page 78, the geodesic curvature $\kappa_\gamma(u)$ of γ is calculated as

$$\kappa_\gamma(u) = \left\| \frac{\partial}{\partial u} \right\|^{-1} \left\langle e_2, \nabla_{\frac{\partial}{\partial u}} e_1 \right\rangle.$$

Using standard properties of the connection ∇, we have

$$\nabla_{\frac{\partial}{\partial u}} e_1 = \left(\frac{\partial}{\partial u} \left\| \frac{\partial}{\partial u} \right\|^{-1} \right) \cdot \frac{\partial}{\partial u} + \left\| \frac{\partial}{\partial u} \right\|^{-1} \cdot \nabla_{\frac{\partial}{\partial u}} \frac{\partial}{\partial u}.$$

Thus

$$\kappa_\gamma(u) = \left\| \frac{\partial}{\partial u} \right\|^{-1} \left\langle e_2, \nabla_{\frac{\partial}{\partial u}} e_1 \right\rangle = \left\| \frac{\partial}{\partial u} \right\|^{-2} \left\| \frac{\partial}{\partial v} \right\|^{-1} \left\langle \frac{\partial}{\partial v}, \nabla_{\frac{\partial}{\partial u}} \frac{\partial}{\partial u} \right\rangle. \qquad (11.5.5)$$

By the Koszul formula (see, for example, [151], page 61) we have

$$2 \left\langle \nabla_{\frac{\partial}{\partial u}} \frac{\partial}{\partial u}, \frac{\partial}{\partial v} \right\rangle = \frac{\partial}{\partial u} \left\langle \frac{\partial}{\partial u}, \frac{\partial}{\partial v} \right\rangle + \frac{\partial}{\partial u} \left\langle \frac{\partial}{\partial v}, \frac{\partial}{\partial u} \right\rangle - \frac{\partial}{\partial v} \left\langle \frac{\partial}{\partial u}, \frac{\partial}{\partial u} \right\rangle$$
$$- \left\langle \frac{\partial}{\partial u}, \left[\frac{\partial}{\partial u}, \frac{\partial}{\partial v} \right] \right\rangle + \left\langle \frac{\partial}{\partial u}, \left[\frac{\partial}{\partial v}, \frac{\partial}{\partial u} \right] \right\rangle + \left\langle \frac{\partial}{\partial v}, \left[\frac{\partial}{\partial u}, \frac{\partial}{\partial u} \right] \right\rangle,$$

where $[X_1, X_2]$ is the commutator of the vector fields X_1 and X_2.

Taking into account that

$$\left\langle \frac{\partial}{\partial u}, \frac{\partial}{\partial v} \right\rangle = 0, \quad \left[\frac{\partial}{\partial u}, \frac{\partial}{\partial v} \right] = -\left[\frac{\partial}{\partial v}, \frac{\partial}{\partial u} \right] = 0, \quad \left[\frac{\partial}{\partial u}, \frac{\partial}{\partial u} \right] = 0,$$

we have

$$\left\langle \nabla_{\frac{\partial}{\partial u}} \frac{\partial}{\partial u}, \frac{\partial}{\partial v} \right\rangle = -\frac{1}{2} \frac{\partial}{\partial v} \left\langle \frac{\partial}{\partial u}, \frac{\partial}{\partial u} \right\rangle. \qquad (11.5.6)$$

Finally, from (11.3.4), (11.5.4), (11.5.5), and (11.5.6) it follows that $\kappa_\gamma \equiv 0$, and thus (see, for example, [118], Proposition 4.3.2) the system of gradient lines consists of geodesics. $\qquad \square$

11.6   Curves with constant geodesic curvature

As was shown in the previous section, the geodesic curvature turns out to be an important invariant to study the geometry of the gradient lines. We show that it is also fundamental to understand the nature of the level lines.

We start with the description of the curves in $\mathbb{D}$ whose geodesic curvature is constant and their characterization by the curvature value.

Recall that the (canonical) Euclidean or flat metric on $\mathbb{D}$ is given by

$$(ds^2)^E = \langle \cdot, \cdot \rangle^E = dx^2 + dy^2,$$

while the hyperbolic metric is given by the expression

$$ds^2 = \frac{dx^2 + dy^2}{\pi(1 - (x^2 + y^2))^2},$$

as introduced in (9.1.1). Considering these two metrics on $\mathbb{D}$, we have two Riemannian manifolds with $\mathbb{D}$ as their base manifold. As a corollary, we need to distinguish the usual, Euclidean, norm in $\mathbb{C}$ and the hyperbolic norm of vectors tangent to $\mathbb{D}$. To do this we use $|\cdot|$ for the former and $\|\cdot\|$ for the later, as it was already used at the end of the previous section.

Furthermore, each one of the above manifolds has a different way to compute the acceleration of a curve in $\mathbb{D}$ which, with the hyperbolic metric and when defined as in [151], for a curve γ is given by the covariant derivative $\nabla_{\gamma'}\gamma'$, where ∇ is the Levi-Civita or Riemannian connection of the hyperbolic metric (see Theorem 11 of [151]). Similarly, the acceleration of a curve γ for the Euclidean geometry is given by $\nabla^E_{\gamma'}\gamma'$, where ∇^E is the Levi-Civita connection of the Euclidean metric. Since the Euclidean and hyperbolic metrics have different Levi-Civita connections, the corresponding accelerations are not the same. Because of this, if γ is a C^2 curve in $\mathbb{D}$, we will denote by γ'' and $\ddot{\gamma}$ the accelerations of γ for the Euclidean and hyperbolic metrics, respectively. Lemma 14 in [151] provides the Schwarz-Christoffel symbols of the Euclidean metric, which can be used to obtain the general formula for the corresponding acceleration of a curve. From this it follows that

$$\gamma'' = (x'', y''),$$

where $\gamma = (x, y)$ and x'', y'' simply denote the second derivatives of the real-valued one-variable functions x, y. On the other hand, we have for every t in the domain of γ,

$$\ddot{\gamma}(t) \quad = \quad (x''(t) + \Gamma^1_{11}(\gamma(t))x'(t)^2 + 2\Gamma^1_{12}(\gamma(t))x'(t)y'(t) + \Gamma^1_{22}(\gamma(t))y'(t)^2,$$
$$y''(t) + \Gamma^2_{11}(\gamma(t))x'(t)^2 + 2\Gamma^2_{12}(\gamma(t))x'(t)y'(t) + \Gamma^2_{22}(\gamma(t))y'(t)^2),$$

where the functions Γ^l_{jk} denote the Schwarz-Christoffel symbols of the hyperbolic metric on $\mathbb{D}$ with respect to the coordinates given by the real and imaginary part of complex numbers. From now on, we will refer to these coordinates as the natural coordinates in $\mathbb{D}$.

The next result states that for a curve in $\mathbb{D}$ that passes through the origin, the Euclidean and hyperbolic geodesic curvatures at the origin differ only by a multiplicative constant which does not depend on the curve. Of course the constant would vary if we move from the origin, since the Euclidean and hyperbolic geodesic curvatures are not the same. However, in our computations we will be dealing with curves through the origin. This can be done since the unit disk $\mathbb{D}$ is homogeneous as a Riemannian manifold with respect to the Möbius transformations.

We denote by κ_γ and κ_γ^E the geodesic curvatures of γ for the hyperbolic and Euclidean metrics, respectively.

Lemma 11.6.1. *Let $\gamma : I \to \mathbb{D}$ be a C^2 curve, where I is an open interval of $\mathbb{R}$. Suppose that there exists $t_0 \in I$ such that $\gamma(t_0) = 0$, the origin in $\mathbb{D}$. Then*

$$\kappa_\gamma(t_0) = \sqrt{\pi}\,\kappa_\gamma^E(t_0).$$

Proof. Using our expression for the hyperbolic metric g and Proposition 13 on page 62 from [151] we find, with respect to the natural coordinates in $\mathbb{D}$, the following expressions for the components of the metric and the Schwarz-Christoffel symbols:

$$g_{jk}(0) = \frac{1}{\pi}\delta_{jk}, \tag{11.6.1}$$
$$\Gamma^l_{jk}(0) = 0,$$

for every $j, k, l = 1, 2$. In particular at the origin of $\mathbb{D}$ we have

$$\|\cdot\| = \frac{1}{\sqrt{\pi}}|\cdot|. \tag{11.6.2}$$

Thus the formulas for the acceleration of a curve in $\mathbb{D}$, for both the hyperbolic and the Euclidean metrics, imply

$$\gamma''(t_0) = \ddot{\gamma}(t_0). \tag{11.6.3}$$

On the other hand, by Definition 4.2. from [118], it follows that the Frenet frame for γ at t_0 with respect to the Euclidean metric is given by

$$e_1^E(t_0) = \frac{\gamma'(t_0)}{|\gamma'(t_0)|},$$
$$e_2^E(t_0) = \frac{(-y'(t_0), x'(t_0))}{|\gamma'(t_0)|} = \frac{i\gamma'(t_0)}{|\gamma'(t_0)|},$$

and by (11.6.1) the Frenet frame for γ at t_0 with respect to the hyperbolic metric is

$$e_1(t_0) = \frac{\gamma'(t_0)}{\|\gamma'(t_0)\|} = \sqrt{\pi}\,e_1^E(t_0),$$
$$e_2(t_0) = \frac{(-y'(t_0), x'(t_0))}{\|\gamma'(t_0)\|} = \frac{i\gamma'(t_0)}{\|\gamma'(t_0)\|} = \sqrt{\pi}\,e_2^E(t_0).$$

Hence, the formula for the geodesic curvature considered in Section 11.5 (see also [118]) yields for the Euclidean metric

$$
\begin{aligned}
\kappa_\gamma^E(t_0) &= \frac{1}{|\gamma'(t_0)|} \left\langle e_2^E(t_0), (\nabla^E_{\gamma'(t_0)} \frac{1}{|\gamma'|} \gamma')(t_0) \right\rangle^E \\
&= \frac{1}{|\gamma'(t_0)|^2} \left\langle i\gamma'(t_0), \frac{1}{|\gamma'(t_0)|} (\nabla^E_{\gamma'(t_0)} \gamma')(t_0) \right\rangle^E \\
&= \frac{1}{|\gamma'(t_0)|^3} \langle i\gamma'(t_0), \gamma''(t_0) \rangle^E \\
&= \frac{1}{|\gamma'(t_0)|^3} \langle i\gamma'(t_0), \gamma''(t_0) \rangle^E ,
\end{aligned}
$$

where we have used for the second identity the derivation properties of ∇^E together with the fact that $e_2(t_0)$ and $\gamma'(t_0)$ are perpendicular. Correspondingly, we have for the hyperbolic metric:

$$
\begin{aligned}
\kappa_\gamma(t_0) &= \frac{1}{\|\gamma'(t_0)\|} \left\langle e_2(t_0), (\nabla_{\gamma'(t_0)} \frac{1}{\|\gamma'\|} \gamma')(t_0) \right\rangle \\
&= \frac{1}{\|\gamma'(t_0)\|^2} \left\langle i\gamma'(t_0)), \frac{1}{\|\gamma'(t_0)\|} (\nabla_{\gamma'(t_0)} \gamma')(t_0) \right\rangle \\
&= \frac{1}{\|\gamma'(t_0)\|^3} \langle i\gamma'(t_0), \ddot\gamma(t_0) \rangle \\
&= \frac{\sqrt{\pi}}{|\gamma'(t_0)|^3} \langle i\gamma'(t_0), \gamma''(t_0) \rangle^E ,
\end{aligned}
$$

where the fourth identity follows from equations (11.6.1), (11.6.2) and (11.6.3). The result is then a consequence of the formulas for these two geodesic curvatures. $\qquad\square$

The previous lemma allows us to obtain a formula for the geodesic curvature of curves in $\mathbb{D}$ with the hyperbolic metric. The following formula was first stated, to the best of our knowledge, by Carathéodory [31].

Theorem 11.6.2. *Let $\gamma : I \to \mathbb{D}$ be a C^2 curve, where I is an open interval in $\mathbb{R}$. Then the geodesic curvature of γ with respect to the hyperbolic metric in $\mathbb{D}$ is given by the expression:*

$$
\kappa_\gamma = \sqrt{\pi} \left(2 \frac{\langle i\gamma, \gamma' \rangle^E}{|\gamma'|} + \frac{(1 - |\gamma|^2) \langle i\gamma', \gamma'' \rangle^E}{|\gamma'|^3} \right).
$$

Proof. Let $t_0 \in I$ be given. Consider the Möbius transformation given by

$$
\phi(z) = \frac{z - \gamma(t_0)}{1 - \overline{\gamma(t_0)}z},
$$

which maps $\mathbb{D}$ onto $\mathbb{D}$ taking $\gamma(t_0)$ into the origin. Let us denote $\alpha = \phi \circ \gamma$. From Lemma 11.6.1 and the fact that ϕ (being a hyperbolic isometry) preserves geodesic curvatures, it follows that

$$\kappa_\gamma(t_0) = \kappa_\alpha(t_0) = \sqrt{\pi}\kappa_\alpha^E(t_0). \tag{11.6.4}$$

On the other hand, using the arguments from the proof of Lemma 11.6.1 we conclude that

$$\kappa_\alpha^E = \frac{\langle i\alpha', \alpha''\rangle^E}{|\alpha'|^3}. \tag{11.6.5}$$

Easy calculations show that

$$\alpha'(t_0) = \frac{\gamma'(t_0)}{1 - |\gamma(t_0)|^2},$$

$$\alpha''(t_0) = \frac{\gamma''(t_0)(1 - |\gamma(t_0)|^2) + 2\overline{\gamma(t_0)}\gamma'(t_0)^2}{(1 - |\gamma(t_0)|^2)^2}.$$

Substituting in (11.6.5) we have

$$\kappa_\alpha^E(t_0) = \frac{(1 - |\gamma(t_0)|^2)^3}{|\gamma'(t_0)|^3} \left\langle \frac{i\gamma'(t_0)}{1 - |\gamma(t_0)|^2}, \frac{\gamma''(t_0)(1 - |\gamma(t_0)|^2) + 2\overline{\gamma(t_0)}\gamma'(t_0)^2}{(1 - |\gamma(t_0)|^2)^2} \right\rangle^E$$

$$= \frac{1}{|\gamma'(t_0)|^3} \left\langle i\gamma'(t_0), \gamma''(t_0)(1 - |\gamma(t_0)|^2) + 2\overline{\gamma(t_0)}\gamma'(t_0)^2 \right\rangle^E$$

$$= 2\frac{\langle i\gamma(t_0), \gamma'(t_0)\rangle^E}{|\gamma'(t_0)|} + \frac{(1 - |\gamma(t_0)|^2)\,\langle i\gamma'(t_0), \gamma''(t_0)\rangle^E}{|\gamma'(t_0)|^3}.$$

The final result now follows from (11.6.4). $\square$

In what follows we will use the notion of circular arcs, understanding them in the extended sense of the Riemann sphere geometry. In particular, every segment of a straight line is considered to be a circular arc.

Theorem 11.6.3. *Every circular arc contained in $\mathbb{D}$ has constant hyperbolic geodesic curvature.*

Proof. Every circular arc contained in $\mathbb{D}$ can be parameterized by either one of the following curves:

$$\gamma_1 : I_1 \to \mathbb{D} \qquad\qquad \gamma_2 : I_2 \to \mathbb{D}$$
$$t \mapsto r_1 e^{r_2 it} + z \qquad\qquad t \mapsto tw + ir_3 w,$$

where $r_1, r_2, r_3 \in \mathbb{R}$ $(r_1, r_2 \neq 0)$, $z, w \in \mathbb{C}$ $(w \neq 0)$ and I_1, I_2 are suitable open intervals in $\mathbb{R}$. For the expression of γ_2 we have used the fact that we can translate the origin in the parameter t to assume that at time $t = 0$ the curve γ_2 passes through the point in the line segment nearest to the origin. Also observe that we require the conditions

$$|z| + |r_1| \leq 1, \qquad \text{or} \qquad ||z| - |r_1|| < 1 \tag{11.6.6}$$

$$|r_3 w| \;\; < \;\; 1, \tag{11.6.7}$$

for the circular arcs to have non-empty intersection with $\mathbb{D}$.

Then a straightforward application of the formula from Theorem 11.6.2 shows that the (hyperbolic) geodesic curvatures of the circular arcs defined by γ_1 and γ_2 are given by

$$\kappa_{\gamma_1}(t) \;\; = \;\; \frac{\text{sign}(r_2)\sqrt{\pi}(r_1^2 - |z|^2 + 1)}{|r_1|}, \tag{11.6.8}$$

$$\kappa_{\gamma_2}(t) \;\; = \;\; -2\sqrt{\pi}\,|w|\,r_3, \tag{11.6.9}$$

for every t in I_1, I_2, respectively. In particular, the geodesic curvatures do not depend on t and thus the geodesic curvature is constant for every circular arc. $\quad\square$

The above formulas for the geodesic curvature of circular arcs allow us to describe the cycles in $\mathbb{D}$ by the value of their curvature as follows.

1. Horocycles: These are given by circular arcs contained in $\mathbb{D}$ which are tangent to the boundary of $\mathbb{D}$. It is easy to check that such circular arcs are obtained from γ_1 as in the proof of Theorem 11.6.3 precisely when $|z| = 1 - |r_1|$. Hence equation (11.6.8) shows that every horocycle has (constant) geodesic curvature with value either $2\sqrt{\pi}$ or $-2\sqrt{\pi}$ according to whether r_2 is positive or negative, respectively. It is easy to check that the sign of such curvature corresponds to the orientation of the horocycle.

2. Elliptic cycles: These correspond to circles completely contained in $\mathbb{D}$. Such curves are obtained from γ_1 in the proof of Theorem 11.6.3 precisely when $|z| < 1 - |r_1|$. By using this condition and equation (11.6.8) it follows that every elliptic cycle has (constant) geodesic curvature whose value lies in the set $(-\infty, -2\sqrt{\pi}) \cup (2\sqrt{\pi}, +\infty)$. For each given elliptic cycle, the interval of such set in which the value of the geodesic curvature lies depends on the sign of r_2. Again, this reflects the orientation of the elliptic cycle. Also, from equation (11.6.8) it is easy to see that for every $\kappa_0 \in (-\infty, -2\sqrt{\pi}) \cup (2\sqrt{\pi}, +\infty)$ there is an elliptic cycle whose geodesic curvature has (constant) value κ_0.

3. Hypercycles: These are given by curves γ_1 precisely when $|z| > 1 - |r_1|$ and by all curves γ_2 as in the proof of Theorem 11.6.3. Such choices correspond to circular arcs that intersect the boundary of $\mathbb{D}$ in two different points. Using equations (11.6.8) and (11.6.9) it is easy to see that every hypercycle has (constant) geodesic curvature whose value lies in the interval $(-2\sqrt{\pi}, 2\sqrt{\pi})$. Moreover, from such equations it is easy to see that for every $\kappa_0 \in (-2\sqrt{\pi}, 2\sqrt{\pi})$ there is a hypercycle whose geodesic curvature has (constant) value κ_0.

Recall that for a 1-dimensional manifold C in the complex plane (e.g., a circular arc in $\mathbb{D}$ or a line in the complex plane), an orientation is an equivalence class of parameterizations where two of them γ_1, γ_2 are considered equivalent if there is a strictly increasing function h such that $\gamma_2 = \gamma_1 \circ h$. A curve is called oriented if it is endowed with an orientation. With these definitions, we say that two oriented curves C_1 and C_2 are tangent with the same orientation at a point z if they have parameterizations γ_1 and γ_2, respectively, such that for some t_1, t_2 we have $\gamma_1(t_1) = z = \gamma_2(t_2)$ and $\gamma_1'(t_1) = c\gamma_2'(t_2)$, for some $c \in (0, +\infty)$. For a given oriented curve C we say that an oriented straight line L is the oriented tangent for C at z if both are tangent with the same orientation at z.

Given $r \in (-1, 1)$, a simple computation shows that all oriented circular arcs that pass through $r = r + i0 \in \mathbb{D}$ with oriented tangent at r given by $\mathbb{R}i$ (oriented in the direction of i) are

$$
\begin{aligned}
\gamma_1(t) &= (r - \mu)e^{\operatorname{sign}(r-\mu)it} + \mu, \\
\gamma_2(t) &= ti + r,
\end{aligned}
$$

where $\mu \in \mathbb{R} \setminus \{r\}$. From equations (11.6.8) and (11.6.9) we conclude that the (hyperbolic) geodesic curvatures of these curves through r are given by

$$
\begin{aligned}
\kappa_{\gamma_1} &= \frac{\sqrt{\pi}(r^2 - 2\mu r + 1)}{r - \mu}, \\
\kappa_{\gamma_2} &= 2\sqrt{\pi}r.
\end{aligned}
$$

A straightforward computation shows that κ_{γ_1}, as a function of μ, has strictly positive derivative and so it is injective. Moreover, we observe that κ_{γ_1} has limits $+\infty$, $2\sqrt{\pi}r$ and $-\infty$ for $\mu \to r^-$, $|\mu| \to +\infty$ and $\mu \to r^+$, respectively. From these remarks it follows that for every $r \in (-1, 1)$ and $\kappa_0 \in \mathbb{R}$ there is exactly one circular arc passing through r with oriented tangent line $\mathbb{R}i$ (oriented in the direction of i) and (constant) geodesic curvature κ_0. In particular, for fixed r, no circular arc given by the curves γ_1 has geodesic curvature equal to that of γ_2.

On the other hand, it is well known (both from geometry and complex analysis) that the Möbius transformations fixing $\mathbb{D}$ and the reflections through geodesics in $\mathbb{D}$ generate a group G that acts transitively on the (hyperbolic) unit tangent bundle of $\mathbb{D}$. In other words, for every $z_1, z_2 \in \mathbb{D}$ and $v_1 \in T_{z_1}\mathbb{D}, v_2 \in T_{z_2}\mathbb{D}$ with $\|v_1\| = \|v_2\|$, there is a transformation $\phi \in G$ that maps $\phi(z_1) = z_2$ and $d\phi(v_1) = v_2$. It is also known that the transformations in G map circular arcs into circular arcs and (being hyperbolic isometries) preserve the geodesic curvature. These facts and the above remarks concerning circular arcs passing through a point $r \in (-1, 1)$ with tangent $\mathbb{R}i$ show that, for every $z \in \mathbb{D}, v \in T_z\mathbb{D} \setminus \{0\}$ and $\kappa_0 \in \mathbb{R}$, there is exactly one oriented circular arc passing through z with (constant) geodesic curvature κ_0 and oriented tangent line $\mathbb{R}v$ (with orientation given by v).

The above observations ensure the existence of curves with constant geodesic curvature and prescribed initial conditions. We also obtain uniqueness when we restrict ourselves to the family of circular arcs in $\mathbb{D}$. The next result proves that the uniqueness part holds for the larger family of C^2 curves in $\mathbb{D}$. This will play an important role in our main results.

Theorem 11.6.4. *Let $\alpha : I \to \mathbb{D}$ and $\beta : J \to \mathbb{D}$ be C^2 curves ($I, J \subset \mathbb{R}$ open intervals) with constant geodesic curvature such that*

1. *$\kappa_\alpha \equiv \kappa_\beta$,*

2. *both $\|\alpha'\|$ and $\|\beta'\|$ are constant,*

3. *$\alpha(t_0) = \beta(t_0)$ and $\alpha'(t_0) = \beta'(t_0)$,*

for some $t_0 \in I \cap J$. Then, $\alpha|_{I \cap J} = \beta|_{I \cap J}$.

Proof. Let $\gamma : I \to \mathbb{D}$ be a C^2 curve, where $I \subset \mathbb{R}$. To require $\|\gamma'\|$ to be constant is equivalent to

$$\langle \gamma', \ddot{\gamma} \rangle = 0. \tag{11.6.10}$$

On the other hand, by the formula of Theorem 11.6.2 it follows that to require from γ to have constant geodesic curvature is equivalent to

$$\sqrt{\pi} \left(2\frac{\langle i\gamma, \gamma' \rangle^E}{|\gamma'|} + \frac{(1 - |\gamma|^2)\,\langle i\gamma', \gamma'' \rangle^E}{|\gamma'|^3} \right) = \kappa_0, \tag{11.6.11}$$

for some κ_0.

Hence, the curves α and β in the statement satisfy the second-order system of ordinary differential equations given by (11.6.10) and (11.6.11) with same initial conditions by (3). Then the conclusion follows from standard results on ordinary differential equations, see for example the section "Integral Curves" of [151]. $\qquad\square$

As a consequence of the previous result we obtain the following characterization of curves in $\mathbb{D}$ with constant hyperbolic geodesic curvature.

Corollary 11.6.5. *Let $\alpha : I \to \mathbb{D}$ be a C^2 curve with constant hyperbolic geodesic curvature, where I is an open interval in $\mathbb{R}$. Then, there is a curve $\gamma : J \to \mathbb{D}$ that parameterizes a circular arc such that $I \subset J$ and $\alpha = \gamma|_I$. Moreover, no open circular arc in $\mathbb{D}$ contains the image of γ as a proper subset.*

Proof. We observe that the discussion in page 27 of [151] implies that a system of ordinary differential equations defines a vector field whose integral curves (see Definition 48 in page 27 of [151]) are precisely the solution of the system. Applying this observation to the system defined by equations (11.6.10), (11.6.11) in the proof of Theorem 11.6.4 together with the remarks that follow Corollary 50 in page 28 of [151], we conclude that there is a unique maximal integral $\gamma : J \to \mathbb{D}$ of the vector field associated to the system of equations (11.6.10), (11.6.11). In particular, γ has constant hyperbolic geodesic curvature and $\alpha = \gamma|_I$.

On the other hand, γ is a curve with constant hyperbolic geodesic curvature and for its initial conditions at a given parameter (value and derivative as well as the value of its geodesic curvature) our remarks concerning the curvature of circular arcs ensure the existence of a circular arc C sharing those same initial conditions. Then the uniqueness property stated by Theorem 11.6.4 implies that γ coincides with the circular arc C in the intersection of their domains. By moving along the parameter of γ we conclude that γ can be seen as obtained from circular arcs glued together in a C^2 manner, which clearly implies that γ is itself a circular arc. The last claim follows from the fact that γ contains all open circular arcs with the initial condition data that γ itself provides. $\qquad\square$

Since the (open) cycles in $\mathbb{D}$ are clearly the maximal (with respect to inclusion) circular arcs, we can sum up the previous results and remarks in the following statement.

Theorem 11.6.6. *A C^2 curve contained in $\mathbb{D}$ has constant geodesic curvature if and only if its image is a segment of a cycle.*

11.7 Third term: level lines are cycles

Let $\mathcal{A}(\mathbb{D})$ be a linear space of C^3 functions in $\mathbb{D}$ which is 2-rich and such that $\mathcal{T}_h(\mathcal{A}(\mathbb{D}))$ is commutative for each $h \in (0,1)$. Given a real-valued function $a \in \mathcal{A}(\mathbb{D})$ which is non-singular in some open set, we keep considering the orthogonal coordinate system (u, v) introduced in Section 11.4. Denote with k_ℓ the geodesic curvature of the level curves of a as a function of (u, v).

Our next result provides an expression for the first derivative of the geodesic curvature k_ℓ in terms of Poisson brackets.

Theorem 11.7.1. *Let $a \in \mathcal{A}(\mathbb{D})$ be a non-constant real-valued function, and let $z \in \mathbb{D}$ be a non-singular point of a. Then in a neighborhood of z we have*

$$\frac{\partial k_\ell}{\partial v} = \frac{g\sqrt{\widetilde{g}^{11}}}{(a')^2 D}\{a, \Delta a\},$$

where g, D and $\widetilde{g}^{11}$ are as in Section 11.4, and a' denotes the partial derivative of a with respect to u to emphasize the independence of a with respect to v.

Proof. By the definition of the geodesic curvature and our previous notation we have

$$k_\ell = \left\|\frac{\partial}{\partial v}\right\|^{-2}\left\|\frac{\partial}{\partial u}\right\|^{-1}\left\langle \nabla_{\frac{\partial}{\partial v}}\frac{\partial}{\partial v}, \frac{\partial}{\partial u}\right\rangle = \frac{\widetilde{\Gamma}^1_{22}\widetilde{g}_{11}}{\widetilde{g}_{22}\sqrt{\widetilde{g}_{11}}} = \widetilde{g}^{22}\widetilde{\Gamma}^1_{22}\sqrt{\widetilde{g}_{11}},$$

from which we conclude the expression

$$\frac{\partial k_\ell}{\partial v} = \frac{\partial}{\partial v}(\widetilde{g}^{22}\widetilde{\Gamma}^1_{22})\sqrt{\widetilde{g}_{11}}, \tag{11.7.1}$$

where we used the fact that $\widetilde{g}_{11}$ does not depend on v (see (11.5.3)). We also observe that

$$\widetilde{\Gamma}^1_{11} = \frac{1}{2}\widetilde{g}^{11}\frac{\partial \widetilde{g}_{11}}{\partial u},$$

and so $\widetilde{\Gamma}^1_{11}$ does not depend on v either.

As our function a depends only on u, by (11.5.2) we have

$$\Delta a = a''\widetilde{g}_{11} - a'(\widetilde{g}^{11}\widetilde{\Gamma}^1_{11} + \widetilde{g}^{22}\widetilde{\Gamma}^1_{22}).$$

Calculating the Poisson brackets in the (u, v) coordinates (see (11.4.1)) we obtain

$$\{a, \Delta a\} = -g^{-1}Da'\frac{\partial \Delta a}{\partial v} = g^{-1}D(a')^2\frac{\partial}{\partial v}(\widetilde{g}^{22}\widetilde{\Gamma}^1_{22}), \qquad (11.7.2)$$

and our formula follows from (11.7.1) and (11.7.2) by eliminating $\frac{\partial}{\partial v}(\widetilde{g}^{22}\widetilde{\Gamma}^1_{22})$. $\square$

We will need the following result that computes the dimension of the jet spaces of rich symbol sets.

Lemma 11.7.2. *Let $\mathcal{A}(\mathbb{D})$ be a k-rich complex vector space of smooth functions, where $k \geq 2$, and let S be the set of common singular points of the elements in $\mathcal{A}(\mathbb{D})$. If $\{a, b\} = 0$ for every $a, b \in \mathcal{A}(\mathbb{D})$, then the complex dimension of $J^l_z(\mathcal{A}(\mathbb{D}))$ is $l + 1$ at every $z \in \mathbb{D} \setminus S$ and for every $l = 0, \dots, k$.*

Proof. First observe that, by the definition of richness, S is finite. Now choose $z \in \mathbb{D} \setminus S$. From the arguments in the first paragraphs of Section 11.4 and those found in the proof of Lemma 11.4.5, our hypothesis ensures the existence of a smooth real coordinate system (u, v) such that in a neighborhood of z, every $a \in \mathcal{A}(\mathbb{D})$ is a (complex) function of u, i.e., it is independent of v.

Fix any $a \in \mathcal{A}(\mathbb{D})$. From the above remarks, there is a complex-valued function of one real variable such that $a = h \circ u$. In particular, all partial derivatives of a at z that involve at least one with respect to v vanish. In other words, the only possibly non-vanishing partial derivatives of a at z are those of the form

$$\frac{\partial^j a}{\partial u^j}(z).$$

Hence, for every $l \geq 0$, the l-jet at z of a is of the form

$$j^l_z(a) = \left(a(z), \frac{\partial a}{\partial u}(z), \frac{\partial^2 a}{\partial u^2}(z), \dots, \frac{\partial^l a}{\partial u^l}(z)\right).$$

Such representation clearly defines a linear inclusion of $J^l_z(\mathcal{A}(\mathbb{D}))$ into $\mathbb{C}^{l+1}$, thus showing that the complex dimension of $J^l_z(\mathcal{A}(\mathbb{D}))$ is at most $l + 1$. Then the result follows by the k-richness condition. $\square$

In the proof of the previous result, it turns out that the natural linear map $J_z^l(\mathcal{A}(\mathbb{D})) \to \mathbb{C}^{l+1}$ given by

$$j_z^l(a) \mapsto \left(a(z), \frac{\partial a}{\partial u}(z), \frac{\partial^2 a}{\partial u^2}(z), \dots, \frac{\partial^l a}{\partial u^l}(z) \right)$$

is in fact an isomorphism. Hence, if $\mathcal{A}(\mathbb{D})$ satisfies the hypotheses of Lemma 11.7.2, then for every $0 \le l \le k$ and $z \in \mathbb{D} \setminus S$, there exist some $a \in \mathcal{A}(\mathbb{D})$ such that $j_z^k(a) = (\delta_{jl})_{j=0,\dots,k}$, i.e., each of its partial derivatives at z vanish up to order k except for $\frac{\partial^l a}{\partial u^l}(z) = 1$. The same sort of result holds for the real-valued functions in $\mathcal{A}(\mathbb{D})$. More precisely, we have the following result.

Lemma 11.7.3. *Let $\mathcal{A}(\mathbb{D})$ and S satisfy the conditions of Lemma 11.7.2 and consider coordinates (u, v) as in the proof of Lemma 11.7.2. If $\{a, b\} = 0$ for every $a, b \in \mathcal{A}(\mathbb{D})$, then at every $z \in \mathbb{D} \setminus S$ and for every $l = 0, \dots, k$ there is some real-valued $a \in \mathcal{A}(\mathbb{D})$ all of whose partial derivatives with respect to (u, v) at z of order $\le k$ vanish except for $\frac{\partial^l a}{\partial u^l}(z)$.*

Proof. The complex linear isomorphism

$$J_z^k(\mathcal{A}(\mathbb{D})) \to \mathbb{C}^{k+1}$$

described above preserves conjugation and so induces an isomorphism

$$J_z^k(\mathcal{A}^{\mathbb{R}}(\mathbb{D})) \to \mathbb{R}^{k+1},$$

where $\mathcal{A}^{\mathbb{R}}(\mathbb{D})$ is the subspace of real-valued functions in $\mathcal{A}(\mathbb{D})$. Then the jets that are mapped into the canonical base in $\mathbb{R}^{k+1}$ satisfy the required conclusion. $\square$

We now prove that the vanishing of the third term of a commutator, or equivalently the condition (11.3.6), implies that the level lines of real-valued defining symbols are cycles.

Theorem 11.7.4. *Let $\mathcal{A}(\mathbb{D})$ be a 3-rich vector space of smooth functions such that $T_h(\mathcal{A}(\mathbb{D}))$ is commutative for each $h \in (0, 1)$. Then for each point $z \in \mathbb{D} \setminus S$ there is a real-valued function $a \in \mathcal{A}(\mathbb{D})$ which has z as a non-singular point and*

$$\{a, \Delta a\}(z) = 0.$$

Proof. The condition (11.3.6) states that

$$\{\Delta a, \Delta b\} + \{a, \Delta^2 b\} + \{\Delta^2 a, b\} = 0, \tag{11.7.3}$$

for all $a, b \in \mathcal{A}(\mathbb{D})$.

The identities

$$
\begin{aligned}
\Delta(ab) &= a\Delta b + b\Delta a + 2\langle \operatorname{grad} a, \operatorname{grad} b \rangle, & (11.7.4) \\
\Delta(h \circ a) &= (h' \circ a)\Delta a + (h'' \circ a)\|\operatorname{grad} a\|^2, \\
\{a, h \circ b\} &= (h' \circ b)\{a, b\}, \\
\{a, bc\} &= b\{a, c\} + c\{a, b\}, \\
\operatorname{grad}(h \circ a) &= (h' \circ a)\operatorname{grad} a,
\end{aligned}
$$

which hold for arbitrary (smooth enough) functions on the unit disk, are straight-forward consequences of the definitions of the operators involved.

Consider any point $z \in \mathbb{D} \setminus S$, with S as in the definition of 3-richness for $\mathcal{A}(\mathbb{D})$. By the 1-richness of $\mathcal{A}(\mathbb{D})$ there is an element $a \in \mathcal{A}(\mathbb{D})$ that has z as a non-singular point. Since $\mathcal{A}(\mathbb{D})$ is closed under complex conjugation we can assume that a is real-valued.

Let $h : \mathbb{R} \to \mathbb{R}$ be a smooth function such that $h \circ a \in \mathcal{A}(\mathbb{D})$. We observe that the vanishing of the Poisson brackets $\{a, b\} = 0$ and the arguments from Section 11.4 imply that every real-valued $b \in \mathcal{A}(\mathbb{D})$ can be written, in a neighborhood of z, as $b = h \circ a$ for some h.

By (11.7.3) it follows that

$$\{\Delta a, \Delta(h \circ a)\} + \{a, \Delta^2(h \circ a)\} + \{\Delta^2 a, (h \circ a)\} = 0. \tag{11.7.5}$$

Let us now compute each one of the three terms in the latter expression. By using the identities in (11.7.4) we obtain

$$
\begin{aligned}
\{\Delta a, \Delta(h \circ a)\} &= \{\Delta a, (h' \circ a)\Delta a\} + \{\Delta a, (h'' \circ a)\|\operatorname{grad} a\|^2\} \quad (11.7.6)\\
&= \Delta a\{\Delta a, (h' \circ a)\} + \{\Delta a, (h'' \circ a)\|\operatorname{grad} a\|^2\}\\
&= (h'' \circ a)\Delta a\{\Delta a, a\} + \{\Delta a, (h'' \circ a)\|\operatorname{grad} a\|^2\}.
\end{aligned}
$$

For the second term in (11.7.3) we have

$$
\begin{aligned}
\{a, \Delta^2(h \circ a)\} &= \{a, \Delta((h' \circ a)\Delta a)\} + \{a, \Delta((h'' \circ a)\|\operatorname{grad} a\|^2)\} \quad (11.7.7)\\
&= \{a, (h' \circ a)\Delta^2 a\} + \{a, \Delta a \Delta(h' \circ a)\}\\
&\quad + 2\{a, \langle \operatorname{grad}(h' \circ a), \operatorname{grad}(\Delta a)\rangle\} + \{a, \Delta((h'' \circ a)\|\operatorname{grad} a\|^2)\}\\
&= (h' \circ a)\{a, \Delta^2 a\} + \Delta a\{a, \Delta(h' \circ a)\} + \Delta(h' \circ a)\{a, \Delta a\}\\
&\quad + 2\{a, (h'' \circ a)\,\langle \operatorname{grad} a, \operatorname{grad}(\Delta a)\rangle\} + \{a, \Delta((h'' \circ a)\|\operatorname{grad} a\|^2)\}\\
&= (h' \circ a)\{a, \Delta^2 a\} + \Delta a\{a, (h'' \circ a)\Delta a\}\\
&\quad + \Delta a\{a, (h''' \circ a)\|\operatorname{grad} a\|^2\}\\
&\quad + ((h'' \circ a)\Delta a + (h''' \circ a)\|\operatorname{grad} a\|^2)\{a, \Delta a\}\\
&\quad + 2(h'' \circ a)\{a, \langle \operatorname{grad} a, \operatorname{grad}(\Delta a)\rangle\} + \{a, \Delta((h'' \circ a)\|\operatorname{grad} a\|^2)\}\\
&= (h' \circ a)\{a, \Delta^2 a\} + 2(h'' \circ a)\Delta a\{a, \Delta a\}\\
&\quad + (h''' \circ a)\|\operatorname{grad} a\|^2\{a, \Delta a\}\\
&\quad + 2(h'' \circ a)\{a, \langle \operatorname{grad} a, \operatorname{grad}(\Delta a)\rangle\} + \{a, \Delta((h'' \circ a)\|\operatorname{grad} a\|^2)\},
\end{aligned}
$$

where we used the fact that $\{a, \|\operatorname{grad} a\|^2\} = 0$.

Finally, by the third equality in (11.7.4), the last term in (11.7.3) can be written as

$$\{\Delta^2 a, (h \circ a)\} = (h' \circ a)\{\Delta^2 a, a\}. \tag{11.7.8}$$

By replacing (11.7.6), (11.7.7) and (11.7.8) into (11.7.3) and after canceling out terms we obtain

$$\{\Delta a, (h'' \circ a)\|\operatorname{grad} a\|^2\} + \{a, \Delta((h'' \circ a)\|\operatorname{grad} a\|^2)\} + (h'' \circ a)\Delta a\{a, \Delta a\}$$
$$+(h''' \circ a)\|\operatorname{grad} a\|^2\{a, \Delta a\} + 2(h'' \circ a)\{a, \langle \operatorname{grad} a, \operatorname{grad}(\Delta a)\rangle\} = 0.$$

Let us now consider the first two terms of this equation. In a neighborhood of z, we can write $(h'' \circ a)\|\operatorname{grad} a\|^2 = f \circ a$ for some function f. This can be proved by writing $\|\operatorname{grad} a\|^2$ in coordinates and using the properties obtained in the proof of Theorem 11.5.1 to conclude that it only depends on a in a neighborhood of z. But then a straightforward computation using the equations in (11.7.4) yields

$$
\begin{aligned}
\{\Delta a, f \circ a\} + \{a, \Delta(f \circ a)\} \;=\;& (f' \circ a)\{\Delta a, a\}\\
&+\{a, (f' \circ a)\Delta a + (f'' \circ a)\|\operatorname{grad} a\|^2\}\\
=\;& (f' \circ a)\{\Delta a, a\} + (f' \circ a)\{a, \Delta a\}\\
&+\{a, (f'' \circ a)\|\operatorname{grad} a\|^2\}\\
=\;& (f' \circ a)\{\Delta a, a\} + (f' \circ a)\{a, \Delta a\}\\
&+(f'' \circ a)\{a, \|\operatorname{grad} a\|^2\}\\
=\;& 0.
\end{aligned}
$$

After such computations we are left with the identity

$$((h'' \circ a)\Delta a + (h''' \circ a)\|\operatorname{grad} a\|^2)\{a, \Delta a\} + 2(h'' \circ a)\{a, \langle \operatorname{grad} a, \operatorname{grad}(\Delta a)\rangle\} = 0. \tag{11.7.9}$$

Now choose a local coordinate system in a neighborhood of z such that a is one of the components of our coordinate map. Then, as remarked before, every $b \in \mathcal{A}(\mathbb{D})$ can be written in a neighborhood of z, as $b = h \circ a$. On the other hand, the 3-richness of $\mathcal{A}(\mathbb{D})$, our hypotheses and Lemma 11.7.3 imply that there is some real-valued $b \in \mathcal{A}(\mathbb{D})$ all of whose partial derivatives at z up to order 3 vanish except for $\frac{\partial^3 a}{\partial u^3}(z)$, where (u, v) is a suitable coordinate system near a neighborhood of z. If we write $b = h \circ a$ as above, then a straightforward application of the chain rule yields

$$
\begin{aligned}
0 \;=\;& \frac{\partial b}{\partial u}(z) = h' \circ a(z)\frac{\partial a}{\partial u}(z),\\[4pt]
0 \;=\;& \frac{\partial^2 b}{\partial u^2}(z) = h'' \circ a(z)\left(\frac{\partial a}{\partial u}(z)\right)^2 + h' \circ a(z)\frac{\partial^2 a}{\partial u^2}(z),\\[4pt]
0 \;\neq\;& \frac{\partial^3 b}{\partial u^3}(z) = h''' \circ a(z)\left(\frac{\partial a}{\partial u}(z)\right)^3 + 3h'' \circ a(z)\frac{\partial a}{\partial u}(z)\frac{\partial^2 a}{\partial u^2}(z)\\[4pt]
&+h' \circ a(z)\frac{\partial^3 a}{\partial u^3}(z).
\end{aligned}
$$

Since $\frac{\partial a}{\partial u}(z) \neq 0$ (because a is non-singular at z and depends only on u) this implies $h'(a(z)) = h''(a(z)) = 0$, $h'''(a(z)) \neq 0$. If we replace such h in the above computation, we reduce equation (11.7.9) at z to

$$\|(\operatorname{grad} a)(z)\|^2 \{a, \Delta a\}(z) = 0.$$

To end the proof we recall that z is a non-singular point of a. $\square$

Corollary 11.7.5. *Let $\mathcal{A}(\mathbb{D})$ be a 3-rich vector space of smooth functions $\mathcal{A}(\mathbb{D})$ which generates for each $h \in (0, 1)$ the commutative C^*-algebra $\mathcal{T}_h(\mathcal{A}(\mathbb{D}))$ of Toeplitz operators. Then the common level lines of all real-valued functions in $\mathcal{A}(\mathbb{D})$ are cycles.*

Proof. Recall first that by Lemma 11.4.5 all functions from $\mathcal{A}(\mathbb{D})$ have the same set of level lines. Given any level line ℓ, Theorem 11.7.4 ensures that for each point $z \in \mathbb{D} \setminus S$ there is a function $a \in \mathcal{A}(\mathbb{D})$ such that $\{a, \Delta a\}(z) = 0$. Repeating the same arguments in a neighborhood V_z of z, each point of which remains non-singular for a, we have that $\{a, \Delta a\} \equiv 0$ in V_z. Thus by Theorem 11.7.1 the line ℓ has a constant geodesic curvature, and by Theorem 11.6.6 is a segment of a cycle.

On the other hand, the line ℓ is the inverse image of a regular value of some function $a \in \mathcal{A}(\mathbb{D})$, and thus is a closed subset of $\mathbb{D}$ (considered as the (open) hyperbolic plane without the boundary). Since the only segment of a cycle that defines a closed subset of $\mathbb{D}$ is the cycle itself, we have that ℓ is a cycle. $\square$

11.8 Commutative Toeplitz operator algebras and pencils of geodesics

To achieve the main result of this chapter, we want to understand the structure of the real-valued functions on $\mathbb{D}$ whose gradient lines define a pencil of geodesics. The following result provides a geometric characterization of such functions.

Theorem 11.8.1. *A non-constant C^3 real-valued function a in $\mathbb{D}$ defines a pencil if and only if the following two conditions are satisfied:*

(i) *The gradient lines of a are geodesics.*

(ii) *Each level line of a is a cycle.*

Proof. By the definition of a pencil and by Theorem 11.6.3 it is clear that the two above conditions are necessary for a to define a pencil.

To prove the sufficiency, assume that a satisfies the two given conditions. Let ℓ be a level line of a. We know that the gradient lines of a are geodesics. This fact and the uniqueness of geodesics imply that all geodesics perpendicular to ℓ are gradient lines of a. Moreover, since ℓ is a closed submanifold of $\mathbb{D}$, for every point in $\mathbb{D}$ there is at least one geodesic perpendicular to ℓ passing through that point (see [27]). This implies that the gradient lines of a are precisely the geodesics perpendicular to ℓ.

Now observe that every horocycle, elliptic cycle or hypercycle is a level curve in a pencil. To prove this for a horocycle we take the pencil defined by all horocycles with the same center as the given one. For an elliptic cycle we first recall that every such cycle has a *hyperbolic center*, i.e., it is the set of points at a fixed hyperbolic distance to a point in $\mathbb{D}$; then we take the pencil given by elliptic cycles with the same hyperbolic center. For a hypercycle we consider the two accumulation points of the hypercycle in the boundary of $\mathbb{D}$, and we take the pencil defined by the hypercycles with the same pair of accumulation points.

With the above facts we complete the proof as follows. The level line ℓ of a, being either a horocycle, an elliptic cycle or a hypercycle, is a level line in a pencil $\mathcal{P}$. Then the gradient lines of both a and $\mathcal{P}$ are the same, since such lines are precisely the geodesics perpendicular to ℓ. Also, the level lines of a and $\mathcal{P}$ are the same since they are given as the leaves of the foliation obtained by integrating the normal bundle to the common set of gradient lines. This shows that the gradient and level lines of a define the pencil $\mathcal{P}$. $\square$

Now Lemma 11.4.5, Theorem, 11.5.1, Corollary 11.7.5, and Theorem 11.8.1 lead directly to the following result.

Corollary 11.8.2. *Let $\mathcal{A}(\mathbb{D})$ be a 3-rich vector space of smooth functions such that $\mathcal{T}_h(\mathcal{A}(\mathbb{D}))$ is commutative for each $h \in (0,1)$. Then there exists a pencil $\mathcal{P}$ of hyperbolic geodesics in $\mathbb{D}$ such that all functions in $\mathcal{A}(\mathbb{D})$ are constant on the cycles of $\mathcal{P}$.*

Now the main result of the chapter reads as follows.

Corollary 11.8.3. *Let $\mathcal{A}(\mathbb{D})$ be a 3-rich vector space of smooth functions. Then the following three statements are equivalent:*

(i) *there is a pencil $\mathcal{P}$ of hyperbolic geodesics in $\mathbb{D}$ such that all functions in $\mathcal{A}(\mathbb{D})$ are constant on the cycles of $\mathcal{P}$;*

(ii) *the C^*-algebra generated by Toeplitz operators with $\mathcal{A}(\mathbb{D})$ defining symbols is commutative on each weighted Bergman space $\mathcal{A}_h^2(\mathbb{D})$, $h \in (0,1)$ (or $\mathcal{A}_\lambda^2(\mathbb{D})$, $\lambda \in (-1, +\infty)$);*

(iii) *the C^*-algebra generated by Toeplitz operators with $\mathcal{A}(\mathbb{D})$ defining symbols is commutative on each weighted Bergman space $\mathcal{A}_h^2(\mathbb{D})$ (or $\mathcal{A}_\lambda^2(\mathbb{D})$) for a sequence of parameters h_n with $\lim_{n\to\infty} h_n = 0$ (or λ_n with $\lim_{n\to\infty} \lambda_n = +\infty$).*

Chapter 12

Dynamics of Properties of Toeplitz Operators with Radial Symbols

Given a smooth defining symbol $a = a(z)$, the family of Toeplitz operators $T_a = \{T_a^{(h)}\}$, where $h \in (0,1)$, was considered in the previous chapter under the Berezin quantization procedure. For a fixed h the Toeplitz operator $T_a^{(h)}$ acts on the weighted Bergman space $\mathcal{A}_h^2(\mathbb{D})$, where the parameter h characterizes the weight (10.1.5) on $\mathcal{A}_h^2(\mathbb{D})$. In the sequel we will consider another form of presentation of the weighted Bergman spaces, see (10.1.1), the space $\mathcal{A}_\lambda^2(\mathbb{D})$ which is parameterized by $\lambda \in (-1, +\infty)$ being connected with $h \in (0,1)$ by the rule $\lambda + 2 = \frac{1}{h}$, see Section 10.1.

In this and the next two chapters we study the behavior of several properties (boundedness, compactness, spectral properties, etc.) of $T_a^{(\lambda)}$ in dependence on λ, and compare their limit behavior under $\lambda \to \infty$ ($h \to 0$) with corresponding properties of the initial defining symbol a.

The word "dynamics" is used to emphasize the main theme of these chapters: *what happens to properties of Toeplitz operators acting on weighted Bergman spaces when the weight parameter varies?*

It seems to be quite impossible to get a reasonably complete answer to the above problem for general (even smooth) defining symbols. At the same time the described classes of commutative C^*-algebras of Toeplitz operators suggest the classes of symbols for with a satisfactory complete answer can be given. Indeed, the key feature of defining symbols constant on cycles, which permits us to get much more complete information than one obtained studying general symbols, is that in each such case the corresponding Toeplitz operators admit a spectral type representation, i.e., they are unitary equivalent to multiplication operators, by a certain sequence in the elliptic case and by certain functions on $\mathbb{R}_+$ and $\mathbb{R}$ in the parabolic and hyperbolic cases, respectively. That is, in each such case we have the exact formula that governed the boundedness and spectral properties.

Many statements in this and the next two chapters are quite similar, and we have at least two alternative ways to organize them: either to group them according to the property under study (boundedness, spectral properties, etc), or to group according to the defining symbol class, originated by a model pencil of a certain type (elliptic, parabolic, or hyperbolic). We select the second way, with the advantage that during each chapter we are dealing with the same class of defining symbols and with the same formula for $\gamma_{a,\lambda}$, permitting one to accumulate a certain intuition in each chapter.

We mention that in this and the next two chapters the term "symbol" used without any adjective will always mean the "defining symbol".

12.1   Boundedness and compactness properties

In this chapter we consider the radial symbols $a = a(|z|) = a(r)$ only. By Theorem 10.4.2 each Toeplitz operator $T_a^{(\lambda)}$ with radial symbols $a = a(r)$, acting on the weighted Bergman space $\mathcal{A}_\lambda^2(\mathbb{D})$, is unitary equivalent to the multiplication operator $\gamma_{a,\lambda} I$, acting on the one-sided l_2. The sequence $\gamma_{a,\lambda} = \{\gamma_{a,\lambda}(n)\}$ is given by

$$\gamma_{a,\lambda}(n) = \frac{1}{B(n+1, \lambda+1)} \int_0^1 a(\sqrt{r})\,(1-r)^\lambda\, r^n dr, \quad n \in \mathbb{Z}_+ = \mathbb{N} \cup \{0\}. \quad (12.1.1)$$

Thus, in particular, the Toeplitz operator $T_a^{(\lambda)}$ with radial symbols $a = a(r)$ is bounded on $\mathcal{A}_\lambda^2(\mathbb{D})$ if and only if the sequence $\gamma_{a,\lambda}$ is bounded, and it is compact on $\mathcal{A}_\lambda^2(\mathbb{D})$ if and only if the sequence $\gamma_{a,\lambda}$ has the zero limit.

As we already know (see Example 6.1.7) in the classical (weightedless) Bergman space there exist symbols *unbounded near the boundary* which generate *bounded* and even *compact* Toeplitz operators.

For every bounded symbol $a(r)$ the Toeplitz operator $T_a^{(\lambda)}$ is bounded on all spaces $\mathcal{A}_\lambda^2(\mathbb{D})$, where $\lambda \in (-1, \infty)$, and the corresponding norms are uniformly bounded by $\sup_r |a(r)|$. As examples show, a Toeplitz operator with unbounded symbol can be bounded (compact) for one value of λ and unbounded (non-compact) for another. At this stage the natural question appears: *given an unbounded symbol, describe the set of λ for which the corresponding Toeplitz operator is bounded (compact)*.

As has been already mentioned, all spaces $\mathcal{A}_\lambda^2(\mathbb{D})$, where $\lambda \in (-1, \infty)$, are natural and appropriate for Toeplitz operators with *bounded* symbols. Admitting *unbounded* symbols and wishing to have a sufficiently large class of them common for all admissible λ, it is convenient to study corresponding Toeplitz operators on $\mathcal{A}_\lambda^2(\mathbb{D})$ with $\lambda \in [0, \infty)$.

In the sequel we consider radial symbols $a(r)$ satisfying the condition $a(\sqrt{r}) \in L_1(0, 1)$, and assume that $\lambda \in [0, \infty)$.

We start with sufficient conditions for boundedness of Toeplitz operators with unbounded radial symbols.

Given a symbol $a(\sqrt{r}) \in L_1(0, 1)$, introduce $B_a^{(0)}(r) = a(\sqrt{r})$. Then

$$
\begin{aligned}
\gamma_{a,\lambda}(n) &= \frac{1}{B(n+1, \lambda+1)} \int_0^1 a(\sqrt{r})(1-r)^\lambda r^n dr \\
&= \frac{1}{B(n+1, \lambda+1)} \int_0^1 a(\sqrt{r}) dr \int_0^r \omega_\lambda(n, s) ds \\
&= \frac{1}{B(n+1, \lambda+1)} \int_0^1 \omega_\lambda(n, s) ds \int_s^1 a(\sqrt{r}) dr \\
&= \frac{1}{B(n+1, \lambda+1)} \int_0^1 B_a^{(1)}(s) \omega_\lambda(n, s) ds.
\end{aligned}
$$

We use Fubini's theorem to change the order of integration. Here we have written

$$
B_a^{(1)}(s) = \int_s^1 a(\sqrt{r}) dr \tag{12.1.2}
$$

and

$$
\omega_\lambda(n, s) = n(1-s)^\lambda s^{n-1} - \lambda(1-s)^{\lambda-1} s^n.
$$

Analogously integrating by parts and neglecting the boundary terms, we get

$$
\gamma_{a,\lambda}(n) = \frac{1}{B(n+1, \lambda+1)} \int_0^1 B_a^{(j)}(s) \frac{d^j}{ds^j} \omega_\lambda(n, s) \, ds, \quad j = 0, 1, \dots, \tag{12.1.3}
$$

where

$$
B_a^{(j)}(s) = \int_s^1 B_a^{(j-1)}(r) \, dr, \qquad j = 1, 2, \dots . \tag{12.1.4}
$$

Theorem 12.1.1. *If there exists $j \in \mathbb{N}$ such that*

$$
B_a^{(j)}(r) = O((1-r)^j), \quad r \to 1, \tag{12.1.5}
$$

then the Toeplitz operator $T_a^{(\lambda)}$ is bounded on each $\mathcal{A}_\lambda^2(\mathbb{D})$, with $\lambda \geq 0$.

If for some $j \in \mathbb{N}$,

$$
B_a^{(j)}(r) = o((1-r)^j), \quad r \to 1, \tag{12.1.6}
$$

then the Toeplitz operator $T_a^{(\lambda)}$ is compact on each $\mathcal{A}_\lambda^2(\mathbb{D})$, with $\lambda \geq 0$.

Proof. To prove the boundedness we show that the sequence (12.1.1) is bounded.

Consider first the case $j = 1$. Integrating by parts we have

$$
\begin{aligned}
|\gamma_{a,\lambda}(n)| &= \left| \frac{1}{B(n+1,\lambda+1)} \int_0^1 B_a^{(1)}(r)\,(\lambda r + n(1-r))\,(1-r)^{\lambda-1}\,r^{n-1}dr \right| \\
&\leq \operatorname{const} \frac{1}{B(n+1,\lambda+1)} \left(\lambda \int_0^1 (1-r)^{\lambda}\,r^n dr \right. \\
&\qquad \left. + n \int_0^1 (1-r)^{\lambda+1}\,r^{n-1}dr \right) \\
&= \operatorname{const} \left(\lambda \frac{B(n+1,\lambda+1)}{B(n+1,\lambda+1)} + n \frac{B(n,\lambda+2)}{B(n+1,\lambda+1)} \right) \\
&= \operatorname{const} (\lambda + (\lambda+1)) = \operatorname{const} (2\lambda + 1).
\end{aligned}
$$

The case $j > 1$ is considered analogously applying consecutive integration by parts.

Sufficiency of the condition (12.1.6) for compactness requires more delicate but similar arguments. $\qquad\qquad\square$

Repeating the arguments of Example 6.1.7 we have

Example 12.1.2. Consider unbounded symbol

$$
a(r) = (1 - r^2)^{-\beta} \sin(1 - r^2)^{-\alpha}, \tag{12.1.7}
$$

where $\alpha > 0$, and $0 < \beta < 1$. By Theorem 12.1.1 with $j = 1$ the corresponding Toeplitz operator

- is bounded for $\alpha \geq \beta$,

- is compact for $\alpha > \beta$

on each weighted Bergman space $\mathcal{A}_\lambda^2(\mathbb{D})$, with $\lambda \geq 0$.

The conditions of Theorem 12.1.1 are only sufficient for oscillating symbols, and the above conclusions are in fact just preliminary. The complete information about the symbol (12.1.7) is given in Example 12.2.6 below and says that the corresponding Toeplitz operator is bounded and compact for all $\lambda \geq 0$, $\alpha > 0$, and $0 < \beta < 1$.

The next theorem deals with necessary conditions for boundedness and compactness.

Theorem 12.1.3. *Let* $a(\sqrt{r}) \in L_1(0,1)$, *and either* $a(r) \geq 0$, *or* $B_a^{(j)}(r) \geq 0$ *for a certain* $j \in \mathbb{N}$. *Then the conditions* (12.1.5), (12.1.6) *are also necessary for the boundedness and compactness of the corresponding Toeplitz operator* $T_a^{(\lambda)}$ *on* $\mathcal{A}_\lambda^2(\mathbb{D})$ *with* $\lambda \geq 0$, *respectively. That is, the conditions* (12.1.5), (12.1.6) *are necessary and sufficient for boundedness and respectively compactness of the Toeplitz operator* $T_a^{(\lambda)}$ *with the symbol* $a = a(r)$ *as specified on* $\mathcal{A}_\lambda^2(\mathbb{D})$, *where* $\lambda \geq 0$.

Proof. We prove that these conditions are necessary when $a(r) \geq 0$. The general case of $B_a^{(j)}(r) \geq 0$ for a certain $j \in \mathbb{N}$ can be treated analogously.

For $\lambda = 0$, setting $s_n = 1 - B(n+1, \lambda+1)$ we have

$$\gamma_{a,0}(n) \geq \text{const } B^{-1}(n+1,1) \int_{s_n}^{1} a(\sqrt{r})dr = \text{const } (1 - s_n)^{-1} B_a^{(1)}(s_n),$$

which proves the statement for $\lambda = 0$. Let now $\lambda = \lambda_0 > 0$. Below we show (Theorem 12.1.7) that if γ_{a,λ_0} is bounded (compact) for some λ_0, then it is also bounded (compact) for each $\lambda \in [0, \lambda_0)$. $\qquad\square$

Corollary 12.1.4. *If $a(r) \geq 0$, and $\lim_{\varepsilon \to 0} \inf_{r \in [1-\varepsilon, 1]} a(r) = +\infty$, then the Toeplitz operator $T_a^{(\lambda)}$ is unbounded on each $\mathcal{A}_\lambda^2(\mathbb{D})$, $\lambda \geq 0$.*

Corollary 12.1.5. *Let $a(\sqrt{r}) \in L_1(0,1)$, and let $a(r) \geq 0$, or $B_a^{(j)}(r) \geq 0$ for some $j \in \mathbb{N}$. Then the Toeplitz operator $T_a^{(\lambda)}$ is bounded (compact), or unbounded (not compact) on each $\mathcal{A}_\lambda^2(\mathbb{D})$ simultaneously.*

To prove the corollary it is sufficient to note that the above conditions for boundedness and compactness do not depend on λ.

We would like to mention in this connection a result by K. Zhu (see [238]) who proved the statement for non-negative symbols (considering operators acting on the weighted Bergman space over the unit ball in $\mathbb{C}^n$). Our statement is not limited to non-negative symbols.

Remark 12.1.6. For general $a(\sqrt{r}) \in L_1(0,1)$ symbols the conditions (12.1.5), (12.1.6) fail to be necessary. Further, while j becomes large these conditions get weaker. But the hypothesis that each operator $T_a^{(\lambda)}$ $(a = a(r))$ bounded (or compact) on $\mathcal{A}_\lambda^2(\mathbb{D})$ will satisfy (12.1.5) (or (12.1.6)) for some large j appears to be false.

Indeed, there exists (see Theorem 12.1.8 below) a Toeplitz operator $T_a^{(\lambda)}$, bounded (compact) on $\mathcal{A}_{\lambda_1}^2(\mathbb{D})$ and unbounded (not compact) on $\mathcal{A}_{\lambda_2}^2(\mathbb{D})$, for some $\lambda_1 < \lambda_2$, while the conditions (12.1.5) and (12.1.6) do not depend on λ. Furthermore, Example 12.2.6 below illustrates dependence of the conditions (12.1.5) and (12.1.6) on j.

We pass now to the analysis of the boundedness and compactness properties in their dependence on $\lambda \in [0, \infty)$.

Theorem 12.1.7. *The following statements hold:*

(i) *if for any $\lambda_0 > 0$, the sequence γ_{a,λ_0} is bounded, then the sequence $\gamma_{a,\lambda}$ is bounded for all $\lambda \in [0, \lambda_0)$;*

(ii) *if for any $\lambda_0 > 0$, $\lim_{n \to \infty} \gamma_{a,\lambda_0}(n) = 0$, then $\lim_{n \to \infty} \gamma_{a,\lambda_0}(n) = 0$ for all $\lambda \in [0, \lambda_0)$.*

Proof. We introduce

$$\beta_{a,\lambda}(n) = \int_0^1 a(\sqrt{r})\,(1-r)^\lambda\,r^n dr,$$

and the power series

$$(1-r)^\sigma = \sum_{j=0}^\infty C_j(\sigma)\,r^j,$$

whose coefficients $C_j(\sigma)$ we represent (see, for example, [1] p. 256) in the form

$$
\begin{aligned}
C_j(\sigma) &= (-1)^j \frac{\Gamma(1+\sigma)}{\Gamma(j+1)\Gamma(1+\sigma-j)} = \frac{(-1)^j}{\pi}\,\frac{\Gamma(1+\sigma)\Gamma(j-\sigma)}{\Gamma(j+1)}\,\sin\pi(j-\sigma)\\
&= -\frac{1}{\pi}\sin\pi\sigma\,\frac{\Gamma(1+\sigma)\Gamma(j-\sigma)}{\Gamma(j+1)} = -\frac{1}{\pi}\sin\pi\sigma\,B(j-\sigma,1+\sigma). \quad (12.1.8)
\end{aligned}
$$

Thus for $\lambda < \lambda_0$ we have

$$
\begin{aligned}
\beta_{a,\lambda}(n) &= \int_0^1 a(\sqrt{r})\,(1-r)^{\lambda_0}\,(1-r)^{\lambda-\lambda_0}\,r^n dr\\
&= -\sum_{j=0}^\infty \frac{\sin\pi(\lambda-\lambda_0)}{\pi}\,B(j-\sigma,1+\sigma)\int_0^1 a(\sqrt{r})\,(1-r)^{\lambda_0}\,r^{n+j}dr\\
&= \frac{\sin\pi(\lambda_0-\lambda)}{\pi}\sum_{j=0}^\infty B(j-\sigma,1+\sigma)\,\beta_{a,\lambda_0}(n+j),
\end{aligned}
$$

where $\sigma = \lambda - \lambda_0 < 0$.

Now it is easy to see that

$$\gamma_{a,\lambda}(n) = \frac{\sin\pi|\sigma|}{\pi}\sum_{j=0}^\infty \frac{B(j-\sigma,1+\sigma)\,B(n+j+1,\lambda_0+1)}{B(n+1,\lambda+1)}\,\gamma_{a,\lambda_0}(n+j). \quad (12.1.9)$$

We need the following asymptotic representation of the Beta function for fixed δ and $L \to \infty$,

$$
\begin{aligned}
B(L,\delta) &= \frac{\Gamma(L)\,\Gamma(\delta)}{\Gamma(L+\delta)} = \Gamma(\delta)\,\frac{e^{-L}L^{L-\frac{1}{2}}(1+O(L^{-1}))}{e^{-(L+\delta)}(L+\delta)^{L+\delta-\frac{1}{2}}(1+O(L^{-1}))}\\
&= \Gamma(\delta)\,e^\delta\left(1-\frac{\delta}{L+\delta}\right)^{L-\frac{1}{2}}(L+\delta)^{-\delta}\,(1+O(L^{-1}))\\
&= \Gamma(\delta)\,L^{-\delta}\,(1+O(L^{-1})). \quad\quad (12.1.10)
\end{aligned}
$$

Assume now that the sequence γ_{a,λ_0} is bounded. Then

$$
\begin{aligned}
|\gamma_{a,\lambda}(n)| &\leq \text{const} \sum_{j=0}^{\infty} \frac{(n+1)^{\lambda+1}}{(j-\sigma)^{1+\sigma}(n+j+1)^{\lambda_0+1}} \\
&\leq \text{const}\,(n+1)^{\lambda+1} \int_0^{\infty} \frac{ds}{(s+1)^{1+\sigma}(n+s+1)^{\lambda_0+1}} \\
&\leq \text{const}\,\frac{(n+1)^{\lambda+1}}{n^{\lambda+1}} \int_0^{\infty} \frac{dv}{v^{1+(\lambda-\lambda_0)}(1+v)^{\lambda_0+1}},
\end{aligned}
$$

where $s+1 = nv$.

Since $1+(\lambda-\lambda_0) < 1$ and $1+\lambda_0 > 1$, the last integral converges, and we have (i). To prove (ii) it is sufficient to observe that if $\lim_{n\to\infty}\gamma_{a,\lambda_0}(n) = 0$, then $\text{const} = c(n)$ tends to 0 when $n \to \infty$ as well. $\qquad\square$

Given a symbol $a = a(r)$, denote by $B(a)$ the set of all $\lambda \in [0,\infty)$ for which the Toeplitz operator $T_a^{(\lambda)}$ is bounded on $\mathcal{A}_\lambda^2(\mathbb{D})$, that is, the sequence $\gamma_{a,\lambda}$ is bounded; and denote by $K(a)$ the set of all $\lambda \in [0,\infty)$ for which the Toeplitz operator $T_a^{(\lambda)}$ is compact on $\mathcal{A}_\lambda^2(\mathbb{D})$, that is the sequence $\gamma_{a,\lambda}$ has the zero limit.

Theorem 12.1.7 shows that the sets $B(a)$ and $K(a)$ may have one of the following forms:

$$
(i)\ \ [0,\infty), \qquad (ii)\ \ [0,\lambda_0), \qquad (iii)\ \ [0,\lambda_0].
$$

We show now that all three possibilities can be realized. The first one is true, for example, for any bounded symbol, in the case of $B(a)$, and for any continuous bounded symbol with $a(1) = 0$, in the case of $K(a)$.

To realize (ii) and (iii) introduce the sequence

$$
\gamma(n) = e^{\frac{i}{5\pi}\ln^2(n+2)}\,\ln^{-\nu}(n+2)\,\ln^{\beta}\ln(n+2),
$$

where $0 < \nu \leq \nu_0 < 1$ and $\beta \in \mathbb{R}$. We show first that there exists a radial symbol $a_{\nu,\beta} = a_{\nu,\beta}(r)$ such that the corresponding sequence (12.1.1) for $\lambda = 0$ has the above form, i.e.,

$$
\gamma_{a_{\nu,\beta},0}(n) = \gamma(n), \qquad n \in \mathbb{Z}_+. \tag{12.1.11}
$$

Given a symbol $a = a(r)$, we have

$$
\gamma_{a,0}(n) = (n+1) \int_0^1 a(\sqrt{r})r^n dr,
$$

or, changing the variable $r = e^{-\xi}$,

$$
\gamma_{a,0}(n) = (n+1) \int_{\mathbb{R}_+} a(e^{-\frac{\xi}{2}})e^{-(n+1)\xi} d\xi.
$$

Introduce the function $b(\xi) = \sqrt{2\pi} a(e^{-\frac{\xi}{2}}) e^{-\xi}$; we assume that $b \in L_2(\mathbb{R}_+)$, which is equivalent to $a \in L_2((0,1), r^3 dr)$. Now we have

$$\frac{1}{n+1}\, \gamma_{a,0}(n) = \frac{1}{\sqrt{2\pi}} \int_{\mathbb{R}_+} b(\xi)\, e^{-n\xi}\, d\xi. \qquad (12.1.12)$$

The right-hand side of the last equation can be treated as the inverse Fourier transform of the L_2-function b, supported on $\mathbb{R}_+$ and evaluated at the point $z = in \in \Pi$.

In this connection we recall (see the beginning of Chapter 3) that the Fourier integral transform provides the isometric isomorphism between the Hardy space $H^2_+(\mathbb{R})$ on the upper half-plane and $L_2(\mathbb{R}_+)$. Recall that the Hardy space $H^2_+(\mathbb{R})$ coincides with the set of all $L_2(\mathbb{R})$-functions which admit analytic continuations φ in the upper half-plane Π and such that the functions $g_y(x) = \varphi(x + iy)$ are square integrable on the real axis ($x \in \mathbb{R}$) for each fixed y with uniformly bounded L_2-norms. In what follows we will identify the functions from $H^2_+(\mathbb{R})$, defined in $\mathbb{R}$, with their analytic continuations φ, defined in Π. Now given a function $b \in L_2(\mathbb{R}_+)$, the corresponding analytic function φ is recovered by the inverse Fourier transform

$$\varphi(z) = \frac{1}{\sqrt{2\pi}} \int_{\mathbb{R}_+} b(\xi)\, e^{iz\xi}\, d\xi, \qquad z \in \Pi.$$

Thus, returning to (12.1.12), each analytic function $\varphi \in H^2_+(\mathbb{R})$ determines the function $b \in L_2(\mathbb{R}_+)$, and thus the radial symbol

$$a(r) = \frac{1}{\sqrt{2\pi}}\, r^{-2}\, b(-2\ln r) \in L_2((0,1), r^3 dr),$$

such that the correspondence sequence $\gamma_{a,0}$ has the form

$$\gamma_{a,0}(n) = (n+1)\, \varphi(in), \qquad n \in \mathbb{Z}_+.$$

That is, to achieve our goal (to prove the existence of radial symbol $a_{\nu,\beta}$ for which we have (12.1.11)) we need to check that the function

$$\varphi(z) = \frac{1}{1 - iz}\, e^{\frac{i}{5\pi} \ln^2(2 - iz)}\, \ln^{-\nu}(2 - iz)\, \ln^\beta \ln(2 - iz)$$

belongs to $H^2(\Pi)$. We define the single-valued branch of the above multi-valued function φ by setting $\arg(2 - iz) \in [-\pi, \pi)$. For $z = x + iy \in \Pi$, we have that $\arg(2 - iz) = \arg(2 + y - ix) \in (-\frac{\pi}{2}, \frac{\pi}{2})$ and the function $\varphi(z)$ is analytic in the upper half-plane Π. Moreover, for $y \geq 0$,

$$|g_y(x)|^2 = |\varphi(x + iy)|^2 \leq \text{const}\, \frac{1}{x^2 + (1 + y)^2}\, (x^2 + (2 + y)^2)^{-\frac{2}{5\pi} \arg(2 + y - ix)}$$

$$\leq \text{const}\, \frac{1}{x^2 + (1 + y)^2}\, (x^2 + (2 + y)^2)^{\frac{1}{5}},$$

which implies that φ belongs to $H^2_+(\mathbb{R})$.

The following theorem gives an example of different behavior of $B(a)$ and $K(a)$ realizing the cases (ii) and (iii).

Theorem 12.1.8. *Let $0 < \nu < 1$. Then*

a) $\qquad B(a_{\nu,0}) = [0,\nu], \qquad\qquad K(a_{\nu,0}) = [0,\nu), \qquad\quad \beta = 0,$

b) $\qquad B(a_{\nu,\beta}) = [0,\nu), \qquad\qquad K(a_{\nu,\beta}) = [0,\nu), \qquad\quad \beta > 0,$

c) $\qquad B(a_{\nu,\beta}) = [0,\nu], \qquad\qquad K(a_{\nu,\beta}) = [0,\nu], \qquad\quad \beta < 0.$

Proof. Let $0 < \lambda < 1$. Then

$$
\begin{aligned}
\gamma_{a_{\nu,\beta},\lambda}(n) \;&=\; \frac{1}{B(n+1,\lambda+1)} \int_0^1 a_{\nu,\beta}(\sqrt{r})((1-r)(1-r)^{\lambda-1} r^n\,dr \\
&=\; \frac{1}{B(n+1,\lambda+1)} \sum_{j=0}^{\infty} C_j(\lambda-1) \int_0^1 a_{\nu,\beta}(\sqrt{r})(1-r)r^{n+j}\,dr \\
&=\; \frac{1}{B(n+1,\lambda+1)} \sum_{j=0}^{\infty} C_j(\lambda-1)(\beta_{a_{\nu,\beta},0}(n+j) - \beta_{a_{\nu,\beta},0}(n+j+1))
\end{aligned}
$$

where

$$
\beta_{a_{\nu,\beta},0}(n) = \frac{\gamma_{a_{\nu,\beta},0}(n)}{n+1}. \tag{12.1.13}
$$

Further

$$
\gamma_{a_{\nu,\beta},\lambda}(n) = \frac{1}{B(n+1,\lambda+1)} \sum_{j=0}^{\infty} C_j(\lambda-1) \int_j^{j+1} \beta'_{a_{\nu,\beta},0}(n+s)\,ds. \tag{12.1.14}
$$

It is easy to see that

$$
\begin{aligned}
\beta'_{a_{\nu,\beta},0}(s) \;&=\; \frac{2i}{5\pi} \cdot \frac{e^{\frac{i}{5\pi}\ln^2(n+s+2)}}{(n+s+1)^2} \ln^{1-\nu}(n+s+2)\ln^{\beta}\ln(n+s+2) \\
&\quad\; + O\left(\frac{\ln^{-\nu}(n+s+2)}{(n+s+1)^2} \ln^{\beta}\ln(n+s+2) \right) \\
&:=\; \delta_0(n+s) + \delta_1(n+s). \tag{12.1.15}
\end{aligned}
$$

By (12.1.8)

$$
C_j(\lambda-1) = \frac{1}{\pi} \sin(\pi(1-\lambda))B(j+1-\lambda,\lambda),
$$

and thus (12.1.10) implies that

$$
C_j(\lambda-1) = \sin \pi(1-\lambda) \frac{\Gamma(\lambda)}{\pi} (j+1-\lambda)^{-\lambda} (1+O((j+1)^{-1})). \tag{12.1.16}
$$

Thus from (12.1.14)–(12.1.16) we get

$$
\begin{aligned}
\gamma_{a_\nu,\beta,\lambda}(n) \;=\;& \frac{c_\lambda}{B(n+1,\lambda+1)} \sum_{j=0}^{\infty} \int_j^{j+1} \frac{\delta_0(n+s)}{(s+1)^\lambda}\,ds \\
&+ \frac{1}{B(n+1,\lambda+1)} \sum_{j=0}^{\infty} O\left(\frac{\delta_1(n+j+1)}{(j+1)^\lambda}\right) \\
&+ \frac{1}{B(n+1,\lambda+1)} \sum_{j=0}^{\infty} O\left(\frac{\delta_0(n+j+1)}{(j+1)^{1+\lambda}}\right) \\
=\;& \frac{c_\lambda}{B(n+1,\lambda+1)} \int_0^{\infty} \frac{\delta_0(s+n)}{(s+1)^\lambda}\,ds + (n+1)^{\lambda+1} \\
&\cdot \sum_{j=0}^{\infty} O\left(\frac{\delta_1(n+j+1)}{(j+1)^\lambda}\right) \\
&+ (n+1)^{\lambda+1} \sum_{j=0}^{\infty} O\left(\frac{\delta_0(n+s+1)}{(j+1)^{1+\lambda}}\right) := I_1 + I_2 + I_3
\end{aligned}
$$

where $c_\lambda = \frac{\sin \pi(1-\lambda)}{\pi}\Gamma(\lambda)$.

We estimate now summands I_2 and I_3.

$$
\begin{aligned}
I_2 \;\le\;& \operatorname{const}(n+1)^{1+\lambda} \int_0^{\infty} \frac{\ln^{-\nu}(n+s+2)\,\ln^\beta \ln(n+s+2)}{(n+s+1)^2\,(s+1)^\lambda}\,ds \\
\asymp\;& \operatorname{const}(n+1)^{1+\lambda} \int_0^{\infty} \frac{\ln^{-\nu}(n+s+1)\,\ln^\beta \ln(n+s+1)}{(n+s+1)^2\,(s+1)^\lambda}\,ds.
\end{aligned}
$$

Changing the variable $s+1 = nv$ we have

$$
\begin{aligned}
I_2 \;\le\;& \operatorname{const} \int_0^{\infty} \frac{\ln^{-\nu} n(v+1)\,\ln^\beta \ln n(v+1)}{(1+v)^2\,v^\lambda}\,dv \\
\le\;& \operatorname{const} \ln^{-\nu} n \, \ln^\beta \ln n \int_0^{\infty} \frac{\left(1+\frac{\ln(1+v)}{\ln n}\right)^{-\nu}\left(1+\frac{\ln\left(1+\frac{\ln(1+v)}{\ln n}\right)}{\ln \ln n}\right)^\beta}{(1+v)^2\,v^\lambda}\,dv \\
\le\;& \operatorname{const} \ln^{-\nu} n \, \ln^\beta \ln n \int_0^{\infty} \frac{(1+\ln(1+v))^{|\beta|}\,dv}{(1+v)^2\,v^\lambda} \\
\le\;& \operatorname{const} \ln^{-\nu} n \, \ln^\beta \ln n, \hspace{5cm} (12.1.17)
\end{aligned}
$$

noting that since $0 \le \lambda < 1$ then $2+\lambda > 1$.

Now we pass to the evaluation of I_3:

$$
\begin{aligned}
I_3 \;\le\;& \operatorname{const}(n+1)^{\lambda+1} \int_0^{\infty} \frac{\ln^{1-\nu}(n+s+2)\,\ln^\beta \ln(n+s+1)}{(n+s+2)^2\,(s+1)^{1+\lambda}}\,ds \\
\asymp\;& \operatorname{const}(n+1)^{\lambda+1} \int_0^{\infty} \frac{\ln^{1-\nu}(n+s+1)\,\ln^\beta \ln(n+s+1)}{(n+s+1)^2\,(s+1)^{1+\lambda}}\,ds.
\end{aligned}
$$

Again changing the variable $s + 1 = nv$ we have

$$
\begin{aligned}
I_3 &\leq \text{const} \, \frac{1}{n} \int_{n-1}^{\infty} \frac{\ln^{1-\nu} n(1+v) \, \ln^{\beta} \ln n(1+v)}{(1+v)^2 v^{1+\lambda}} \, dv \\
&\leq \text{const} \, \frac{\ln^{1-\nu} n \, \ln^{\beta}(\ln n)}{n} \int_{n-1}^{\infty} \frac{dv}{v^{1+\lambda}} \\
&\leq \text{const} \, \frac{\ln^{1-\nu} n \, \ln^{\beta} \ln n}{n^{1-\lambda}}.
\end{aligned}
\tag{12.1.18}
$$

Thus we have (see (12.1.17) and (12.1.18))

$$
I_2 = O(\ln^{-\nu} n \, \ln^{\beta} \ln n)
\tag{12.1.19}
$$

and

$$
I_3 = O\left(\frac{\ln^{1-\nu} n \, \ln^{\beta} \ln n}{n^{1-\lambda}} \right).
\tag{12.1.20}
$$

Now we consider the "main" term

$$
\begin{aligned}
I_1 &= \frac{\widetilde{c}_\lambda}{B(n+1, \lambda+1)} \int_0^{\infty} \frac{e^{\frac{i}{5\pi} \ln^2 (n+s+2)} \ln^{1-\nu}(n+s+2) \ln^{\beta} \ln(m+s+2)}{(m+s+1)^2 (s+1)^{\lambda}} \, ds \\
&\asymp \frac{\widetilde{c}_\lambda}{B(n+1, \lambda+1)} \int_0^{\infty} \frac{e^{\frac{i}{5\pi} \ln^2 (n+s+1)} \ln^{1-\nu}(n+s+1) \ln^{\beta} \ln(m+s+1)}{(m+s+1)^2 (s+1)^{\lambda}} \, ds \\
&= \frac{\widetilde{c}_\lambda}{B(n+1, \lambda+1) n^{1+\lambda}} \int_0^{\infty} \frac{e^{\frac{i}{5\pi} \ln^2 n(v+1)} \ln^{1-\nu} n(1+v) \ln^{\beta} \ln(n(1+v))}{(1+v)^2 v^{\lambda}} \, dv,
\end{aligned}
$$

where $\widetilde{c}_\lambda = \frac{2i}{5\pi} c_\lambda$.

Using the asymptotic of the B-function we have

$$
\begin{aligned}
I_1 &= \widetilde{c}_\lambda \Gamma(1+\lambda)(1 + O(n^{-1})) e^{\frac{i}{5\pi} \ln^2 n} \ln^{1-\nu} n \, \ln^{\beta}(\ln n) \\
&\quad \cdot \int_0^{\infty} \frac{e^{i\frac{2\pi}{5}(\ln n)\ln(1+v)} e^{\frac{i}{5\pi}\ln^2(v+1)}}{(v+1)^2 v^{\lambda}} \\
&\quad \cdot \frac{\left(1 + \frac{\ln(1+v)}{\ln n}\right)^{1-\nu} \left(1 + \frac{\ln\left(1 + \frac{\ln(1+v)}{\ln n}\right)}{\ln \ln n}\right)^{\beta}}{(v+1)^2 v^{\lambda}} \, dv.
\end{aligned}
\tag{12.1.21}
$$

Introduce a large parameter $\Lambda = \frac{2\pi}{5} \ln n$. We have the oscillatory integral with two points of singularity: $v_0 = 0$ and $v_0 = \infty$. Let

$$
1 = \chi_1(v) + \chi_2(v)
\tag{12.1.22}
$$

with $\chi_{1,2}(v) \in C^{\infty}(0, \infty)$, $\operatorname{supp} \chi_1 \subset (0, 1)$ and $\operatorname{supp} \chi_2 \subset (1/2, \infty)$. According to this representation set

$$
I_1 := I_{1,1} + I_{1,2}.
$$

Now $I_{1,1}$ can be written in the form

$$
\begin{aligned}
I_{1,1} &= \widetilde{c}_\lambda \Gamma(1+\lambda)(1+O(n^{-1})) e^{\frac{i}{5\pi} \ln^2 n} \ln^{1-\nu} n \ln^\beta(\ln n) \\
&\quad \cdot \left(\int_0^1 \frac{e^{i\Lambda \ln(1+v)} F(v)}{v^\lambda} \, dv + \int_0^1 K(v,n)\, dv \right),
\end{aligned}
\tag{12.1.23}
$$

where $F(v) = \frac{e^{\frac{i}{5\pi} \ln^2(v+1)}}{(1+v)^2} \chi_1(v)$, and the function $K(v,n)$ admits the estimate

$$
|K(v,n)| \le \mathrm{const}\, \frac{v}{(1+v)^2 v^\lambda \ln n (\ln \ln n)} = \frac{\mathrm{const}\, v^{1-\lambda}}{\ln n (\ln \ln n)}.
$$

Thus

$$
\int_0^1 |K(v,n)|\, dv \le \frac{\mathrm{const}}{\ln n (\ln \ln n)}.
\tag{12.1.24}
$$

According to the stationary phase method (see, for example, [76], p.97) we have

$$
\begin{aligned}
\int_0^1 \frac{e^{i\Lambda \ln(1+v)}}{v^\lambda} F(v)\, dv &= \frac{F(0)\Gamma(1-\lambda) e^{i\frac{\pi}{2}(1-\lambda)}}{\Lambda^{1-\lambda}} \left(1 + \frac{1}{\Lambda}\right) \\
&= \frac{d_\lambda}{\ln^{1-\lambda} n} \left(1 + O\left(\frac{1}{\ln n}\right)\right)
\end{aligned}
$$

where

$$
d_\lambda = \frac{\Gamma(1-\lambda) e^{i\frac{\pi}{2}(1-\lambda)}}{(2\pi/5)^{1-\lambda}}.
\tag{12.1.25}
$$

Thus from (12.1.24) and (12.1.25) we have

$$
I_{1,1} = \widetilde{c}_\lambda d_\lambda \Gamma(1+\lambda) e^{\frac{i}{5\pi} \ln^2 n} \ln^{\lambda-\nu} n \ln^\beta(\ln n) \left(1 + O\left(\frac{1}{\ln^\lambda n (\ln \ln \lambda)}\right)\right).
\tag{12.1.26}
$$

Pass now to the term $I_{1,2}$. Represent it in the form

$$
I_{1,2} = \widetilde{c}_\lambda \Gamma(1+\lambda)(1+O(n^{-1})) e^{\frac{i}{5\pi} \ln^2 n} \ln^{1-\nu} n \ln^\beta(\ln n) \int_{1/2}^\infty e^{i\Lambda \ln(1+v)} \Phi(v,n)\, dv.
$$

Integrating by parts we have

$$
\begin{aligned}
I_{1,2} &= \widetilde{c}_\lambda \Gamma(1+\lambda)(1+O(n^{-1})) e^{\frac{i}{5\pi} \ln^2 n} \ln^{1-\nu} n \ln^\beta(\ln n) \\
&\quad \cdot \frac{1}{i\Lambda} \int_{1/2}^\infty e^{i\Lambda \ln(1+v)} \left[(1+v) \frac{\partial \Phi}{\partial v}(v,n) + \Phi(v,n) \right] dv.
\end{aligned}
$$

Since

$$
|\Phi| + \left| (1+v)\frac{\partial \Phi}{\partial v} \right| \le M \frac{\ln^{1-\nu} v \ln^{|\beta|}(\ln v)}{v^{2+\lambda}},
$$

where M is independent of n, we have

$$I_{1,2} = O(\ln^{-\nu} n \, \ln^{\beta}(\ln n)). \tag{12.1.27}$$

Thus from (12.1.26) and (12.1.27) we get

$$I_1 = \widetilde{c}_\lambda d_\lambda \Gamma(1+\lambda) e^{\frac{i}{5\pi} \ln^2 n} \ln^{\lambda-\nu} n \, \ln^{\beta}(\ln n) \left(1 + O\left(\frac{1}{\ln^\lambda n}\right)\right). \tag{12.1.28}$$

Thus finally (12.1.19), (12.1.20) and (12.1.28) show that

$$\gamma_{a_{\nu,\beta},\lambda}(n) = \widetilde{c}_\lambda d_\lambda \Gamma(1+\lambda) e^{i \ln^2 n} \ln^{\lambda-\nu} n \, \ln^{\beta}(\ln n) \left(1 + O\left(\frac{1}{\ln^\lambda n}\right)\right),$$

which implies all the statements of the theorem. $\qquad\square$

12.2 Schatten classes

In many questions of operator theory, separation in the ideal of all compact operators of certain specific classes plays an important role. A special role is played by the so-called Schatten classes, and among them, the trace and Hilbert-Schmidt classes.

It is evident that Toeplitz operator $T_a^{(\lambda)}$ with radial symbol $a(r)$ belongs to Schatten class $K_p(\lambda)$, $1 \le p < \infty$, on the space $\mathcal{A}_\lambda^2(\mathbb{D})$ if and only if

$$\|T_a^{(\lambda)}\|_{p,\lambda} = \left(\sum_{n=1}^{\infty} |\gamma_{a,\lambda}(n)|^p\right)^{1/p} < \infty.$$

In the context of this paper the following question is very natural: *given a radial symbol $a(r)$, what is the structure of the pairs (p, λ) for which $T_a^{(\lambda)} \in K_p(\lambda)$?*

We start with the sufficient conditions on a symbol $a(r)$ for $T_a^{(\lambda)} \in K_p(\lambda)$.

Theorem 12.2.1. *Let $a(\sqrt{r}) \in L_1(0,1)$ and let for some $j = 0, 1, \ldots$, the function $B_a^{(j)}(r)$ (see (12.1.4)) satisfy one of the conditions*

$$\int_0^1 |B_a^{(j)}(r)|(1-r)^{-(1+j+\frac{1}{p})} \, dr \; < \; \infty, \qquad p \ge 1, \tag{12.2.1}$$

$$\int_0^1 |B_a^{(j)}(r)|^p (1-r)^{-(2+j-\varepsilon)} \, dr \; < \; \infty, \qquad p > 1, \tag{12.2.2}$$

where $\varepsilon > 0$ can be arbitrarily small. Then $T_a^{(\lambda)} \in K_p(\lambda)$.

Proof. Assume first that $j = 0$, $p > 1$, and the condition (12.2.1) holds. Then

$$\|T_a^{(\lambda)}\|_{p,\lambda} = \left(\sum_{n=0}^{\infty} \left|\frac{1}{B(n+1,\lambda+1)} \int_0^1 a(\sqrt{r})(1-r)^\lambda r^n \, dr\right|^p\right)^{\frac{1}{p}}.$$

Applying the Hölder inequality we have

$$
\begin{aligned}
\|T_a^{(\lambda)}\|_{p,\lambda} \;&=\; \left(\sum_{n=0}^{\infty} \left| \frac{1}{B(n+1,\lambda+1)} \int_0^1 \left(a^{\frac{1}{p}}(\sqrt{r})(1-r)^{(\lambda+1-\frac{1}{p^2})} r^n \right) \right. \right. \\
&\qquad \left. \left. \cdot \left(a^{\frac{1}{q}}(\sqrt{r})(1-r)^{-(1-\frac{1}{p^2})} dr \right) \right|^p \right)^{\frac{1}{p}} \\
&\leq\; \left(\sum_{n=0}^{\infty} \frac{1}{B^p(n+1,\lambda+1)} \int_0^1 |a(\sqrt{r})|(1-r)^{(\lambda+1-\frac{1}{p^2})p} r^{np} dr \right. \\
&\qquad \left. \cdot \left(\int_0^1 |a(\sqrt{r})|(1-r)^{-q(1-\frac{1}{p^2})} dr \right)^{\frac{p}{q}} \right)^{\frac{1}{p}} \\
&\leq\; \left(\int_0^1 |a(\sqrt{r})|(1-r)^{-(1+\frac{1}{p})} dr \right)^{\frac{1}{q}} \left(\int_0^1 |a(\sqrt{r})|(1-r)^{(\lambda+1-\frac{1}{p^2})p} \right. \\
&\qquad \left. \cdot \left(\sum_{n=0}^{\infty} \frac{r^{np}}{B^p(n+1,\lambda+1)} \right) dr \right)^{\frac{1}{p}}.
\end{aligned}
\tag{12.2.3}
$$

Estimate now the last sum using the asymptotic representation (12.1.10),

$$
\begin{aligned}
\left| \sum_{n=0}^{\infty} \frac{r^{np}}{B^p(n+1,\lambda+1)} \right| \;&\leq\; \text{const} \sum_{n=0}^{\infty} n^{p(\lambda+1)} r^{np} \\
&\leq\; \text{const} \int_0^{\infty} u^{p(\lambda+1)} r^{up} du \\
&=\; \text{const} \int_0^{\infty} e^{-up \ln r^{-1}} u^{p(\lambda+1)} du \\
&=\; \text{const} \, (p \ln r^{-1})^{-p(\lambda+1)-1} \int_0^{\infty} e^{-v} v^{p(\lambda+1)} dv \\
&=\; \text{const} \, (p \ln r^{-1})^{-p(\lambda+1)-1} \, \Gamma(p(\lambda+1)+1) \\
&\leq\; \text{const} \, (1-r)^{-p(\lambda+1)-1}.
\end{aligned}
\tag{12.2.4}
$$

Note that the estimate (12.2.4) holds for all $p \geq 1$. Since

$$
\left(\lambda + 1 - \frac{1}{p^2} \right) p - p(\lambda+1) - 1 = -\left(1 + \frac{1}{p} \right),
$$

from (12.2.3) and (12.2.4) it follows that

$$
\|T_a^{(\lambda)}\|_{p,\lambda} \leq \operatorname{const} \left(\int_0^\infty |a(\sqrt{r})|(1-r)^{-(1+\frac{1}{p})}\,dr \right)^{\frac{1}{q}}
$$

$$
\cdot \left(\int_0^\infty |a(\sqrt{r})|(1-r)^{-(1+\frac{1}{p})}\,dr \right)^{\frac{1}{p}}
$$

$$
= \operatorname{const} \int_0^\infty |a(\sqrt{r})|(1-r)^{-(1+\frac{1}{p})}\,dr.
$$

Thus according to (12.2.1) with $j=0$ we get $T_a^{(\lambda)} \in K_p(\lambda)$.

If $p=1$ then

$$
\|T_a^{(\lambda)}\|_{1,\lambda} \leq \sum_{n=0}^\infty \frac{1}{B(n+1,\lambda+1)} \int_0^1 |a(\sqrt{r})|(1-r)^\lambda r^n\,dr
$$

$$
\leq \int_0^1 |a(\sqrt{r})|(1-r)^\lambda \left(\sum_{n=0}^\infty \frac{r^n}{B(n+1,\lambda+1)} \right) dr.
$$

Since

$$
\sum_{n=0}^\infty \frac{r^n}{B(n+1,\lambda+1)} = (\lambda+1)(1-r)^{-(\lambda+2)} \tag{12.2.5}
$$

we have

$$
\|T_a^{(\lambda)}\|_{1,\lambda} \leq (\lambda+1) \int_0^1 |a(\sqrt{r})|(1-r)^{-2}\,dr,
$$

and thus $T_a^{(\lambda)} \in K_1(\lambda)$ according to (12.2.1) with $j=0$.

Let now $p>1$ and let the condition (12.2.2) hold with $j=0$. Applying the Hölder inequality in another way and using (12.2.4) we have

$$
\|T_a^{(\lambda)}\|_{p,\lambda} = \left(\sum_{n=0}^\infty \left| \frac{1}{B(n+1,\lambda+1)} \int_0^1 \left(a(\sqrt{r})(1-r)^{(\lambda+\frac{1-\sigma}{q})} r^n \right) \right. \right.
$$

$$
\left. \left. \cdot \left((1-r)^{-\frac{1-\sigma}{q}} \right) dr \right|^p \right)^{\frac{1}{p}}
$$

$$
\leq \operatorname{const} \left(\int_0^1 (1-r)^{-(1-\sigma)}\,dr \right)^{\frac{1}{q}}
$$

$$
\cdot \left(\int_0^1 |a(\sqrt{r})|^p (1-r)^{(\lambda+\frac{1-\sigma}{q})p-(\lambda+1)p-1}\,dr \right)^{\frac{1}{p}}.
$$

Taking $\sigma = \frac{\varepsilon}{p-1} > 0$ we have

$$
\|T_a^{(\lambda)}\|_{p,\lambda} \leq \operatorname{const} \left(\int_0^1 |a(\sqrt{r})|^p (1-r)^{-(2-\varepsilon)}\,dr \right)^{\frac{1}{p}},
$$

and thus $T_a^{(\lambda)} \in K_p(\lambda)$ according to (12.2.2).

The case $j > 0$ is considered by integrating by parts (as in the proof of Theorem 12.1.1) and by repeating the above arguments. $\qquad\square$

The next theorem shows that for a non-negative symbol or for a symbol having any non-negative mean $B_a^{(j)}(r)$, condition (12.2.1) is necessary as well for trace class operators, i.e., for $p = 1$.

Theorem 12.2.2. *Given a symbol $a(r)$, let for some $j \in \mathbb{Z}_+$, $B_a^{(j)}(r) \geq 0$ a.e. and $T_a^{(\lambda)} \in K_1(\lambda)$. Then*

$$\int_0^1 B_a^{(j)}(r)\,(1-r)^{-(2+j)}\,dr < \infty. \qquad (12.2.6)$$

Proof. Let first $j = 0$, then according to (12.2.5) we have

$$\|T_a^{(\lambda)}\|_{1,\lambda} = (\lambda+1)\int_0^1 a(\sqrt{r})\,(1-r)^{-2}\,dr < \infty,$$

and the condition (12.2.6) holds.

Let now $j = 1$ and thus $B_a^{(1)}(r) \geq 0$ a.e. Integrating by parts we have

$$\|T_a^{(\lambda)}\|_{1,\lambda} = \sum_{n=0}^{\infty} \frac{1}{B(n+1,\lambda+1)} \int_0^1 B_a^{(1)}(r)\,(\lambda(1-r)^{\lambda-1}r^n + n(1-r)^\lambda r^{n-1})dr.$$

Using (12.2.5) and the representation

$$\sum_{n=0}^{\infty} \frac{nr^{n-1}}{B(n+1,\lambda+1)} = (\lambda+1)(\lambda+2)(1-r)^{-(\lambda+3)}$$

we get (12.2.6):

$$\begin{aligned}
\|T_a^{(\lambda)}\|_{1,\lambda} &= \lambda(\lambda+1)\int_0^1 B_a^{(1)}(r)\,(1-r)^{-3}\,dr \\
&\quad + (\lambda+1)(\lambda+2)\int_0^1 B_a^{(1)}(r)\,(1-r)^{-3}\,dr \\
&= 2(\lambda+1)^2 \int_0^1 B_a^{(1)}(r)\,(1-r)^{-3}\,dr.
\end{aligned}$$

The cases $j > 1$ are considered analogously. $\qquad\square$

We give now several examples for the above theorems.

Example 12.2.3. For the symbol $a(r) = (1-r^2)^\alpha$ we have

$$\gamma_{a,\lambda}(n) = \frac{1}{B(n+1,\lambda+1)} \int_0^1 (1-r)^{\lambda+\alpha}\,r^n\,dr = \frac{B(n+1,\lambda+\alpha+1)}{B(n+1,\lambda+1)}.$$

The asymptotic representation (12.1.10) implies

$$\gamma_{a,\lambda}(n) = \frac{\Gamma(\lambda + \alpha + 1)}{\Gamma(\lambda + 1)}(n+1)^{-\alpha}\left(1 + O\left(\frac{1}{n+1}\right)\right).$$

Thus there exists C_1, $C_2 > 0$ such that

$$C_1\left(\sum_{n=0}^{\infty}(n+1)^{-\alpha p}\right)^{\frac{1}{p}} \le \|T_a^{(\lambda)}\|_{p,\lambda} \le C_2\left(\sum_{n=0}^{\infty}(n+1)^{-\alpha p}\right)^{\frac{1}{p}},$$

and $T_a^{(\lambda)} \in K_p(\lambda)$ if and only if $\alpha \in (\frac{1}{p}, \infty)$.

Note, that for $\alpha \in (\frac{1}{p}, \infty)$ the above symbol satisfies both conditions (12.2.1) and (12.2.2).

Example 12.2.4 (Unbounded increasing sequence). Consider the following non-negative symbol

$$a(\sqrt{r}) = \begin{cases} a_k, & r \in I_k = [r_k, r_k + \varepsilon_k], \quad k \in \mathbb{N} \\ 0 & r \in [0,1] \setminus \cup_{k=1}^{\infty} I_k \end{cases},$$

where $\{r_k\}$ is a positive increasing sequence tending to 1, $r_k + \varepsilon_k < r_{k+1}$, and a_k, $\varepsilon_k > 0$.

The conditions (12.2.1) and (12.2.2) for $j = 0$ for such a symbol are

$$p\sum_{k=1}^{\infty} a_k\left((1 - (r_k + \varepsilon_k))^{-\frac{1}{p}} - (1 - r_k)^{-\frac{1}{p}}\right) < \infty,$$

$$(1 - \varepsilon)^{-1}\sum_{k=1}^{\infty} a_k^p\left((1 - (r_k + \varepsilon_k))^{-(1-\varepsilon)} - (1 - r_k)^{-(1-\varepsilon)}\right) < \infty.$$

Setting

$$a_k = k^a, \quad a > 0, \quad r_k = 1 - \frac{1}{k}, \quad \varepsilon_k = \frac{k^{-b}}{2}, \quad b > 2 + \frac{1}{p},$$

the above conditions are equivalent to

$$\sum_{k=1}^{\infty} k^a \cdot k^{\frac{1}{p} - b + 1} < \infty, \tag{12.2.7}$$

$$\sum_{k=1}^{\infty} k^{ap} \cdot k^{(1-\varepsilon)-b+1} < \infty. \tag{12.2.8}$$

Thus (12.2.7) holds if

$$a < b - 2 - \frac{1}{p},$$

while (12.2.8) holds if

$$a < \frac{b-3}{p},$$

recalling that $\varepsilon > 0$ can be chosen arbitrarily small.

For our choice of b we have (for $p > 1$)

$$b - 2 - \frac{1}{p} > \frac{b-3}{p},$$

thus the first condition gives a bigger region for $T_a^{(\lambda)} \in K_p(\lambda)$, and thus the first condition of Theorem 12.2.1 is better adapted for increasing sequences $\{a_k\}$ than the second one.

Example 12.2.5 (Decreasing sequence). Consider the symbol from the previous example, but setting now

$$a_k = k^{-a}, \quad a > 0, \quad r_k = 1 - \frac{1}{k}, \quad \varepsilon_k = \frac{k^{-b}}{2}, \quad b \in [2, 2 + \frac{1}{p}].$$

In this case after a short calculation we come to the following conditions:

$$2 + \frac{1}{p} - b \ > \ a,$$

$$\frac{3-b}{p} \ > \ a.$$

For the current choice of b we have (for $p > 1$)

$$\frac{3-b}{p} > 2 + \frac{1}{p} - b,$$

and thus now the second condition gives a bigger region for $T_a^{(\lambda)} \in K_p(\lambda)$; that is, the second condition of Theorem 12.2.1 is better adapted for decreasing sequences $\{a_k\}$ than the first one.

In the three previous examples the symbols were non-negative. Consider now

Example 12.2.6 (Unbounded oscillating symbol). Let

$$a(r) = (1 - r^2)^{-\beta} \sin(1 - r^2)^{-\alpha}, \qquad \alpha > 0, \ \ 0 < \beta < 1.$$

Recall (see (6.1.5)) that we have the following asymptotic representation in a neighborhood of $r = 1$,

$$B_a^{(1)}(r) = \frac{\cos(1-r)^{-\alpha}}{\alpha} (1-r)^{\alpha-\beta+1} + O((1-r)^{2\alpha-\beta+1}). \tag{12.2.9}$$

Thus by condition (12.2.1) for $j = 1$ of Theorem 12.2.1 we have $T_a^{(\lambda)} \in K_p(\lambda)$ if

$$(\alpha - \beta + 1) - (1 + 1 + \tfrac{1}{p}) > -1 \qquad \text{or} \qquad \alpha - \beta > \tfrac{1}{p}.$$

Setting $\beta_1 = \alpha - \beta + 1$, analogously to (12.2.9) we can obtain the representation

$$B_a^{(2)}(r) = -\frac{\sin(1 - r)^{-\alpha}}{\alpha^2}(1 - r)^{\alpha + \beta_1 + 1} + O((1 - r)^{2\alpha + \beta_1 + 1}).$$

Apply now condition (12.2.7) for $j = 2$; we have that $T_a^{(\lambda)} \in K_p(\lambda)$ if

$$(\alpha + \beta_1 + 1) - (1 + 2 + \tfrac{1}{p}) > -1 \qquad \text{or} \qquad 2\alpha - \beta > \tfrac{1}{p}.$$

In the same manner, for any $j \in \mathbb{N}$ using the asymptotic representation of the corresponding mean $B^{(j)}(r)$ we can get that $T_a^{(\lambda)} \in K_p(\lambda)$ if

$$j\alpha - \beta > \frac{1}{p}. \tag{12.2.10}$$

Since for any $\alpha > 0$ and $0 < \beta < 1$ there exists $j \in \mathbb{N}$ such that (12.2.10) holds, the operator $T_a^{(\lambda)}$ always belongs to $K_p(\lambda)$, for any $\lambda \geq 0$, $p \geq 1$, $\alpha > 0$, and $0 < \beta < 1$.

Theorem 12.2.7. *Let $T_a^{(\lambda_0)} \in K_p(\lambda_0)$ for some $\lambda_0 > 0$, and let $1 \leq p < \infty$. Then for all $\lambda \in [0, \lambda_0]$ we have*

$$T_a^{(\lambda)} \in K_p(\lambda).$$

Proof. Using the representation (12.1.9) we have

$$\|T_a^{(\lambda)}\|_{p,\lambda} = \left(\sum_{n=1}^{\infty} \left| \frac{\sin \pi(\lambda_0 - \lambda)}{\pi} \sum_{j=0}^{\infty} \frac{B(j - (\lambda - \lambda_0), 1 + (\lambda - \lambda_0))}{B(n + 1, \lambda + 1)} \right. \right.$$
$$\left. \left. \times B(n + j + 1, \lambda_0 + 1)\gamma_{a\lambda_0}(n + j) \right|^p \right)^{1/p}.$$

Then by (12.1.10)

$$\|T_a^{(\lambda)}\|_{p,\lambda} \leq \text{const} \left(\sum_{n=1}^{\infty} \left| \sum_{j=0}^{\infty} \frac{n^{\lambda+1} \gamma_{a,\lambda_0}(n + j)}{(n + j + 1)^{\lambda_0 + 1}(j + 1)^{1-(\lambda_0 - \lambda)}} \right|^p \right)^{1/p}$$

$$= \text{const} \left(\sum_{n=1}^{\infty} \left| \sum_{\nu=0}^{\infty} \sum_{\nu \leq \frac{j+1}{n} < \nu+1} \frac{\gamma_{a,\lambda_0}(n + j)}{n \left(1 + \frac{j+1}{n}\right)^{\lambda+1} \left(\frac{j+1}{n}\right)^{1-(\lambda_0 - \lambda)}} \right|^p \right)^{1/p}$$

$$=: \text{const} \left(\sum_{n=1}^{\infty} \left| \sum_{\nu=0}^{\infty} \theta_\nu(n) \right|^p \right)^{1/p}.$$

Using the Minkowski inequality we have

$$\|T_a^{(\lambda)}\|_{p,\lambda} \leq \mathrm{const} \sum_{\nu=0}^{\infty} \left(\sum_{n=1}^{\infty} |\theta_\nu(n)|^p \right)^{1/p}.$$

It is easy to see that

$$|\theta_\nu(n)| \leq \frac{1}{(1+\nu)^{\lambda+1}\,\nu^{1-(\lambda_0-\lambda)}} \sum_{\nu \leq \frac{j+1}{n} < \nu+1} \frac{|\gamma_{a,\lambda_0}(n+j)|}{n}$$

for each $\nu = 1, 2, \ldots$.

Suppose first that $p > 1$. According to the Hölder inequality

$$|\theta_\nu(n)|^p \leq \frac{1}{(1+\nu)^{p(\lambda+1)}\,\nu^{p(1-(\lambda_0-\lambda))}} \left(\frac{n}{n^q} \right)^{p/q} \sum_{\nu \leq \frac{j+1}{n} < \nu+1} |\gamma_{a,\lambda_0}(n+j)|^p,$$

where $1/p + 1/q = 1$. So we have

$$|\theta_\nu(n)|^p \leq \frac{\mathrm{const}}{(1+\nu)^{(\lambda+1)p}\,\nu^{(1-(\lambda-\lambda_0))p}} \sum_{\nu \leq \frac{j+1}{n} < \nu+1} \frac{|\gamma_{a,\lambda_0}(n+j)|^p}{n},$$

for $\nu = 1, 2, \ldots$. Thus

$$\|T_a^{(\lambda)}\|_{p,\lambda} \leq \mathrm{const} \left[\left(\sum_{n=1}^{\infty} |\theta_0(n)|^p \right)^{1/p} \right.$$

$$\left. + \sum_{\nu=1}^{\infty} \frac{1}{(1+\nu)^{\lambda+1}\,\nu^{1-(\lambda-\lambda_0)}} \left(\sum_{n=1}^{\infty} \frac{1}{n} \sum_{\nu \leq \frac{j+1}{n} < \nu+1} |\gamma_{a,\lambda_0}(n+j)|^p \right)^{1/p} \right].$$

Consider first the series

$$\sum_{n=1}^{\infty} \frac{1}{n} \sum_{\nu \leq \frac{j+1}{n} < \nu+1} |\gamma_{a,\lambda_0}(n+j)|^p = \sum_{n=1}^{\infty} \frac{1}{n} \sum_{\nu+1 \leq \frac{m+1}{n} < \nu+2} |\gamma_{a,\lambda_0}(m)|^p$$

$$= \sum_{m=\nu}^{\infty} |\gamma_{a,\lambda_0}(m)|^p \sum_{\frac{m+1}{\nu+2} < n \leq \frac{m+1}{\nu+1}} \frac{1}{n} \leq \mathrm{const} \sum_{m=\nu}^{\infty} |\gamma_{a,\lambda_0}(m)|^p \int_{\frac{m+1}{\nu+2}}^{\frac{m+1}{\nu+1}} \frac{du}{u}$$

$$= \mathrm{const} \sum_{m=\nu}^{\infty} |\gamma_{a,\lambda_0}(m)|^p \left(\ln \frac{m+1}{\nu+1} - \ln \frac{m+1}{\nu+2} \right)$$

$$= \mathrm{const} \ln \frac{\nu+2}{\nu+1} \cdot \|T_a^{(\lambda_0)}\|_{p,\lambda_0}^p.$$

Thus

$$\|T_a^{(\lambda)}\|p, \lambda \ \leq \ \text{const} \left[\left(\sum_{n=1}^{\infty} n^{-(\lambda_0 - \lambda)p} \right.\right.$$

$$\left. \cdot \left| \sum_{\frac{j+1}{n} < 1} \frac{\gamma_{a,\lambda_0}(n+j)}{(1 + \frac{j+1}{n})^{\lambda+1} (j+1)^{1-(\lambda_0-\lambda)}} \right|^p \right)^{1/p}$$

$$\left. + \|T_a^{(\lambda_0)}\|_{p,\lambda_0} \sum_{\nu=1}^{\infty} \frac{\ln^{1/p}(1 + \frac{1}{\nu+1})}{(1+\nu)^{\lambda+1} \nu^{1-(\lambda-\lambda_0)}} \right] . \qquad (12.2.11)$$

The second term of (12.2.11) is bounded as

$$\lambda + 1 + (1 - (\lambda - \lambda_0)) + \frac{1}{p} = 2 + \lambda_0 + \frac{1}{p} > 1.$$

Applying the Hölder inequality to the first term of (12.2.11) we get

$$I_1 \ = \ \left(\sum_{n=1}^{\infty} n^{-(\lambda_0-\lambda)p} \left| \sum_{j+1<n} \frac{\gamma_{a,\lambda_0}(n+j)}{\left(1 + \frac{j+1}{n}\right)^{\lambda+1} (j+1)^{1-(\lambda_0-\lambda)}} \right|^p \right)^{1/p}$$

$$\leq \ \left(\sum_{n=1}^{\infty} n^{-(\lambda_0-\lambda)p} \left| \sum_{j+1<n} \left(\frac{\gamma_{a,\lambda_0}(n+j)}{(j+1)^{\frac{1-(\lambda_0-\lambda)}{p}}} \right) \cdot \frac{1}{(j+1)^{\frac{1-(\lambda_0-\lambda)}{q}}} \right|^p \right)^{1/p}$$

$$\leq \ \left(\sum_{n=1}^{\infty} n^{-(\lambda_0-\lambda)p} \sum_{j+1<n} \frac{|\gamma_{a,\lambda_0}(n+j)|^p}{(j+1)^{1-(\lambda_0-\lambda)}} \cdot \left(\sum_{j+1<n} \frac{1}{(j+1)^{1-(\lambda_0-\lambda)}} \right)^{p/q} \right)^{1/p}$$

$$\leq \ \text{const} \left(\sum_{n=1}^{\infty} n^{-(\lambda_0-\lambda)p} n^{(\lambda_0-\lambda)\frac{p}{q}} \sum_{j+1<n} \frac{|\gamma_{a,\lambda_0}(n+j)|^p}{(j+1)^{1-(\lambda_0-\lambda)}} \right)^{1/p}$$

$$= \ \text{const} \left(\sum_{n=1}^{\infty} n^{-(\lambda_0-\lambda)} \sum_{n+1 \leq m+1 < 2n} \frac{|\gamma_{a,\lambda_0}(m)|^p}{(m-n+1)^{1-(\lambda_0-\lambda)}} \right)^{1/p}$$

$$\leq \ \text{const} \left(\sum_{m=1}^{\infty} |\gamma_{a,\lambda_0}(m)|^p \sum_{\frac{m+1}{2} < n \leq m} \frac{n^{-(\lambda_0-\lambda)}}{(m-n+1)^{1-(\lambda_0-\lambda)}} \right)^{1/p}$$

$$\leq \ \text{const} \left(\sum_{m=1}^{\infty} |\gamma_{a,\lambda_0}(m)|^p \int_{\frac{m+1}{2}}^{m+1} \frac{u^{-(\lambda_0-\lambda)} \, du}{(m+1-u)^{1-(\lambda_0-\lambda)}} \right)^{1/p} .$$

Changing $u = (m+1)v$ in the last integral we have

$$I_1 \leq \text{const} \left(\sum_{m=1}^{\infty} |\gamma_{a,\lambda_0}(m)|^p \int_{\frac{1}{2}}^{1} \frac{v^{-(\lambda_0 - \lambda)} \, dv}{(1-v)^{1-(\lambda_0 - \lambda)}} \right)^{1/p}.$$

Since $1 - (\lambda_0 - \lambda) < 1$, the last integral exists, and the first summand in (12.2.11) is estimated as

$$I_1 \leq \text{const} \, \|T_a^{(\lambda_0)}\|_{p,\lambda_0}.$$

Thus (for $p > 1$) we have

$$\|T_a^{(\lambda)}\|_{p,\lambda} \leq \text{const} \, \|T_a^{(\lambda_0)}\|_{p,\lambda_0}.$$

The case $p = 1$ is considered analogously. $\qquad\qquad\square$

Note that the above result admits the following reformulation. Given a radial symbol $a(r)$, let $T_a^{(\lambda_0)} \in K_{p_0}(\lambda_0)$. Then $T_a^{(\lambda)} \in K_p(\lambda)$ for all $(p, \lambda) \in [p_0, \infty) \times [0, \lambda_0]$.

12.3　Spectra of Toeplitz operators, continuous symbols

In this and the next two chapters we will use the following notion of the limit of a family of sets. Let E be a subset of $\mathbb{R}$ having $+\infty$ as a limit point (normally $E = [0, +\infty)$), and let there be for each $\lambda \in E$ a set $M_\lambda \subset \mathbb{C}$. Define the set M_∞ as the set of all $z \in \mathbb{C}$ for which there exists a sequence of complex numbers $\{z_n\}_{n \in \mathbb{N}}$ such that

   (i)　for each $n \in \mathbb{N}$ there exists $\lambda_n \in E$ such that $z_n \in M_{\lambda_n}$,

  (ii)　$\lim_{n \to \infty} \lambda_n = +\infty$,

 (iii)　$z = \lim_{n \to \infty} z_n$.

We will write

$$M_\infty = \lim_{\lambda \to +\infty} M_\lambda,$$

and call M_∞ the (partial) limit set of a family $\{M_\lambda\}_{\lambda \in E}$ when $\lambda \to +\infty$.

For the case when E is a discrete set with a unique limit point at infinity, the above notion coincides with the partial limiting set introduced in [97], Section 3.1.1. Following the arguments of Proposition 3.5 in [97] one can show that

$$M_\infty = \bigcap_\lambda \text{clos} \left(\bigcup_{\mu \geq \lambda} M_\mu \right).$$

Note that obviously

$$\lim_{\lambda \to +\infty} M_\lambda = \lim_{\lambda \to +\infty} \overline{M}_\lambda = M_\infty.$$

The *a priori* spectral information for Toeplitz operators with L_∞-symbols is given by Lemma A.2.3 and says that for each $a \in L_\infty(\mathbb{D})$ and each $\lambda > -1$,

$$\operatorname{sp} T_a^{(\lambda)} \subset \operatorname{conv}(\text{ess-Range } a). \tag{12.3.1}$$

For a radial symbol $a = a(r)$ the Toeplitz operator $T_a^{(\lambda)}$ is unitary equivalent to the multiplication operator $\gamma_{a,\lambda} I$, where the sequence $\gamma_{a,\lambda} = \{\gamma_{a,\lambda}(n)\}$ is given by (12.1.1). Thus we have obviously

$$\operatorname{sp} T_a^{(\lambda)} = \overline{M_\lambda(a)},$$

where $M_\lambda(a) = \operatorname{Range} \gamma_{a,\lambda}$.

An interesting question here is: what is the limit behavior of $\operatorname{sp} T_a^{(\lambda)}$ when $\lambda \to \infty$?

Theorem 12.3.1. *Let $a = a(r) \in C[0,1]$. Then*

$$\lim_{\lambda \to +\infty} \operatorname{sp} T_a^{(\lambda)} = \lim_{\lambda \to +\infty} M_\lambda(a) = \operatorname{Range} a. \tag{12.3.2}$$

Obviously the set $\operatorname{Range} a$ coincides with the spectrum $\operatorname{sp} aI$ of the operator of multiplication by $a = a(r)$ acting, say, on any of $L_2(\mathbb{D}, d\mu_\lambda)$, thus another form of (12.3.2) is

$$\lim_{\lambda \to +\infty} \operatorname{sp} T_a^{(\lambda)} = \operatorname{sp} aI.$$

Proof. We find the asymptotic of (12.1.1) when $\lambda \to +\infty$ using the Laplace method. Introduce a large parameter $L = \sqrt{\lambda^2 + n^2}$, and represent (12.1.1) in the form

$$\gamma_{a,\lambda}(n) = \frac{1}{B(n+1, \lambda+1)} \int_0^1 a(\sqrt{r}) e^{-LS(r,\varphi)} \, dr, \tag{12.3.3}$$

where $S(r,\varphi) = \sin\varphi \cdot \ln(1-r)^{-1} + \cos\varphi \cdot \ln r^{-1}$ and $\sin\varphi = \frac{\lambda}{L}$, $\cos\varphi = \frac{n}{L}$ with $\varphi \in [0, \frac{\pi}{2})$.

Find the maximum point of $S(r,\varphi)$; we have

$$S'_r(r,\varphi) = \frac{\sin\varphi}{1-r} - \frac{\cos\varphi}{r},$$

and obviously $S'_r(r,\varphi) = 0$ for

$$r_0 = \frac{\cos\varphi}{\sin\varphi + \cos\varphi} \in [0,1],$$

moreover for $\varphi = 0$ $r_0 = 1$ and for $\varphi = \frac{\pi}{2}$ $r_0 = 0$.

Rewrite now (12.3.3) in the form

$$
\begin{aligned}
\gamma_{a,\lambda}(n) \;=\; & \frac{1}{B(n+1,\lambda+1)} \left(\int_0^1 a(\sqrt{r_0})\, e^{-LS(r,\varphi)}\, dr \right. \\
& + \int_{U(r_0)} \left(a(\sqrt{r}) - a(\sqrt{r_0}) \right) e^{-LS(r,\varphi)}\, dr \\
& \left. + \int_{[0,1]\setminus U(r_0)} \left(a(\sqrt{r}) - a(\sqrt{r_0}) \right) e^{-LS(r,\varphi)}\, dr \right),
\end{aligned}
$$

where $U(r_0)$ is a neighborhood of r_0 such that for a sufficiently small ε,

$$
\sup_{r\in U(r_0)} |a(\sqrt{r}) - a(\sqrt{r_0})| < \varepsilon.
$$

We have

$$
\int_0^1 a(\sqrt{r_0})\, e^{-LS(r,\varphi)}\, dr = a(\sqrt{r_0})\, B(n+1,\lambda+1).
$$

Further

$$
\left| \int_{U(r_0)} \left(a(\sqrt{r}) - a(\sqrt{r_0}) \right) e^{-LS(r,\varphi)}\, dr \right| \le \varepsilon \int_{U(r_0)} e^{-LS(r,\varphi)}\, dr \le \varepsilon\, B(n+1,\lambda+1).
$$

Finally,

$$
\begin{aligned}
& \left| \int_{[0,1]\setminus U(r_0)} \left(a(\sqrt{r}) - a(\sqrt{r_0}) \right) e^{-LS(r,\varphi)}\, dr \right| \\
& \le \; 2 \sup_{r\in[0,1]} |a(\sqrt{r})| \int_{[0,1]\setminus U(r_0)} e^{-LS(r,\varphi)}\, dr \\
& \le \; \left(2 \sup_{r\in[0,1]} |a(\sqrt{r})|\, e^{-L\delta(\varepsilon)} \right) B(n+1,\lambda+1),
\end{aligned}
$$

where $\delta(\varepsilon) > 0$ can be taken independently on $r_0 \in [0,1]$.

Since ε can be taken arbitrarily small uniformly on $r_0 \in [0,1]$, we have the representation

$$
\gamma_{a,\lambda}(n) = a(\sqrt{r_0})(1 + \alpha(L)), \tag{12.3.4}
$$

where $\lim_{L\to\infty} \alpha(L) = 0$ uniformly on $r_0 \in [0,1]$, or $\varphi \in [0,\frac{\pi}{2}]$.

Finally, the last representation guarantees the statement of the theorem. $\qquad\square$

Recall that for a continuous symbol $a(r)$ and for each fixed λ the spectrum $\operatorname{sp} T_a^{(\lambda)}$ coincides with the closure of the set $\gamma_{a,\lambda} = \{\gamma_{a,\lambda}(n)\}$, i.e., is a discrete set with the unique limit point $a(1)$, and, in general, seems to have no strict connection with the range of $a(r)$. The definite tendency in the behavior of elements of the sequence $\gamma_{a,\lambda}$ starts to arise as λ tends to ∞, and the limit set $M_\infty(a)$ of these sequences coincides with the range of the initial symbol $a(r)$.

We illustrate the theorem on the symbol (hypocycloid)

$$a_1(r) = \frac{3}{4}(r + i\sqrt{1 - r^2})^8 + (r - i\sqrt{1 - r^2})^4,$$

presenting the images of the sequence $\gamma_{a_1,\lambda}(n)$ for the following values of λ: 0, 5, 12, and 200.

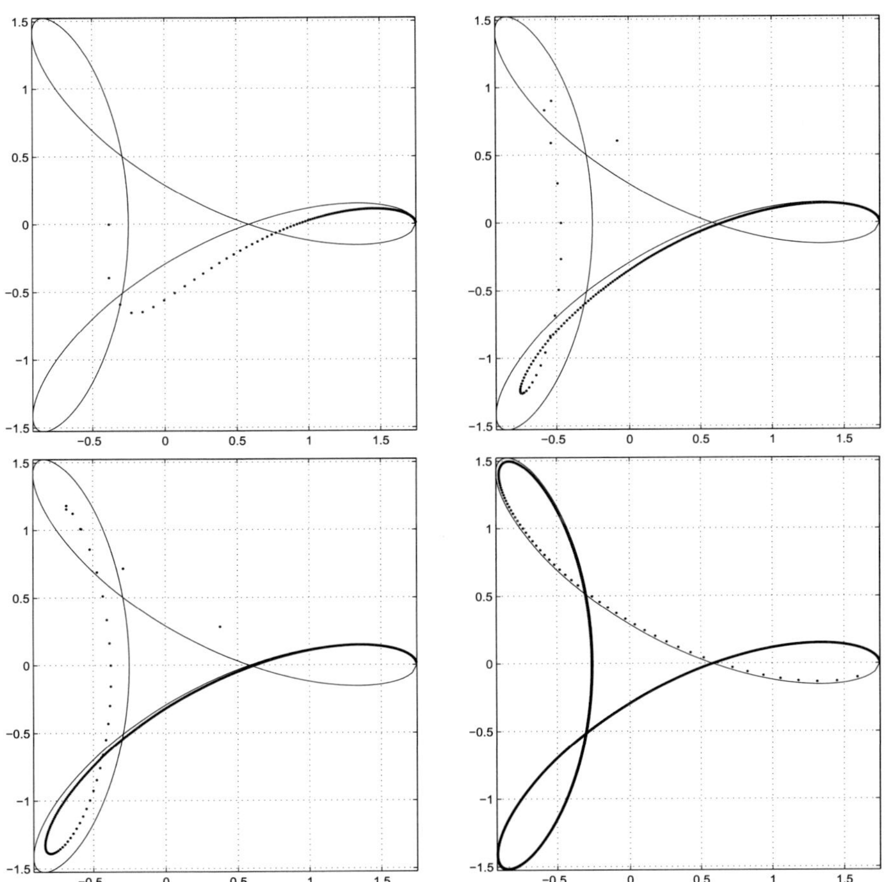

Figure 12.1: The images of $\gamma_{a_1,\lambda}$ for $\lambda = 0$, $\lambda = 5$, $\lambda = 12$, and $\lambda = 200$.

12.4 Spectra of Toeplitz operators, piece-wise continuous symbols

Let $b(r) = a(\sqrt{r})$ be a piece-wise continuous function which has jumps at a finite set of points

$$0 < r_1 < r_2 < \ldots < r_m < 1.$$

Introduce the sets

$$J_j(a) = \{z \in \mathbb{C} : z = a(\sqrt{r}), r \in [r_j + 0, r_{j+1} - 0]\},$$

where $j = 0, 1, \ldots, m$, $r_0 = 0$ and $r_{m+1} = 1$, and $I_j(a)$ being the straight line segment with the endpoints $a(\sqrt{r_j - 0})$ and $a(\sqrt{r_j + 0})$, for all $j = 0, 1, \ldots, m$.

Introduce now

$$\widetilde{R}(a) = \left(\bigcup_{j=0}^{m} J_j(a)\right) \cup \left(\bigcup_{j=1}^{m} I_j(a)\right),$$

that is,

$$\widetilde{R}(a) = \operatorname{Range} a \cup \left(\bigcup_{j=1}^{m} I_j(a)\right).$$

Theorem 12.4.1. *Let* $b(r) = a(\sqrt{r})$ *be a piece-wise continuous function as above. Then*

$$\lim_{\lambda \to +\infty} \operatorname{sp} T_a^{(\lambda)} = M_\infty(a) = \widetilde{R}(a).$$

Proof. We use the Laplace method as in Theorem 12.3.1. For any $\varepsilon > 0$ we take $\delta > 0$ such that for each interval $I \subset (r_j, r_{j+1})$ with length less than δ, $j = 1, 2, \ldots, m$, the following holds:

$$\sup_{s_1, s_2 \in I} |a(\sqrt{s_1}) - a(\sqrt{s_2})| < \varepsilon.$$

Suppose first that the maximum point $s_0 = \dfrac{\cos \varphi}{\sin \varphi + \cos \varphi}$ satisfies the condition

$$\sup_{j=1,2,\ldots,m} |s_0 - r_j| \geq \delta.$$

We have

$$
\begin{aligned}
\gamma_{a,\lambda}(n) &= a(\sqrt{s_0}) + \frac{1}{B(n+1, \lambda+1)} \int_{s_0-\delta}^{s_0+\delta} (a(\sqrt{r}) - a(\sqrt{s_0}))\, e^{-LS(r,\varphi)} dr \\
&\quad + \frac{1}{B(n+1, \lambda+1)} \int_{[0,1]\backslash(s_0-\delta, s_0+\delta)} (a(\sqrt{r}) - a(\sqrt{s_0}))\, e^{-LS(r,\varphi)} dr \\
&= a(\sqrt{s_0}) + O(\varepsilon) + O(e^{-\sigma L}),
\end{aligned}
\tag{12.4.1}
$$

where $\sigma = \min_{[0,1]\setminus(s_0-\delta,s_0+\delta)}(S(r,\varphi)-S(s_0,\varphi))$. Thus, varying φ and δ, we have that

$$J_j(a) \subset M_\infty(a), \qquad j = 0,1,\ldots,m.$$

Now suppose that there exists j such that $|s_0 - r_j| < \delta$. Then we have

$$
\begin{aligned}
\gamma_{a,\lambda}(n) &= \frac{1}{B(n+1,\lambda+1)}\left(a(\sqrt{r_j-0}) \int_{s_0-\delta}^{r_j} e^{-LS(r,\varphi)}\,dr \right.\\
&\qquad \left. + a(\sqrt{r_j+0}) \int_{r_j}^{s_0+\delta} e^{-LS(r,\varphi)}\,dr \right)\\
&\quad + \frac{1}{B(n+1,\lambda+1)}\left(\int_{s_0-\delta}^{r_j} \left(a(\sqrt{r}) - a(\sqrt{r_j-0}) \right) e^{-LS(r,\varphi)}\,dr \right.\\
&\qquad + \int_{r_j}^{s_0+\delta} \left(a(\sqrt{r}) - a(\sqrt{r_j+0}) \right) e^{-LS(r,\varphi)}\,dr\\
&\qquad \left. + \int_{[0,1]\setminus(s_0-\delta,s_0+\delta)} a(\sqrt{r})\, e^{-LS(r,\varphi)}\,dr \right)\\
&= \frac{1}{B(n+1,\lambda+1)}\left(a(\sqrt{r_j-0}) \int_{s_0-\delta}^{r_j} e^{-LS(r,\varphi)}\,dr \right.\\
&\qquad \left. + a(\sqrt{r_j+0}) \int_{r_j}^{s_0+\delta} e^{-LS(r,\varphi)}\,dr \right) + O(\varepsilon) + O(e^{-L\sigma}). \quad (12.4.2)
\end{aligned}
$$

Taking δ small enough we can suppose that

$$\frac{r_1}{2} < s_0 < \frac{1+r_n}{2}.$$

Thus the functions $(S''(s_0,\varphi))^{\pm 1}$ are uniformly bounded (on s_0), and then the following asymptotic calculations are uniform on s_0:

$$
\begin{aligned}
B(n+1,\lambda+1) &= \int_0^1 e^{-LS(r,\varphi)}\,dr\\
&= e^{-LS(s_0,\varphi)} \int_0^1 e^{-L\frac{S''(s_0,\varphi)(r-s_0)^2}{2}}\,dr\,(1+o(1))\\
&= e^{-LS(s_0,\varphi)} \int_{-s_0}^{1-s_0} e^{-L\frac{S''(s_0,\varphi)u^2}{2}}\,du\,(1+o(1))\\
&= \sqrt{\frac{2}{S''(s_0,\varphi)}}\, \frac{e^{-LS(s_0,\varphi)}}{L^{\frac{1}{2}}} \int_{-\infty}^{+\infty} e^{-v^2}\,dv\,(1+o(1)). \quad (12.4.3)
\end{aligned}
$$

Analogously,

$$\int_{r_j}^{s_0+\delta} e^{-LS(r,\varphi)}\,dr = \sqrt{\frac{2}{S''(s_0,\varphi)}}\, \frac{e^{-LS(s_0,\varphi)}}{L^{\frac{1}{2}}} \int_{x_j}^{+\infty} e^{-v^2}\,dv\,(1+o(1)), \quad (12.4.4)$$

and

$$\int_{s_0-\delta}^{r_j} e^{-LS(r,\varphi)} dr = \sqrt{\frac{2}{S''(s_0,\varphi)}} \, \frac{e^{-LS(s_0,\varphi)}}{L^{\frac{1}{2}}} \int_{-\infty}^{x_j} e^{-v^2} dv \, (1+o(1)), \qquad (12.4.5)$$

where

$$x_j = (r_j - s_0) \left(\frac{LS''(s_0,\varphi)}{2} \right)^{\frac{1}{2}}. \qquad (12.4.6)$$

Thus from (12.4.2)–(12.4.5) we have

$$\gamma_{a,\lambda}(n) = \left(a(\sqrt{r_j - 0})t + a(\sqrt{r_j + 0})\tau \right) (1 + o(1)) + O(\varepsilon) + O(e^{-L\sigma}), \qquad (12.4.7)$$

where

$$t = \frac{\int_{-\infty}^{x_j} e^{-v^2} dv}{\int_{-\infty}^{+\infty} e^{-v^2} dv} \qquad \text{and} \qquad \tau = \frac{\int_{x_j}^{+\infty} e^{-v^2} dv}{\int_{-\infty}^{+\infty} e^{-v^2} dv}.$$

Now it is evident that $t, \tau \in [0,1]$ and $t + \tau = 1$. Moreover, according to (12.4.6) for each $x \in \mathbb{R}$ there exist s_0 and L such that

$$x_j = (r_j - s_0) \left(\frac{LS''(s_0,\varphi)}{2} \right)^{\frac{1}{2}} = x.$$

Thus t runs over the whole segment $[0,1]$, which implies $I_j(a) \subset M_\infty(a)$. Thus

$$\widetilde{R}(a) \subset M_\infty(a).$$

The representations (12.4.1) and (12.4.7) imply the inverse inclusion

$$\widetilde{R}(a) \supset M_\infty(a). \qquad \qquad \square$$

Remark 12.4.2. The appearance of line segments which connect the one-sided limit values at the points of discontinuity of symbols is quite typical in the theory of Toeplitz operators with piece-wise continuous symbols acting either on the Hardy, or on the Bergman space (see, for example, [83, 202]). We stress the principal difference between our case and the cases just mentioned. In the case of Toeplitz operators with piece-wise continuous symbols the line segments appear in the *essential spectrum* of the Toeplitz operator. In our case any Toeplitz operator is a compact perturbation of a multiple of the identity, i.e., $T_{a(r)} = a(1)I + K$, and its essential spectrum consists of a single point $a(1)$ for all λ. For each fixed λ the spectrum of a Toeplitz operator coincides with the union of the *discrete set* (the sequence $\gamma_{a,\lambda}(n)$) with its limit point $a(1)$. The tendency of the line segment to form stars appears for large values of λ, while the line segments themselves appear only in the *limit set of spectra*.

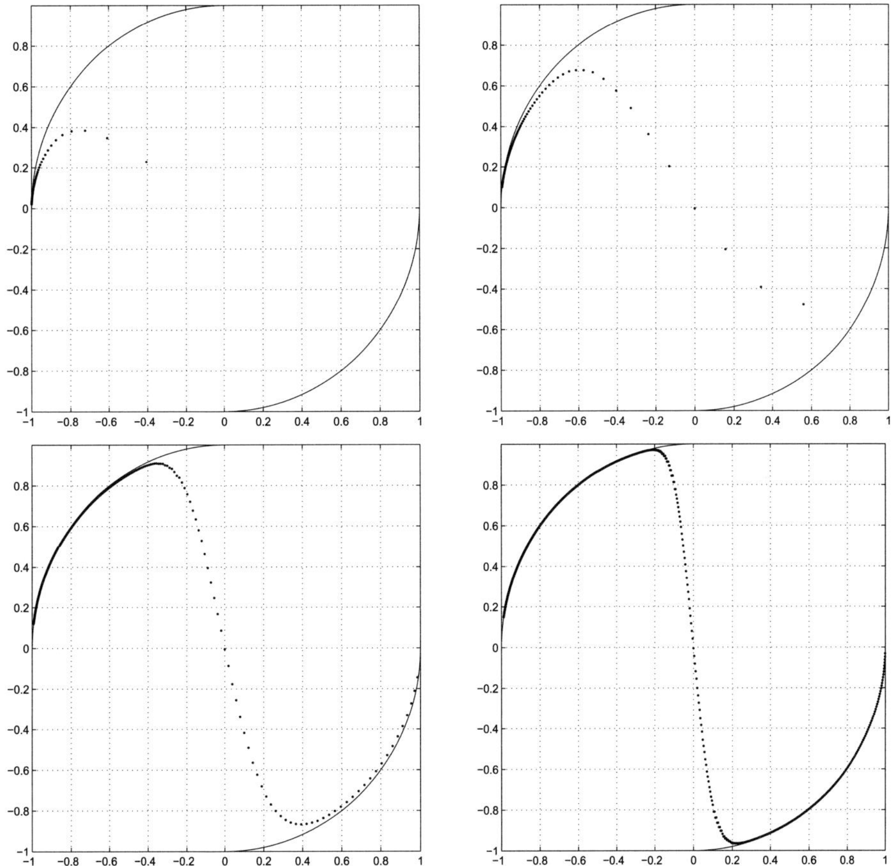

Figure 12.2: The images of $\gamma_{a_1,\lambda}$ for $\lambda = 0$, $\lambda = 4$, $\lambda = 40$, and $\lambda = 200$.

We illustrate this effect on the piece-wise continuous symbol

$$a(r) = \begin{cases} e^{-i\pi r^2}, & r \in [0, 1/\sqrt{2}] \\ e^{i\pi r^2}, & r \in (1/\sqrt{2}, 1] \end{cases},$$

for λ taking the values 0, 4, 40, and 200.

For L_∞-symbols apart from the a priori information (12.3.1), we have obviously

$$\lim_{\lambda \to +\infty} \operatorname{sp} T_a^{(\lambda)} = M_\infty(a) \subset \operatorname{conv}(\text{ess-Range}\, a). \tag{12.4.8}$$

Let us give a number of examples illustrating possible interrelations between these sets.

Example 12.4.3. Let $a = a(r) \in C[0,1]$. Then according to Theorem 12.3.1

$$M_\infty(a) = \operatorname{Range} a \ (= \text{ess-Range}\, a).$$

Example 12.4.4. Let

$$a(r) = \left\{ \begin{array}{ll} z_1, & r \in [0, \tfrac{1}{2}) \\ z_2, & r \in [\tfrac{1}{2}, 1] \end{array} \right. ,$$

where $z_1, z_2 \in \mathbb{C}$, and $z_1 \neq z_2$. Then according to Theorem 12.4.1 $M_\infty(a)$ coincides with the straight line segment $[z_1, z_2]$ joining the points z_1 and z_2, and thus

$$M_\infty(a) = \operatorname{conv}(\text{ess-Range}\, a) \ (= \operatorname{conv}(\operatorname{Range} a)).$$

Example 12.4.5. Let

$$a(r) = \left\{ \begin{array}{ll} z_1, & r \in [0, \tfrac{1}{4}) \\ z_2, & r \in [\tfrac{1}{4}, \tfrac{1}{2}) \\ z_3, & r \in [\tfrac{1}{2}, \tfrac{3}{4}) \\ z_4, & r \in [\tfrac{3}{4}, 1] \end{array} \right. ,$$

where z_1, z_2, z_3, z_4 are different points on the complex plane. By Theorem 12.4.1 we have

$$M_\infty(a) = [z_1, z_2] \cup [z_2, z_3] \cup [z_3, z_1] \cup [z_3, z_4],$$

and the set $M_\infty(a)$ is a part of the boundary of the convex hull of ess-Range $a = \operatorname{Range} a$:

$$M_\infty(a) \subset \partial \operatorname{conv}(\operatorname{Range} a).$$

If $z_4 = z_1$ then the set $M_\infty(a)$ coincides with the triangle having the vertices z_1, z_2, z_3, and

$$M_\infty(a) = \partial \operatorname{conv}(\operatorname{Range} a).$$

Example 12.4.6. Let $\{r_j\}_{j\in\mathbb{Z}_+}$ be a sequence of increasing positive numbers with $\lim_{j\to\infty} r_j = 1$. Define the symbol $a(\sqrt{r})$ as follows

$$a(\sqrt{r}) = \left\{ \begin{array}{ll} e^{i\theta_j}, & r \in [r_{2j}, r_{2j+1}) \\ -e^{i\theta_j}, & r \in [r_{2j+1}, r_{2j+2}), \quad j = 0, 1, \ldots \end{array} \right. ,$$

where $\{\theta_j\}_{j\in\mathbb{Z}_+}$ is a dense set in $[0, \pi]$.

As in the proof of Theorem 12.4.1, it is easy to check that each diameter $[e^{i\theta_j}, -e^{i\theta_j}]$ of the unit disk $\mathbb{D}$ with endpoints $e^{i\theta_j}$ and $-e^{i\theta_j}$ belongs to $M_\infty(a)$, which implies $\overline{\mathbb{D}} \subset M_\infty(a)$.

We have that $\operatorname{clos} \operatorname{Range} a = \partial \mathbb{D} = \mathbb{T}$, thus finally

$$M_\infty(a) = \overline{\mathbb{D}} = \operatorname{conv}(\operatorname{Range} a).$$

In all examples above the set $M_\infty(a)$ is connected. But this is not so in general.

Example 12.4.7. Let $a(r) = r^{2i}$. Then $\operatorname{Range} a = S^1$, and for a fixed λ the elements

$$
\begin{aligned}
\gamma_{a,\lambda}(n) &= \frac{1}{B(n+1, \lambda+1)} \int_0^1 (1-r)^\lambda \, r^{n+i} dr \\
&= \frac{B(n+1+i, \lambda+1)}{B(n+1, \lambda+1)} = \frac{\Gamma(n+\lambda+2)}{\Gamma(n+\lambda+2+i)} \cdot \frac{\Gamma(n+1+i)}{\Gamma(n+1)}
\end{aligned}
$$

form the sequence tending to $\lim_{r \to 1} a(r) = 1 \in S^1$, as $n \to \infty$.

Let $\lambda \to \infty$. By the asymptotic formulas for the Gamma function (see, for example, [1], p. 257) we have

$$
\gamma_{a,\lambda}(n) = \frac{\Gamma(n+1+i)}{\Gamma(n+1)} (n+\lambda+2)^{-i} \left(1 + O(\frac{1}{n+\lambda})\right).
$$

Thus the set $M_\infty(a)$ is the union of the circles $\mathbb{T}_n$ centered in the origin and having radii

$$
r_n = \left| \frac{\Gamma(n+1+i)}{\Gamma(n+1)} \right| = \left| \frac{(n+i)(n-1+i)\ldots i\Gamma(i)}{n!} \right|, \qquad n \in \mathbb{Z}_+,
$$

that is

$$
M_\infty(a) = \bigcup_{n=0}^\infty \mathbb{T}_n.
$$

According to formula 6.1.29 from [1] we have

$$
|\Gamma(i)| = \left(\frac{\pi}{\sinh \pi} \right)^{\frac{1}{2}} = \left(\frac{2\pi}{e^\pi - e^{-\pi}} \right)^{\frac{1}{2}} \approx 0.52.
$$

Thus

$$
r_n = |\Gamma(i)| \prod_{j=1}^n \left(1 + \frac{1}{j^2}\right)^{\frac{1}{2}}.
$$

That is $r_n < r_{n+1}$ for all $n \in \mathbb{Z}_+$. Moreover by the asymptotic formula for the Gamma function we have

$$
\lim_{n \to \infty} r_n = 1.
$$

Note that the values of a few first radii are

$$
r_0 \approx 0.52, \qquad r_1 \approx 0.73, \qquad r_2 \approx 0.82, \qquad r_3 \approx 0.86.
$$

The process of forming the limit set $M_\infty(a)$ here is somewhat different compared to the previous examples. This is caused by an oscillatory type of discontinuity of the symbol at the point $r = 0$. For each fixed and sufficiently large λ the corresponding set $\gamma_{a,\lambda}$ is a part of a spiral approaching the unit circle and ending at $a(1)$. For increasing λ, these spirals become longer and compress to the unit circle; for very large values of λ the points of the sequence $\gamma_{a,\lambda}(n)$ are practically positioned on the corresponding circles of the limit set, moving along these circles as λ changes. To make a more informative picture we draw the values of $\gamma_{a,\lambda}(n)$ with fixed n for different values of λ. We illustrate below the forming of the first, second and fourth circles of the limit set $M_\infty(a)$ with λ changing from 0 to 20000. Note that in the picture of the limit set the part between the last circle drawn and the unit circle is omitted.

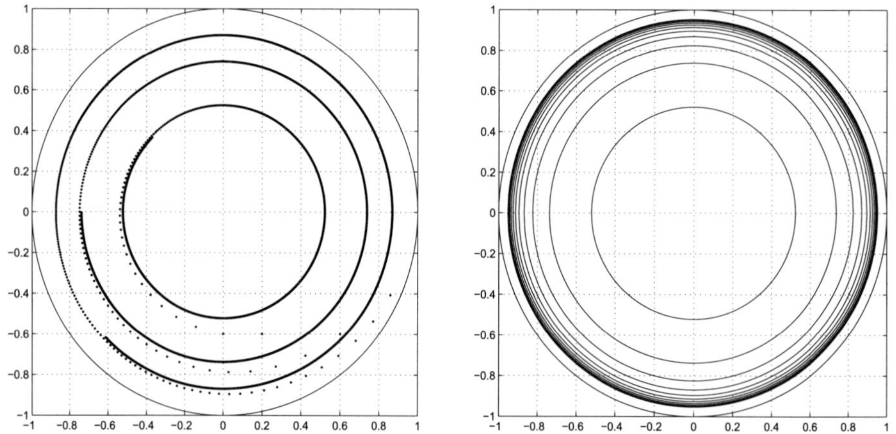

Figure 12.3: Formation of the circles $n = 0, 1, 3$ for $\lambda = 0, 1, \ldots, 20000$, and the limit set $M_\infty(a)$.

12.5   Spectra of Toeplitz operators, unbounded symbols

As has been already mentioned, there exist unbounded radial symbols for which the corresponding Toeplitz operators are bounded on each weighted Bergman space $\mathcal{A}_\lambda^2(\mathbb{D})$, with $\lambda \geq 0$.

     Note that in spite of the fact that for each fixed λ the sequence $\{\gamma_{a,\lambda}(n)\}$ is bounded for symbols satisfying, for example, the conditions of Theorem 12.1.1, the estimates given in the proof are not uniform on λ. This suggests that in spite of having $\operatorname{sp} T_a$ bounded for each λ, the limit set $M_\infty(a)$ may be unbounded. The next theorem shows that this observation is not accidental.

Theorem 12.5.1. *Let $a(\sqrt{r}) \in L_1(0,1) \cap C[0,1]$. Then*

$$\operatorname{Range} a \subset M_\infty(a).$$

Proof. We apply the Laplace method as in Theorem 12.3.1. Fix any point $r_0 \in (0,1)$, and for each $n \in \mathbb{N}$ take λ_n such that

$$\frac{\lambda_n}{n} = r_0^{-1} - 1.$$

Then by (12.3.4) we have

$$\gamma_{a,\lambda}(n) = a(\sqrt{r_0})(1 + \alpha(\sqrt{\lambda_n^2 + n^2})),$$

where $\lim_{L \to \infty} \alpha(L) = 0$. Thus if $n \to \infty$ then $\lambda_n \to \infty$ as well, and we have

$$a(\sqrt{r_0}) \in M_\infty(a).$$

$\square$

Let us show now that the property (12.4.8), previously established for bounded symbols, remains valid for unbounded radial symbols as well.

Theorem 12.5.2. *Let $a(\sqrt{r}) \in L_1(0,1)$. Then*

$$M_\infty(a) \subset \mathrm{conv}(\mathrm{ess\text{-}Range}\, a). \tag{12.5.1}$$

Proof. For each $M > 0$ consider the bounded symbol

$$a_M(r) = \begin{cases} a(r), & \text{if} \quad |a(r)| \leq M \\ 0, & \text{if} \quad |a(r)| > M \end{cases},$$

and the corresponding sequence

$$\gamma_{a_M,\lambda}(n) = \frac{1}{B(n+1,\lambda+1)} \int_0^1 a_M(\sqrt{r})\,(1-r)^\lambda\, r^n dr, \quad n \in \mathbb{Z}_+.$$

By (12.3.1) we have obviously

$$\{\gamma_{a_M,\lambda}(n)\} \subset \mathrm{conv}(\mathrm{ess\text{-}Range}\, a_M) \subset \mathrm{conv}(\mathrm{ess\text{-}Range}\, a).$$

The equality

$$\lim_{M \to \infty} \gamma_{a_M,\lambda}(n) = \gamma_{a,\lambda}(n),$$

verified by the Lebesgue dominated convergence theorem, implies that

$$\{\gamma_{a,\lambda}(n)\} \subset \mathrm{conv}(\mathrm{ess\text{-}Range}\, a),$$

and we have (12.5.1). $\square$

Note that for functions $a(\sqrt{r}) \in L_1(0,1) \cap C(0,1)$ Theorems 12.5.1 and 12.5.2 imply that

$$\mathrm{Range}\, a \subset M_\infty(a) \subset \mathrm{conv}(\mathrm{Range}\, a).$$

We now show that each of these inclusions can be an equality.

Example 12.5.3. For each $j \in \mathbb{N}$ define $I_j = [1 - j^{-1} - j^{-3}, 1 - j^{-1}]$, and let $\{\theta_j\}_{j \in \mathbb{N}} \subset [0, 2\pi)$ be a sequence such that $\mathrm{clos}\{\theta_j\}_{j \in \mathbb{N}} = [0, 2\pi]$. Define the symbol by

$$a(\sqrt{r}) = \begin{cases} je^{i\theta_j}, & r \in I_j, \quad j \in \mathbb{N} \\ 0, & r \in [0,1] \setminus \cup_{j=1}^{\infty} I_j \end{cases}.$$

This symbol satisfies the condition (12.1.5) for $j = 1$, and thus the corresponding Toeplitz operator $T_a^{(\lambda)}$ is bounded for every $\lambda > -1$. Indeed,

$$
\begin{aligned}
|B^{(1)}(r)| &= \left| \int_r^1 a(\sqrt{s}) ds \right| \le \int_r^1 |a(\sqrt{s})| \, ds \\
&\le \sum_{1-j^{-1} > r} j \cdot j^{-3} = \sum_{j > (1-r)^{-1}} j^{-2} \le \mathrm{const}\, (1 - r).
\end{aligned}
$$

Further, we have that $a(\sqrt{r}) = je^{i\theta_j}$ for $r \in I_j$, and $a(\sqrt{r}) = 0$ for $r \in (1 - j^{-1}, 1 - (j+1)^{-1} - (j+1)^{-3}]$, which implies just in the same way as in Theorem 12.4.1, that the straight line segment $[0, je^{i\theta_j}]$ belongs to $M_\infty(a)$.

    Thus

$$\mathrm{Range}\, a \subset M_\infty(a) = \mathbb{C} = \mathrm{conv}(\mathrm{Range}\, a).$$

Example 12.5.4. Given $\alpha \in (0, 1)$ introduce $a(\sqrt{r}) = r^{i-\alpha}$. Calculate

$$
\begin{aligned}
\gamma_{a,\lambda}(n) &= \frac{1}{B(n+1, \lambda+1)} \int_0^1 (1 - r)^\lambda \, r^{n+i-\alpha} dr \\
&= \frac{B(n+1+i-\alpha, \lambda+1)}{B(n+1, \lambda+1)} \\
&= \frac{\Gamma(n+\lambda+2)}{\Gamma(n+\lambda+2+i-\alpha)} \cdot \frac{\Gamma(n+1+i-\alpha)}{\Gamma(n+1)},
\end{aligned}
$$

and it is easy to see that for each fixed λ the sequence $\{\gamma_{a,\lambda}(n)\}$ is bounded, which implies the boundedness of the corresponding Toeplitz operator.

    Let now $\lambda \to \infty$. Using the asymptotic formulas for the Gamma function (see, for example, [1], p. 257) we have

$$\gamma_{a,\lambda}(n) = \frac{\Gamma(n+1+i-\alpha)}{\Gamma(n+1)} (n+\lambda+2)^{\alpha-i} \left(1 + O(\frac{1}{n+\lambda})\right).$$

If n is fixed or bounded, this expression gives us only one point from the set $M_\infty(a)$, namely ∞. Let now both n and λ tend to infinity. Then we have

$$
\begin{aligned}
\gamma_{a,\lambda}(n) &= \left(\frac{n+\lambda+2}{n+1}\right)^{\alpha-i} \left(1 + O(\frac{1}{n+1})\right)\left(1 + O(\frac{1}{n+\lambda})\right) \\
&= \left(1 + \frac{\lambda+1}{n+1}\right)^{\alpha-i} \left(1 + O(\frac{1}{n+1}) + O(\frac{1}{n+\lambda})\right).
\end{aligned}
$$

Given arbitrary $u \in (0,1)$, take λ and n in such a form that

$$\left(1 + \frac{\lambda+1}{n+1}\right) = u^{-1}.$$

Thus

$$\gamma_{a,\lambda}(n) = u^{i-\alpha}\left(1 + O(\frac{1}{n+1}) + O(\frac{1}{n+\lambda})\right),$$

and thus in this case

$$\operatorname{Range} a = M_\infty(a) \subset \operatorname{conv}(\operatorname{Range} a).$$

Let us mention some peculiarities of this example. In spite of the *unbound-edness* of the symbol, the Toeplitz operator $T_a^{(\lambda)}$ is *bounded* for each λ, and the sequence $\gamma_{a,\lambda}$ is *bounded* by $\|T_a^{(\lambda)}\|$ for each λ. The rough estimate here is

$$\sup_n |\gamma_{a,\lambda}(n)| = \|T_a^{(\lambda)}\| \approx \operatorname{const} \lambda^\alpha.$$

That is, being bounded for each λ, the norm and $\sup_n |\gamma_{a,\lambda}(n)|$ grow as λ tends to infinity. The tendency of $\gamma_{a,\lambda}$ to approach $M_\infty(a)$ here is developed first of all in the growth of the size (i.e., $\sup_n |\gamma_{a,\lambda}(n)|$) of $\operatorname{sp} T_a^{(\lambda)}$ while λ tends to infinity.

In the next figure, setting $\alpha = 0.1$ we present the sequence $\gamma_{a,\lambda}$ for $\lambda = 100000$ and the limit set $M_\infty(a) = \operatorname{Range} a$.

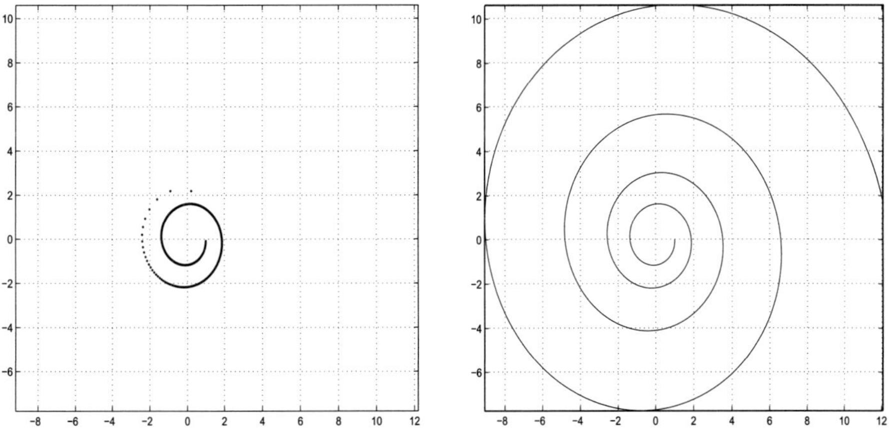

Figure 12.4: The sequence $\gamma_{a,\lambda} = \{\gamma_{a,\lambda}(n)\}$ for $\lambda = 100000$ and the limit set $M_\infty(a)$.

Chapter 13

Dynamics of Properties of Toeplitz Operators on the Upper Half-Plane: Parabolic Case

13.1 Boundedness of Toeplitz operators with symbols depending on $y = \operatorname{Im} z$

Recall that by Theorems 10.4.9 and 10.5.1, the function

$$
\begin{aligned}
\gamma_{a,\lambda}(t) &= \frac{t^{\lambda+1}}{\Gamma(\lambda+1)} \int_0^\infty a\!\left(\frac{\eta}{2}\right) \eta^\lambda\, e^{-t\eta}\, d\eta \\
&= \frac{1}{\Gamma(\lambda+1)} \int_0^\infty a\!\left(\frac{\eta}{2t}\right) \eta^\lambda\, e^{-\eta}\, d\eta
\end{aligned}
\tag{13.1.1}
$$

is responsible for the boundedness of a Toeplitz operator with symbol $a = a(y)$. If $a = a(y) \in L_\infty(\mathbb{R}_+)$, then the operator $T_a^{(\lambda)}$ is obviously bounded on all spaces $\mathcal{A}_\lambda^2(\Pi)$, where $\lambda \in (-1, \infty)$, and the corresponding norms are uniformly bounded by $\sup_z |a(z)|$. That is, all spaces $\mathcal{A}_\lambda^2(\Pi)$, where $\lambda \in (-1, \infty)$, are natural and appropriate for Toeplitz operators with *bounded* symbols. One of our aims is a systematic study of *unbounded* symbols. To avoid unnecessary technicalities in this chapter we will always assume that $\lambda \in [0, \infty)$.

As is easy to see, the major contribution to the integral (13.1.1) for "very big t" ($t \to \infty$) is determined by values of $a(y)$ at a neighborhood of the point 0, and the major contribution for "very small t" ($t \to 0$) is determined by values of $a(y)$ at a neighborhood of ∞. In particular, as it follows directly from (13.1.1), if

$a(y)$ has limits on the points 0 and ∞, then

$$\lim_{t\to\infty} \gamma_{a,\lambda}(t) = \lim_{y\to 0} a(y),$$

$$\lim_{t\to 0} \gamma_{a,\lambda}(t) = \lim_{y\to\infty} a(y).$$

As a matter of fact, 0 and ∞ are the only worrisome points for unbounded symbols $a(y) \in L_1(\mathbb{R}_+, 0)$. Moreover the behaviour of certain means of a symbol, rather than the behaviour of a symbol itself, plays a crucial role under the study of boundedness properties.

Given $\lambda \in [0, +\infty)$ and a locally summable function $a(y)$ introduce the means

$$B_{a,\lambda}^{(1)}(\xi) = \int_0^\xi a(t/2)t^\lambda dt,$$

$$B_{a,\lambda}^{(j)}(\xi) = \int_0^\xi B_{a,\lambda}^{(j-1)}(t)dt, \quad j = 2, 3, \ldots.$$

We note that contrary to the radial case (12.1.2) the above functions depend now on the weight parameter.

Theorem 13.1.1. *Let $a(y) \in L_1(\mathbb{R}_+, 0)$. If for any $\lambda_0 \in [0, +\infty)$ and any $j \in \mathbb{N}$ the function $B_{a,\lambda_0}^{(j)}(\xi)$ has the following asymptotic behaviors in the neighborhoods of the points $\xi = 0$ and $\xi = \infty$,*

$$B_{a,\lambda_0}^{(j)}(\xi) = O(\xi^{j+\lambda_0}), \quad \xi \to 0, \tag{13.1.2}$$

and

$$B_{a,\lambda_0}^{(j)}(\xi) = O(\xi^{j+\lambda_0}), \quad \xi \to \infty, \tag{13.1.3}$$

then for every $\lambda \in [\lambda_0, \infty)$,

$$\sup_{x\in\mathbb{R}_+} |\gamma_{a,\lambda}(x)| < \infty,$$

and the corresponding Toeplitz operator $T_a^{(\lambda)}$ is bounded on $\mathcal{A}_\lambda^2(\Pi)$ for every $\lambda \in [\lambda_0, \infty)$.

Proof. Let $\lambda \geq \lambda_0$. Assume first that $j = 1$. Then the conditions (13.1.2) and (13.1.3) imply that for all $\xi \in \mathbb{R}_+$ the estimate

$$|B_{a,\lambda_0}^{(1)}(\xi)| \leq \text{const } \xi^{1+\lambda_0} \tag{13.1.4}$$

holds, where "const" does not depend on $\xi \in \mathbb{R}_+$. Integrating by parts we have for all $x \in \mathbb{R}_+$,

$$
\begin{aligned}
|\gamma_{a,\lambda}(x)| &= \frac{x^{\lambda+1}}{\Gamma(\lambda+1)} \left| \int_0^\infty t^{\lambda-\lambda_0} e^{-xt} dB_{a,\lambda_0}^{(1)}(t) \right| \\
&= \frac{x^{\lambda+1}}{\Gamma(\lambda+1)} \left| \int_0^\infty B_{a,\lambda_0}^{(1)}(t)[(\lambda-\lambda_0)t^{\lambda-\lambda_0-1} - xt^{\lambda-\lambda_0}]e^{-xt} dt \right| \\
&\leq \mathrm{const}\, \frac{x^{\lambda+1}}{\Gamma(\lambda+1)} \int_0^\infty ((\lambda-\lambda_0)t^\lambda + xt^{\lambda+1})e^{-xt} dt \\
&\leq \mathrm{const}\, [(\lambda-\lambda_0) + (\lambda+1)] = \mathrm{const}\, (2\lambda - \lambda_0 + 1)
\end{aligned}
$$

and the case $j = 1$ is done.

For $j \geq 2$ we use the inequalities

$$
|B_{a,\lambda_0}^{(j)}(\xi)| \leq \mathrm{const}\, \xi^{j+\lambda_0} \tag{13.1.5}
$$

(where $\xi \in \mathbb{R}_+$ and "const" does not depend on ξ) and integration by parts j-times. $\qquad\square$

Remark 13.1.2. The condition (13.1.2) guarantees the boundedness of the function $\gamma_{a,\lambda}(x)$ in a neighborhood of $x = \infty$ while the condition (13.1.3) guarantees the boundedness of the functions $\gamma_{a,\lambda}(x)$ in a neighborhood of $x = 0$.

The next statement sets a partial order on the family of sufficient conditions for boundedness of Toeplitz operators given by Theorem 13.1.1.

Theorem 13.1.3. 1. *Let the conditions* (13.1.2) *and* (13.1.3) *hold for* $j = j_0$ *and some* λ_0. *Then these conditions hold for* $j = j_0 + 1$ *and the same* λ_0.

2. *Let the conditions* (13.1.2) *and* (13.1.3) *hold for* $j = j_0$ *and some* λ_0. *Then these conditions hold for* $j = j_0$ *and* λ_0 *replaced by any* $\lambda_1 \geq \lambda_0$.

Proof. Assume that we have (13.1.2) and (13.1.3) for $j = j_0$. Then according to (13.1.5) we have

$$
\begin{aligned}
|B_{a,\lambda_0}^{(j_0+1)}(\xi)| &\leq \int_0^\xi |B_{a,\lambda_0}^{(j_0)}(t)| dt \\
&\leq \mathrm{const}\, \int_0^\xi t^{j_0+\lambda_0} dt \leq \mathrm{const}\, \xi^{j_0+1+\lambda_0}.
\end{aligned}
$$

Thus the first statement is proved.

Assume we have (13.1.2) and (13.1.3) for $j = 1$ and $\lambda = \lambda_0$. If $\lambda_1 > \lambda_0$ then

$$
\begin{aligned}
|B^{(1)}_{a,\lambda_1}(\xi)| &\leq \left| \int_0^\xi t^{\lambda_1 - \lambda_0} dB^{(1)}_{a,\lambda_0}(t) \right| \\
&= \left| B^{(1)}_{a,\lambda_0}(\xi)\xi^{\lambda_1 - \lambda_0} - (\lambda_1 - \lambda_0) \int_0^\xi B^{(1)}_{a,\lambda_0}(t) t^{\lambda_1 - \lambda_0 - 1} dt \right| \\
&\leq \text{const} \left(\left| \xi^{1+\lambda_0} \xi^{\lambda_1 - \lambda_0} + \int_0^\xi t^{1+\lambda_0} t^{\lambda_1 - \lambda_0 - 1} dt \right| \right) \\
&\leq \text{const}\, \xi^{1+\lambda_1}.
\end{aligned}
$$

Assume that (13.1.2) and (13.1.3) hold for $j = 2$ and $\lambda = \lambda_0$. Then for each $\lambda_1 > \lambda_0$ we have

$$
\begin{aligned}
|B^{(2)}_{a,\lambda_1}(\xi)| &= \left| \int_0^\xi \int_0^u a(t/2) t^{\lambda_1} dt\, du \right| \\
&= \left| \int_0^\xi \int_0^u t^{\lambda_1 - \lambda_0} dB^{(1)}_{a,\lambda_0}(t)\, du \right| \\
&= \left| \int_0^\xi B^{(1)}_{a,\lambda_0}(u) u^{\lambda_1 - \lambda_0} du - (\lambda_1 - \lambda_0) \int_0^\xi \int_0^u B^{(1)}_{a,\lambda_0}(t) t^{\lambda_1 - \lambda_0 - 1} dt\, du \right| \\
&= \left| B^{(2)}_{a,\lambda_0}(\xi)\xi^{\lambda_1 - \lambda_0} - (\lambda_1 - \lambda_0) \int_0^\xi B^{(2)}_{a,\lambda_0}(u) u^{\lambda_1 - \lambda_0 - 1} du \right. \\
&\qquad - (\lambda_1 - \lambda_0) \int_0^\xi B^{(2)}_{a,\lambda_0}(u) u^{\lambda_1 - \lambda_0 - 1} du \\
&\qquad \left. + (\lambda_1 - \lambda_0)(\lambda_1 - \lambda_0 - 1) \int_0^\xi \int_0^u B^{(2)}_{a,\lambda_0}(t) t^{\lambda_1 - \lambda_0 - 2} dt\, du \right| \\
&\leq \text{const} \left(\xi^{2+\lambda_1} + \frac{2(\lambda_1 - \lambda_0)}{\lambda_1 + 2} \xi^{2+\lambda_1} + \frac{(\lambda_1 - \lambda_0)(\lambda_1 - \lambda_0 - 1)}{(\lambda_1 + 1)(\lambda_1 + 2)} \xi^{2+\lambda_1} \right) \\
&\leq \text{const}\, \xi^{2+\lambda_1}.
\end{aligned}
$$

The cases $j_0 > 2$ for the second statement are considered analogously. $\qquad\square$

Example 13.1.4. Consider the unbounded symbol

$$
a(t/2) = t^{-\beta} \sin t^{-\alpha},
$$

where $0 < \beta < 1$, $\alpha > 0$. Applying Theorem 13.1.1 for $j = 1$ and $\lambda_0 = 0$ we have

$$
B^{(1)}_{a,0}(\xi) = \int_0^\xi t^{-\beta} \sin t^{-\alpha} dt = \frac{1}{\alpha} \int_{\xi^{-\alpha}}^\infty y^{\frac{\beta-1}{\alpha} - 1} \sin y\, dy. \tag{13.1.6}
$$

As in the proof of Theorem 11.1.1, integrating by parts twice we get

$$B_{a,0}^{(1)}(\xi) \;=\; \frac{\xi^{\alpha-\beta+1}}{\alpha}\cos\xi^{-\alpha} - \frac{(\beta-\alpha-1)}{\alpha^2}\xi^{2\alpha-\beta+1}\sin\xi^{-\alpha}$$
$$-\frac{(\beta-\alpha-1)(\beta-2\alpha-1)}{\alpha^3}\int_{\xi^{-\alpha}}^{\infty} y^{\frac{\beta-1}{\alpha}-3}\sin y\,dy.$$

Thus we have

$$B_{a,0}^{(1)}(\xi) = \frac{\xi^{\alpha-\beta+1}}{\alpha}\cos\xi^{-\alpha} + O(\xi^{2\alpha-\beta+1}), \quad \xi \to 0. \tag{13.1.7}$$

To get the asymptotic at infinity we use again the representation (13.1.6):

$$B_{a,0}^{(1)}(\xi) = \frac{1}{\alpha}\int_{\xi^{-\alpha}}^{1} y^{\frac{\beta-\alpha-1}{\alpha}}\sin y\,dy + \frac{1}{\alpha}\int_{1}^{\infty} y^{\frac{\beta-\alpha-1}{\alpha}}\sin y\,dy.$$

Since $(\beta-\alpha-1)/\alpha < 0$ the second integral converges. Integrating the first integral by parts (another way) we get

$$B_{a,0}^{(1)}(\xi) = c_0 + c_1\xi^{1-\beta-\alpha} + O(\xi^{1-\beta-2\alpha}), \quad c_0, c_1 \in \mathbb{C}. \tag{13.1.8}$$

Thus if

$$\alpha \geq \beta, \tag{13.1.9}$$

then the conditions (13.1.2) and (13.1.3) hold for $j = 1$, $\lambda_0 = 0$, and the operator $T_a^{(\lambda)}$ is bounded for each $\lambda \geq 0$.

Now apply Theorem 13.1.1 for $j = 2$, $\lambda_0 = 0$. Let $\alpha < \beta$. Using the inequality (13.1.7) (for $\beta := \beta - \alpha - 1$) and (13.1.8) we get

$$B_{a,0}^{(2)}(\xi) = O(\xi^{2\alpha-\beta+2}), \quad \xi \to 0,$$

and

$$B_{a,0}^{(2)}(\xi) = O(\xi) + O(\xi^{2-\beta+\alpha}), \quad \xi \to \infty.$$

Thus the operator $T_a^{(\lambda)}$ is bounded if

$$\alpha \geq \frac{\beta}{2}. \tag{13.1.10}$$

Analogously apply Theorem 13.1.1 for $j = 3, 4, \ldots$ ($\lambda_0 = 0$), we have that for

$$\alpha \geq \frac{\beta}{j} \tag{13.1.11}$$

operator $T_a^{(\lambda)}$ is bounded. Since there exists j large enough for (13.1.11) to hold, we have that for arbitrary $0 < \beta < 1$ and $\alpha > 0$ the operator $T_a^{(\lambda)}$ is bounded for each $\lambda \geq 0$.

Remark 13.1.5. Example 13.1.4 shows that the conditions (13.1.2) and (13.1.3) for $j = j_1$, $\lambda_0 = 0$, compared with those for $j = j_2$, $\lambda_0 = 0$, and $j_1 > j_2$, widen in fact the class of symbols for which the boundedness of the corresponding Toeplitz operators can be justified.

The sufficient conditions of Theorem 13.1.1 provide the simultaneous boundedness of an operator $T_a^{(\lambda)}$ for all $\lambda \in [\lambda_0, \infty)$ at once. We pass now to the more delicate question of the boundedness of a Toeplitz operator $T_a^{(\lambda)}$ on the space $\mathcal{A}_\lambda^2(\Pi)$ in its dependence on λ. The following result plays a central role here.

Theorem 13.1.6. *Let $a(y)$ belong to $L_1(\mathbb{R}_+, 0)$ and the operator $T_a^{(\lambda_0)}$ be bounded on $\mathcal{A}_{\lambda_0}^2(\Pi)$ for a certain $\lambda_0 > 0$. Then $T_a^{(\lambda)}$ is bounded on $\mathcal{A}_\lambda^2(\Pi)$ for each $\lambda \in [0, \lambda_0]$.*

Proof. Let the operator $T_a^{(\lambda)}$ be bounded on $\mathcal{A}_{\lambda_0}^2(\Pi)$, that is $\sup_{x>0} |\gamma_{a,\lambda_0}(x)| < \infty$. For $\lambda < \lambda_0$, we have

$$
\begin{aligned}
\gamma_{a,\lambda}(x) &= \frac{x^{\lambda+1}}{\Gamma(\lambda+1)} \int_0^\infty a(t/2) t^\lambda e^{-xt} dt \\
&= \frac{x^{\lambda+1}}{\Gamma(\lambda+1)\Gamma(\lambda_0-\lambda)} \int_0^\infty a(t/2) t^{\lambda_0} e^{-xt} dt \int_0^\infty y^{\lambda_0-\lambda-1} e^{-yt} dy \\
&= \frac{x^{\lambda+1}}{\Gamma(\lambda+1)\Gamma(\lambda_0-\lambda)} \int_0^\infty y^{\lambda_0-\lambda-1} dy \int_0^\infty a(t/2) t^{\lambda_0} e^{-(x+y)t} dt \\
&= \frac{\Gamma(1+\lambda_0)}{\Gamma(\lambda+1)\Gamma(\lambda_0-\lambda)} \int_0^\infty y^{\lambda_0-\lambda-1} (1+y)^{-\lambda_0-1} \gamma_{a,\lambda_0}(x(1+y)) dy.
\end{aligned}
$$

Thus

$$
|\gamma_{a,\lambda}(x)| \leq \sup_{x>0} |\gamma_{a,\lambda_0}(x)| \frac{\Gamma(1+\lambda_0)}{\Gamma(\lambda+1)\Gamma(\lambda_0-\lambda)} \int_0^\infty y^{\lambda_0-\lambda-1} (1+y)^{-\lambda_0-1} dy. \qquad \square
$$

The next theorem extends the range of λ, given by Theorem 13.1.1, for simultaneous boundedness of Toeplitz operators on $\mathcal{A}_\lambda^2(\Pi)$.

Theorem 13.1.7. *Under the hypothesis of Theorem 13.1.1 the Toeplitz operator $T_a^{(\lambda)}$ is bounded on $\mathcal{A}_\lambda^2(\Pi)$ for each $\lambda \in [0, \infty)$.*

Proof. Follows directly from Theorems 13.1.1 and 13.1.6. $\qquad \square$

Theorem 13.1.6 allows us to obtain in particular the necessity of the hypothesis of Theorem 13.1.1 in the case of non-negative symbols or non-negative means.

Theorem 13.1.8.   1. *Assume that $a(y) \in L_1(\mathbb{R}_+, 0)$ and $a(y) \geq 0$ almost everywhere. Let the operator $T_a^{(\lambda')}$ be bounded on $\mathcal{A}_{\lambda'}^2(\Pi)$ for some $\lambda' \geq 0$. Then the conditions (13.1.2) and (13.1.3) hold for $j = 1$ and $\lambda_0 = 0$ and consequently the operator $T_a^{(\lambda)}$ is bounded on $\mathcal{A}_\lambda^2(\Pi)$ for each $\lambda \in [0, \infty)$.*

2. *Assume that $B_{a,\mu}^{(j)}(y) \geq 0$ almost everywhere for some $j = j_0 \geq 1$ and $\mu \geq 0$, and that the operator $T_a^{(\lambda')}$ is bounded on $\mathcal{A}_{\lambda'}^2(\Pi)$ for some $\lambda' \geq 0$. Then the conditions (13.1.2) and (13.1.3) hold for $j = j_0 + 1$ and $\lambda_0 = \mu$, and consequently the operator $T_a^{(\lambda)}$ is bounded on $\mathcal{A}_\lambda^2(\Pi)$ for each $\lambda \in [0, \infty)$.*

Proof. 1. If $T_a^{(\lambda')}$ is bounded on $\mathcal{A}_{\lambda'}^2(\Pi)$, then according to Theorem 13.1.6 the operator $T_a^{(0)}$ is bounded on $\mathcal{A}_0^2(\Pi)$. We have

$$\gamma_{a,0}(x) = x \int_0^\infty a(t/2)e^{-xt}dt \geq x \int_0^{x^{-1}} a(t/2)e^{-xt}dt \geq \frac{x}{e} B_{a,0}^{(1)}(x^{-1}).$$

Thus denoting $\xi = x^{-1}$ we have

$$B_{a,0}^{(1)}(\xi) \leq (e \sup_{x \in \mathbb{R}_+} |\gamma_{a,0}(x)|)\xi.$$

2. Assume first that $j_0 = 1$ and $\mu \geq \lambda'$. We have

$$\gamma_{a,\lambda'}(x) = \frac{x^{\lambda'+1}}{\Gamma(\lambda' + 1)} \int_0^\infty a(t/2)t^{\lambda'} e^{-xt}\, dt.$$

Integrating by parts we get

$$\begin{aligned}
\gamma_{a,\lambda'}(x) &= \frac{x^{\lambda'+1}}{\Gamma(\lambda' + 1)} \int_0^\infty B_{a,\mu}^{(1)}(t)[(\mu - \lambda') + xt]t^{\lambda'-\mu-1}e^{-xt}dt \\
&\geq \frac{x^{\lambda'+1}}{\Gamma(\lambda' + 1)} \left(\int_0^{x^{-1}} B_{a,\mu}^{(1)}(t)dt \right) [(\mu - \lambda') + 1]x^{-(\lambda'-\mu-1)}e^{-1} \\
&= \frac{x^{\mu+2}(\mu - \lambda' + 1)}{e\,\Gamma(\lambda' + 1)} B_{a,\mu}^{(2)}(x^{-1}).
\end{aligned}$$

Again denoting $\xi = x^{-1}$ we have

$$B_{a,\mu}^{(2)}(\xi) \leq \frac{e\,\Gamma(\lambda' + 1)}{\mu - \lambda' + 1} \sup_{x \in \mathbb{R}_+} |\gamma_{a,\lambda'}(x)| \, \xi^{\mu+2}.$$

The above integration by parts is correct because for arbitrary $a(t) \in L_1(\mathbb{R}_+, 0)$ we have

$$|B_{a,\mu}^{(1)}(\xi)| = o(\xi^\mu), \qquad \xi \to 0.$$

Let now $j_0 = 1$ and $\mu < \lambda'$. Then according to Theorem 13.1.6 the operator $T_a^{(\mu)}$ is bounded on $\mathcal{A}_\mu^2(\Pi)$. Repeating the above reasonings for the function $\gamma_{a,\mu}(x)$ we complete the consideration of the case $j_0 = 1$.

The cases $j_0 > 1$ are considered analogously. $\qquad\square$

Remark 13.1.9. Simultaneous boundedness of the operators $T_a^{(\lambda)}$ for all λ in the case of arbitrary (depending on both variables) non-negative symbol was shown in [240]. We extend this result for a class of not necessarily non-negative symbols depending only on y.

For a non-negative function $a(t)$ we set

$$m_{a,0}(x) = \inf_{(0,x)} a(t/2) \quad\text{and}\quad m_{a,\infty}(x) = \inf_{(x/2,x)} a(t/2).$$

Corollary 13.1.10. *Given a non-negative symbol* $a(y)$*, if either*

$$\lim_{x\to 0} m_{a,0}(x) = \infty \tag{13.1.12}$$

or

$$\lim_{x\to\infty} m_{a,\infty}(x) = \infty, \tag{13.1.13}$$

then Toeplitz operator $T_a^{(\lambda)}$ *is unbounded on each* $\mathcal{A}_\lambda^2(\Pi)$*,* $\lambda \in [0, +\infty)$*.*

Proof. If the condition (13.1.12) holds then

$$B_{a,0}^{(1)}(\xi) = \int_0^\xi a(t/2)dt \geq \xi m_{a,0}(\xi)$$

and

$$\xi^{-1}B^{(1)}(\xi) \to \infty \quad\text{as}\quad \xi \to 0.$$

Now let the condition (13.1.13) hold. Then

$$\xi^{-1}B_{a,0}^{(1)}(\xi) > \xi^{-1}\int_{\xi/2}^\xi a(t/2)dt \geq \frac{1}{2}m_{a,\infty}(\xi) \to \infty \quad\text{as}\quad \xi \to \infty. \qquad \square$$

Note, that Corollary 13.1.10 shows that infinitely growing positive symbols cannot generate bounded Toeplitz operators. To generate a bounded Toeplitz operator, its unbounded symbol must necessarily have (see Example 13.1.4) sufficiently sophisticated oscillating behaviour at neighborhoods of the "critical" points 0 and ∞.

Given a symbol $a(y) \in L_1(\mathbb{R}_+, 0)$, denote by $B(a)$ the set of values $\lambda \in [0, \infty)$ for which the corresponding Toeplitz operator $T_a^{(\lambda)}$ is bounded. Theorem 13.1.6 suggests that the set $B(a)$, being non-empty, may have only one of the following three types:

$$[0, \infty), \qquad [0, \nu), \qquad [0, \nu].$$

We show that all of these possibilities can be realized. Indeed, the first case is true for bounded symbols. To realize the two remaining cases we give the following theorem.

Theorem 13.1.11. *There exists a family of symbols* $a_{\nu,\beta}(y)$*, with* $\nu \in (0,1)$*,* $\beta \geq 0$*, such that for the corresponding Toeplitz operators* $T_{a_{\nu,\beta}}^{(\lambda)}$ *we have*

a) $B(a_{\nu,0}) = [0,\nu], \quad \beta = 0,$

b) $B(a_{\nu,\beta}) = [0,\nu), \quad \beta > 0.$

Proof. To prove the above statement we show that the asymptotic behavior of the corresponding function $\gamma_{a_{\nu,\beta},\lambda}(x)$ when $x \to \infty$ is

$$\gamma_{a_{\nu,\beta},\lambda}(x) = c_\lambda e^{\frac{i}{5\pi}\ln^2(1+x)} \ln^{\lambda-\nu}(1+x)\ln^\beta \ln(1+x) + o(\ln^{\lambda-\nu}(1+x)\ln^\beta \ln(1+x)),$$
$$(13.1.14)$$

where $c_\lambda \neq 0$, and

$$\lim_{x \to 0} \gamma_{a_{\nu,\beta},\lambda}(x) = 0. \qquad (13.1.15)$$

To introduce the function $a_{\nu,\beta}(y)$ we consider

$$f_{\nu,\beta}(z) = e^{\frac{5\pi}{4}i} \exp\{\frac{i}{5\pi}\ln^2(z+i)\}[\ln(z+i) - i5\pi/2]^{-\nu}\ln^\beta(\ln(z+i) - i5\pi/2),$$
$$(13.1.16)$$

where the branch of the function $f_{\nu,\beta}(z)$ is fixed by $\arg z \in [3\pi/2, 7\pi/2]$. We set now

$$a_{\nu,\beta}(t/2) = \frac{1}{\sqrt{2\pi}}\int_{\mathbb{R}} f_{\nu,\beta}(x)e^{-ixt}dx.$$

The function $f_{\nu,\beta}(z)$ belongs to the Hardy space $H^2_+(\mathbb{R})$, thus $a_{\nu,\beta}(t) \in L_2(\mathbb{R}_+)$ and the formula

$$f_{\nu,\beta}(z) = \int_{\mathbb{R}_+} a_{\nu,\beta}(t/2)e^{izt}dt \qquad (13.1.17)$$

holds. Thus, $\gamma_{a_{\nu,\beta},0}(x) = x f_{\nu,\beta}(ix)$.

Recall that $t^\alpha e^{-xt} = \mathcal{D}^\alpha e^{-xt}$, where the Liouville fractional derivative is given, as usual, by

$$\mathcal{D}^\alpha \varphi(x) = \frac{1}{d_{1,1}(\alpha)}\int_{\mathbb{R}_+} \frac{\varphi(x+t) - \varphi(x)}{t^{1+\alpha}}dt,$$

where

$$d_{1,1}(\alpha) = \int_{\mathbb{R}_+} \frac{e^{-\xi} - 1}{\xi^{1+\alpha}}d\xi, \quad 0 < \alpha < 1.$$

Therefore, setting $c(\lambda) = \frac{1}{d_{1,1}(\lambda)\Gamma(\lambda+1)}$, we have

$$
\begin{aligned}
\gamma_{a_{\nu,\beta},\lambda}(x) &= c(\lambda)x^{\lambda+1} \int_{\mathbb{R}_+} \frac{f_{\nu,\beta}(i(x+t)) - f_{\nu,\beta}(ix)}{t^{1+\lambda}}\,dt \\[2mm]
&= c(\lambda)\frac{x^{\lambda+1}}{(x+1)^\lambda} \int_{\mathbb{R}_+} \frac{f_{\nu,\beta}(i(x+xt+t)) - f_{\nu,\beta}(ix)}{t^{1+\lambda}}\,dt \\[2mm]
&= c(\lambda,x)(x+1) \int_{\mathbb{R}_+} \frac{dt}{t^{1+\lambda}}\,dt \int_0^t \frac{d}{d\xi} f_{\nu,\beta}(i(x+x\xi+\xi))\,d\xi \\[2mm]
&= c(\lambda,x)(x+1) \int_{\mathbb{R}_+} \frac{d}{d\xi} f_{\nu,\beta}(i(x+x\xi+\xi))\,d\xi \int_\xi^\infty \frac{dt}{t^{1+\lambda}}\,dt \\[2mm]
&= \lambda^{-1} c(\lambda,x)(x+1) \int_{\mathbb{R}_+} \frac{1}{\xi^\lambda}\frac{d}{d\xi} f_{\nu,\beta}(i(x+x\xi+\xi))\,d\xi,
\end{aligned}
$$

where $c(\lambda,x) = c(\lambda)x^{\lambda+1}/(1+x)^{\lambda+1}$. Note that

$$
f_{\nu,\beta}(iy) = \frac{\exp\{\frac{i}{5\pi}\ln^2(1+y)\}}{1+y} \ln^{-\nu}(1+y) \ln^\beta \ln(1+y), \tag{13.1.18}
$$

hence we have

$$
\begin{aligned}
\gamma_{a_{\nu,\beta},\lambda}(x) = -\lambda^{-1} c(\lambda,x) \int_{\mathbb{R}_+} \frac{\exp\{\frac{i}{5\pi}\ln^2(1+x)(1+\xi)\}}{\xi^\lambda(1+\xi)^2} \Bigg(\frac{2i}{5\pi}\omega_{\nu,\beta}(x,\xi) \\
- \omega_{\nu+1,\beta}(x,\xi) - \nu\omega_{\nu+2,\beta}(x,\xi) - \beta\omega_{\nu+2,\beta-1}(x,\xi) \Bigg)\,d\xi, \tag{13.1.19}
\end{aligned}
$$

where

$$
\omega_{\nu,\beta}(x,\xi) = \ln^{1-\nu}(1+x)(1+\xi) \ln^\beta \ln(1+x)(1+\xi).
$$

We split the above integral into four integrals according to the sum of four terms in the brackets. These integrals are of the same type, and differ (up to a constant) only by parameters ν, β. Obviously, the principal term of the behavior of $\gamma_{a,\lambda}(x)$ when $x \to \infty$ is determined by the integral corresponding to the first summand, i.e.,

$$
\begin{aligned}
I(x,\lambda,\nu,\beta) &= \int_{\mathbb{R}_+} \frac{\exp\{\frac{i}{5\pi}\ln^2(1+x)(1+\xi)\}}{\xi^\lambda(1+\xi)^2} \omega_{\nu,\beta}(x,\xi)\,d\xi \\[2mm]
&= \int_{\mathbb{R}_+} \frac{\exp\{\frac{i}{5\pi}\ln^2(1+x)(1+\xi)\}}{\xi^\lambda(1+\xi)^2} \omega_{\nu,\beta}(x,\xi)\,(\chi_0(\xi) + \chi_\infty(\xi))\,d\xi \\[2mm]
&= I_0(x,\lambda,\nu,\beta) + I_\infty(x,\lambda,\nu,\beta).
\end{aligned}
$$

Here $\chi_0(\xi)$ is a smooth function on $\mathbb{R}_+$, satisfying the conditions $\chi_0(\xi) = 1$ for $0 \le \xi \le 1$ and $\chi_0(\xi) = 0$ for $\xi \ge 2$; and $\chi_\infty(\xi) = 1 - \chi_0(\xi)$.

Integrating the second integral by parts we have

$$
\begin{aligned}
I_\infty(x, \lambda, \nu, \beta) &= -\frac{5\pi i}{2} \int_1^\infty \frac{\omega_{\nu+1,\beta}(x,\xi)}{\xi^\lambda(1+\xi)} \chi_\infty(\xi)\, d\exp\{\frac{i}{5\pi}\ln^2(1+x)(1+\xi)\} \\
&= \frac{5\pi i}{2} \int_1^\infty \exp\{\frac{i}{5\pi}\ln^2(1+x)(1+\xi)\} \\
&\quad \cdot \frac{\partial}{\partial\xi}\left(\frac{\omega_{\nu+1,\beta}(x,\xi)}{\xi^\lambda(1+\xi)}\chi_\infty(\xi)\right) d\xi.
\end{aligned}
$$

For $\xi > 1$ and large enough x the following inequality holds:

$$
\left|\frac{\partial}{\partial\xi}\left(\frac{\omega_{\nu+1,\beta}(x,\xi)}{\xi^\lambda(1+\xi)}\chi_\infty(\xi)\right)\right| \le \text{const}\, \frac{\omega_{\nu+1,\beta}(x,0)}{\xi^\lambda(1+\xi)^2}.
$$

Thus we have

$$
|I_\infty(x, \lambda, \nu, \beta)| = O(\omega_{\nu+1,\beta}(x,0)) = O(\ln^{-\nu}(1+x)\ln^\beta\ln(1+x)).
$$

For $I_0(x, \lambda, \nu, \beta)$ according to a Lemma of Erdelyi (see [76]), we have

$$
I_0(x, \lambda, \nu, \beta) = (1 + O(\ln^{-1}(1+x)))\widetilde{I}_0(x, \lambda, \nu, \beta),
$$

where

$$
\begin{aligned}
\widetilde{I}_0(x, \lambda, \nu, \beta) &= \omega_{\nu,\beta}(x,0)e^{\frac{i}{5\pi}\ln^2(1+x)}\int_{\mathbb{R}_+} \frac{e^{i\frac{2\ln(1+x)}{5\pi}\xi}}{\xi^\lambda(1+\xi)^2}\chi_0(\xi)e^{\frac{i}{5\pi}\ln^2(1+\xi)}d\xi \\
&= \omega_{\nu,\beta}(x,0)e^{\frac{i}{5\pi}\ln^2(1+x)}\int_0^2 \frac{e^{i\frac{2\ln(1+x)}{5\pi}\xi}}{\xi^\lambda}F(\xi)d\xi.
\end{aligned}
$$

Applying Erdelyi's lemma ([76]) once again, we have

$$
\begin{aligned}
\int_0^2 \frac{e^{i\frac{2\ln(1+x)}{5\pi}\xi}}{\xi^\lambda}F(\xi)d\xi &= \frac{5\pi}{2i}\Gamma(1-\lambda)e^{i\pi(1-\lambda)/2}e^{\frac{i}{5\pi}\ln^2(1+x)}\ln^{\lambda-1}(1+x) \\
&\quad + o(\ln^{\lambda-1}(1+x)), \quad x \to \infty.
\end{aligned}
$$

This and the above considerations prove (13.1.14). Finally, it is easy to see that (13.1.19) implies (13.1.15). $\qquad\square$

13.2 Continuous symbols

Given a symbol $a = a(y)$, the Toeplitz operator $T_a^{(\lambda)}$ acting on the space $\mathcal{A}_\lambda^2(\Pi)$ is unitary equivalent to the multiplication operator $\gamma_{a,\lambda}I$, where the function $\gamma_{a,\lambda}(x)$, $x \in \mathbb{R}_+$, is given by (10.4.4). Thus we have obviously

$$
\text{sp}\, T_a^{(\lambda)} = \overline{M_\lambda(a)},
$$

where $M_\lambda(a) = \text{Range}\,\gamma_{a,\lambda}$.

Theorem 13.2.1. *Let $a = a(y) \in C(\overline{\mathbb{R}}_+) = C[0, +\infty]$. Then*

$$\lim_{\lambda \to +\infty} \operatorname{sp} T_a^{(\lambda)} = M_\infty(a) = \operatorname{Range} a. \tag{13.2.1}$$

Note, that $\operatorname{Range} a$ coincides with the spectrum $\operatorname{sp} aI$ of the operator of multiplication by $a = a(y)$ acting, say, on any of $L_2(\Pi, d\mu_\lambda)$; thus another form of (13.2.1) is

$$\lim_{\lambda \to +\infty} \operatorname{sp} T_a^{(\lambda)} = \operatorname{sp} aI.$$

Proof. Introduce the large parameter $L = \sqrt{x^2 + \lambda^2}$ and represent $\gamma_{a,\lambda}(x)$ in the form

$$\gamma_{a,\lambda}(x) = \frac{x^{\lambda+1}}{\Gamma(\lambda+1)} \int_0^\infty a(t/2) e^{-LS(t,\varphi)} dt, \tag{13.2.2}$$

where

$$S(t, \varphi) = \frac{x}{L} t - \frac{\lambda}{L} \ln t = (\sin \varphi) t + (\cos \varphi) \ln 1/t, \quad \text{with} \quad \varphi \in [0, \pi/2].$$

The function $S(t, \varphi)$, as a function of t, has a minimum at the point

$$t_\varphi = \frac{\cos \varphi}{\sin \varphi} \in (0, \infty).$$

Write (13.2.2) in the form

$$\begin{aligned}
\gamma_{a,\lambda}(x) - a(t_\varphi/2) &= \frac{x^{\lambda+1}}{\Gamma(\lambda+1)} \left[\int_{\mathbb{R}_+ \cap U(t_\varphi)} (a(t/2) - a(t_\varphi/2)) e^{-LS(t,\varphi)} dt \right. \\
&\quad \left. + \int_{\mathbb{R}_+ \setminus U(t_\varphi)} (a(t/2) - a(t_\varphi/2)) e^{-LS(t,\varphi)} dt \right] \\
&\equiv I_1(L) + I_2(L),
\end{aligned} \tag{13.2.3}$$

where $U(t_\varphi)$ is a neighborhood of the point t_φ such that

$$\sup_{t \in U(t_\varphi)} |a(t/2) - a(t_\varphi/2)| < \varepsilon,$$

with $\varepsilon > 0$ sufficiently small. We have,

$$I_1(L) \leq \varepsilon$$

uniformly in φ. Next, $I_2(L) \leq \varepsilon$ uniformly on φ as well. Indeed rewrite the integral $I_2(L)$ in the form

$$\begin{aligned}
I_2(L) &= \frac{x^{\lambda+1}}{\Gamma(\lambda+1)} \int_0^{t_\varphi - \sigma} (a(t/2) - a(t_\varphi/2)) e^{-LS(t,\varphi)} dt \\
&\quad + \frac{x^{\lambda+1}}{\Gamma(\lambda+1)} \int_{t_\varphi + \sigma}^\infty (a(t/2) - a(t_\varphi/2)) e^{-LS(t,\varphi)} dt \\
&\equiv I_{2,1}(L) + I_{2,2}(L)
\end{aligned}$$

where $\sigma > 0$ is small enough.

Let us use the asymptotic Euler formula for the Γ-function (see, [86], formula 8.327)

$$\Gamma(\lambda + 1) = \lambda\Gamma(\lambda) = \frac{\lambda e^{-\lambda}\lambda^{\lambda-1/2}}{\sqrt{2\pi}}(1 + O(\lambda^{-1/2})), \quad \lambda \to \infty,$$

where we set $\lambda = xt_\varphi$. Then the integral $I_{2,2}(L)$ admits the estimate

$$|I_{2,2}(L)| \leq \text{const } x^{1/2} \int_{t_\varphi+\sigma}^{\infty} |a(t/2) - a(t_\varphi/2)|e^{-x\widetilde{S}(t,\varphi)}\,dt$$

where

$$\widetilde{S}(t,\varphi) = (t - t_\varphi) - t_\varphi(\ln t - \ln t_\varphi).$$

It is evident that there exists $\Delta\,(> 0)$ which does not depend on φ and such that for $t \geq t_\varphi + \delta$ the following inequality holds,

$$\widetilde{S}(t,\varphi) > \Delta(t - t_\varphi), \quad t > t_\varphi.$$

Thus we have

$$\begin{aligned}
|I_{2,2}(L)| &\leq \text{const } x^{1/2} \int_{t_\varphi+\sigma} |a(t/2) - a(t\varphi/2)|e^{-x\Delta(t-t_\varphi)}\,dt \\
&\leq \text{const } x^{1/2}e^{-(x-1)\Delta\sigma} \int_{t_\varphi+\sigma} |a(t/2) - a(t\varphi/2)|e^{-\Delta(t-t_\varphi)}\,dt.
\end{aligned}$$

According to the definition of the class $L_1(\mathbb{R}_+, 0)$ the last integral is finite and we have, uniformly in φ,

$$\lim_{L\to\infty} I_{2,2}(L) = 0.$$

Analogously one can get that, uniformly in φ,

$$\lim_{L\to\infty} I_{2,1}(L) = 0$$

and consequently $\lim_{L\to\infty} I_2(L) = 0$.

Since ε can be arbitrarily small, from the above we get

$$\gamma_{a,\lambda}(x) = a(t_\varphi/2)\,(1 + \alpha(L)), \tag{13.2.4}$$

where $\alpha(L) \to 0$, when $L \to \infty$, uniformly in φ. $\qquad\square$

13.3 Piece-wise continuous symbols

Let $b(t) = a(t/2)$ be a piece-wise continuous function on $[0, +\infty]$ having jumps at a finite set of points $\{t_j\}_{j=1}^{m}$,

$$0 = t_0 < t_1 < t_2 < \ldots < t_m < t_{m+1} = +\infty,$$

and such that $a(t_j/2 \pm 0), j = 1, \ldots, m$, exist. Introduce the sets

$$J_j(a) := \{z \in \mathbb{C} : \ z = a(t/2), \ \ t \in (t_j, t_{j+1})\}$$

where $j = 0, \ldots, m$, and let $I_j(a)$ be the straight line segment with the endpoints $a(t_j/2 - 0)$ and $a(t_j/2 + 0)$, $j = 1, 2, \ldots m$.

Introduce now

$$\widetilde{R}(a) = \left(\cup_{j=0}^m J_j(a)\right) \cup \left(\cup_{j=1}^m I_j(a)\right).$$

Theorem 13.3.1. *Let $a(t/2)$ be a piece-wise continuous function on $[0, +\infty]$. Then*

$$\lim_{\lambda \to \infty} \mathrm{sp}_\lambda \, T_a^{(\lambda)} = M_\infty(a) = \widetilde{R}(a).$$

Proof. The proof is quite analogous to that of Theorem 14.3.1, see also Theorem 12.4.1. $\qquad\qquad\qquad\qquad\qquad\qquad\qquad\qquad\qquad\qquad\qquad\qquad$ $\square$

For L_∞-symbols apart from the *a priori* information (12.3.1) we have obviously

$$\lim_{\lambda \to \infty} \mathrm{sp}_\lambda \, T_a^{(\lambda)} = M_\infty(a) \subset \mathrm{conv}(\mathrm{ess\,Range}\, a). \tag{13.3.1}$$

At the same time the position of $M_\infty(a)$ inside $\mathrm{conv}(\mathrm{ess\,Range}\, a)$ may essentially vary. We give a number of examples illustrating possible interrelations between these sets.

Example 13.3.2. Let $a(t) \in C[0, +\infty]$. Then according to Theorem 13.2.1,

$$M_\infty(a) = \mathrm{Range}\, a \ (= \mathrm{ess\,Range}\, a).$$

Example 13.3.3. Let

$$a(t/2) = \left\{ \begin{array}{ll} \alpha_1, & t \in (0, 1) \\ \alpha_2, & t \in [1, \infty] \end{array} \right. .$$

where $\alpha_1, \alpha_2 \in \mathbb{C}$ and $\alpha_1 \neq \alpha_2$. Then according to Theorem 13.3.1, $M_\infty(a)$ coincides with the straight line segment $[\alpha_1, \alpha_2]$ joining the points α_1 and α_2, whence

$$M_\infty(a) = \mathrm{conv}(\mathrm{ess\,Range}\, a) \ \ (= \mathrm{conv}(\mathrm{Range}\, a)).$$

Example 13.3.4. Let

$$a(t/2) = \left\{ \begin{array}{ll} \alpha_1, & t \in [0, 1) \\ \alpha_2, & t \in [1, 2) \\ \alpha_3, & t \in [2, \infty] \end{array} \right. .$$

where $\alpha_1, \alpha_2, \alpha_3$ are different points from $\mathbb{C}$. Then by Theorem 13.3.1 we have

$$M_\infty(a) = [\alpha_1, \alpha_2] \cup [\alpha_2, \alpha_3]$$

and in this case the set $M_\infty(a)$ is a part of the boundary of the convex hull $\mathrm{ess\,Range}\, a = \mathrm{Range}\, a$, that is

$$M_\infty(a) \subset \partial\,\mathrm{conv}(\mathrm{Range}\, a).$$

Example 13.3.5. Let $\alpha_1, \alpha_2, \alpha_3$ be as above, and

$$a(t/2) = \begin{cases} \alpha_1, & t \in [0,1) \\ \alpha_2, & t \in [1,2) \\ \alpha_3, & t \in [2,3) \\ \alpha_1, & t \in [3,\infty] \end{cases}.$$

By Theorem 13.3.1 the set $M_\infty(a)$ coincides with a triangle having the vertices α_1, α_2, and α_3,

$$M_\infty(a) = [\alpha_1, \alpha_2] \cup [\alpha_2, \alpha_3] \cup [\alpha_3, \alpha_4].$$

Thus in this case

$$M_\infty(a) = \partial \operatorname{conv}(\operatorname{Range} a).$$

Example 13.3.6. Let $\{t_j\}_{j\in\mathbb{Z}_+}$ be a sequence of increasing positive numbers with $\lim_{j\to\infty} t_j = \infty$ and $t_0 = 0$. Define the symbol $a(t)$ as follows,

$$a(t/2) = \begin{cases} e^{i\xi_j}, & t \in [t_{2j}, t_{2j+1}) \\ -e^{i\xi_j}, & t \in [t_{2j+1}, t_{2j+2}) \end{cases}.$$

where $\{\xi_j\}_{j\in\mathbb{Z}_+} \subset [0,\pi]$ with the closure $\overline{\{\xi_j\}_{j\in\mathbb{Z}_+}} = [0,\pi]$.

As in Theorem 13.3.1 one can show that each diameter $[e^{i\xi_j}, -e^{i\xi_i}]$ of the unit disk $\mathbb{D}$ having $e^{i\xi_j}$ and $-e^{i\xi_j}$ as endpoints, belongs to $M_\infty(a)$, which implies $\overline{\mathbb{D}} \subset M_\infty(a)$. We have that $\overline{\operatorname{Range} a} = \partial\mathbb{D} = \mathbb{T}$. Finally,

$$M_\infty(a) = \overline{\mathbb{D}} = \operatorname{conv}(\operatorname{Range} a).$$

13.4 Oscillating symbols

We consider here a case of a discontinuity of the second kind, the oscillating symbols. To be more precise, the following two model situations will be considered: a strong oscillation and a slow oscillation. In spite of their qualitative sameness, an oscillation type discontinuity, the results differ drastically.

Theorem 13.4.1 (Strong oscillation). *Let* $a(t) = e^{2it}$, *then* $\operatorname{Range} a = \mathbb{T}$ *and* $M_\infty(a) = \mathbb{D}$.

Proof. For $a(t/2) = e^{it}$ we have

$$
\begin{aligned}
\gamma_{a,\lambda}(x) &= \frac{x^{\lambda+1}}{\Gamma(\lambda+1)} \int_0^\infty t^\lambda e^{-(x-i)t}\, dt \\
&= \frac{x^{\lambda+1}}{(x-i)^{\lambda+1}} \cdot \frac{1}{\Gamma(\lambda+1)} \int_0^\infty s^\lambda e^{-s}\, ds \\
&= \left(\frac{x}{x-i}\right)^{\lambda+1} \\
&= \exp\left[\frac{\lambda+1}{2} \ln\left(1 - \frac{1}{x^2+1}\right)\right] \cdot \exp\left[(\lambda+1)i \arctan(x^{-1})\right]. \quad (13.4.1)
\end{aligned}
$$

Given a point $z_0 \in \mathbb{D}$, we represent it in the form

$$z_0 = \exp(-\alpha_0 + i\beta_0),$$

where $\alpha_0 > 0$ and $\beta_0 \in [0, 2\pi)$.

Introduce the sequences

$$x_k = \frac{\beta_0 + 2\pi k}{2\alpha_0} \quad \text{and} \quad \lambda_k = \frac{(\beta_0 + 2\pi k)^2}{2\alpha_0} - 1 = 2\alpha_0 x_k^2 - 1, \quad k \in \mathbb{N}.$$

Then for large values of k we have

$$
\begin{aligned}
\gamma_{a,\lambda_k}(x_k) &= \exp\left[\frac{\lambda_k + 1}{2} \ln\left(1 - \frac{1}{x_k^2 + 1}\right)\right] \cdot \exp\left[(\lambda_k + 1)i \arctan(x_k^{-1})\right] \\
&= \exp\left[-\frac{\lambda_k + 1}{2x_k^2} + (\lambda_k + 1)O(x_k^{-4})\right] \\
&\quad \cdot \exp\left[i\frac{\lambda_k + 1}{x_k} + (\lambda_k + 1)O(x_k^{-3})\right] \\
&= \exp\left[-\alpha_0 + O(k^{-2}) + i(\beta_0 + 2\pi k) + O(k^{-1})\right].
\end{aligned}
$$

It is easy to see now that

$$\lim_{k \to \infty} \gamma_{a,\lambda_k}(x_k) = z_0;$$

that is, $z_0 \in M_\infty(a)$, and $\mathbb{D} \subset M_\infty(a)$. The inverse inclusion follows from (13.3.1). $\qquad\square$

We note that formula (13.4.1) permits us to understand the form of the image of $\gamma_{a,\lambda}$ for each fixed (and sufficiently large) value of λ. First of all, it is easy to see that

$$\lim_{x \to \infty} \gamma_{a,\lambda}(x) = 1 \quad \text{and} \quad \lim_{x \to 0} \gamma_{a,\lambda}(x) = 0.$$

If $0 < m < x < M < +\infty$, then the absolute value of $\gamma_{a,\lambda}(x)$ changes much more slowly than its argument. That is for each fixed λ, the image of $\gamma_{a,\lambda}$ looks like a spiral starting at the point $z = 1$ and tending to $z = 0$ as x tends to 0. Moreover, when λ is growing the branches of a spiral became closer to each other.

Theorem 13.4.2 (Slow oscillation). *Let $a(t) = (2t)^i$, then we have that* Range $a = \mathbb{T}$ *and $M_\infty(a) = \mathbb{T}$.*

Proof. For $a(t/2) = t^i$ we have

$$
\begin{aligned}
\gamma_{a,\lambda}(x) &= \frac{x^{\lambda+1}}{\Gamma(\lambda+1)} \int_0^\infty t^{\lambda+i} e^{-xt} dt \\
&= \frac{1}{\Gamma(\lambda+1)} \int_0^\infty s^{\lambda+i} e^{-s} ds = x^i \frac{\Gamma(\lambda+1+i)}{\Gamma(\lambda+1)}.
\end{aligned}
$$

That is for a fixed λ the image of $\gamma_{a,\lambda}$ coincides with the circle centered at the origin and having radius equal to $\left|\frac{\Gamma(\lambda+1+i)}{\Gamma(\lambda+1)}\right|$.

By [86], formula 8.328.2, we have

$$\lim_{\lambda\to\infty}\left|\frac{\Gamma(\lambda+1+i)}{\Gamma(\lambda+1)}\right|=1. \qquad \square$$

We note that Theorems 13.4.1 and 13.4.2 can be generalized for a wide class of strong and slowly oscillating symbols. For example, if $a_1(t)=(2t+1)^i$, then $M_\infty(a_1)=\mathbb{T}$, as in Theorem 13.4.2. The function $a_1(t)$ is continuous at the point $t=0$, thus $\gamma_{a_1,\lambda}(\infty)=a_1(0)=1$, for all λ. For a fixed λ the image of $\gamma_{a_1,\lambda}$ is a spiral outgoing starting at point $z=1$ and tending to the limit circle with radius equal to $\left|\frac{\Gamma(\lambda+1+i)}{\Gamma(\lambda+1)}\right|$ and centered at the origin (the same circle as in Theorem 13.4.2).

We illustrate the above on figures presenting images of functions $\gamma_{a,\lambda}$ for the two oscillating symbols

$$a_1(t)=(1+2t)^i=e^{i\ln(1+2t)} \quad \text{and} \quad a_2(t)=e^{i2t}, \quad t\in[0,\infty),$$

and for the following values of λ: 0, 10, and 1000.

We note that both symbols are continuous at the point $t=0$ and have oscillation type discontinuity at infinity, and both of them are of the same form

$$a_k(t)=e^{i\varphi_k(t)}, \qquad k=1,2,$$

where the corresponding functions $\varphi_k(t)$ are continuous and increasing on $[0,+\infty]$ with $\varphi_k(0)=0$ and $\varphi_k(+\infty)=+\infty$. The only difference between them is the speed of growth at infinity. This difference leads to a drastic difference between the spectrum behaviour of the corresponding Toeplitz operators.

13.5 Unbounded symbols

Theorem 13.5.1. *Let $a(t)\in L_1(\mathbb{R}_+,0)\cap C(\mathbb{R}_+)$. Then*

$$\text{Range}\, a\subset M_\infty(a).$$

Proof. Analogous to that of Theorem 13.2.1. $\qquad \square$

We show now that the property (13.3.1), previously established for bounded symbols, still remains valid for our unbounded symbols.

Theorem 13.5.2. *Let $a(t)\in L_1(\mathbb{R}_+,0)$. Then*

$$M_\infty(a)\subset \text{conv}(\text{ess Range}\, a).$$

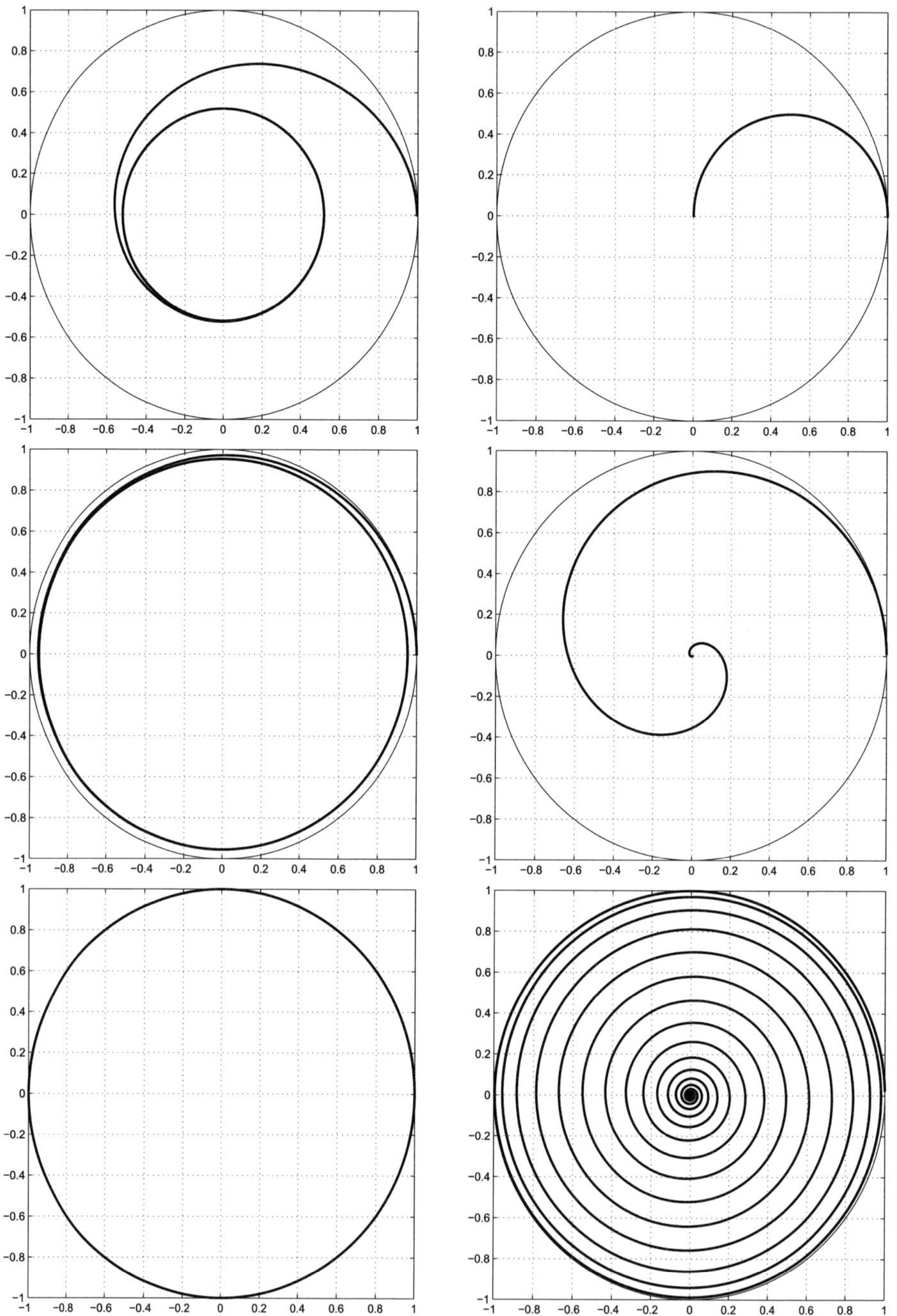

Figure 13.1: The functions $\gamma_{a_1,\lambda}(x)$ and $\gamma_{a_2,\lambda}(x)$ for $\lambda = 0$, $\lambda = 10$, and $\lambda = 1000$.

Proof. For each $M > 0$ consider the function

$$a_M(t) = \begin{cases} a(t), & \text{if } |a(t)| \leq M \\ 0, & \text{if } |a(t)| > M \end{cases}.$$

The function $a_M(t)$ is bounded, whence

$$\operatorname{Range} \gamma_{a_M,\lambda} \subset \operatorname{conv}(\operatorname{ess Range} a_M) \subset \operatorname{conv}(\operatorname{ess Range} a).$$

The equality

$$\lim_{M \to \infty} \gamma_{a_M,\lambda}(x) = \gamma_{a,\lambda}(x),$$

verified by the Lebesgue dominated convergence theorem, implies

$$\operatorname{Range} \gamma_{a,\lambda}(x) \subset \operatorname{conv}(\operatorname{Range} a). \qquad \square$$

Corollary 13.5.3. *For functions* $a(t) \in L_1(\mathbb{R}_+, 0) \cap C(\mathbb{R}_+)$,

$$\operatorname{Range} a \subset M_\infty(a) \subset \operatorname{conv}(\operatorname{Range} a).$$

Example 13.5.4. For each $j \in \mathbb{N}$ define $I_j = [j-1, j-1+1/j^3]$ and let $\{\xi_j\}_{j \in \mathbb{N}}$ be a sequence such that $\overline{\{\xi_j\}_{j \in \mathbb{N}}} = [0, 2\pi]$. Define the symbol by

$$a(t/2) = \begin{cases} je^{i\xi_j}, & t \in I_j, \ j \in \mathbb{N}, \\ 0, & \text{otherwise.} \end{cases}$$

Obviously, $B_a^{(1)}(\xi) \leq \sum_{j \in \mathbb{N}} \frac{1}{j^2}$, and the corresponding Toeplitz operator $T_a^{(\lambda)}$ is bounded for every $\lambda > 0$. Theorem 13.3.1 implies that the straight line segment $[0, je^{i\xi_j}]$ belongs to $M_\infty(a)$. Thus

$$M_\infty(a) = \mathbb{C} = \operatorname{conv}(\operatorname{Range} a).$$

Example 13.5.5. For given $\alpha \in (0,1)$ introduce $a(t/2) = t^{i-\alpha}$ and calculate

$$\gamma_{a,\lambda}(x) = \frac{x^{\lambda+1}}{\Gamma(\lambda+1)} \int_0^\infty t^{\lambda+i-\alpha} e^{-xt} dt = \frac{x^{\alpha-i}\Gamma(\lambda+1+\alpha-i)}{\Gamma(\lambda+1)}.$$

By the asymptotics of Γ-function (see, [86], formula 8.327)

$$\gamma_{a,\lambda}(x) = x^{\alpha-i}(\lambda+1)^{i-\alpha}(1+o(1)), \quad \lambda \to \infty.$$

Given arbitrary $\eta > 0$, one can take x and λ such that $\frac{\lambda+1}{x} = \eta$. Thus,

$$\gamma_{a,\lambda}(x) = \eta^{i-\alpha}(1+o(1)), \quad \lambda \to \infty,$$

and in this case,

$$\operatorname{Range} a = M_\infty(a).$$

Chapter 14

Dynamics of Properties of Toeplitz Operators on the Upper Half-Plane: Hyperbolic Case

14.1 Boundedness of Toeplitz operators with symbols depending on $\theta = \arg z$.

Recall that by Theorems 10.4.16 and 10.5.1, the function

$$\gamma_{a,\lambda}(\xi) = \left(\int_0^\pi e^{-2\xi\theta} \, \sin^\lambda \theta \, d\theta \right)^{-1} \int_0^\pi a(\theta) \, e^{-2\xi\theta} \, \sin^\lambda \theta \, d\theta, \qquad \xi \in \mathbb{R} \quad (14.1.1)$$

is responsible for the boundedness of a Toeplitz operator with symbol $a(\theta)$ ($\in L_1(0,\pi)$). If $a(\theta) \in L_\infty(0,\pi)$, then the operator $T_a^{(\lambda)}$ is obviously bounded on all spaces $\mathcal{A}_\lambda^2(\Pi)$, for $\lambda \in (-1,\infty)$, and the corresponding norms are uniformly bounded by $\sup_z |a(z)|$. That is, all spaces $\mathcal{A}_\lambda^2(\Pi)$, where $\lambda \in (-1,\infty)$, are natural and appropriate for Toeplitz operators with *bounded* symbols. Studying *unbounded* symbols, we wish to have a sufficiently large class of them common to all admissible λ; moreover, we are especially interested in properties of Toeplitz operators for large values of λ. Thus it is convenient for us to consider λ belonging only to $[0,\infty)$, which we will always assume in what follows.

For $a(\theta) \in L_1(0,\pi)$ the function $\gamma_{a,\lambda}(\xi)$ is continuous at all finite points $\xi \in \mathbb{R}$. For a "very large ξ" ($\xi \to +\infty$) the exponent $e^{-2\xi\theta}$ has a very sharp maximum at the point $\theta = 0$, and thus the major contribution to the integral containing $a(\theta)$ in (14.1.1) for these "very large ξ" is determined by values of

$a(\theta)$ in a neighborhood of the point 0. The major contribution for a "very large negative ξ" ($\xi \to -\infty$) is determined by values of $a(\theta)$ in a neighborhood of π, due to a very sharp maximum of $e^{-2\xi\theta}$ at $\theta = \pi$ for these values of ξ. In particular, a minor modification of the proof of Lemma 7.2.3 shows that if $a(\theta)$ has limits on the points 0 and π, then

$$\lim_{\xi \to +\infty} \gamma_{a,\lambda}(\xi) = \lim_{\theta \to 0} a(\theta),$$
$$\lim_{\xi \to -\infty} \gamma_{a,\lambda}(\xi) = \lim_{\theta \to \pi} a(\theta).$$

As a matter of fact, 0 and π are the only worrisome points for unbounded symbols $a(\theta) \in L_1(0, \pi)$. Moreover the behaviour of certain means of a symbol, rather than the behaviour of a symbol itself, plays a crucial role in the study of boundedness properties.

Given $\lambda \in [0, \infty)$ and a function $a(\theta) \in L_1(0, \pi)$ introduce the following means:

$$C_{a,\lambda}^{(1)}(\sigma) = \int_0^\sigma a(\theta) \sin^\lambda \theta d\theta,$$
$$D_{a,\lambda}^{(1)}(\sigma) = \int_\sigma^\pi a(\theta) \sin^\lambda \theta d\theta,$$
$$C_{a,\lambda}^{(j)}(\sigma) = \int_0^\sigma C_{a,\lambda}^{(j-1)}(\theta) d\theta, \quad j = 2, 3, \ldots,$$
$$D_{a,\lambda}^{(j)}(\sigma) = \int_\sigma^\pi D_{a,\lambda}^{(j-1)}(\theta) d\theta, \quad j = 2, 3, \ldots.$$

We note that, as in the previous chapter, the above means depend on the weight parameter.

Theorem 14.1.1. *Let $a(\theta) \in L_1(0, \pi)$. If for certain $\lambda_0 \in [0, \infty)$ and $j_0, j_1 \in \mathbb{N}$ the following conditions hold,*

$$C_{a,\lambda_0}^{(j_0)}(\sigma) = O(\sigma^{j_0 + \lambda_0}), \quad \sigma \to 0, \tag{14.1.2}$$

$$D_{a,\lambda_0}^{(j_1)}(\sigma) = O((\pi - \sigma)^{j_1 + \lambda_0}), \quad \sigma \to \pi, \tag{14.1.3}$$

then the corresponding Toeplitz operator $T_a^{(\lambda)}$ is bounded on $\mathcal{A}_\lambda^2(\Pi)$ for each $\lambda \in [\lambda_0, \infty)$.

Proof. Note that the function $\gamma_{a,\lambda}(\xi)$ is continuous at finite points. We assume now that $\xi \to +\infty$ and the condition (14.1.2) holds with $j_0 = 1$. Then

$$
\begin{aligned}
\gamma_{a,\lambda}(\xi) &= 2^\lambda(\lambda+1)\vartheta_\lambda^2(\xi)\int_0^\pi \sin^{\lambda-\lambda_0}(\theta)e^{-2\xi\theta}dC_{a,\lambda_0}^{(1)}(\theta) \\
&= 2^\lambda(\lambda+1)\vartheta_\lambda^2(\xi)\left|\int_0^\pi C_{a,\lambda_0}^{(1)}(\theta)[(\lambda-\lambda_0)\sin^{\lambda-\lambda_0-1}\theta\cos\theta\right.\\
&\quad \left. -\,2\xi\sin^{\lambda-\lambda_0}\theta]e^{-2\xi\theta}d\theta\right| \\
&\leq \text{const}\,2^\lambda(\lambda+1)\vartheta_\lambda^2(\xi)\left[(\lambda-\lambda_0)\int_0^\infty \theta^\lambda e^{-2\xi\theta}d\theta\right.\\
&\quad \left. +\,2\xi\int_0^\infty \theta^{\lambda+1}e^{-2\xi\theta}d\theta\right] \\
&\leq \text{const}\,\vartheta_\lambda^2(\xi)\left[(\lambda-\lambda_0)(2\xi)^{-(\lambda+1)}\Gamma(\lambda+1)+(2\xi)^{-(\lambda+1)}\Gamma(\lambda+2)\right] \\
&\leq \text{const}\,(2\lambda-\lambda_0+1)2^\lambda(\lambda+1)\vartheta_\lambda^2(\xi)(2\xi)^{-(\lambda+1)}\Gamma(\lambda+1).
\end{aligned}
$$

It is easy to get the asymptotic representation of the function $\vartheta_\lambda^2(\xi)$. According to (10.3.10) we have

$$
\begin{aligned}
2^{-\lambda}(\lambda+1)^{-1}\vartheta_\lambda^{-2}(\xi) &= \int_0^\pi e^{-2\xi\theta}\sin^\lambda\theta d\theta \\
&= \int_0^\pi \theta^\lambda e^{-2\xi\theta}d\theta[1+\theta(\xi^{-1})] \\
&= (2\xi)^{-(\lambda+1)}\Gamma(\lambda+1)[1+O(\xi^{-1})]. \qquad (14.1.4)
\end{aligned}
$$

Thus we have finally

$$
|\gamma_{a,\lambda}(\xi)| \leq \text{const}\,(2\lambda-\lambda_0+1).
$$

The case $\xi \to -\infty$ (and $j_1 = 1$) is reduced to the one considered using the change of variable $\theta = \pi - \theta'$ in the integral for $\gamma_{a,\lambda}(\xi)$, see Remark 8.5.1.

The cases $j_{0,1} > 1$ are considered analogously using integration by parts. $\square$

The proof of the following statement is analogous to that of Theorem 13.1.3.

Theorem 14.1.2. 1. *Let conditions (14.1.2), (14.1.3) hold for $j_0 = j_0'$, $j_1 = j_1'$, and some λ_0. Then these conditions hold for $j_0 = j_0' + 1$, $j_1 = j_1' + 1$, and the same λ_0.*

2. *Let conditions (14.1.2), (14.1.3) hold for $j_0 = j_0'$, $j_1 = j_1'$, and some λ_0. Then these conditions hold for $j_0 = j_0'$, $j_1 = j_1'$, and λ_0 replaced by any $\lambda_1 \geq \lambda_0$.*

Example 14.1.3. Consider the family of unbounded symbols

$$
a(\theta) = (\sin\theta)^{-\beta}\sin[(\sin\theta)^{-\alpha}].
$$

As in Example 13.1.4 it can be proved that for all $\lambda \geq 0$ the operator $T_a^{(\lambda)}$ is bounded for each $\beta \in (0,1)$ and $\alpha > 0$.

Theorem 14.1.4. *Let the Toeplitz operator $T_a^{(\lambda)}$, with $a(\theta) \in L_1(0,\pi)$, be bounded on some $\mathcal{A}_{\lambda_0}^2(\Pi)$. Then it is bounded on each $\mathcal{A}_\lambda^2(\Pi)$, where $\lambda \in [0, \lambda_0]$.*

Proof. Let $\sup_{\xi \in \mathbb{R}} |\gamma_{a,\lambda_0}(\xi)| < \infty$. We split $a(\theta)$ into two functions which vanish in neighborhoods of 0 and π, respectively. The study of either of these two cases is quite similar, thus we suppose that $a(\theta)$ vanishes in a neighborhood of π, for example. Suppose also that $\xi \to \infty$. A similar argument is applicable for the study of the behavior of $\gamma_{a,\lambda}(\xi)$ under $\xi \to -\infty$. For $\lambda \in [0, \lambda_0)$, write

$$\gamma_{a,\lambda}(\xi) = \frac{2^{2\lambda - \lambda_0}(\lambda+1)\vartheta_\lambda^2(\xi)}{\Gamma(\lambda_0 - \lambda)} \int_0^\infty y^{\lambda_0 - \lambda - 1} dy \int_0^\pi a(\theta) e^{-2\theta(\xi + \frac{\sin\theta}{\theta} y)} \sin^{\lambda_0}\theta d\theta.$$

Using $\frac{\sin\theta}{\theta} = 1 + O(\theta^2)$, as $\theta \to 0$, for some $c_\lambda \neq 0$, we have

$$\begin{aligned}
\gamma_{a,\lambda}(\xi) &= (c_\lambda + o(1))\vartheta_\lambda^2(\xi) \int_0^\infty y^{\lambda_0 - \lambda - 1} dy \int_0^\pi a(\theta) e^{-2\theta(\xi + y)} \sin^{\lambda_0}\theta d\theta \\
&= \frac{(c_\lambda + o(1))\vartheta_\lambda^2(\xi)}{2^{\lambda_0}(\lambda_0 + 1)} \int_0^\infty y^{\lambda_0 - \lambda - 1} \frac{\gamma_{a,\lambda_0}(\xi + y)}{\vartheta_{\lambda_0}^2(\xi + y)} dy.
\end{aligned}$$

Using (14.1.4) and $\sup_{\xi \in \mathbb{R}} |\gamma_{a,\lambda_0}(\xi)| < \infty$ we have

$$\begin{aligned}
|\gamma_{a,\lambda}(\xi)| &\leq \text{const } \xi^{\lambda+1} \int_0^\infty y^{\lambda_0 - \lambda - 1}(\xi + y)^{-(\lambda_0 + 1)} dy \\
&= \text{const } \int_0^\infty u^{\lambda_0 - \lambda - 1}(1 + u)^{-(\lambda_0 + 1)} du < \infty,
\end{aligned}$$

since $\lambda < \lambda_0$ and $\lambda + 2 > 1$. $\qquad\qquad\qquad\qquad\qquad\qquad\qquad\qquad\qquad\square$

As an immediate corollary of Theorems 14.1.1 and 14.1.4 we have now

Theorem 14.1.5. *Under the hypothesis of Theorem 14.1.1 the Toeplitz operator $T_a^{(\lambda)}$ is bounded on $\mathcal{A}_\lambda^2(\Pi)$ for each $\lambda \in [0, \infty)$.*

The proof of the next theorem is analogous to that of Theorem 13.1.8.

Theorem 14.1.6.   1. *Assume that $a(\theta) \in L_1(0,\pi)$ and $a(\theta) \geq 0$ almost everywhere. Let the operator $T_a^{(\lambda')}$ be bounded on $\mathcal{A}_{\lambda'}^2(\Pi)$ for some $\lambda' > 0$. Then the conditions (14.1.2) and (14.1.3) hold for $j_0 = j_1 = 1$, $\lambda_0 = 0$ and consequently the operator $T_a^{(\lambda)}$ is bounded on $\mathcal{A}_\lambda^2(\Pi)$ for arbitrary $\lambda \in [0, \infty)$.*

2. *Assume that $C_{a,\mu_0}^{(j_0)}(\sigma) \geq 0$ and $D_{a,\mu_1}^{(j_1)}(\sigma) \geq 0$ almost everywhere for some $j_0 \geq 1$, $j_1 \geq 1$ and $\mu_0 \geq 0$, $\mu_1 \geq 0$, and that the operator $T_a^{(\lambda')}$ is bounded on $\mathcal{A}_{\lambda'}^2(\Pi)$ for some $\lambda' \geq 0$. Then the operator $T_a^{(\lambda)}$ is bounded on $\mathcal{A}_\lambda^2(\Pi)$ for arbitrary $\lambda \in [0, \infty)$.*

For a non-negative $a(\theta)$ we set

$$\begin{aligned}
m_{a,0}(\sigma) &= \operatorname{ess-inf}_{\theta \in (0,\sigma)} a(\theta), \\
m_{a,\pi}(\sigma) &= \operatorname{ess-inf}_{\theta \in (\sigma,\pi)} a(\theta).
\end{aligned}$$

Corollary 14.1.7. *Given a non-negative symbol, if either $\lim_{\sigma \to 0} m_{a,0}(\sigma) = \infty$ or $\lim_{\sigma \to \pi} m_{a,\pi}(\sigma) = \infty$, then the Toeplitz operator $T_a^{(\lambda)}$ is unbounded on each $\mathcal{A}_\lambda^2(\Pi)$, with $\lambda \in [0,\infty)$.*

For a symbol $a(\theta) \in L_1(0,\pi)$ we denote by $\widetilde{B}(a)$ the set of points $\lambda \in [0,\infty)$ for which the corresponding Toeplitz operator $T_a^{(\lambda)}$ is bounded on $\mathcal{A}_\lambda^2(\Pi)$. As in the parabolic case we have the following result, the proof of which is analogous to that of Theorem 13.1.11.

Theorem 14.1.8. *There exists a family of symbols $a_{\nu,\beta}(\theta)$, where $\nu \in (0,1)$, $\beta \in \mathbb{R}$, such that*

a) $\widetilde{B}(a_{\nu,0}) = [0,\nu], \quad \beta = 0,$

b) $\widetilde{B}(a_{\nu,\beta}) = [0,\nu), \quad \beta > 0.$

14.2 Continuous symbols

Given a symbol $a = a(\theta)$, the Toeplitz operator $T_a^{(\lambda)}$ acting on the space $\mathcal{A}_\lambda^2(\Pi)$ is unitary equivalent to the multiplication operator $\gamma_{a,\lambda} I$, where the function $\gamma_{a,\lambda}(\xi)$, $\xi \in \mathbb{R}$, is given by (10.4.8). Thus we have obviously

$$\operatorname{sp} T_a^{(\lambda)} = \overline{M_\lambda(a)},$$

where $M_\lambda(a) = \operatorname{Range} \gamma_{a,\lambda}$.

Theorem 14.2.1. *Let $a = a(\theta) \in C[0,\pi]$. Then*

$$\lim_{\lambda \to \infty} \operatorname{sp} T_a^{(\lambda)} = \operatorname{Range} a. \tag{14.2.1}$$

Proof. We find the asymptotics of the function $\gamma_{a,\lambda}(\xi)$ when $\lambda \to \pm\infty$ using the Laplace method. Introduce the large parameter $L = \sqrt{\lambda^2 + (2\xi)^2}$ and represent $\gamma_{a,\lambda}(\xi)$ in the form

$$\widetilde{\gamma}_{a,\lambda}(\xi) = 2^\lambda (\lambda+1) \vartheta_\lambda^2(\xi) \int_0^\pi a(\theta) e^{-LS(\theta,\varphi)} d\theta,$$

where

$$S(\theta,\varphi) = \sin\varphi \ln(\sin\theta)^{-1} + (\cos\varphi)\theta,$$

$$\sin\varphi = \lambda/L, \quad \cos\varphi = 2\xi/L \quad \text{with } \varphi \in [0,\pi).$$

To find the point of minimum of $S(\theta, \varphi)$ calculate

$$S'_\theta(\theta, \varphi) = -(\sin \varphi) \cot \theta + \cos \varphi.$$

It is obvious that $S'_\theta(\theta_\varphi, \varphi) = 0$, for $\theta_\varphi \in (0, \pi)$, if and only if $\theta_\varphi = \varphi$.

Rewrite (14.2.1) in the form

$$\gamma_{a,\lambda}(\xi) - a(\varphi) = 2^\lambda(\lambda+1)\vartheta_\lambda^2(\xi)\left[\int_{U(\varphi)\cap[0,\pi]}(a(\theta) - a(\varphi))e^{-LS(\theta,\varphi)}d\theta\right.$$
$$\left. + \int_{[0,\pi]\setminus U(\varphi)}(a(\theta) - a(\varphi))e^{-LS(\theta,\varphi)}d\theta\right] \equiv I_1(L) + I_2(L)$$

where $U(\varphi)$ is a neighborhood of φ such that $\sup_{\theta \in U(\varphi)}|a(\theta) - a(\varphi)| < \varepsilon$ for sufficiently small ε. We have used

$$2^\lambda(\lambda+1)\vartheta_\lambda^2(\xi)\int_0^\pi a(\varphi)e^{-LS(\theta,\varphi)}d\theta = a(\varphi).$$

Further

$$\left|\int_{U(\varphi)}(a(\theta) - a(\varphi))e^{-LS(\theta,\varphi)}d\theta\right| \leq \varepsilon \int_{U(\varphi)} e^{-LS(\theta,\varphi)}d\theta \leq \varepsilon\left(2^\lambda(\lambda+1)\vartheta_\lambda^2(\xi)\right)^{-1}.$$

Finally,

$$\left|\int_{[0,\pi]\setminus U(\varphi)}(a(\theta) - a(\varphi))e^{-LS(\theta,\varphi)}d\theta\right| \leq 2\sup_{\theta \in [0,\pi]}|a(\theta)|\int_{[0,\pi]\setminus U(\varphi)} e^{-LS(\theta,\varphi)}d\theta$$
$$\leq \left(2M \sup_{\theta \in [0,\pi]}|a(\theta)|e^{-L\sigma(\varepsilon)}\right)$$
$$\cdot \left(2^\lambda(\lambda+1)\vartheta_\lambda^2(\xi)\right)^{-1},$$

where $\sigma(\varepsilon) = \min_{\theta \in [0,\pi]\setminus U(\varphi)}(S(\theta, \varphi) - S(\varphi, \varphi))$. We note that $\sigma(\varepsilon)$ and M can be taken independent of $\varphi \in (0, \pi)$.

Since ε can be arbitrary small uniformly as $\varphi \in (0, \pi)$, we have

$$\gamma_{a,\lambda}(u) = a(\varphi)(1 + \alpha(L)) \tag{14.2.2}$$

where $\lim_{L\to\infty} \alpha(L) = 0$ uniformly as $\varphi \in (0, \pi)$, which proves the theorem. $\qquad\square$

We illustrate the theorem on the continuous symbol (hypocycloid)

$$a(\theta) = \frac{3}{4}e^{4i\theta} + e^{-2i\theta},$$

presenting the image of $\gamma_{a,\lambda}$ for the following values of λ: 0, 5, 12, and 200.

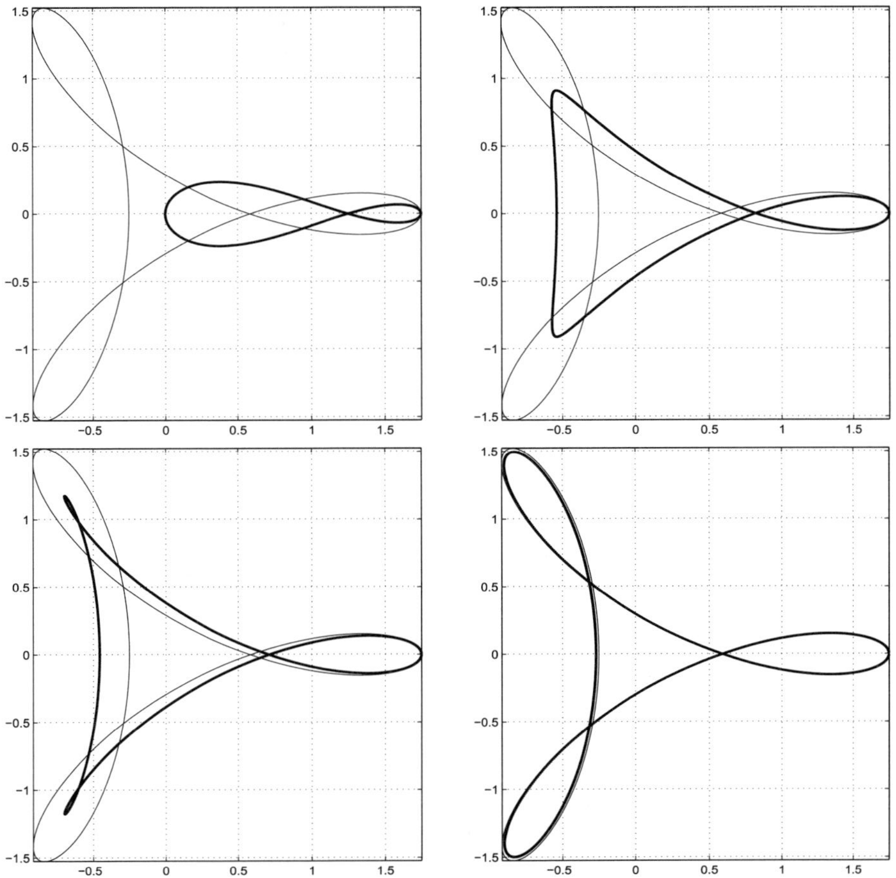

Figure 14.1: The function $\gamma_{a,\lambda}$ for $\gamma_{a,\lambda}$ for $\lambda = 0$, $\lambda = 5$, $\lambda = 12$, and $\lambda = 200$.

14.3 Piece-wise continuous symbols

Let $a(\theta)$ be a piece-wise continuous function having jumps at a finite set of points $\{\theta_j\}_{j=1}^m$ where

$$\theta_0 = 0 < \theta_1 < \theta_2 < \ldots < \theta_m < \pi = \theta_{m+1},$$

and $a(\theta_j \pm 0), j = 1, \ldots, m,$ exist. Introduce the sets

$$J_j(a) := \{z \in \mathbb{C} : z = a(\theta), \ \theta \in (\theta_j, \theta_{j+1})\}$$

where $j = 0, \ldots, m$, and let $I_j(a)$ be the segment with the endpoints $a(\theta_j - 0)$ and $a(\theta_j + 0)$, $j = 1, 2, \ldots m$. We set

$$\widetilde{R}(a) = \left(\bigcup_{j=0}^{m} J_j(a) \right) \cup \left(\bigcup_{j=1}^{m} I_j(a) \right).$$

Theorem 14.3.1. *Let $a(\theta)$ be a piece-wise continuous function. Then*

$$\lim_{\lambda \to \infty} \mathrm{sp}\, T_a^{(\lambda)} = M_\infty(a) = \widetilde{R}(a).$$

Proof. We use the Laplace method as in Theorem 14.2.1. For any $\varepsilon > 0$ we take $\delta > 0$ such that for each interval $I \subset (\theta_j, \theta_{j+1})$ with length less then δ, $j = 1, 2, \ldots, m$, the following inequality holds,

$$\sup_{s_1, s_2 \in I} |a(s_1) - a(s_2)| < \varepsilon.$$

Suppose first that the minimum point $s_\varphi = \varphi$ satisfies the condition

$$\inf_{j=1,2,\ldots,m} |\varphi - \theta_j| > \delta.$$

We have

$$
\begin{aligned}
\gamma_{a,\lambda}(\xi) \;=\;& a(\varphi) + 2^\lambda(\lambda + 1)\vartheta_\lambda^2(\xi) \int_{\varphi-\delta}^{\varphi+\delta} (a(\theta) - a(\varphi))e^{-LS(\theta,\varphi)}d\theta \\
&+ 2^\lambda(\lambda + 1)\vartheta_\lambda^2(\xi) \int_{[0,\pi]\backslash(\varphi-\delta,\varphi+\delta)} (a(\theta) - a(\varphi))e^{-LS(\theta,\varphi)}d\theta \\
\;=\;& a(\varphi) + O(\varepsilon) + O(e^{-\sigma L}) \qquad\qquad\qquad\qquad\qquad (14.3.1)
\end{aligned}
$$

where

$$\sigma = \min_{[0,\pi]\backslash(\varphi-\delta,\varphi+\delta)} (S(\theta, \varphi) - S(\varphi, \varphi)).$$

Thus varying $\varphi \in \cup_{j=0}^{m}(\theta_j, \theta_{j+1})$ we have that

$$J_j(a) \subset M_\infty(a), \quad j = 0, 1, \ldots, m.$$

Now suppose that there exist j such that $|\varphi - \theta_j| < \delta$. Then we have

$$\gamma_{a,\lambda}(\xi) \qquad\qquad\qquad\qquad\qquad\qquad\qquad\qquad\qquad\qquad (14.3.2)$$

$$
= 2^\lambda(\lambda + 1)\vartheta_\lambda^2(\xi) \left(a(\theta_j - 0) \int_{\varphi-\delta}^{\theta_j} e^{-LS(\theta,\varphi)}d\theta + a(\theta_j + 0) \int_{\theta_j}^{\varphi+\delta} e^{-LS(\theta,\varphi)}d\theta \right)
$$

$$
+ 2^\lambda(\lambda + 1)\vartheta_\lambda^2(\xi) \left(\int_{\varphi-\delta}^{\theta_j} (a(\theta) - a(\theta_j - 0))e^{-LS(\theta,\varphi)}d\theta \right.
$$

$$
\left. + \int_{\theta_j}^{\varphi+\delta} (a(\theta) - a(\theta_j + 0))e^{-LS(\theta,\varphi)}d\theta + \int_{(0,\pi)\backslash(\varphi-\delta,\varphi+\delta)} a(\theta)e^{-LS(\theta,\varphi)}d\theta \right).
$$

Taking δ small enough we suppose that

$$\frac{\theta_1}{2} < s_\varphi(=\varphi) < \frac{\pi + \theta_m}{2}.$$

Thus the function

$$(S''_{\theta,\theta}(\varphi,\varphi))^{-1} = -\sin\varphi$$

is uniformly bounded on φ and the following asymptotic calculations are uniform on φ:

$$
\begin{aligned}
[2^\lambda(\lambda+1)\vartheta_\lambda^2(\xi)]^{-1} &= \int_0^\pi e^{-LS(\theta,\varphi)}d\theta \\
&= e^{-LS(\varphi,\varphi)}\int_0^\pi e^{-\frac{L}{2}(\sin^{-1}\varphi)(\theta-\varphi)^2}d\theta(1+O(1)) \\
&= e^{-LS(\varphi,\varphi)}\int_{-\varphi}^{\pi-\varphi} e^{-\frac{L}{2}(\sin^{-1}\varphi)u^2}du(1+O(1)) \\
&= \sqrt{2\sin\varphi}\,\frac{e^{-LS(\varphi,\varphi)}}{L^{1/2}}\int_{-\infty}^{\infty} e^{-v^2}dv(1+O(1)). \quad (14.3.3)
\end{aligned}
$$

Analogously

$$\int_{\theta_j}^{\varphi+\delta} e^{-LS(\theta,\varphi)}d\theta = \sqrt{2\sin\varphi}\,\frac{e^{-LS(\varphi,\varphi)}}{L^{1/2}}\int_{x_j}^{\infty} e^{-v^2}dv(1+O(1)) \qquad (14.3.4)$$

and

$$\int_{\varphi-\delta}^{\theta_j} e^{-LS(\theta,\varphi)}d\theta = \sqrt{2\sin\theta}\,\frac{e^{-LS(\varphi,\varphi)}}{L^{1/2}}\int_{-\infty}^{x_j} e^{-v^2}dv(1+O(1)), \qquad (14.3.5)$$

where

$$x_j = \left(\frac{L}{2\sin\varphi}\right)^{1/2}(\theta_j - \varphi).$$

Thus from (14.3.3)–(14.3.5) we have

$$\gamma_{a,\lambda}(\xi) = (a(\theta_j - 0)t + a(\theta_j + 0)\tau)(1 + O(1) + O(\varepsilon) + O(e^{-i\sigma})), \qquad (14.3.6)$$

where

$$t = \left(\int_{-\infty}^{x_j} e^{-v^2}dv\right)\Big/\left(\int_{-\infty}^{\infty} e^{-v^2}dv\right) \quad \text{and} \quad \tau = \left(\int_{x_j}^{\infty} e^{-v^2}dv\right)\Big/\left(\int_{-\infty}^{\infty} e^{-v^2}dv\right).$$

Now it is evident that $t,\tau \in [0,1]$ and $\tau + t = 1$, which implies $I_j(a) \subset M_\infty(a)$. Thus

$$\widetilde{R}(a) \subset M_\infty(a).$$

The representations (14.3.1) and (14.3.6) imply the inverse inclusion

$$\widetilde{R}(a) \supset M_\infty(a). \qquad \square$$

We illustrate the theorem on the following piece-wise continuous symbol which has six jump points,

$$
a(\theta) = \begin{cases}
\exp i\left[-\frac{\pi}{6} + \frac{2\pi}{3}\cdot\frac{7\theta}{\pi}\right], & \theta \in \left[0, \frac{\pi}{7}\right) \\
\frac{1}{3}\exp i\left[\frac{\pi}{6} + \frac{2\pi}{3}\cdot\left(\frac{7\theta}{\pi} - 1\right)\right], & \theta \in \left[\frac{\pi}{7}, \frac{2\pi}{7}\right) \\
\exp i\left[-\frac{\pi}{6} + \frac{2\pi}{3}\cdot\left(\frac{7\theta}{\pi} - 2\right)\right], & \theta \in \left[\frac{2\pi}{7}, \frac{3\pi}{7}\right) \\
\frac{1}{3}\exp i\left[-\frac{\pi}{6} + \frac{2\pi}{3}\cdot\left(\frac{7\theta}{\pi} - 3\right)\right], & \theta \in \left[\frac{3\pi}{7}, \frac{4\pi}{7}\right) \\
\exp i\left[-\frac{\pi}{6} + \frac{2\pi}{3}\cdot\left(\frac{7\theta}{\pi} - 4\right)\right], & \theta \in \left[\frac{4\pi}{7}, \frac{5\pi}{7}\right) \\
\frac{1}{3}\exp i\left[-\frac{\pi}{6} + \frac{2\pi}{3}\cdot\left(\frac{7\theta}{\pi} - 5\right)\right], & \theta \in \left[\frac{5\pi}{7}, \frac{6\pi}{7}\right) \\
\exp\left(-i\frac{\pi}{6}\right), & \theta \in \left[\frac{6\pi}{7}, \pi\right]
\end{cases}
$$

We present the image of the symbol $a = a(\theta)$, the image of $\gamma_{a,\lambda}$ for the following values of λ: 1, 10, 70, and 500, as well as the limit set $M_\infty(a)$.

We have obviously

$$\lim_{\lambda\to\infty} \operatorname{sp} T_a^{(\lambda)} = M_\infty(a) \subset \operatorname{conv}(\operatorname{ess}\operatorname{Range} a). \tag{14.3.7}$$

Wishing to illustrate the possible interrelations between these sets we can repeat the arguments of Examples 13.3.2–13.3.5 and construct the (piece-wise continuous) symbols $a = a(\theta)$ to realize the following possibilities:

$$
\begin{aligned}
M_\infty(a) &= \operatorname{Range} a \quad (= \operatorname{ess}\operatorname{Range} a), \\
M_\infty(a) &= \operatorname{conv}(\operatorname{ess}\operatorname{Range} a) \quad (= \operatorname{conv}(\operatorname{Range} a)), \\
M_\infty(a) &\subset \partial\operatorname{conv}(\operatorname{Range} a), \\
M_\infty(a) &= \partial\operatorname{conv}(\operatorname{Range} a).
\end{aligned}
$$

14.4　Unbounded symbols

Theorem 14.4.1. *Let* $a(\theta) \in L_1(0, \pi) \cap C(0, 1)$. *Then*

$$\operatorname{Range} a \subset M_\infty(a).$$

Proof. We apply the Laplace method as in Theorem 14.2.1. Fix any point $\varphi \in (0, \pi)$ and consider for each ξ large enough the value $\lambda = 2\xi\arctan\varphi$. Then by (14.2.2) we have

$$\gamma_{a,\lambda}(\xi) = a(\varphi)(1 + \alpha(\lambda\sqrt{1 + (2\arctan\varphi)^{-2}})$$

where $\lim_{L\to\infty} \alpha(L) = 0$. Thus if $\xi \to \infty$ then $\lambda \to \infty$ as well and we have

$$a(\varphi) \in M_\infty(a). \qquad \square$$

The next theorem, whose proof is analogous to that of Theorem 13.5.2, shows that the property (14.3.7), previously established for bounded symbols, remains valid for summable symbols.

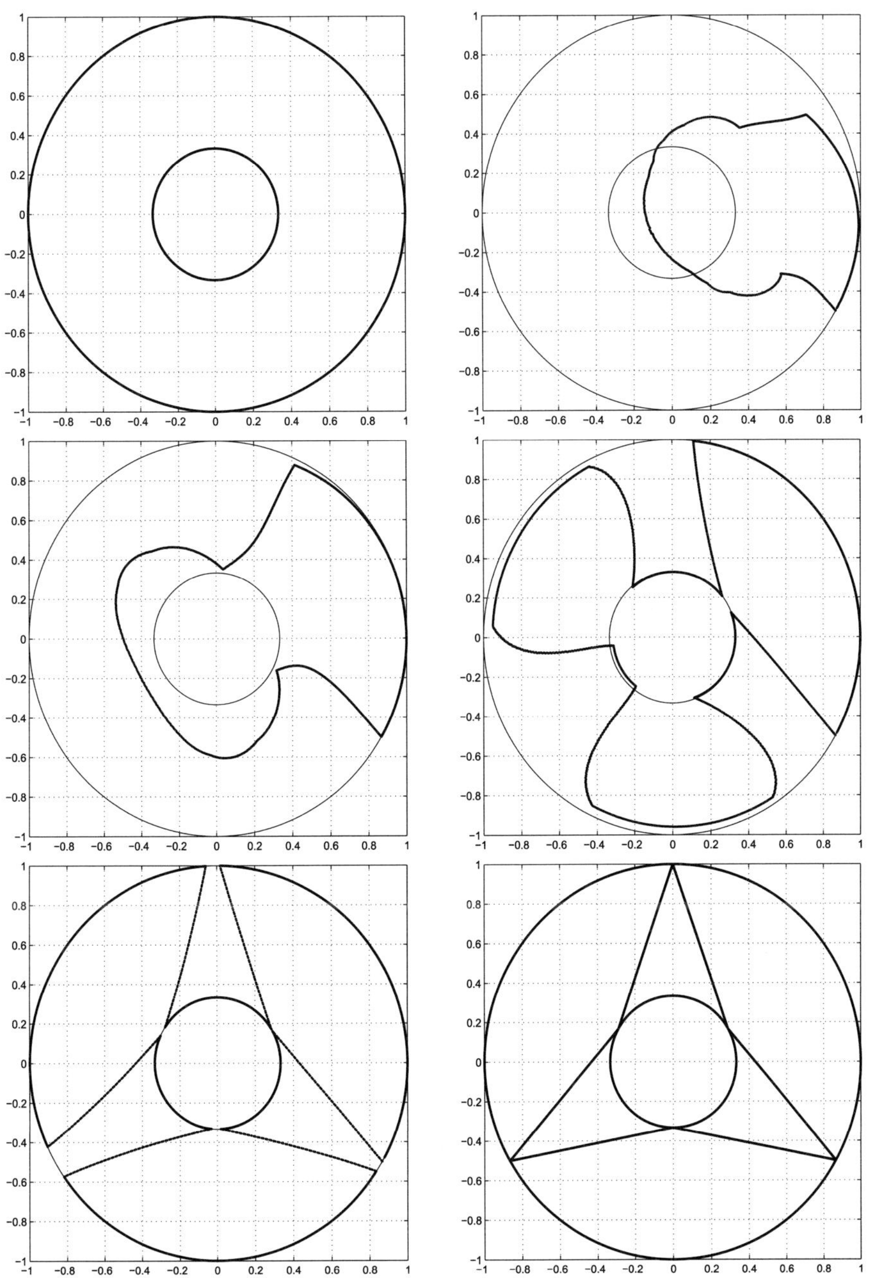

Figure 14.2: The symbol $a(\theta)$, the function $\gamma_{a,\lambda}$ for $\lambda = 1$, $\lambda = 10$, $\lambda = 70$, $\lambda = 500$, and the limit set $M_\infty(a)$.

Theorem 14.4.2. *Let $a(\theta) \subset L_1(0, \pi)$. Then*

$$M_\infty(a) \subset \operatorname{conv}(\operatorname{ess} \operatorname{Range} a).$$

Note that for functions $a(\theta) \in L_1(0, \pi) \cap C(0, \pi)$, Theorems 14.4.1 and 14.4.2 imply that

$$\operatorname{Range} a \subset M_\infty(a) \subset \operatorname{conv}(\operatorname{Range} a),$$

and we show that $\operatorname{Range} a$ can coincide with each of these extreme sets.

Example 14.4.3. For each $j \in \mathbb{N}$ define $I_j = [j^{-1} - j^{-3}, j^{-1}]$ and let $\overline{\{\xi_j\}_{j \in \mathbb{N}}} = [0, 2\pi]$. Define the symbol as

$$a(\theta) = \left\{ \begin{array}{ll} je^{i\xi_j}, & \theta \in I_j, \quad j \in \mathbb{N} \\ 0, & \theta \in (0, \pi) \setminus \bigcup_{j=1}^{\infty} I_j \end{array} \right. .$$

It can be easily shown that

$$M_\infty(a) = \mathbb{C} = \operatorname{conv}(\operatorname{Range} a).$$

Example 14.4.4. Given $\alpha \in [0, 1)$, introduce $a(\theta) = (\sin \theta)^{i-\alpha}$, which is unbounded for $\alpha \in (0, 1)$, while it is bounded and oscillating for $\lambda = 0$. Calculate, using [86], formula 3.892.1,

$$
\begin{aligned}
\gamma_{a,\lambda}(\xi) &= \frac{\int_0^\pi (\sin \theta)^{\lambda + i - \alpha} e^{-2\xi\theta}\, d\theta}{\int_0^\pi (\sin \theta)^{\lambda} e^{-2\xi\theta}\, d\theta} \\
&= \frac{2^{\alpha - i}(\lambda + 1)}{\lambda + i - \alpha + 1} \frac{B\left(\frac{\lambda}{2} + 1 + i\xi, \frac{\lambda}{2} + 1 - i\xi\right)}{B\left(\frac{\lambda + i - \alpha}{2} + 1 + i\xi, \frac{\lambda + i - \alpha}{2} + 1 - i\xi\right)} \\
&= \frac{2^{\alpha - i}(\lambda + 1)}{\lambda + i - \alpha + 1} \frac{\Gamma(\lambda + 2 + i - \alpha)}{\Gamma(\lambda + 2)} \frac{\Gamma\left(\frac{\lambda}{2} + 1 + i\xi\right)}{\Gamma\left(\frac{\lambda + i - \alpha}{2} + 1 + i\xi\right)} \\
&\quad \cdot \frac{\Gamma\left(\frac{\lambda}{2} + 1 - i\xi\right)}{\Gamma\left(\frac{\lambda + i - \alpha}{2} + 1 - i\xi\right)} .
\end{aligned}
$$

Applying the asymptotic formulas (see [86], formulas 8.327 and 8.328.2) for the Γ-function we have

$$\gamma_{a,\lambda}(\xi) = \left[\left(\frac{(\lambda + 2)^2}{(\lambda + 2)^2 + 4\xi^2}\right)^{\frac{1}{2}}\right]^{i-\alpha} \left(1 + O\left(\frac{1}{\lambda + 1}\right)\right).$$

Given any $v \in (0, \pi)$, we can take ξ and λ such that

$$\left(\frac{(\lambda + 2)^2}{(\lambda + 2)^2 + 4\xi^2}\right)^{\frac{1}{2}} = \sin v.$$

Thus

$$\gamma_{a,\lambda}(\xi) = (\sin v)^{i-\alpha} \left(1 + O\left(\frac{1}{\lambda + 1}\right)\right),$$

and in this case $M_\infty(a) = \operatorname{Range} a$.

Appendix A

Coherent states and Berezin transform

A.1 General approach to coherent states

Let H be a Hilbert space, and let $\{\varphi_g\}_{g \in G}$ be a subset of elements of H parameterized by elements g of some set G with measure $d\mu$.

Then $\{\varphi_g\}_{g \in G}$ is called a *system of coherent states* if for all $\varphi \in H$,

$$\|\varphi\|^2 = (\varphi, \varphi) = \int_G |(\varphi, \varphi_g)|^2 d\mu,$$

or, equivalently, if for all $\varphi_1, \varphi_2 \in H$,

$$(\varphi_1, \varphi_2) = \int_G (\varphi_1, \varphi_g)\overline{(\varphi_2, \varphi_g)} d\mu. \tag{A.1.1}$$

We define the isomorphic inclusion

$$V : H \longrightarrow L_2(G)$$

by the rule

$$V : \varphi \in H \longmapsto f = f(g) = (\varphi, \varphi_g) \in L_2(G).$$

By (A.1.1) we have

$$(\varphi_1, \varphi_2) = \langle f_1, f_2 \rangle,$$

where $(\cdot, \cdot)$ and $\langle \cdot, \cdot \rangle$ are the scalar products on H and $L_2(G)$, respectively.

Further

$$f_h(g) = V\varphi_h = (\varphi_h, \varphi_g) = (\overline{\varphi}_g, \overline{\varphi}_h) = \overline{V\varphi_g} = \overline{f_g(h)}.$$

Let $H_2(G) = V(H) \subset L_2(G)$.

Theorem A.1.1. *A function $f \in L_2(G)$ is an element of $H_2(G)$ if and only if for all $h \in G$,*
$$\langle f, f_h \rangle = f(h).$$

Proof. Let $f(h) = (\varphi, \varphi_h) \in H_2(G)$, then
$$\langle f, f_h \rangle = f(h).$$

Conversely, let $f \in L_2(G)$ with
$$\langle f, f_h \rangle = f(h),$$

for all $h \in G$. Consider
$$L_2(G) = H_2(G) \oplus H_2^\perp(G)$$

and correspondingly
$$f = f_0 + f^\perp.$$

For all $h \in G$ $f_h = (\varphi_h, \varphi_g) \in H_2(G)$, thus we have
$$0 = \langle f^\perp, f_h \rangle = \langle f, f_h \rangle - \langle f_0, f_h \rangle = f(h) - f_0(h) = f^\perp(h),$$

hence $f^\perp \equiv 0$, or $f = f_0 + 0 \in H_2(G)$. $\square$

Theorem A.1.2. *The operator*

$$(Pf)(g) = \int_G (\varphi_t, \varphi_g) f(t) d\mu(t)$$

is the orthogonal projection of $L_2(G)$ onto $H_2(G)$.

Proof. We have

$$
\begin{aligned}
\langle (Pf)(g), f_h(g) \rangle &= \int_G \left(\int_G f(t)\overline{(\varphi_g, \varphi_t)} d\mu(t) \right) \overline{(\varphi_h, \varphi_g)} d\mu(g) \\
&= \int_G f(t) d\mu(t) \int_G \overline{(\varphi_g, \varphi_t)} \, \overline{(\varphi_h, \varphi_g)} d\mu(g) \\
&= \int_G f(t) d\mu(t) \int_G (\varphi_t, \varphi_g) \overline{(\varphi_h, \varphi_g)} d\mu(g) \\
&= \int_G f(t)(\varphi_t, \varphi_h) d\mu(t) = (Pf)(h).
\end{aligned}
$$

Thus $\operatorname{Im} P = H_2(G)$. Let $f(g) = (\varphi, \varphi_g) \in H_2(G)$, then

$$(Pf)(g) = \int_G (\varphi, \varphi_t) \overline{(\varphi_g, \varphi_t)} d\mu(t) = (\varphi, \varphi_g) = f(g).$$

Thus $P|_{H_2(G)} = I$, or $P^2 = P$. Finally, the operator P is obviously self-adjoint. $\square$

Remark A.1.3. Let $f_g(t) = V\varphi_g = (\varphi_g, \varphi_t)$, $g \in G$, be the image in $H_2(G)$ of the system of coherent states $\{\varphi_g\}_{g \in G}$, then

$$(Pf)(g) = \langle f, f_g \rangle.$$

The function $a_A(g)$, $g \in G$, is called the *anti-Wick* (or *contravariant*) symbol of an operator $A : H \to H$ if

$$V A V^{-1}|_{H_2(G)} = P a_A(g) P = P a_A(g) I|_{H_2(G)} : H_2(G) \longrightarrow H_2(G),$$

or, in other terminology, the operator $V A V^{-1}|_{H_2(G)}$ is the Toeplitz operator

$$T_{a_A(g)} = P a_A(g) I|_{H_2(G)} : H_2(G) \longrightarrow H_2(G),$$

with the defining symbol $a_A(g)$.

Given an operator $A : H \to H$, introduce the (Wick) function

$$\widetilde{a}(g, h) = \frac{(A\varphi_h, \varphi_g)}{(\varphi_h, \varphi_g)}, \qquad g, h \in G. \tag{A.1.2}$$

If an operator A has an anti-Wick symbol, that is $V A V^{-1} = T_{a(g)}$ for some function $a = a(g)$, then

$$\widetilde{a}(g, h) = \frac{\langle T_a f_h, f_g \rangle}{\langle f_h, f_g \rangle} = f_h^{-1}(g)\langle P a f_h, f_g \rangle = f_h^{-1}(g)\langle a f_h, P f_g \rangle = f_h^{-1}(g)\langle a f_h, f_g \rangle, \tag{A.1.3}$$

where $g, h \in G$, and the operator T_a admits the following representation in the term of its Wick symbol:

$$
\begin{aligned}
(T_a f)(g) &= \int_G a(t) f(t) f_t(g) d\mu(t) = \int_G a(t) f_t(g) d\mu(t) \int_G f(h) f_h(t) d\mu(h) \\
&= \int_G f(h) d\mu(h) \int_G a(t) f_t(g) f_h(t) d\mu(t) \\
&= \int_G f(h) d\mu(h) \frac{f_h(g)}{\langle f_h, f_g \rangle} \int_G a(t) f_h(t) \overline{f_g(t)} d\mu(t) \\
&= \int_G \widetilde{a}(g, h) f(h) f_h(g) d\mu(h).
\end{aligned}
\tag{A.1.4}
$$

Interchanging the integrals above, we understand them in a weak sense.

Note that

$$(T_a f_h)(g) = \langle T_a f_h, f_g \rangle = \langle a f_h, f_g \rangle = \widetilde{a}(g, h) f_h(g),$$

and

$$(T_a^* f_h)(g) = \langle T_a^* f_h, f_g \rangle = \langle f_h, T_a f_g \rangle = \overline{\langle T_a f_g, f_h \rangle} = \overline{\widetilde{a}(h, g)} f_g(h) = \overline{\widetilde{a}(h, g)} f_h(g).$$

Let now T_{a_1} and T_{a_2} be two Toeplitz operators with the Wick symbols $\widetilde{a}_1(g,h)$ and $\widetilde{a}_2(g,h)$, correspondingly, then the Wick symbol $\widetilde{a}(g,h)$ of the composition $T_{a_1}T_{a_2}$ is calculated by the formula

$$
\begin{aligned}
\widetilde{a}(g,h) = (\widetilde{a}_1 \star \widetilde{a}_2)(g,h) &= f_h^{-1}(g)\langle T_{a_1}T_{a_2}f_h, f_g\rangle = f_h^{-1}(g)\langle T_{a_2}f_h, T_{a_1}^*f_g\rangle \\
&= f_h^{-1}(g)\langle \widetilde{a}_2(\zeta,h)f_h(\zeta), \overline{\widetilde{a}_1(g,\zeta)}f_g(\zeta)\rangle. \qquad \text{(A.1.5)}
\end{aligned}
$$

The restriction of the function $\widetilde{a}(g,h)$ onto the diagonal

$$
\widetilde{a}_A(g) = \widetilde{a}(g) = \widetilde{a}(g,g) = \frac{(A\varphi_g, \varphi_g)}{(\varphi_g, \varphi_g)}, \qquad g \in G, \qquad \text{(A.1.6)}
$$

is called the *Wick* (or *covariant*, or *Berezin*) symbol of the operator $A : H \to H$, and the formula (A.1.4) gives the representation of the operator T_a in terms of the Wick symbol.

Theorem A.1.4. *The Wick and anti-Wick symbols of an operator $A : H \to H$ are connected by the Berezin transform*

$$
\widetilde{a}_A(g) = \int_G a_A(t)\frac{(\varphi_g, \varphi_t)(\varphi_t, \varphi_g)}{(\varphi_g, \varphi_g)}d\mu(t).
$$

Proof. Indeed,

$$
\begin{aligned}
\widetilde{a}_A(g) &= \frac{(A\varphi_g, \varphi_g)}{(\varphi_g, \varphi_g)} = \frac{\langle P a_A f_g, f_g\rangle}{(\varphi_g, \varphi_g)} = \frac{\langle a_A f_g, P f_g\rangle}{(\varphi_g, \varphi_g)} \\
&= \frac{\langle a_A f_g, f_g\rangle}{(\varphi_g, \varphi_g)} = \frac{1}{(\varphi_g, \varphi_g)}\int_G a_A(t)f_g(t)\overline{f_g(t)}d\mu(t) \\
&= \int_G a_A(t)\frac{(\varphi_g, \varphi_t)(\varphi_t, \varphi_g)}{(\varphi_g, \varphi_g)}d\mu(t).
\end{aligned}
$$
$\qquad\qquad\square$

Introduce the normalized system

$$
\widetilde{f}_g(t) = \frac{f_g}{\|f_g\|}, \qquad g \in G.
$$

Remark A.1.5. In terms of $\{\widetilde{f}_g\}_{g \in G}$ the Wick and anti-Wick symbols of an operator A are connected by the formula

$$
\widetilde{a}_A(g) = \int_G a_A(t)|\widetilde{f}_g(t)|^2 d\mu(t).
$$

Let H_2 be a closed subspace of a Hilbert space $L_2 = L_2(G, d\mu)$, and let P be the orthogonal projection on L_2 with image H_2. Given $a \in L_\infty(G, d\mu)$, consider the Toeplitz operator

$$
T_a = Pa|_{H_2} : H_2 \longrightarrow H_2.
$$

To have the function a as the anti-Wick symbol of the operator T_a we need a system $\{e_x\}_{x \in G}$ of elements of H_2 with the property

$$\forall \; \varphi(y) \in H_2 \quad (\varphi(y), e_x(y)) = \varphi(x).$$

Then

$$\|\varphi\|^2 = \int_G |(\varphi(y), e_x(y))|^2 d\mu(x) = \int_G |\varphi(x)|^2 d\mu(x),$$

thus the system $\{e_x\}_{x \in G}$ is a system of coherent states, and the mapping $V : H_2 \to H_2$ is the identical operator I.

Remark A.1.6. The above system of coherent states $\{e_x\}_{x \in G}$ is unique.

Indeed, let $\{e'_x\}_{x \in G}$ be another system of coherent states. Then for each x and t from G we have

$$e_x(t) = (e_x, e'_t) = \overline{(e'_t, e_x)} = \overline{e'_t(x)} = e'_x(t).$$

Now the orthogonal projection P on L_2 with image H_2 is given by

$$(P\varphi)(x) = \int_G (e_t, e_x)\varphi(t)d\mu(t) = \int_G \varphi(t)e_t(x)d\mu(t) = (\varphi(t), e_x(t)).$$

Finally, the Wick symbol of the Toeplitz operator T_a is given by

$$\widetilde{a}_{T_a}(x) = \int_G a(t)\frac{(e_x, e_t)(e_t, e_x)}{(e_x, e_x)}d\mu(t) = \int_G a(t)\frac{|e_x(t)|^2}{e_x(x)}d\mu(t),$$

or normalizing the system $\{e_x\}_{x \in G}$:

$$\widetilde{e}_x(t) = \frac{e_x(t)}{\sqrt{e_x(x)}}, \qquad \|\widetilde{e}_x(t)\|_{L_2(G)} = 1,$$

we have

$$\widetilde{a}_{T_a}(x) = \int_G a(t)|\widetilde{e}_x(t)|^2 d\mu(t).$$

A.2 Numerical range and spectra

Let A be a bounded operator acting on a Hilbert space H. Its *numerical range* $W(A)$ is the set

$$W(A) = \{(A\varphi, \varphi) : \|\varphi\| = 1, \; \varphi \in H\}.$$

Theorem A.2.1. *The following properties hold:*

1. *The numerical range $W(A)$ is always convex.*

2. *The numerical range $W(A)$ is not necessarily closed.*

3. *The closure of the numerical range includes the spectrum of the operator*

$$\operatorname{sp} A \subset \overline{W(A)}.$$

4. *If the operator A is normal, then*

$$\operatorname{conv}(\operatorname{sp} A) = \overline{W(A)}.$$

Proof. See, for example, [99]. $\qquad\qquad\qquad\qquad\qquad\qquad\qquad\qquad\square$

Corollary A.2.2. *The closure of the numerical range of a multiplication operator $A = aI$, $a \in L_\infty$, coincides with the convex hull of the essential range of the multiplier a:*

$$\operatorname{conv}(\text{ess-Range}\, a) = \overline{W(A)}.$$

Consider now the following general situation. Let H_2 be a closed subspace of a Hilbert space $L_2 = L_2(G, d\mu)$, and let P be the orthogonal projection on L_2 with image H_2. Given $a \in L_\infty(G, d\mu)$, consider the Toeplitz operator

$$T_a = Pa|_{H_2} : H_2 \longrightarrow H_2.$$

Lemma A.2.3. *The following inclusion holds,*

$$\operatorname{sp} T_a \subset \operatorname{conv}(\operatorname{sp} T_a) \subset \operatorname{conv}(\text{ess-Range}\, a).$$

Proof. For $\varphi \in H_2$ we have obviously

$$(T_a\varphi, \varphi) = (Pa\varphi, \varphi) = (a\varphi, P\varphi) = (a\varphi, \varphi).$$

Thus, $W(T_a) \subset W(aI)$, the operator aI is normal, thus

$$\operatorname{sp} T_a \subset \operatorname{conv}(\operatorname{sp} T_a) \subset \overline{W(T_a)} \subset \overline{W(aI)} = \operatorname{conv}(\text{ess-Range}\, a). \qquad \square$$

Corollary A.2.4. *For an operator A with anti-Wick symbol a_A we have*

$$\operatorname{sp} A \subset \operatorname{conv}(\operatorname{sp} A) \subset \operatorname{conv}(\text{ess-Range}\, a_A).$$

Lemma A.2.5. *Let A be a normal operator with anti-Wick symbol a_A and Wick symbol $\widetilde{a}_A$. Then*

$$\text{Range}\, \widetilde{a}_A \subset \operatorname{conv}(\operatorname{sp} A) \subset \operatorname{conv}(\text{ess-Range}\, a_A).$$

Proof. Indeed,

$$\text{Range}\, \widetilde{a}_A \subset \overline{W(A)} = \operatorname{conv}(\operatorname{sp} A) \subset \operatorname{conv}(\text{ess-Range}\, a_A). \qquad \square$$

A.3 Coherent states in the Bergman space

The reproducing property Bergman kernel function shows that the system of functions $k_{h,\zeta}(z) = K_h(z,\overline{\zeta})$, $\zeta \in \mathbb{D}$, forms a system of coherent states in the space $\mathcal{A}_h^2(\mathbb{D})$. That is, now we have $G = \mathbb{D}$, $d\mu = d\mu_h(\zeta)$, $H = H_2(G) = \mathcal{A}_h^2(\mathbb{D})$, $L_2(G) = L_2(\mathbb{D}, d\mu_h)$, $\varphi_g = f_g = k_{h,g}$, where $g = \zeta \in \mathbb{D}$. Thus, given Toeplitz operator T_a with the defining symbol $a = a(z)$ on the weighted Bergman space $\mathcal{A}_h^2(\mathbb{D})$, the corresponding Wick function is calculated by the formula, see (A.1.3), (following tradition we put the "bar" over the second argument)

$$
\begin{aligned}
\widetilde{a}_h(z,\overline{\zeta}) &= f_\zeta^{-1}(z)\langle a(w)f_\zeta(w), f_z(w)\rangle \\
&= (1 - z\overline{\zeta})^{\frac{1}{h}} \int_{\mathbb{D}} \frac{a(w)}{(1 - z\overline{w})^{\frac{1}{h}}(1 - w\overline{\zeta})^{\frac{1}{h}}} \, d\mu_h(w) \\
&= (\frac{1}{h} - 1) \int_{\mathbb{D}} a(w) \left(\frac{(1 - z\overline{\zeta})(1 - w\overline{w})}{(1 - z\overline{w})(1 - w\overline{\zeta})} \right)^{\frac{1}{h}} d\mu(w).
\end{aligned}
$$

Thus the Wick (A.1.4) form of the Toeplitz operator T_a is

$$
\begin{aligned}
(T_a)(z) &= \int_{\mathbb{D}} \frac{\widetilde{a}_h(z,\overline{\zeta})f(\zeta)}{(1 - z\overline{\zeta})^{\frac{1}{h}}} \, d\mu_h(\zeta) \\
&= (\frac{1}{h} - 1) \int_{\mathbb{D}} \widetilde{a}_h(z,\overline{\zeta})f(\zeta) \left(\frac{1 - |\zeta|^2}{1 - z\overline{\zeta}} \right)^{\frac{1}{h}} d\mu(\zeta),
\end{aligned}
$$

the Berezin transform has the form

$$
\widetilde{a}_h(z,\overline{z}) = (\frac{1}{h} - 1) \int_{\mathbb{D}} a(\zeta) \left(\frac{(1 - |z|^2)(1 - |\zeta|^2)}{(1 - z\overline{\zeta})(1 - \zeta\overline{z})} \right)^{\frac{1}{h}} d\mu(\zeta),
$$

and the composition formula for Wick symbols by (A.1.5) is

$$
(\widetilde{a}_h \star \widetilde{b}_h)(z,\overline{z}) = (\frac{1}{h} - 1) \int_{\mathbb{D}} \widetilde{a}_h(z,\overline{\zeta}) \widetilde{b}_h(\zeta,\overline{z}) \left(\frac{(1 - |z|^2)(1 - |\zeta|^2)}{(1 - z\overline{\zeta})(1 - \zeta\overline{z})} \right)^{\frac{1}{h}} d\mu(\zeta).
$$

Given a function a, we denote by $S_h a$, where

$$
\begin{aligned}
(S_h a)(z,\overline{z}) &= (\frac{1}{h} - 1) \int_{\mathbb{D}} a(\zeta) \left(\frac{(1 - |z|^2)(1 - |\zeta|^2)}{(1 - z\overline{\zeta})(1 - \zeta\overline{z})} \right)^{\frac{1}{h}} d\mu(\zeta) \\
&= (\frac{1}{h} - 1)\frac{1}{\pi} \int_{\mathbb{D}} a(\zeta) \left(\frac{(1 - |z|^2)(1 - |\zeta|^2)}{(1 - z\overline{\zeta})(1 - \zeta\overline{z})} \right)^{\frac{1}{h}} \frac{dv(\zeta)}{(1 - |\zeta|^2)^2}, \quad (A.3.7)
\end{aligned}
$$

its Berezin transform.

A.4   Berezin transform

The Berezin transform also appears naturally in the context of problems of function theory and operator theory. As customary in operator theory we will label here the weighted Bergman spaces by the parameter $\lambda \in (-1, +\infty)$, see Section 10.1. Recall that the parameter λ is connected with the parameter h, used in quantization, by the formula $\lambda + 2 = \frac{1}{h}$.

To motivate the appearance of the Berezin transform we start with the weightless case and the Poisson formula.

Let h be a harmonic function in the unit disk $\mathbb{D}$ and continuous on the closed disk $\overline{\mathbb{D}}$. By the mean value property we have

$$h(0) = \frac{1}{2\pi} \int_0^{2\pi} h(e^{i\theta}) d\theta.$$

Denote by φ_z the Möbius transformation which interchanges 0 and z,

$$\varphi_z(\zeta) = \frac{z - \zeta}{1 - \overline{z}\zeta}. \tag{A.4.8}$$

Now replacing h with $h \circ \varphi_z$ and changing of variables in the above integral we obtain the usual Poisson formula

$$h(z) = \frac{1}{2\pi} \int_0^{2\pi} \frac{1 - |z|^2}{|1 - ze^{-i\theta}|^2} h(e^{i\theta}) d\theta.$$

We show now that the Berezin transform appears as an area version of the above Poisson formula. Given a bounded harmonic function h, consider the area version of the mean value property

$$h(0) = \frac{1}{\pi} \int_{\mathbb{D}} h(\zeta) dv(\zeta).$$

Let us do the same as in the previous case replacing h with $h \circ \varphi_z$ and changing of variables in the above integral we obtain

$$h(z) = \frac{1}{\pi} \int_{\mathbb{D}} \frac{(1 - |z|^2)^2}{|1 - z\overline{\zeta}|^4} h(\zeta) dv(\zeta).$$

Now the operator defined on $L_\infty(\mathbb{D})$ by the formula

$$(S_0 a)(z) = \frac{1}{\pi} \int_{\mathbb{D}} \frac{(1 - |z|^2)^2}{|1 - z\overline{\zeta}|^4} a(\zeta) dv(\zeta)$$

is nothing but the Berezin transform for the weightless case $\lambda = 0$.

We pass now to the weighted case. Recall that the weight parameter $\lambda \in (-1, +\infty)$ and that the measure (10.1.1) given by

$$
\begin{aligned}
d\mu_\lambda(z) &= \frac{\lambda + 1}{\pi}(1 - |z|^2)^\lambda dv(z) \\
&= (\lambda + 1)(1 - |z|^2)^{\lambda+2} d\mu(z)
\end{aligned}
$$

is a probability measure.

Consider again a bounded harmonic function h in $\mathbb{D}$. Then the mean value property together with the rotation invariance of the measure $d\mu_\lambda$ yields

$$
h(0) = \int_{\mathbb{D}} h(\zeta) d\mu_\lambda(\zeta) = \frac{\lambda + 1}{\pi} \int_{\mathbb{D}} h(\zeta)(1 - |\zeta|^2)^\lambda dv(\zeta).
$$

For the third time we replace h by $h \circ \varphi_z$ and change variables. Then

$$
\begin{aligned}
h(z) &= \frac{\lambda + 1}{\pi} \int_{\mathbb{D}} \left(\frac{(1 - |z|^2)(1 - |\zeta|^2)}{(1 - z\overline{\zeta})(1 - \zeta\overline{z})} \right)^{\lambda+2} h(\zeta) \frac{dv(\zeta)}{(1 - |\zeta|^2)^2} \\
&= (\lambda + 1) \int_{\mathbb{D}} h(\zeta) \left(\frac{(1 - |z|^2)(1 - |\zeta|^2)}{(1 - z\overline{\zeta})(1 - \zeta\overline{z})} \right)^{\lambda+2} d\mu(\zeta). \qquad \text{(A.4.9)}
\end{aligned}
$$

Now for any $a \in L_\infty(\mathbb{D})$ the operator

$$
(S_\lambda a)(z) = (\lambda + 1) \int_{\mathbb{D}} a(\zeta) \left(\frac{(1 - |z|^2)(1 - |\zeta|^2)}{(1 - z\overline{\zeta})(1 - \zeta\overline{z})} \right)^{\lambda+2} d\mu(\zeta)
$$

is nothing but the Berezin transform (A.3.7), where $\frac{1}{h} = \lambda + 2$.

We note that a change of variables provides us with another formula for the Berezin transform

$$
(S_\lambda a)(z) = \int_{\mathbb{D}} a \circ \varphi_z(\zeta) \, d\mu_\lambda(\zeta). \qquad \text{(A.4.10)}
$$

From the last formula, the fact that $d\mu_\lambda$ is a probability measure, and formula (A.4.9) it follows immediately that the Berezin transform S_λ is bounded on $L_\infty(\mathbb{D})$ for all $\lambda \in (-1, +\infty)$ and $\|S_\lambda\| = 1$.

We mention as well the following result.

Theorem A.4.1. ([102], Proposition 2.2) *Let $\beta, \lambda \in (-1, +\infty)$ and $p \in [1, +\infty)$. Then the operator S_λ is bounded on $L_p(\mathbb{D}, d\mu_\beta)$ if and only if*

$$
-(\lambda + 2) < \beta + 1 < (\lambda + 1)p.
$$

Moreover [102], the Berezin transform S_λ is uniformly bounded on $L_1(\mathbb{D}, d\mu_\beta)$ as $\lambda \to +\infty$.

Theorem A.4.2. *For every $\lambda \in (-1, +\infty)$ the Berezin transform S_λ is Möbius invariant; i.e., let φ be a Möbius transformation in $\mathbb{D}$, then*

$$(S_\lambda a) \circ \varphi = S_\lambda(a \circ \varphi).$$

for every $a \in L_1(\mathbb{D}, d\mu_\beta)$, where $\beta < \lambda$.

Proof. Given a Möbius transformation φ, it is easy to see that the point $0 \in \mathbb{D}$ is a fixed point of the Möbius transformation $\varphi_{\varphi(z)} \circ \varphi \circ \varphi_z$; here φ_z is given by (A.4.8). Thus by Schwarz' lemma, there is w, with $|w| = 1$, such that

$$\varphi_{\varphi(z)} \circ \varphi \circ \varphi_z(\zeta) = w\zeta,$$

or

$$\varphi \circ \varphi_z(\zeta) = \varphi_{\varphi(z)}(w\zeta),$$

for all $\zeta \in \mathbb{D}$. Using (A.4.10) and the rotation invariance of the measure $d\mu_\lambda$ we have

$$
\begin{aligned}
S_\lambda(a \circ \varphi)(z) &= \int_{\mathbb{D}} a \circ \varphi \circ \varphi_z(\zeta)\, d\mu_\lambda(\zeta) \\
&= \int_{\mathbb{D}} a \circ \varphi_{\varphi(z)}(w\zeta)\, d\mu_\lambda(\zeta) \\
&= (S_\lambda a)(\varphi(z)).
\end{aligned}
$$

$\square$

Theorem A.4.3. *Let $\lambda \in (-1, +\infty)$ and $a \in C(\overline{\mathbb{D}})$. Then $S_\lambda a \in C(\overline{\mathbb{D}})$ and $a - S_\lambda a \in C_0(\mathbb{D})$.*

Proof. Given any $z_0 \in \partial\mathbb{D}$, we have that $\varphi_z(\zeta) \to z_0$ as $z \to z_0$ for all $\zeta \in \mathbb{D}$. Thus by the dominated convergence theorem

$$(S_\lambda a)(z_0) = \lim_{z \to z_0} \int_{\mathbb{D}} a \circ \varphi_z(\zeta)\, d\mu_\lambda(\zeta) = a(z_0).$$

$\square$

We mention as well the following properties of the Berezin transform.

Theorem A.4.4. *([102], Proposition 2.5) Let $\beta \in (-1, +\infty)$ and $a \in L_1(\mathbb{D}, d\mu_\beta)$. Then $S_\lambda a \to a$ in $L_1(\mathbb{D}, d\mu_\beta)$ as $\lambda \to +\infty$.*

Theorem A.4.5. *([102], Proposition 2.6) For each $\lambda \in (-1, +\infty)$ the operator S_λ is injective in $L_1(\mathbb{D}, d\mu_\lambda)$.*

Lemma A.4.6. *([102], Corollary of Proposition 2.7) For each $\lambda \in (-1, +\infty)$ we have*

$$S_{\lambda+1} = \left(1 - \frac{1}{(\lambda + 1)(\lambda + 2)}\, \Delta\right) S_\lambda,$$

where the Laplace-Beltrami operator Δ is given by

$$\Delta = (1 - z\bar{z})^2\, \frac{\partial^2}{\partial z \partial \bar{z}}. \tag{A.4.11}$$

Iterating the above formula and using Theorem A.4.4 we have

Theorem A.4.7. *For each $\lambda \in (-1, +\infty)$ and each $n \in \mathbb{N}$,*

$$S_{\lambda+n} = \prod_{k=1}^{n} \left(1 - \frac{1}{(\lambda + k)(\lambda + k + 1)} \Delta \right) S_{\lambda}.$$

For each $\lambda \in (-1, +\infty)$ the Berezin transform admits the following representation in terms of the Laplace-Beltrami operator:

$$S_{\lambda} = \prod_{k=1}^{\infty} \left(1 - \frac{1}{(\lambda + k)(\lambda + k + 1)} \Delta \right)^{-1}.$$

Appendix B

Berezin Quantization on the Unit Disk

B.1 Definition of the quantization

Given a symplectic manifold (M, ω) and a subalgebra $\mathcal{A}(M)$ of $C^\infty(M)$ which is closed under the Poisson structure, by the (special) Berezin quantization we mean an associative algebra $\mathfrak{A}$ with involution having the following description:

1. There exists a family of unital *-algebras $\mathcal{A}_\hbar$ such that

    (i) the parameter $\hbar$ belongs to a set $E \subset \mathbb{R}_+$ having 0 as a limit point $(0 \notin E)$,

    (ii) $\mathfrak{A}$ is a subalgebra of $\bigoplus_{\hbar \in E} \mathcal{A}_\hbar$, with component-wise operations.

The elements $A \in \mathfrak{A}$ are obviously of the form $A = \{A_\hbar\}_{\hbar \in E}$, where $A_\hbar \in \mathcal{A}_\hbar$.

2. There exists a *-homomorphism $\kappa : \mathfrak{A} \to \mathcal{A}(M)$, where the algebra $\mathcal{A}(M)$ is considered with respect to the point-wise operations and the involution given by the complex conjugation, $a^* = \bar{a}$, such that

    (i) for each two points $x_1, x_2 \in M$ there is a function $a \in \kappa(\mathfrak{A})$ such that $a(x_1) \neq a(x_2)$.

3. for each $\hbar \in E$ there exists a *-isomorphism $\sigma_\hbar : \mathcal{A}_\hbar \to \widetilde{\mathcal{A}}_\hbar$ of the algebra $\mathcal{A}_\hbar$ onto an algebra $\widetilde{\mathcal{A}}_\hbar$ of smooth functions on M with point-wise linear operations, a special "$\star = \star_\hbar$" multiplication operation, the so-called star product, and complex conjugation as involution, such that

    (i) the homomorphism κ and the isomorphisms $\sigma_\hbar$ are connected by

$$\sigma_\hbar(A_\hbar) = \kappa(A) + O(\hbar), \quad \text{for all } A = \{A_\hbar\}_{\hbar \in E} \in \mathfrak{A},$$

(ii) for every $A = \{A_\hbar\}$, $B = \{B_\hbar\} \in \mathfrak{A}$

$$\sigma_\hbar\left([A_\hbar, B_\hbar]\right) = [\sigma_\hbar(A_\hbar), \sigma_\hbar(B_\hbar)] = i\hbar\left\{\kappa(A), \kappa(B)\right\} + O(\hbar^2),$$

where $\{\cdot, \cdot\}$ is the Poisson bracket on M.

The last two conditions can be formulated as well in the form

(i$'$) the homomorphism κ and the isomorphisms $\sigma_\hbar$, $\hbar \in E$, are connected by

$$\kappa(A) = \lim_{\hbar \to 0} \sigma_\hbar(A_\hbar), \quad \text{for all} \quad A = \{A_\hbar\}_{\hbar \in E} \in \mathfrak{A},$$

(ii$'$) for every $A = \{A_\hbar\}$, $B = \{B_\hbar\} \in \mathfrak{A}$,

$$\lim_{\hbar \to 0} \frac{1}{\hbar}[\sigma_\hbar(A_\hbar), \sigma_\hbar(B_\hbar)] = i\left\{\kappa(A), \kappa(B)\right\}.$$

In many cases, especially when the Toeplitz operators involved (see, for example, next section), more details can be added to the above description. In particular the algebras $\mathcal{A}_\hbar$ often appear as certain *-algebras of bounded linear operators on corresponding Hilbert spaces. In this case item 1 in the above description reads as follows:

1. There exists a family of Hilbert spaces $H_\hbar$ and a corresponding family of unital *-algebras $\mathcal{A}_\hbar$ of bounded linear operators acting on $H_\hbar$ such that

    (i) the parameter $\hbar$ belongs to a set $E \subset \mathbb{R}_+$ having 0 as a limit point $(0 \notin E)$,

    (ii) $\mathfrak{A}$ is a subalgebra of $\bigoplus_{\hbar \in E} \mathcal{A}_\hbar$, with component-wise operations.

Then we add

4. There exists a linear map $\iota : \mathcal{A}(M) \longrightarrow \mathfrak{A}$ such that

    (i) the algebra $\mathfrak{A}$ is generated by elements $\iota(a)$, with $a \in \mathcal{A}(M)$,

    (ii) for every $a \in \mathcal{A}(M)$　$\kappa(\iota(a)) = a$,

    (iii) let $e(x) \equiv 1$, then $\iota(e) = I = \{I_\hbar\}_{\hbar \in E}$.

    (iv) for every $a \in \mathcal{A}(M)$　$\iota(\overline{a}) = \iota(a)^*$.

From 4.(ii) it follows that the linear map ι is injective. Thus if the algebra $\mathcal{A}(M)$ separates the points of M, then we obviously have property 2.(i).

B.2 Quantization on the unit disk

As classical mechanics we consider here the pair $(\mathbb{D}, \omega)$, where $\mathbb{D}$ is the unit disk and the symplectic form ω is given by

$$\omega = d\mu(z) = \frac{1}{\pi} \frac{dx \wedge dy}{(1 - (x^2 + y^2))^2} = \frac{1}{2\pi i} \frac{d\bar{z} \wedge dz}{(1 - |z|^2)^2}. \tag{B.2.1}$$

Given two functions $a, b \in C^\infty(\mathbb{D})$, their Poisson bracket is

$$\begin{aligned}
\{a, b\} &= \pi(1 - (x^2 + y^2))^2 \left(\frac{\partial a}{\partial y} \frac{\partial b}{\partial x} - \frac{\partial a}{\partial x} \frac{\partial b}{\partial y} \right) \\
&= 2\pi i (1 - z\bar{z})^2 \left(\frac{\partial a}{\partial z} \frac{\partial b}{\partial \bar{z}} - \frac{\partial a}{\partial \bar{z}} \frac{\partial b}{\partial z} \right).
\end{aligned} \tag{B.2.2}$$

Recall that the Laplace-Beltrami operator has the form

$$\begin{aligned}
\Delta &= \pi(1 - (x^2 + y^2))^2 \left(\frac{\partial^2}{\partial x^2} + \frac{\partial^2}{\partial y^2} \right) \\
&= 4\pi(1 - z\bar{z})^2 \frac{\partial^2}{\partial z \partial \bar{z}}.
\end{aligned} \tag{B.2.3}$$

We describe the algebra $\mathfrak{A}$ and the mapping $\iota : \mathcal{A}(\mathbb{D}) \to \mathfrak{A}$.

Let $E = (0, \frac{1}{2\pi})$, for each $\hbar \in E$, and consequently $h \in (0, 1)$. Introduce $H_\hbar$ as the weighted Bergman space $\mathcal{A}_h^2(\mathbb{D})$. For each (smooth) function $a = a(z) \in \mathcal{A}(\mathbb{D})$ consider the family of Toeplitz operators $T_a^{(h)}$ with (anti-Wick) symbol a acting on $\mathcal{A}_h^2(\mathbb{D})$, for $h \in (0, 1)$, and denote by $\mathcal{T}_h(\mathcal{A}(\mathbb{D}))$ the *-algebra generated by Toeplitz operators $T_a^{(h)}$ with defining symbols $a \in \mathcal{A}(\mathbb{D})$.

For each $\hbar \in E$ define $\mathcal{A}_\hbar$ as the algebra $\mathcal{T}_h(\mathcal{A}(\mathbb{D}))$. Then the algebra $\mathfrak{A}$ is a subalgebra of

$$\bigoplus_{\hbar \in E} \mathcal{A}_\hbar = \bigoplus_{h \in (0,1)} \mathcal{T}_h(\mathcal{A}(\mathbb{D})),$$

which is generated by all operator families of the form

$$T_a = \{T_a^{(h)}\}_{h \in (0,1)}, \quad a \in \mathcal{A}(\mathbb{D}).$$

Now the linear map $\iota : \mathcal{A}(\mathbb{D}) \to \mathfrak{A}$ is given by

$$\iota : a \in \mathcal{A}(\mathbb{D}) \longmapsto T_a = \{T_a^{(h)}\}_{h \in (0,1)} \in \mathfrak{A},$$

and the homomorphism $\kappa : \mathfrak{A} \to \mathcal{A}(\mathbb{D})$ is generated by the following mapping of generators of the algebra $\mathfrak{A}$,

$$\kappa^\sharp : T_a = \{T_a^{(h)}\}_{h \in (0,1)} \longmapsto a.$$

Recall that the Wick symbol of the Toeplitz operator $T_a^{(h)}$ by (10.1.7) and (A.3.7) has the form

$$\widetilde{a}_h(z,\overline{z}) = (S_h a)(z,\overline{z}) = (\frac{1}{h} - 1) \int_{\mathbb{D}} a(\zeta) \left(\frac{(1 - |z|^2)(1 - |\zeta|^2)}{(1 - z\overline{\zeta})(1 - \zeta\overline{z})} \right)^{\frac{1}{h}} d\mu(\zeta).$$

For each $h \in (0,1)$ define the function algebra

$$\widetilde{\mathcal{A}}_h = \{\widetilde{a}_h(z,\overline{z}) : a \in \mathcal{A}(\mathbb{D})\}$$

with point-wise linear operations, and with the multiplication law defined by the product of Toeplitz operators and given by the formula (10.1.8)

$$(\widetilde{a}_h \star \widetilde{b}_h)(z,\overline{z}) = (\frac{1}{h} - 1) \int_{\mathbb{D}} \widetilde{a}_h(z,\overline{\zeta}) \widetilde{b}_h(\zeta,\overline{z}) \left(\frac{(1 - |z|^2)(1 - |\zeta|^2)}{(1 - z\overline{\zeta})(1 - \zeta\overline{z})} \right)^{\frac{1}{h}} d\mu(\zeta).$$

For each $h \in (0,1)$ introduce the isomorphism

$$\sigma_h : \mathcal{A}_h \longrightarrow \widetilde{\mathcal{A}}_h$$

defining it on generators by the rule

$$\sigma_h : T_a^{(h)} \longmapsto \widetilde{a}_h(z,\overline{z}).$$

The statements of the correspondence principle are given by the equalities established in the next two sections (see (B.3.7) and (B.4.14))

$$\begin{aligned}
\widetilde{a}_h(z,\overline{z}) &= a(z,\overline{z}) + O(\hbar), \\
(\widetilde{a}_h \star \widetilde{b}_h - \widetilde{b}_h \star \widetilde{a}_h)(z,\overline{z}) &= i\hbar\,\{a,b\} + O(\hbar^2).
\end{aligned}$$

B.3   Two first terms of asymptotic of the Wick symbol

We consider the Berezin transform (A.3.7)

$$(S_h a)(z,\overline{z}) = \frac{1}{\pi}(\frac{1}{h} - 1) \int_{\mathbb{D}} a(\zeta) \left(\frac{(1 - |z|^2)(1 - |\zeta|^2)}{(1 - z\overline{\zeta})(1 - \zeta\overline{z})} \right)^{\frac{1}{h}} \frac{dv(\zeta)}{(1 - |\zeta|^2)^2} \qquad (B.3.4)$$

and calculate the first two asymptotic terms when $h \to 0$.

Assume that the function $a(\zeta) = a(x,y)$ has four continuous derivatives. First, set $z = 0$. Then

$$(S_h a)(0,0) = \frac{1}{\pi}(\frac{1}{h} - 1) \int_{\mathbb{D}} a(\zeta)(1 - |\zeta|^2)^{\frac{1}{h} - 2} dv(\zeta).$$

Passing to polar coordinates we have

$$(S_h a)(0,0) = H \int_0^1 A(r) r e^{(H-1)\ln(1-r^2)} dr,$$

where $H = \frac{1}{h} - 1$, and

$$A(r) = \frac{1}{\pi} \int_{-\pi}^{\pi} a(r\cos\varphi, r\sin\varphi) d\varphi.$$

The point $r_0 = 0$ is the point of maximum of the phase function $S(r) = \ln(1-r^2)$. Further, $S'(0) = 0$ and $S''(0) = -2 \neq 0$. Thus according to the Laplace integral method (see, for example, [76]) we have

$$(S_h a)(0,0) = H\left(\sum_{j=0}^4 \frac{A^{(j)}(0)}{j!} \int_0^1 r^{j+1} e^{(H-1)\ln(1-r^2)} dr + o(H^{-3})\right).$$

Calculate now consecutive terms

for $j = 0$: $\qquad A(0) = 2a(0),$ and

$$\int_0^1 r(1-r^2)^{H-1} dr = \frac{1}{2}\int_0^1 (1-r^2)^{H-1} dr^2 = \frac{1}{2H};$$

for $j = 1$:

$$A'(0) = \frac{1}{\pi}\int_{-\pi}^{\pi} (a'_x(0)\cos\varphi + a'(0)\sin\varphi) d\varphi = 0;$$

for $j = 2$:

$$\begin{aligned}
A''(0) &= \frac{1}{\pi}\int_{-\pi}^{\pi} (a''_{x,x}(0)\cos^2\varphi + 2a''_{x,y}(0)\sin\varphi\cos\varphi + a''_{y,y}(0)\sin^2\varphi) d\varphi \\
&= a''_{x,x}(0) + a''_{y,y}(0) = (Da)(0),
\end{aligned}$$

where $D = \frac{\partial^2}{\partial x^2} + \frac{\partial^2}{\partial y^2}$, and

$$\int_0^1 r^3(1-r^2)^{H-1} dr = \frac{1}{2H(H+1)};$$

for $j = 3$:

$$A'''(0) = \frac{1}{\pi} \sum_{k+l=3} \frac{3!}{k!l!} \frac{\partial^3 a(0)}{\partial x^k \partial y^l} \int_{-\pi}^{\pi} \cos^k\varphi \sin^l\varphi \, d\varphi = 0;$$

for $j = 4$:

$$A^{(4)}(0) = \frac{1}{\pi} \sum_{k+l=4} \frac{4!}{k!l!} \frac{\partial^4 a(0)}{\partial x^k \partial y^l} \int_{-\pi}^{\pi} \cos^k \varphi \sin^l \varphi \, d\varphi$$

$$= \frac{\partial^4 a(0)}{\partial x^4} \frac{1}{\pi} \int_{-\pi}^{\pi} \cos^4 \varphi \, d\varphi + 6 \frac{\partial^4 a(0)}{\partial x^2 \partial y^2} \frac{1}{\pi} \int_{-\pi}^{\pi} \cos^2 \varphi \sin^2 \varphi \, d\varphi$$

$$+ \frac{\partial^4 a(0)}{\partial y^4} \frac{1}{\pi} \int_{-\pi}^{\pi} \sin^4 \varphi \, d\varphi.$$

Now

$$\int_{-\pi}^{\pi} \cos^4 \varphi \, d\varphi = \int_{-\pi}^{\pi} \sin^4 \varphi \, d\varphi = \frac{3}{4} \pi,$$

and

$$\int_{-\pi}^{\pi} \cos^2 \varphi \sin^2 \varphi \, d\varphi = \frac{\pi}{4}.$$

Thus

$$A^{(4)}(0) = \frac{3}{4} \left(\frac{\partial^4 a(0)}{\partial x^4} + 2 \frac{\partial^4 a(0)}{\partial x^2 \partial y^2} + \frac{\partial^4 a(0)}{\partial y^4} \right) = \frac{3}{4} (D^2 a)(0).$$

Further,

$$\int_0^1 r^5 (1 - r^2)^{H-1} dr = \frac{1}{H(H+1)(H+2)}.$$

Thus finally

$$(S_h a)(0,0) = a(0) + \frac{1}{4H} (Da)(0) + \frac{1}{32(H+1)(H+2)} (D^2 a)(0) + o(H^{-2})$$

$$= a(0) + \frac{h}{4} (Da)(0) + \frac{h^2}{32} (D^2 a)(0) + o(h^2). \tag{B.3.5}$$

Consider now $z \neq 0$. Changing in (B.3.4)

$$\zeta = \frac{z - w}{1 - \bar{z}w} \quad \text{and} \quad \bar{\zeta} = \frac{\bar{z} - \bar{w}}{1 - z\bar{w}}$$

we have

$$(S_h a)(z, \bar{z}) = \frac{1}{\pi} \left(\frac{1}{h} - 1 \right) \int_{\mathbb{D}} a \left(\frac{z - w}{1 - \bar{z}w}, \frac{\bar{z} - \bar{w}}{1 - z\bar{w}} \right) (1 - w\bar{w})^{\frac{1}{h}} \frac{dv(w)}{(1 - w\bar{w})^2}.$$

Thus for the function $c_{z,\bar{z}}(w, \bar{w}) = a \left(\frac{z-w}{1-\bar{z}w}, \frac{\bar{z}-\bar{w}}{1-z\bar{w}} \right)$ we have by (B.3.5)

$$(S_h a)(z, \bar{z}) = c_{z,\bar{z}}(0,0) + \frac{h}{4} (Dc_{z,\bar{z}})(0,0) + \frac{h^2}{32} (D^2 c_{z,\bar{z}})(0,0) + o(h^2). \tag{B.3.6}$$

Now

$$\frac{1}{4} D = \frac{1}{4}\left(\frac{\partial^2}{\partial x^2} + \frac{\partial^2}{\partial y^2}\right) = \frac{\partial^2}{\partial z \partial \bar{z}}.$$

Calculate

$$\frac{\partial^2 c_{z,\bar{z}}(w,\bar{w})}{\partial w \partial \bar{w}}\Big|_{w=\bar{w}=0} = \left(\frac{\partial^2 c(\zeta(w),\overline{\zeta(w)})}{\partial \zeta \partial \bar{\zeta}} \cdot \frac{-(1-z\bar{z})}{(1-\bar{z}w)^2} \cdot \frac{-(1-z\bar{z})}{(1-z\bar{w})^2}\right)_{w=\bar{w}=0}$$

$$= (1-z\bar{z})^2 \frac{\partial^2 a(z,\bar{z})}{\partial z \partial \bar{z}} = \frac{1}{4\pi}(\Delta a)(z,\bar{z}),$$

where the Laplace-Beltrami operator Δ is given by (B.2.3).

Start calculating $(D^2 c_{z,\bar{z}})(0,0)$. We have

$$\frac{\partial^3 c_{z,\bar{z}}(w,\bar{w})}{\partial w \partial \bar{w}^2} = -\frac{\partial^3 a}{\partial \zeta \partial \bar{\zeta}^2} \cdot \frac{1-z\bar{z}}{(1-\bar{z}w)^2} \cdot \frac{(1-z\bar{z})^2}{(1-z\bar{w})^4} + \frac{\partial^2 a}{\partial \zeta \partial \bar{\zeta}} \cdot \frac{1-z\bar{z}}{(1-\bar{z}w)^2} \cdot \frac{2z(1-z\bar{z})}{(1-z\bar{w})^3},$$

and

$$\frac{\partial^4 c_{z,\bar{z}}(w,\bar{w})}{\partial w^2 \partial \bar{w}^2} = \frac{\partial^4 a}{\partial \zeta^2 \partial \bar{\zeta}^2} \cdot \frac{(1-z\bar{z})^2}{(1-\bar{z}w)^4} \cdot \frac{(1-z\bar{z})^2}{(1-z\bar{w})^4}$$

$$- \frac{\partial^3 a}{\partial \zeta \partial \bar{\zeta}^2} \cdot \frac{2\bar{z}(1-z\bar{z})}{(1-\bar{z}w)^3} \cdot \frac{(1-z\bar{z})^2}{(1-z\bar{w})^4}$$

$$- \frac{\partial^3 a}{\partial \zeta^2 \partial \bar{\zeta}} \cdot \frac{(1-z\bar{z})^2}{(1-\bar{z}w)^4} \cdot \frac{2z(1-z\bar{z})}{(1-z\bar{w})^3}$$

$$+ \frac{\partial^2 a}{\partial \zeta \partial \bar{\zeta}} \cdot \frac{2\bar{z}(1-z\bar{z})}{(1-\bar{z}w)^3} \cdot \frac{2z(1-z\bar{z})}{(1-z\bar{w})^3}.$$

Thus

$$\frac{1}{16}(D^2 c_{z,\bar{z}})(0,0) = (1-z\bar{z})^4 \frac{\partial^4 a}{\partial z^2 \partial \bar{z}^2} - 2(1-z\bar{z})^3\left(\bar{z}\frac{\partial^3 a}{\partial z \partial \bar{z}^2} + z\frac{\partial^3 a}{\partial z^2 \partial \bar{z}}\right)$$

$$+ 4z\bar{z}(1-z\bar{z})^2 \frac{\partial^2 a}{\partial z \partial \bar{z}},$$

or, as is easy to see,

$$\frac{1}{16}(D^2 c_{z,\bar{z}})(0,0) = \left((1-z\bar{z})^2\frac{\partial^2}{\partial z \partial \bar{z}}\right)^2 a(z,\bar{z}) + 2(1-z\bar{z})^2\frac{\partial^2}{\partial z \partial \bar{z}}a(z,\bar{z})$$

$$= \frac{1}{16\pi^2}(\Delta^2 a)(z,\bar{z}) + \frac{1}{2\pi}(\Delta a)(z,\bar{z}).$$

Finally,

$$(S_h a)(z,\overline{z}) \;=\; a(z,\overline{z}) + \frac{h}{4\pi}(\Delta a)(z,\overline{z})$$

$$+ \frac{h^2}{32\pi^2}[(\Delta^2 a)(z,\overline{z}) + 8\pi(\Delta a)(z,\overline{z})] + o(h^2)$$

$$=\; a(z,\overline{z}) + \frac{\hbar}{2}(\Delta a)(z,\overline{z}) + \frac{\hbar^2}{8}[(\Delta^2 a)(z,\overline{z}) + 8\pi(\Delta a)(z,\overline{z})] + o(\hbar^2).$$

We summarize the above in the following proposition.

Proposition B.3.1. *For each $h \in (0,1]$ and any four-times continuously differentiable function $a = a(z,\overline{z})$ we have*

$$\widetilde{a}_h(z,\overline{z}) \;=\; (S_h a)(z,\overline{z})$$

$$=\; a(z,\overline{z}) + \frac{\hbar}{2}(\Delta a)(z,\overline{z}) + \frac{\hbar^2}{8}[(\Delta^2 a)(z,\overline{z}) + 8\pi(\Delta a)(z,\overline{z})] + o(\hbar^2)$$

$$=\; a(z,\overline{z}) + O(\hbar), \tag{B.3.7}$$

where the Laplace-Beltrami operator Δ is given by $\Delta = 4\pi(1 - z\overline{z})^2 \dfrac{\partial^2}{\partial z \partial \overline{z}}$.

B.4 Three first terms of asymptotic in a commutator

Given two six-times continuously differentiable functions $a = a(z,\overline{z})$ and $b = b(z,\overline{z})$, by the composition formula (10.1.8) we have

$$(\widetilde{a}_h \star \widetilde{b}_h)(z,\overline{z}) = (\frac{1}{h} - 1) \int_{\mathbb{D}} \widetilde{a}_h(z,\overline{\zeta})\, \widetilde{b}_h(\zeta,\overline{z}) \left(\frac{(1-|z|^2)(1-|\zeta|^2)}{(1-z\overline{\zeta})(1-\zeta\overline{z})} \right)^{\frac{1}{h}} d\mu(\zeta).$$

Substituting the expressions for $\widetilde{a}_h$ and $\widetilde{b}_h$ by (10.1.7) and then changing the order of integration we have

$$(\widetilde{a}_h \star \widetilde{b}_h)(z,\overline{z}) \;=\; (\frac{1}{h} - 1)^3 \int_{\mathbb{D}} \left(\int_{\mathbb{D}} a(w,\overline{w}) \left(\frac{(1-z\overline{\zeta})(1-w\overline{w})}{(1-z\overline{w})(1-w\overline{\zeta})} \right)^{\frac{1}{h}} d\mu(w) \right)$$

$$\times \left(\int_{\mathbb{D}} b(w',\overline{w'}) \left(\frac{(1-\overline{z}\zeta)(1-w'\overline{w'})}{(1-\zeta\overline{w'})(1-w'\overline{z})} \right)^{\frac{1}{h}} d\mu(w') \right)$$

$$\times \left(\frac{(1-|z|^2)(1-|\zeta|^2)}{(1-z\overline{\zeta})(1-\zeta\overline{z})} \right)^{\frac{1}{h}} d\mu(\zeta)$$

$$= \ (\frac{1}{h} - 1)^2 \int_{\mathbb{D}} \int_{\mathbb{D}} a(w, \overline{w})\, b(w', \overline{w'})$$

$$\times \left(\frac{(1 - |z|^2)(1 - |w|^2)(1 - |w'|^2)}{(1 - w'\overline{w'})(1 - z\overline{w})(1 - w'\overline{z})} \right)^{\frac{1}{h}} d\mu(w) d\mu(w')$$

$$\times (\frac{1}{h} - 1) \int_{\mathbb{D}} \left(\frac{(1 - w'\overline{w'})(1 - \zeta\overline{\zeta})}{(1 - w\overline{\zeta})(1 - \zeta\overline{w'})} \right)^{\frac{1}{h}} d\mu(\zeta).$$

The last integral is equal to 1, as the Wick symbol of the identity operator (see, (A.1.6) and (10.1.7)). Now change variables

$$w = \frac{z - s}{1 - \overline{z}s} \quad \text{and} \quad w' = \frac{z - s'}{1 - \overline{z}s'},$$

so

$$\widetilde{a}_h \star \widetilde{b}_h a = (\frac{1}{h} - 1)^2 \int_{\mathbb{D}} \int_{\mathbb{D}} a_1(s, \overline{s})\, b_1(s', \overline{s'}) \left(\frac{(1 - |s|^2)(1 - |s'|^2)}{(1 - s\overline{s'})} \right)^{\frac{1}{h}} d\mu(s) d\mu(s'),$$

where

$$a_1(s, \overline{s}) = a\left(\frac{z - s}{1 - \overline{z}s}, \frac{\overline{z} - \overline{s}}{1 - z\overline{s}} \right) \quad \text{and} \quad b_1(s', \overline{s'}) = b\left(\frac{z - s'}{1 - \overline{z}s'}, \frac{\overline{z} - \overline{s'}}{1 - z\overline{s'}} \right).$$

Set $H = \frac{1}{h} - 1$. Then using the formula

$$(1 - s\overline{s'})^{-(H+1)} = \sum_{j=0}^{\infty} c_j\, s^j\, \overline{s'}^{\,j},$$

where

$$c_0 = 1, \quad c_j = \frac{(H + 1)(H + 2)\ldots(H + j)}{j!}, \quad j \in \mathbb{N},$$

we have (substituting s' for s in the second integral)

$$(\widetilde{a}_h \star \widetilde{b}_h)(z, \overline{z}) \ = \ \sum_{j=0}^{\infty} c_j \left(\frac{H}{\pi} \int_{\mathbb{D}} a_1(s, \overline{s})\,(1 - |s|^2)^{H-1} s^j\, dv(s) \right) \quad (\text{B.4.8})$$

$$\times \left(\frac{H}{\pi} \int_{\mathbb{D}} b_1(s, \overline{s})\,(1 - |s|^2)^{H-1} \overline{s}^{\,j}\, dv(s) \right).$$

Represent the function $a_1(s, \overline{s})$ as

$$a_1(s, \overline{s}) \ = \ a_1(0, 0) + s\, \frac{\partial a_1}{\partial s}(0, 0) + \overline{s}\, \frac{\partial a_1}{\partial \overline{s}}(0, 0)$$

$$+ \sum_{m=2}^{6} \sum_{k+l=m} \frac{s^k \overline{s}^{\,l}}{k!\, l!} \frac{\partial^m a_1}{\partial s^k \partial \overline{s}^{\,l}}(0, 0) + o(|s|^6 + |\overline{s}|^6),$$

and do the same for the function $b_1(s, \bar{s})$.

Substitute the above expressions for a_1 and b_1 into (B.4.8) and calculate consequently for $j = 0$, $j = 1$, $j = 2$, and $j = 3$ integrals which give non-zero contribution to the first four terms of the asymptotic for $(\widetilde{a}_h \star \widetilde{b}_h)(z, \bar{z})$.

Note first that

$$\frac{H}{\pi} \int_{\mathbb{D}} (1 - |s|^2)^{H-1} s^l \bar{s}^k \, ds = \frac{H}{\pi} \int_0^1 (1 - r^2)^{H-1} r^{l+k+1} \, dr \int_{-\pi}^{\pi} e^{i(l-k)\theta} \, d\theta$$

$$= \begin{cases} 2H \int_0^1 (1 - r^2)^{H-1} r^{2k+1} \, dr, & l = k \\ 0, & l \neq k \end{cases}$$

$$= \begin{cases} H \int_0^1 (1 - u)^{H-1} u^k \, du, & l = k \\ 0, & l \neq k \end{cases}$$

$$= \begin{cases} H \frac{\Gamma(H)\Gamma(k+1)}{\Gamma(H+k+1)}, & l = k \\ 0, & l \neq k \end{cases}$$

$$= \begin{cases} \frac{k!}{(H+1)(H+2)\ldots(H+k)}, & l = k \\ 0, & l \neq k \end{cases} .$$

For $j = 0$ we have

$$\frac{H}{\pi} \int_{\mathbb{D}} a_1(0,0) \, (1 - |s|^2)^{H-1} \, ds = a_1(0,0),$$

$$\frac{H}{\pi} \int_{\mathbb{D}} b_1(0,0) \, (1 - |s|^2)^{H-1} \, ds = b_1(0,0),$$

$$\frac{H}{\pi} \int_{\mathbb{D}} \frac{\partial^2 a_1}{\partial s \partial \bar{s}}(0,0) \, (1 - |s|^2)^{H-1} \, s\bar{s} \, ds = \frac{1}{H+1} \frac{\partial^2 a_1}{\partial s \partial \bar{s}}(0,0),$$

$$\frac{H}{\pi} \int_{\mathbb{D}} \frac{\partial^2 b_1}{\partial s \partial \bar{s}}(0,0) \, (1 - |s|^2)^{H-1} \, s\bar{s} \, ds = \frac{1}{H+1} \frac{\partial^2 b_1}{\partial s \partial \bar{s}}(0,0),$$

$$\frac{H}{\pi} \int_{\mathbb{D}} \frac{\partial^4 a_1}{\partial s^2 \partial \bar{s}^2}(0,0) \, (1 - |s|^2)^{H-1} \, s^2\bar{s}^2 \, ds = \frac{2}{(H+1)(H+2)} \frac{\partial^4 a_1}{\partial s^2 \partial \bar{s}^2}(0,0),$$

$$\frac{H}{\pi} \int_{\mathbb{D}} \frac{\partial^4 b_1}{\partial s^2 \partial \bar{s}^2}(0,0) \, (1 - |s|^2)^{H-1} \, s^2\bar{s}^2 \, ds = \frac{2}{(H+1)(H+2)} \frac{\partial^4 b_1}{\partial s^2 \partial \bar{s}^2}(0,0),$$

$$\frac{H}{\pi} \int_{\mathbb{D}} \frac{\partial^6 a_1}{\partial s^3 \partial \bar{s}^3}(0,0) \, (1 - |s|^2)^{H-1} \, s^3\bar{s}^3 \, ds = \frac{3!}{(H+1)(H+2)(H+3)} \frac{\partial^6 a_1}{\partial s^3 \partial \bar{s}^3}(0,0),$$

$$\frac{H}{\pi} \int_{\mathbb{D}} \frac{\partial^6 b_1}{\partial s^3 \partial \bar{s}^3}(0,0) \, (1 - |s|^2)^{H-1} \, s^3\bar{s}^3 \, ds = \frac{3!}{(H+1)(H+2)(H+3)} \frac{\partial^6 b_1}{\partial s^3 \partial \bar{s}^3}(0,0).$$

Thus for $j = 0$ we have the following contribution to the asymptotics,

$$(\widetilde{a}_h \star \widetilde{b}_h)_0(z,\overline{z}) = a_1(0,0)b_1(0,0)$$
$$+ h\left(a_1(0,0)\frac{\partial^2 b_1}{\partial s \partial \overline{s}}(0,0) + b_1(0,0)\frac{\partial^2 a_1}{\partial s \partial \overline{s}}(0,0)\right)$$
$$+ h^2 \frac{\partial^2 a_1}{\partial s \partial \overline{s}}(0,0)\frac{\partial^2 b_1}{\partial s \partial \overline{s}}(0,0)$$
$$+ \frac{h^2}{2}\left(a_1(0,0)\frac{\partial^4 b_1}{\partial s^2 \partial \overline{s}^2}(0,0) + b_1(0,0)\frac{\partial^4 a_1}{\partial s^2 \partial \overline{s}^2}(0,0)\right)$$
$$+ \frac{h^3}{2}\left(\frac{\partial^2 a_1}{\partial s \partial \overline{s}}(0,0)\frac{\partial^4 b_1}{\partial s^2 \partial \overline{s}^2}(0,0) + \frac{\partial^2 b_1}{\partial s \partial \overline{s}}(0,0)\frac{\partial^4 a_1}{\partial s^2 \partial \overline{s}^2}(0,0)\right)$$
$$+ \frac{h^3}{6}\left(a_1(0,0)\frac{\partial^6 b_1}{\partial s^3 \partial \overline{s}^3}(0,0) + b_1(0,0)\frac{\partial^6 a_1}{\partial s^3 \partial \overline{s}^3}(0,0)\right) + o(h^3).$$

Observe that $(\widetilde{a}_h \star \widetilde{b}_h)_0(z,\overline{z}) - (\widetilde{b}_h \star \widetilde{a}_h)_0(z,\overline{z}) = 0$, and thus for $j = 0$ we have no contribution to the first three asymptotic terms of the commutator $(\widetilde{a}_h \star \widetilde{b}_h - \widetilde{b}_h \star \widetilde{a}_h)(z,\overline{z})$.

For $j = 1$ the non-zero integrals are as follows:

$$\frac{H}{\pi}\int_{\mathbb{D}} \frac{\partial a_1}{\partial \overline{s}}(0,0)\,(1 - |s|^2)^{H-1}\,s\overline{s}\,ds = \frac{1}{H+1}\frac{\partial a_1}{\partial \overline{s}}(0,0),$$

$$\frac{H}{\pi}\int_{\mathbb{D}} \frac{\partial b_1}{\partial s}(0,0)\,(1 - |s|^2)^{H-1}\,s\overline{s}\,ds = \frac{1}{H+1}\frac{\partial b_1}{\partial s}(0,0),$$

$$\frac{H}{\pi}\int_{\mathbb{D}} \frac{\partial^3 a_1}{\partial s \partial \overline{s}^2}(0,0)\,(1 - |s|^2)^{H-1}\,s^2\overline{s}^2\,ds = \frac{2}{(H+1)(H+2)}\frac{\partial^3 a_1}{\partial s \partial \overline{s}^2}(0,0),$$

$$\frac{H}{\pi}\int_{\mathbb{D}} \frac{\partial^3 b_1}{\partial s^2 \partial \overline{s}}(0,0)\,(1 - |s|^2)^{H-1}\,s^2\overline{s}^2\,ds = \frac{2}{(H+1)(H+2)}\frac{\partial^3 b_1}{\partial s^2 \partial \overline{s}}(0,0),$$

$$\frac{H}{\pi}\int_{\mathbb{D}} \frac{\partial^5 a_1}{\partial s^2 \partial \overline{s}^3}(0,0)\,(1 - |s|^2)^{H-1}\,s^3\overline{s}^3\,ds = \frac{3!}{(H+1)(H+2)(H+3)}\frac{\partial^5 a_1}{\partial s^2 \partial \overline{s}^3}(0,0),$$

$$\frac{H}{\pi}\int_{\mathbb{D}} \frac{\partial^5 b_1}{\partial s^3 \partial \overline{s}^2}(0,0)\,(1 - |s|^2)^{H-1}\,s^3\overline{s}^3\,ds = \frac{3!}{(H+1)(H+2)(H+3)}\frac{\partial^5 b_1}{\partial s^3 \partial \overline{s}^2}(0,0).$$

Taking into account that $c_1 = H + 1$ we have

$$(\widetilde{a}_h \star \widetilde{b}_h)_1(z,\overline{z}) = h\,\frac{\partial a_1}{\partial \overline{s}}(0,0)\frac{\partial b_1}{\partial s}(0,0) \tag{B.4.9}$$
$$+ h^2\left(\frac{\partial a_1}{\partial \overline{s}}(0,0)\frac{\partial^3 b_1}{\partial \overline{s} \partial s^2}(0,0) + \frac{\partial b_1}{\partial s}(0,0)\frac{\partial^3 a_1}{\partial \overline{s}^2 \partial s}(0,0)\right)$$
$$+ h^3\frac{\partial^3 a_1}{\partial s \partial \overline{s}^2}(0,0)\frac{\partial^3 b_1}{\partial s^2 \partial \overline{s}}(0,0)$$
$$+ \frac{h^3}{2}\left(\frac{\partial a_1}{\partial \overline{s}}(0,0)\frac{\partial^5 b_1}{\partial s^3 \partial \overline{s}^2}(0,0) + \frac{\partial b_1}{\partial s}(0,0)\frac{\partial^5 a_1}{\partial s^2 \partial \overline{s}^3}(0,0)\right) + o(h^3).$$

Consider now the case $j = 2$:

$$\frac{H}{\pi} \int_{\mathbb{D}} \frac{\partial^2 a_1}{\partial \bar{s}^2}(0,0)\,(1 - |s|^2)^{H-1}\, s^2 \bar{s}^2\, ds = \frac{2}{(H+1)(H+2)} \frac{\partial^2 a_1}{\partial \bar{s}^2}(0,0),$$

$$\frac{H}{\pi} \int_{\mathbb{D}} \frac{\partial^2 b_1}{\partial s^2}(0,0)\,(1 - |s|^2)^{H-1}\, s^2 \bar{s}^2\, ds = \frac{2}{(H+1)(H+2)} \frac{\partial^2 b_1}{\partial s^2}(0,0),$$

$$\frac{H}{\pi} \int_{\mathbb{D}} \frac{\partial^4 a_1}{\partial s \partial \bar{s}^3}(0,0)\,(1 - |s|^2)^{H-1}\, s^3 \bar{s}^3\, ds = \frac{3!}{(H+1)(H+2)(H+3)} \frac{\partial^4 a_1}{\partial s \partial \bar{s}^3}(0,0),$$

$$\frac{H}{\pi} \int_{\mathbb{D}} \frac{\partial^4 b_1}{\partial s^3 \partial \bar{s}}(0,0)\,(1 - |s|^2)^{H-1}\, s^3 \bar{s}^3\, ds = \frac{3!}{(H+1)(H+2)(H+3)} \frac{\partial^4 b_1}{\partial s^3 \partial \bar{s}}(0,0).$$

Recall that $c_2 = \frac{1}{2}(H+1)(H+2)$, thus

$$
\begin{aligned}
(\widetilde{a}_h \star \widetilde{b}_h)_2(z, \bar{z}) =\ & \frac{h^2}{2} \frac{\partial^2 a_1}{\partial \bar{s}^2}(0,0) \frac{\partial^2 b_1}{\partial s^2}(0,0) + \frac{h^3}{2}\left(\frac{\partial^2 a_1}{\partial \bar{s}^2}(0,0) \frac{\partial^4 b_1}{\partial s^3 \partial \bar{s}}(0,0) \right. \\
& \left. + \frac{\partial^2 b_1}{\partial s^2}(0,0) \frac{\partial^4 a_1}{\partial s \partial \bar{s}^3}(0,0) \right) + o(h^3).
\end{aligned}
\tag{B.4.10}
$$

Consider finally the case $j = 3$:

$$\frac{H}{\pi} \int_{\mathbb{D}} \frac{\partial^3 a_1}{\partial \bar{s}^3}(0,0)\,(1 - |s|^2)^{H-1}\, s^3 \bar{s}^3\, ds = \frac{3!}{(H+1)(H+2)(H+3)} \frac{\partial^3 a_1}{\partial \bar{s}^3}(0,0),$$

$$\frac{H}{\pi} \int_{\mathbb{D}} \frac{\partial^3 b_1}{\partial s^3}(0,0)\,(1 - |s|^2)^{H-1}\, s^3 \bar{s}^3\, ds = \frac{3!}{(H+1)(H+2)(H+3)} \frac{\partial^3 b_1}{\partial s^3}(0,0).$$

Taking into account that $c_3 = \frac{(H+1)(H+2)(H+3)}{3!}$ we have

$$(\widetilde{a}_h \star \widetilde{b}_h)_3(z, \bar{z}) = \frac{h^3}{6} \frac{\partial^3 a_1}{\partial \bar{s}^3}(0,0) \frac{\partial^3 b_1}{\partial s^3}(0,0) + o(h^3). \tag{B.4.11}$$

Now from (B.4.9), (B.4.10), and (B.4.11) we have

$$
\begin{aligned}
\widetilde{a}_h \star \widetilde{b}_h - \widetilde{b}_h \star \widetilde{a}_h =\ & h\left(\frac{\partial a_1}{\partial \bar{s}}(0,0) \frac{\partial b_1}{\partial s}(0,0) - \frac{\partial a_1}{\partial s}(0,0) \frac{\partial b_1}{\partial \bar{s}}(0,0) \right) \\
& + h^2 \left[\left(\frac{\partial a_1}{\partial \bar{s}}(0,0) \frac{\partial^3 b_1}{\partial \bar{s} \partial s^2}(0,0) - \frac{\partial a_1}{\partial s}(0,0) \frac{\partial^3 b_1}{\partial \bar{s}^2 \partial s}(0,0) \right) \right. \\
& + \left(\frac{\partial b_1}{\partial s}(0,0) \frac{\partial^3 a_1}{\partial \bar{s}^2 \partial s}(0,0) - \frac{\partial b_1}{\partial \bar{s}}(0,0) \frac{\partial^3 a_1}{\partial \bar{s} \partial s^2}(0,0) \right) \\
& + \left. \frac{1}{2}\left(\frac{\partial^2 a_1}{\partial \bar{s}^2}(0,0) \frac{\partial^2 b_1}{\partial s^2}(0,0) - \frac{\partial^2 a_1}{\partial s^2}(0,0) \frac{\partial^2 b_1}{\partial \bar{s}^2}(0,0) \right) \right]
\end{aligned}
\tag{B.4.12}
$$

$$+h^3\left\{\left(\frac{\partial^3 a_1}{\partial s\partial\overline{s}^2}(0,0)\,\frac{\partial^3 b_1}{\partial s^2\partial\overline{s}}(0,0)-\frac{\partial^3 a_1}{\partial s^2\partial\overline{s}}(0,0)\,\frac{\partial^3 b_1}{\partial s\partial\overline{s}^2}(0,0)\right)\right.$$

$$+\frac{1}{6}\left(\frac{\partial^3 a_1}{\partial\overline{s}^3}(0,0)\,\frac{\partial^3 b_1}{\partial s^3}(0,0)-\frac{\partial^3 a_1}{\partial s^3}(0,0)\,\frac{\partial^3 b_1}{\partial\overline{s}^3}(0,0)\right)$$

$$+\frac{1}{2}\left[\left(\frac{\partial a_1}{\partial\overline{s}}(0,0)\,\frac{\partial^5 b_1}{\partial s^3\partial\overline{s}^2}(0,0)-\frac{\partial a_1}{\partial s}(0,0)\,\frac{\partial^5 b_1}{\partial s^2\partial\overline{s}^3}(0,0)\right)\right.$$

$$\left.+\left(\frac{\partial b_1}{\partial s}(0,0)\,\frac{\partial^5 a_1}{\partial s^2\partial\overline{s}^3}(0,0)-\frac{\partial b_1}{\partial\overline{s}}(0,0)\,\frac{\partial^5 a_1}{\partial s^3\partial\overline{s}^2}(0,0)\right)\right]$$

$$+\frac{1}{2}\left[\left(\frac{\partial^2 a_1}{\partial\overline{s}^2}(0,0)\,\frac{\partial^4 b_1}{\partial s^3\partial\overline{s}}(0,0)-\frac{\partial^2 a_1}{\partial s^2}(0,0)\,\frac{\partial^4 b_1}{\partial s\partial\overline{s}^3}(0,0)\right)\right.$$

$$\left.+\left(\frac{\partial^2 b_1}{\partial s^2}(0,0)\,\frac{\partial^4 a_1}{\partial s\partial\overline{s}^3}(0,0)-\frac{\partial^2 b_1}{\partial\overline{s}^2}(0,0)\,\frac{\partial^4 a_1}{\partial s^3\partial\overline{s}}(0,0)\right)\right]$$

$$-\left[\left(\frac{\partial a_1}{\partial\overline{s}}(0,0)\,\frac{\partial^3 b_1}{\partial\overline{s}\partial s^2}(0,0)-\frac{\partial a_1}{\partial s}(0,0)\,\frac{\partial^3 b_1}{\partial\overline{s}^2\partial s}(0,0)\right)\right.$$

$$+\left(\frac{\partial b_1}{\partial s}(0,0)\,\frac{\partial^3 a_1}{\partial\overline{s}^2\partial s}(0,0)-\frac{\partial b_1}{\partial\overline{s}}(0,0)\,\frac{\partial^3 a_1}{\partial\overline{s}\partial s^2}(0,0)\right)$$

$$\left.\left.+\frac{1}{2}\left(\frac{\partial^2 a_1}{\partial\overline{s}^2}(0,0)\,\frac{\partial^2 b_1}{\partial s^2}(0,0)-\frac{\partial^2 a_1}{\partial s^2}(0,0)\,\frac{\partial^2 b_1}{\partial\overline{s}^2}(0,0)\right)\right]\right\}+o(h^3).$$

We note that the last group of terms in h^3, starting with the sign '-', originates from the group in h^2 of (B.4.9) and (B.4.10) because of

$$\frac{1}{(H+1)(H+2)}=\frac{h^2}{1+h}=h^2-h^3+o(h^3).$$

For the function

$$c_1(s,\overline{s})=c\left(\frac{z-s}{1-\overline{z}s},\frac{\overline{z}-\overline{s}}{1-z\overline{s}}\right)$$

calculate

$$\frac{\partial c_1}{\partial s}=\frac{\partial c}{\partial w}\cdot(-1)\frac{1-z\overline{z}}{(1-\overline{z}s)^2},$$

$$\frac{\partial c_1}{\partial\overline{s}}=\frac{\partial c}{\partial\overline{w}}\cdot(-1)\frac{1-z\overline{z}}{(1-z\overline{s})^2},$$

$$\frac{\partial^2 c_1}{\partial s^2}=\frac{\partial^2 c}{\partial w^2}\cdot\frac{(1-z\overline{z})^2}{(1-\overline{z}s)^4}-2\overline{z}\frac{1-z\overline{z}}{(1-\overline{z}s)^3}\cdot\frac{\partial c}{\partial w},$$

$$\frac{\partial^2 c_1}{\partial\overline{s}^2}=\frac{\partial^2 c}{\partial\overline{w}^2}\cdot\frac{(1-z\overline{z})^2}{(1-z\overline{s})^4}-2z\frac{1-z\overline{z}}{(1-z\overline{s})^3}\cdot\frac{\partial c}{\partial\overline{w}},$$

$$\frac{\partial^3 c_1}{\partial \overline{s} \partial s^2} = -\frac{\partial^3 c}{\partial \overline{w} \partial w^2} \cdot \frac{(1 - z\overline{z})^3}{(1 - \overline{z}s)^4 (1 - z\overline{s})^2} + 2\overline{z}\, \frac{(1 - z\overline{z})^2}{(1 - \overline{z}s)^3 (1 - z\overline{s})^2} \cdot \frac{\partial^2 c}{\partial \overline{w} \partial w},$$

$$\frac{\partial^3 c_1}{\partial \overline{s}^2 \partial s} = -\frac{\partial^3 c}{\partial \overline{w}^2 \partial w} \cdot \frac{(1 - z\overline{z})^3}{(1 - \overline{z}s)^2 (1 - z\overline{s})^4} + 2z\, \frac{(1 - z\overline{z})^2}{(1 - \overline{z}s)^2 (1 - z\overline{s})^3} \cdot \frac{\partial^2 c}{\partial \overline{w} \partial w},$$

$$\frac{\partial^3 c_1}{\partial s^3} = -\frac{\partial^3 c}{\partial w^3} \cdot \frac{(1 - z\overline{z})^3}{(1 - \overline{z}s)^6} + 6\overline{z}\, \frac{(1 - z\overline{z})^2}{(1 - \overline{z}s)^5} \cdot \frac{\partial^2 c}{\partial w^2} - 6\overline{z}^2\, \frac{1 - z\overline{z}}{(1 - \overline{z}s)^4} \cdot \frac{\partial c}{\partial w},$$

$$\frac{\partial^3 c_1}{\partial \overline{s}^3} = -\frac{\partial^3 c}{\partial \overline{w}^3} \cdot \frac{(1 - z\overline{z})^3}{(1 - z\overline{s})^6} + 6z\, \frac{(1 - z\overline{z})^2}{(1 - z\overline{s})^5} \cdot \frac{\partial^2 c}{\partial \overline{w}^2} - 6z^2\, \frac{1 - z\overline{z}}{(1 - z\overline{s})^4} \cdot \frac{\partial c}{\partial \overline{w}},$$

$$\frac{\partial^4 c_1}{\partial \overline{s} \partial s^3} = \frac{\partial^4 c}{\partial \overline{w} \partial w^3} \cdot \frac{(1 - z\overline{z})^4}{(1 - \overline{z}s)^6 (1 - z\overline{s})^2} - 6\overline{z}\, \frac{(1 - z\overline{z})^3}{(1 - \overline{z}s)^5 (1 - z\overline{s})^2} \cdot \frac{\partial^3 c}{\partial \overline{w} \partial w^2}$$
$$- 6\overline{z}^2\, \frac{(1 - z\overline{z})^2}{(1 - \overline{z}s)^4 (1 - z\overline{s})^2} \cdot \frac{\partial^2 c}{\partial \overline{w} \partial w},$$

$$\frac{\partial^4 c_1}{\partial \overline{s}^3 \partial s} = \frac{\partial^4 c}{\partial \overline{w}^3 \partial w} \cdot \frac{(1 - z\overline{z})^4}{(1 - \overline{z}s)^2 (1 - z\overline{s})^6} - 6z\, \frac{(1 - z\overline{z})^3}{(1 - \overline{z}s)^2 (1 - z\overline{s})^5} \cdot \frac{\partial^3 c}{\partial \overline{w}^2 \partial w}$$
$$- 6z^2\, \frac{(1 - z\overline{z})^2}{(1 - \overline{z}s)^2 (1 - z\overline{s})^4} \cdot \frac{\partial^2 c}{\partial \overline{w} \partial w},$$

$$\frac{\partial^5 c_1}{\partial \overline{s}^2 \partial s^3} = -\frac{\partial^5 c}{\partial \overline{w}^2 \partial w^3} \cdot \frac{(1 - z\overline{z})^5}{(1 - \overline{z}s)^6 (1 - z\overline{s})^4} + \frac{\partial^4 c}{\partial \overline{w} \partial w^3} \cdot \frac{2z\,(1 - z\overline{z})^4}{(1 - \overline{z}s)^6 (1 - z\overline{s})^3}$$
$$- 6\overline{z}\, \frac{(1 - z\overline{z})^4}{(1 - \overline{z}s)^5 (1 - z\overline{s})^4} \cdot \frac{\partial^4 c}{\partial \overline{w}^2 \partial w^2} - 12 z\overline{z}\, \frac{(1 - z\overline{z})^3}{(1 - \overline{z}s)^5 (1 - z\overline{s})^3} \cdot \frac{\partial^3 c}{\partial \overline{w} \partial w^2}$$
$$- 6\overline{z}^2\, \frac{(1 - z\overline{z})^3}{(1 - \overline{z}s)^4 (1 - z\overline{s})^4} \cdot \frac{\partial^3 c}{\partial \overline{w}^2 \partial w} + 12 z\overline{z}^2\, \frac{(1 - z\overline{z})^2}{(1 - \overline{z}s)^4 (1 - z\overline{s})^3} \cdot \frac{\partial^2 c}{\partial \overline{w} \partial w},$$

$$\frac{\partial^5 c_1}{\partial \overline{s}^3 \partial s^2} = -\frac{\partial^5 c}{\partial \overline{w}^3 \partial w^2} \cdot \frac{(1 - z\overline{z})^5}{(1 - \overline{z}s)^4 (1 - z\overline{s})^6} + \frac{\partial^4 c}{\partial \overline{w}^3 \partial w} \cdot \frac{2\overline{z}\,(1 - z\overline{z})^4}{(1 - \overline{z}s)^3 (1 - z\overline{s})^6}$$
$$- 6z\, \frac{(1 - z\overline{z})^4}{(1 - \overline{z}s)^4 (1 - z\overline{s})^5} \cdot \frac{\partial^4 c}{\partial \overline{w}^2 \partial w^2} - 12 z\overline{z}\, \frac{(1 - z\overline{z})^3}{(1 - \overline{z}s)^3 (1 - z\overline{s})^5} \cdot \frac{\partial^3 c}{\partial \overline{w}^2 \partial w}$$
$$- 6z^2\, \frac{(1 - z\overline{z})^3}{(1 - \overline{z}s)^4 (1 - z\overline{s})^4} \cdot \frac{\partial^3 c}{\partial \overline{w} \partial w^2} + 12 z^2\overline{z}\, \frac{(1 - z\overline{z})^2}{(1 - \overline{z}s)^3 (1 - z\overline{s})^4} \cdot \frac{\partial^2 c}{\partial \overline{w} \partial w},$$

and

$$\frac{\partial c_1}{\partial s}(0,0) = -(1 - z\overline{z})\, \frac{\partial c}{\partial z}(z, \overline{z}),$$

$$\frac{\partial c_1}{\partial \overline{s}}(0,0) = -(1 - z\overline{z})\, \frac{\partial c}{\partial \overline{z}}(z, \overline{z}),$$

$$\frac{\partial^2 c_1}{\partial s^2}(0,0) = (1 - z\overline{z})^2\, \frac{\partial^2 c}{\partial z^2}(z, \overline{z}) - 2\overline{z}(1 - z\overline{z})\, \frac{\partial c}{\partial z}(z, \overline{z}),$$

$$\frac{\partial^2 c_1}{\partial \overline{s}^2}(0,0) = (1 - z\overline{z})^2\, \frac{\partial^2 c}{\partial \overline{z}^2}(z, \overline{z}) - 2z(1 - z\overline{z})\, \frac{\partial c}{\partial \overline{z}}(z, \overline{z}),$$

$$\frac{\partial^3 c_1}{\partial \bar{s}\partial s^2}(0,0) = -(1-z\bar{z})^3 \frac{\partial^3 c}{\partial \bar{z}\partial z^2}(z,\bar{z}) + 2\bar{z}(1-z\bar{z})^2 \frac{\partial^2 c}{\partial \bar{z}\partial z}(z,\bar{z}),$$

$$\frac{\partial^3 c_1}{\partial \bar{s}^2\partial s}(0,0) = -(1-z\bar{z})^3 \frac{\partial^3 c}{\partial \bar{z}^2\partial z}(z,\bar{z}) + 2z(1-z\bar{z})^2 \frac{\partial^2 c}{\partial \bar{z}\partial z}(z,\bar{z}),$$

$$\frac{\partial^3 c_1}{\partial s^3}(0,0) = -(1-z\bar{z})^3 \frac{\partial^3 c}{\partial z^3} + 6\bar{z}(1-z\bar{z})^2 \frac{\partial^2 c}{\partial z^2} - 6\bar{z}^2(1-z\bar{z})\frac{\partial c}{\partial z},$$

$$\frac{\partial^3 c_1}{\partial \bar{s}^3}(0,0) = -(1-z\bar{z})^3 \frac{\partial^3 c}{\partial \bar{z}^3} + 6z(1-z\bar{z})^2 \frac{\partial^2 c}{\partial \bar{z}^2} - 6z^2(1-z\bar{z})\frac{\partial c}{\partial \bar{z}},$$

$$\frac{\partial^4 c_1}{\partial \bar{s}\partial s^3}(0,0) = (1-z\bar{z})^4 \frac{\partial^4 c}{\partial \bar{z}\partial z^3} - 6\bar{z}(1-z\bar{z})^3 \frac{\partial^3 c}{\partial \bar{z}\partial z^2} + 6\bar{z}^2(1-z\bar{z})^2 \frac{\partial^2 c}{\partial \bar{z}\partial z},$$

$$\frac{\partial^4 c_1}{\partial \bar{s}^3\partial s}(0,0) = (1-z\bar{z})^4 \frac{\partial^4 c}{\partial \bar{z}^3\partial z} - 6z(1-z\bar{z})^3 \frac{\partial^3 c}{\partial \bar{z}^2\partial z} + 6z^2(1-z\bar{z})^2 \frac{\partial^2 c}{\partial \bar{z}\partial z},$$

$$\begin{aligned}
\frac{\partial^5 c_1}{\partial \bar{s}^2\partial s^3}(0,0) = {}& -(1-z\bar{z})^5 \frac{\partial^5 c}{\partial \bar{z}^2\partial z^3} + 2z(1-z\bar{z})^4 \frac{\partial^4 c}{\partial \bar{z}\partial z^3} \\
&+ 6\bar{z}(1-z\bar{z})^4 \frac{\partial^4 c}{\partial \bar{z}^2\partial z^2} - 12z\bar{z}(1-z\bar{z})^3 \frac{\partial^3 c}{\partial \bar{z}\partial z^2} \\
&- 6\bar{z}^2(1-z\bar{z})^3 \frac{\partial^3 c}{\partial \bar{z}^2\partial z} + 12z\bar{z}^2(1-z\bar{z})^2 \frac{\partial^2 c}{\partial \bar{z}\partial z},
\end{aligned}$$

$$\begin{aligned}
\frac{\partial^5 c_1}{\partial \bar{s}^3\partial s^2}(0,0) = {}& -(1-z\bar{z})^5 \frac{\partial^5 c}{\partial \bar{z}^3\partial z^2} + 2\bar{z}(1-z\bar{z})^4 \frac{\partial^4 c}{\partial \bar{z}^3\partial z} \\
&+ 6z(1-z\bar{z})^4 \frac{\partial^4 c}{\partial \bar{z}^2\partial z^2} - 12z\bar{z}(1-z\bar{z})^3 \frac{\partial^3 c}{\partial \bar{z}^2\partial z} \\
&- 6z^2(1-z\bar{z})^3 \frac{\partial^3 c}{\partial \bar{z}\partial z^2} + 12z^2\bar{z}(1-z\bar{z})^2 \frac{\partial^2 c}{\partial \bar{z}\partial z}.
\end{aligned}$$

We start with the two-term asymptotic expansion of the commutator. Substituting the first six expressions of the above derivatives into (B.4.12) we have

$$\begin{aligned}
(\widetilde{a}_h \star \widetilde{b}_h - \widetilde{b}_h \star \widetilde{a}_h)(z,\bar{z}) = {}& h\,(1-z\bar{z})^2 \left(\frac{\partial a}{\partial \bar{z}}\frac{\partial b}{\partial z} - \frac{\partial a}{\partial z}\frac{\partial b}{\partial \bar{z}} \right) \\
&+ h^2 \left[(1-z\bar{z})^4 \left[\left(\frac{\partial a}{\partial \bar{z}}\frac{\partial^3 b}{\partial \bar{z}\partial z^2} - \frac{\partial a}{\partial z}\frac{\partial^3 b}{\partial \bar{z}^2\partial z} \right) \right.\right. \\
&\left.\left. + \left(\frac{\partial b}{\partial z}\frac{\partial^3 a}{\partial \bar{z}^2\partial z} - \frac{\partial b}{\partial \bar{z}}\frac{\partial^3 a}{\partial \bar{z}\partial z^2} \right) + \frac{1}{2}\left(\frac{\partial^2 a}{\partial \bar{z}^2}\frac{\partial^2 b}{\partial z^2} - \frac{\partial^2 a}{\partial z^2}\frac{\partial^2 b}{\partial \bar{z}^2} \right) \right] \right. \\
&\left. - 2(1-z\bar{z})^3 \left[\bar{z}\left(\frac{\partial a}{\partial \bar{z}}\frac{\partial^2 b}{\partial \bar{z}\partial z} - \frac{\partial b}{\partial \bar{z}}\frac{\partial^2 a}{\partial \bar{z}\partial z} \right) \right.\right. \\
&\left.\left. + z\left(\frac{\partial b}{\partial z}\frac{\partial^2 a}{\partial \bar{z}\partial z} - \frac{\partial a}{\partial z}\frac{\partial^2 b}{\partial \bar{z}\partial z} \right) \right. \right.
\end{aligned}$$

$$+ \ \frac{1}{2}\left(\bar{z}\left(\frac{\partial b}{\partial z}\frac{\partial^2 a}{\partial \bar{z}^2} - \frac{\partial a}{\partial z}\frac{\partial^2 b}{\partial \bar{z}^2} \right) + z\left(\frac{\partial a}{\partial \bar{z}}\frac{\partial^2 b}{\partial z^2} - \frac{\partial b}{\partial \bar{z}}\frac{\partial^2 a}{\partial z^2} \right) \right) \Big]$$

$$+ \ 2z\bar{z}(1 - z\bar{z})^2 \left(\frac{\partial a}{\partial \bar{z}}\frac{\partial b}{\partial z} - \frac{\partial a}{\partial z}\frac{\partial b}{\partial \bar{z}} \right) \Big] + o(h^2).$$

Recall (see (B.2.2)) that the Poisson bracket on the unit disk is given by

$$\{a, b\} \ = \ \pi(1 - (x^2 + y^2))^2 \left(\frac{\partial a}{\partial y}\frac{\partial b}{\partial x} - \frac{\partial a}{\partial x}\frac{\partial b}{\partial y} \right)$$

$$= \ 2\pi i(1 - z\bar{z})^2 \left(\frac{\partial a}{\partial z}\frac{\partial b}{\partial \bar{z}} - \frac{\partial a}{\partial \bar{z}}\frac{\partial b}{\partial z} \right),$$

and that the Laplace-Beltrami operator (B.2.3) has the form

$$\Delta \ = \ \pi(1 - (x^2 + y^2))^2 \left(\frac{\partial^2}{\partial x^2} + \frac{\partial^2}{\partial y^2} \right)$$

$$= \ 4\pi(1 - z\bar{z})^2 \frac{\partial^2}{\partial z \partial \bar{z}}.$$

Observe now that

$$\frac{i}{8\pi^2}\Delta\{a, b\} \ = \ (1 - z\bar{z})^2 \frac{\partial}{\partial z}\left(\frac{\partial}{\partial \bar{z}}(1 - z\bar{z})^2 \left(\frac{\partial a}{\partial \bar{z}}\frac{\partial b}{\partial z} - \frac{\partial b}{\partial \bar{z}}\frac{\partial a}{\partial z} \right) \right)$$

$$= \ (1 - z\bar{z})^2 \frac{\partial}{\partial z}\left(-2z(1 - z\bar{z})\left(\frac{\partial a}{\partial \bar{z}}\frac{\partial b}{\partial z} - \frac{\partial b}{\partial \bar{z}}\frac{\partial a}{\partial z} \right) \right.$$

$$+ \ (1 - z\bar{z})^2 \left(\frac{\partial^2 a}{\partial \bar{z}^2}\frac{\partial b}{\partial z} + \frac{\partial a}{\partial \bar{z}}\frac{\partial^2 b}{\partial \bar{z}\partial z} - \frac{\partial^2 b}{\partial \bar{z}^2}\frac{\partial a}{\partial z} - \frac{\partial b}{\partial \bar{z}}\frac{\partial^2 a}{\partial \bar{z}\partial z} \right) \right)$$

$$= \ (1 - z\bar{z})^2 \left(-2(1 - z\bar{z})\left(\frac{\partial a}{\partial \bar{z}}\frac{\partial b}{\partial z} - \frac{\partial b}{\partial \bar{z}}\frac{\partial a}{\partial z} \right) \right.$$

$$+ 2z\bar{z}\left(\frac{\partial a}{\partial \bar{z}}\frac{\partial b}{\partial z} - \frac{\partial b}{\partial \bar{z}}\frac{\partial a}{\partial z} \right)$$

$$- 2z(1 - z\bar{z})\left(\frac{\partial^2 a}{\partial \bar{z}\partial z}\frac{\partial b}{\partial z} + \frac{\partial a}{\partial \bar{z}}\frac{\partial^2 b}{\partial z^2} - \frac{\partial^2 b}{\partial \bar{z}\partial z}\frac{\partial a}{\partial z} - \frac{\partial b}{\partial \bar{z}}\frac{\partial^2 a}{\partial z^2} \right)$$

$$- 2\bar{z}(1 - z\bar{z})\left(\frac{\partial^2 a}{\partial \bar{z}^2}\frac{\partial b}{\partial z} + \frac{\partial a}{\partial \bar{z}}\frac{\partial^2 b}{\partial \bar{z}\partial z} - \frac{\partial^2 b}{\partial \bar{z}^2}\frac{\partial a}{\partial z} - \frac{\partial b}{\partial \bar{z}}\frac{\partial^2 a}{\partial \bar{z}\partial z} \right)$$

$$+ (1 - z\bar{z})^2 \left(\frac{\partial^3 a}{\partial \bar{z}^2\partial z}\frac{\partial b}{\partial z} + \frac{\partial^2 a}{\partial \bar{z}^2}\frac{\partial^2 b}{\partial z^2} + \frac{\partial^2 a}{\partial \bar{z}\partial z}\frac{\partial^2 b}{\partial \bar{z}\partial z} + \frac{\partial a}{\partial \bar{z}}\frac{\partial^3 b}{\partial \bar{z}\partial z^2} \right.$$

$$\left. \left. - \ \frac{\partial^3 b}{\partial \bar{z}^2\partial z}\frac{\partial a}{\partial z} - \frac{\partial^2 b}{\partial \bar{z}^2}\frac{\partial^2 a}{\partial z^2} - \frac{\partial^2 b}{\partial \bar{z}\partial z}\frac{\partial^2 a}{\partial \bar{z}\partial z} - \frac{\partial b}{\partial \bar{z}}\frac{\partial^3 a}{\partial \bar{z}\partial z^2} \right) \right)$$

$$
= (1 - z\overline{z})^4 \left[\left(\frac{\partial^2 a}{\partial \overline{z}^2} \frac{\partial^2 b}{\partial z^2} - \frac{\partial^2 b}{\partial \overline{z}^2} \frac{\partial^2 a}{\partial z^2} \right) \right.
$$

$$
\left. + \left(\frac{\partial a}{\partial \overline{z}} \frac{\partial^3 b}{\partial \overline{z} \partial z^2} - \frac{\partial a}{\partial z} \frac{\partial^3 b}{\partial \overline{z}^2 \partial z} \right) + \left(\frac{\partial b}{\partial z} \frac{\partial^3 a}{\partial \overline{z}^2 \partial z} - \frac{\partial b}{\partial \overline{z}} \frac{\partial^3 a}{\partial \overline{z} \partial z^2} \right) \right]
$$

$$
- 2(1 - z\overline{z})^3 \left[z \left(\left(\frac{\partial a}{\partial \overline{z}} \frac{\partial^2 b}{\partial z^2} - \frac{\partial b}{\partial \overline{z}} \frac{\partial^2 a}{\partial z^2} \right) \right. \right.
$$

$$
\left. + \left(\frac{\partial b}{\partial z} \frac{\partial^2 a}{\partial \overline{z} \partial z} - \frac{\partial a}{\partial z} \frac{\partial^2 b}{\partial \overline{z} \partial z} \right) \right)
$$

$$
\left. + \overline{z} \left(\left(\frac{\partial^2 a}{\partial \overline{z}^2} \frac{\partial b}{\partial z} - \frac{\partial^2 b}{\partial \overline{z}^2} \frac{\partial a}{\partial z} \right) + \left(\frac{\partial a}{\partial \overline{z}} \frac{\partial^2 b}{\partial \overline{z} \partial z} - \frac{\partial b}{\partial \overline{z}} \frac{\partial^2 a}{\partial \overline{z} \partial z} \right) \right) \right]
$$

$$
+ 4 z\overline{z}(1 - z\overline{z})^2 \left(\frac{\partial a}{\partial \overline{z}} \frac{\partial b}{\partial z} - \frac{\partial b}{\partial \overline{z}} \frac{\partial a}{\partial z} \right)
$$

$$
- 2(1 - z\overline{z})^2 \left(\frac{\partial a}{\partial \overline{z}} \frac{\partial b}{\partial z} - \frac{\partial b}{\partial \overline{z}} \frac{\partial a}{\partial z} \right),
$$

and

$$
\frac{i}{8\pi^2} \left(\{a, \Delta b\} + \{\Delta a, b\} \right)
$$

$$
= (1 - z\overline{z})^4 \left[\left(\frac{\partial a}{\partial \overline{z}} \frac{\partial^3 b}{\partial \overline{z} \partial z^2} - \frac{\partial a}{\partial z} \frac{\partial^3 b}{\partial \overline{z}^2 \partial z} \right) + \left(\frac{\partial b}{\partial z} \frac{\partial^3 a}{\partial \overline{z}^2 \partial z} - \frac{\partial b}{\partial \overline{z}} \frac{\partial^3 a}{\partial \overline{z} \partial z^2} \right) \right]
$$

$$
- 2(1 - z\overline{z})^3 \left[z \left(\frac{\partial b}{\partial z} \frac{\partial^2 a}{\partial \overline{z} \partial z} - \frac{\partial a}{\partial z} \frac{\partial^2 b}{\partial \overline{z} \partial z} \right) + \overline{z} \left(\frac{\partial a}{\partial \overline{z}} \frac{\partial^2 b}{\partial \overline{z} \partial z} - \frac{\partial b}{\partial \overline{z}} \frac{\partial^2 a}{\partial \overline{z} \partial z} \right) \right].
$$

Thus we have finally

$$
\widetilde{a}_h \star \widetilde{b}_h - \widetilde{b}_h \star \widetilde{a}_h = \frac{ih}{2\pi} \{a, b\} + \frac{h^2}{2} \left(\frac{i}{8\pi^2} \Delta\{a, b\} + \frac{i}{8\pi^2} \{a, \Delta b\} + \frac{i}{8\pi^2} \{\Delta a, b\} \right.
$$

$$
\left. + \frac{i}{\pi} \{a, b\} \right) + o(h^2)
$$

$$
= i\hbar \{a, b\} i \frac{\hbar^2}{4} \left(\Delta\{a, b\} + \{a, \Delta b\} + \{\Delta a, b\} + 8\pi\{a, b\} \right)
$$

$$
+ o(\hbar^2), \tag{B.4.13}
$$

where the Poisson bracket $\{\,,\,\}$ and the Laplace-Beltrami operator Δ in coordinates $(z, \overline{z})$ are given by (B.2.2) and by (B.2.3), respectively.

Note that, in particular, one has

$$
(\widetilde{a}_h \star \widetilde{b}_h - \widetilde{b}_h \star \widetilde{a}_h)(z, \overline{z}) = i\hbar \{a, b\} + O(\hbar^2). \tag{B.4.14}
$$

The three-term asymptotic expansion of the commutator is obtained by substituting of the expressions for partial derivatives of the function c_1 calculated at the origin to (B.4.12) and by doing the symbolic calculations in MatLab 6.5.

The final result is as follows:

$$\widetilde{a}_h \star \widetilde{b}_h - \widetilde{b}_h \star \widetilde{a}_h$$

$$= \frac{ih}{2\pi}\{a,b\} + \frac{h^2}{2}\left[\frac{i}{8\pi^2}\left(\Delta\{a,b\} + \{a,\Delta b\} + \{\Delta a,b\}\right) + \frac{i}{\pi}\{a,b\}\right]$$

$$+ h^3\left[\frac{i}{192\pi^3}\left(\{\Delta a,\Delta b\} + \{a,\Delta^2 b\} + \{\Delta^2 a,b\}\right.\right.$$

$$\left. + \Delta^2\{a,b\} + \Delta\{a,\Delta b\} + \Delta\{\Delta a,b\}\right)$$

$$\left. + \frac{7i}{48\pi^2}\left(\Delta\{a,b\} + \{a,\Delta b\} + \{\Delta a,b\}\right) + \frac{i}{2\pi}\{a,b\}\right] + o(h^3)$$

$$= i\hbar\{a,b\} + i\frac{\hbar^2}{4}\left(\Delta\{a,b\} + \{a,\Delta b\} + \{\Delta a,b\} + 8\pi\{a,b\}\right)$$

$$+ i\frac{\hbar^3}{24}\left[\{\Delta a,\Delta b\} + \{a,\Delta^2 b\} + \{\Delta^2 a,b\} + \Delta^2\{a,b\}\right.$$

$$+ \Delta\{a,\Delta b\} + \Delta\{\Delta a,b\} + 28\pi\left(\Delta\{a,b\} + \{a,\Delta b\} + \{\Delta a,b\}\right)$$

$$\left. + 96\pi^2\{a,b\}\right] + o(\hbar^3),$$

where the Poisson bracket $\{\,,\,\}$ and the Laplace-Beltrami operator Δ in coordinates $(z,\bar{z})$ are given by (B.2.2) and by (B.2.3), respectively.

Bibliographical Remarks

Chapter 1. Preliminaries

The notion of locally equivalent operators and localization theory were introduced and developed by I. Simonenko in [177, 178]. Without pretending to be complete we mention as well different approaches to the local principle by G. R. Allan [7, 8], R. Douglas [58, 59], I. Gohberg and N. Krupnik [85], V. Pilidi [159], S. Roch and B. Silbermann [169]. We mention as well the works by J. Dauns and K. H. Hofmann [54], M. J. Dupré [60], J. M. G. Fell [77], K. H. Hofmann [103], M. Takesaki and J. Tomiyama [195], J. Varela [197], and N. B. Vasiliev [218] dedicated to the description of algebras and rings in terms of continuous sections. These two directions have been developing independently, and there have been no links between these two series of papers. Our approach is based on general constructions of [54, 103] and results of [197]. It has been presented in a short note [204], as a part of Section 2 in [205], and in [198].

The description of the C^*-algebra generated by two orthogonal projections (or by two self-adjoint idempotents) was understood in the last third of the previous century by many authors (see, for example, S. Borac [28], R. Douglas [57], T. Fink, S. Roch, and B. Silbermann [78], P. Halmos [98], G. K. Pedersen [153], I. Raeburn and A. M. Sinclair [165], I. Vaĭs [196], N. Vasilevski [200], N. Vasilevski and I. Spitkovsky [216], Y. Weiss [220]).

We follow here the papers [198, 200]. The C^*-algebra generated by more than two orthogonal projections is wild in general. At the same time under some special relations among the projections the algebra can be described (see, for example, A. Böttcher, I. Gohberg, Yu. Karlovich, N. Krupnik, S. Roch, B. Silbermann, and I. Spitkovsky [29], Yu. I. Karlovich and L. Pessoa [114], Yu. P. Moskaleva and Yu. S. Samoĭlenko [147], V. S. Sunder [194], N. L. Vasilevski [206]). Studying the C^*-algebra generated by more than two orthogonal projections we follow [206].

Chapter 2. Prologue

There are many excellent sources where the Bergman space and the Bergman projection are treated. We mention here the book [26] of S. Bergman himself and the books by R. Courant [45], A. Dzhuraev [64], H. Hedenmalm, B. Korenblum, and

K. Zhu [102], W. Rudin [170], K. Zhu [240]. Describing the Bergman space, the Bergman kernel function, and the Bergman projection we follow mainly A. Dzhuraev [64]. The properties of two-dimensional integral operators mentioned in Section 2.4 are taken from N. Vekua [219]. The description of the algebra generated by the Bergman projection and piece-wise continuous coefficients is taken from the author's papers [202, 203]. The irreducibility of the algebra genetated by Toeplitz operators with continuous defining symbols and the structure of this algebra were established by L. Coburn [42]. Toeplitz operators with piece-wise continuous defining symbols were treated by G. McDonald [141] and N. Vasilevski [200, 201]. The criterion of compactness of Toeplitz operators with bounded defining symbols was obtained, among other results, by S. Axler and D. Zheng [12].

Chapter 3. Bergman and Poly-Bergman Spaces

Studying middle Hankel type operators Q. Jiang and L. Peng [108] obtained a direct sum decomposition of the weighted L_2 space on the upper half-plane. Their study was based on the admissible wavelet technique. Our approach to a direct sum decomposition of the L_2 space onto the poly-Bergman type spaces is based on the author's papers [208, 212]. The definitions of the poly-Bergman type spaces are taken from M. B. Balk [17], M. B. Balk and M. F. Zuev [18], A. Dzhuraev [63, 64]. A connection between the two-dimensional singular integral operators and the poly-Bergman type projections were studied by A. Dzhuraev [63, 64], J. Ramírez and I. Spitkovsky [167], Yu. I. Karlovich and L. Pessoa [115], N. Vasilevski [215]. Our presentation is based on [215].

Chapter 4. Bergman Type Spaces on the Unit Disk

The decomposition of the space $L_2(\mathbb{D})$ into a direct sum of Bergman type spaces was obtained by L. Peng, R. Rochberg, and Z. Wu in [154]. Our presentation is based on [211, 212]. See also the related papers by J. He and L. Peng [101], L. Peng and C. X. Xu [155], L. Peng and G. Zhang [156].

Chapter 5. Toeplitz Operators with Commutative Symbol Algebras

We follow here the author's papers [207, 212]. The maximal C^*-algebra of defining symbols for which the semi-commutator of two Toeplitz operators is compact was described by D. Sarason [172, 173] for the Hardy space setting and by K. Zhu [237, 240] for the Bergman space setting.

Chapter 6. Toeplitz Operators on the Unit Disk with Radial Symbols

Toeplitz operators with bounded radial defining symbols were studied by B. Korenblum and K. Zhu in [120]. They found two important properties of such operators: the diagonal form of Toeplitz operators with respect to the standard monomial basis in $\mathcal{A}^2(\mathbb{D})$ and the criterion for their compactness. Different aspects of the theory of Toeplitz operators with radial defining symbols were studied in the papers by S. Grudsky, A. Karapetyants, and N. Vasilevski [87, 90] S. Grudsky and N. Vasilevski [92], A. Karapetyants and A. Golikov [112], Y. Lu [134], K. Stroethoff [181], N. Vasilevski [211, 212], Q. Xu [224], N. Zorboska [242] in the context of the Bergman space, in Y. Lee [125], J. Miao [144], K. Stroethoff [182] in the context of the harmonic Bergman space, and in S. Grudsky and N. Vasilevski [95], Sangadji and K. Stroethoff [171] in the context of the Fock space. In this chapter we follow the papers by S. Grudsky and N. Vasilevski [92] and N. Vasilevski [212].

Chapter 7. Toeplitz Operators with Homogeneous Symbols

The commutativity property for Toeplitz operators with defining symbols homogeneous on the upper half-plane was established by N. Vasilevski [214]. The C^*-algebra generated by the Bergman projection and the multiplication operators by piece-wise continuous functions having more than two limit values at the boundary points of discontinuity was studied by M. Loaiza [131]. Toeplitz operators with defining symbols having more than two limit values at the boundary points of discontinuity were studied by M. Loaiza [132]. Describing the above we use a distinct approach; some facts presented in this chapter have never been published previously. An analysis of Toeplitz operators with defining symbols homogeneous of zero order was done by N. Vasilevski [199].

Chapter 8. Anatomy of the Algebra Generated by Toeplitz Operators with Piece-wise Continuous Symbols

In this chapter we follow the papers by S. Grudsky and N. Vasilevski [93, 94]. The necessary information on pseudodifferential operators with compound symbols and Theorem 8.6.4 are taken from Yu. Karlovich [113]. Theorem 8.7.2 is taken from V. Dybin and S. Grudsky [61].

Chapter 9. Commuting Toeplitz Operators and Hyperbolic Geometry

This chapter is based on the paper [214]. The notion of the Bergman metric and its elementary properties are taken from the books by S. Bergman [26], B. V. Shabat [176], K. Zhu [240]. Describing Möbius transformations, their properties and required facts from the hyperbolic geometry we follow the book by A. F. Beardon [21].

Chapter 10. Weighted Bergman Spaces

The definition and basic properties of weighted Bergman spaces parameterized by $\lambda \in (-1, \infty)$ and the corresponding Bergman projections are taken from the books

by H. Hedenmalm, B. Korenblum, and K. Zhu [102], K.Zhu [240]. The description
of weighted Bergman spaces parameterized by $h \in (0,1)$ is taken from the book
by F. Berezin [25]. A majority of the results of Sections 10.3 and 10.4 are taken
from the papers by S. Grudsky, A. Karapetyants, and N. Vasilevski [88, 89].

Chapter 11. Commutative Algebras of Toeplitz Operators

We follow here the paper by S. Grudsky, R. Quiriga-Barranco, and N. Vasilevski
[91]. The idea of considering $T_z^{(\lambda)} T_{\bar{z}}^{(\lambda)}$ in the proof of Lemma 11.2.1 was inspired
by the L. Coburn paper [44]. The general concept of quantization and the quan-
tization on the unit disk was introduced and studied by F. Berezin in [23]. The
notion of jet and its elementary properties are taken from the books by I. Kolář,
P. Michor, and J. Slovák [119] and D. Saunders [174]. The definitions, properties
and results from differential geometry used in the chapter are taken from the books
by W. Klingenberg [118] and B. O'Nell [151].

Chapters 12, 13, and 14. Dynamics of Properties of Toeplitz Oper-
ators

In these chapters we follow the papers by S. Grudsky, A. Karapetyants, and
N. Vasilevski [90, 89, 88], respectively. See also [217]. The definition of the limit
set of a family of sets is taken from the book by R. Hagen, S. Roch, and B. Silbermann [97].

Appendix A. Coherent states and Berezin transform

The definition of coherent states and their properties are given, for example,
in the books by S. T. Ali, J.-P. Antoine, and J.-P. Gazeau [6], F. Berezin [25],
J. R. Klauder, B.-S. Skagerstam (editors) [117], A. M. Perelomov [158]. The notion of the numerical spectrum and its properties is taken from P. Halmos [99].
Berezin's definition of the Berezin transform can be found in F. Berezin [25]. The
properties of the Berezin transform are taken from the book by H. Hedenmalm,
B. Korenblum, and K. Zhu [102].

Appendix B. Berezin Quantization on the Unit Disk

The quantization procedure described in this chapter was introduced by F. Berezin
in [22, 24, 23, 25]; see also M. Englis [70]. For quantization on the unit disk see F.
Berezin's works [23, 25].

For the reader's benefit we have included in the Bibliography a substantial
number of papers devoted to aspects of the theory of Toeplitz operators which
are not necessarily connected with the main content of the book. We present our
honest apologies to those whose papers were unintentionally omitted.

We mention as well a number of papers devoted to multidimensional extensions of some results presented. The paper by S. Grudsky, A. Karapetyants, and N. Vasilevski [87] treats the Toeplitz operators with radial defining symbols on the unit ball, R. Quiroga-Barranco and N. Vasilevski [161] describes the commutative algebras of Toeplitz operators on the Reinhardt domains, R. Quiroga-Barranco and N. Vasilevski [162, 163] contain the classification of the commutative algebras of Toeplitz operators on the unit ball, and the author's papers [209, 210] study the commutative algebras of Toeplitz operators on the tube domains.

Bibliography

[1] M. Abramowitz, I. Stegun. *Handbook of mathematical functions: with formulas, graphs, and mathematical tables.* New York : Dover Publications, 1965.

[2] P. Ahern. On the range of the Berezin transform. *J. Funct. Anal.,* 215(1):206–216, 2004.

[3] P. Ahern, Ž. Čučković. Products of Toeplitz operators on the Bergman space. *Illinois J. Math.,* 45(1):113–121, 2001.

[4] P. Ahern, Ž. Čučković. A theorem of Brown-Halmos type for Bergman space Toeplitz operators. *J. Funct. Anal.,* 187(1):200–210, 2001.

[5] P. Ahern, Ž. Čučković. Some examples related to the Brown-Halmos theorem for the Bergman space. *Acta Sci. Math. (Szeged),* 70(1-2):373–378, 2004.

[6] S. T. Ali, J.-P. Antoine, and J.-P. Gazeau. *Coherent states, wavelets and their generalizations.* Graduate Texts in Contemporary Physics. Springer-Verlag, New York, 2000.

[7] G. R. Allan. On one-sided inverses in Banach algebras of holomorphic vector valued functions. *J. London Math. Soc.,* 42(3):463–470, 1967.

[8] G. R. Allan. Ideals of vector valued functions. *Proc. London Math. Soc.,* 18:193–216, 1968.

[9] H. B. An, R. Y. Jian. Slant Toeplitz operators on Bergman spaces. *Acta Math. Sinica (Chin. Ser.),* 47(1):103–110, 2004.

[10] S. Axler, J. Conway, and G. McDonald. Toeplitz operators on Bergman spaces. *Can. J. Math.,* 34:466–483, 1982.

[11] S. Axler, D. Zheng. The Berezin transform on the Toeplitz algebra. *Stud. Math.,* 127(2):113–136, 1998.

[12] S. Axler, D. Zheng. Compact operators via the Berezin transform. *Indiana Univ. Math. J.,* 47(2):387–400, 1998.

[13] S. Axler. Bergman spaces and their operators. In *Surveys of some recent results in operator theory, Vol. I*, volume 171 of *Pitman Res. Notes Math. Ser.*, pages 1–50. Longman Sci. Tech., Harlow, 1988.

[14] S. Axler, Ž. Čučković. Commuting Toeplitz operators with harmonic symbols. *Integral Equations Operator Theory*, 14(1):1–12, 1991.

[15] S. Axler, Ž. Čučković, and N. V. Rao. Commutants of analytic Toeplitz operators on the Bergman space. *Proc. Amer. Math. Soc.*, 128(7):1951–1953, 2000.

[16] S. Axler, D. Zheng. Toeplitz algebras on the disk. *J. Funct. Anal.*, 243(1):67–86, 2007.

[17] M. B. Balk. *Polyanalytic functions and their generalizations, Complex analysis I. Encycl. Math. Sci*, volume 85, pages 195–253. Springer Verlag, 1997.

[18] M. B. Balk, M. F. Zuev. On polyanalytic functions. *Russ. Math. Surveys*, 25(5):201–223, 1970.

[19] H. Bateman. *Table of Integral Transformas, Vol. I*. McGraw-Hill, 1954.

[20] H. Bateman, A. Erdélyi. *Higher transcendental functions, Vol. 2*. McGraw-Hill, 1954.

[21] A. F. Beardon. *The Geometry of Discrete Groups*. Springer-Verlag, Berlin etc., 1983.

[22] F. A. Berezin. Quantization. *Math. USSR Izvestija*, 8:1109–1165, 1974.

[23] F. A. Berezin. General concept of quantization. *Commun. Math. Phys.*, 40:135–174, 1975.

[24] F. A. Berezin. Quantization in complex symmetric spaces. *Math. USSR Izvestija*, 9:341–379, 1975.

[25] F. A. Berezin. *Method of Second Quantization*. "Nauka", Moscow, 1988.

[26] S. Bergman. *The Kernel Function and Conformal Mapping*. American Mathematical Society, New York, 1950.

[27] R. L. Bishop, B. O'Neill. Manifolds of negative curvature. *Trans. of AMS*, 145:1–49, 1969.

[28] S. Borac. On the algebra generated by two projections. *J. Math. Phys.*, 36(2):863–874, 1995.

[29] A. Böttcher, I. Gohberg, Yu. Karlovich, N. Krupnik, S. Roch, B. Silbermann, and I. Spitkovsky. Banach algebras generated by N idempotents and applications. In *Singular integral operators and related topics (Tel Aviv, 1995)*, volume 90 of *Oper. Theory Adv. Appl.*, pages 19–54. Birkhäuser, Basel, 1996.

[30] A. Böttcher, H. Wolf. Large sections of Bergman space Toeplitz operators with piecewise continuous symbols. *Math. Nachr.*, 156:129–155, 1992.

[31] C. Caratheodory. *Conformal representation.* Cambridge University Press, 1969.

[32] X. M. Chen, Q. X. Xu. Commutative relationship of Toeplitz operators on the Bergman space. *Northeast. Math. J.*, 9(4):427–434, 1993.

[33] B. R. Choe, Y. J. Lee. Compact Toeplitz operators with bounded symbols on the Bergman space. *J. Korean Math. Soc.*, 31(2):289–307, 1994.

[34] B. R. Choe, Y. J. Lee. Norm and essential norm estimates of Toeplitz operators on the Bergman space. *Commun. Korean Math. Soc.*, 11(4):937–958, 1996.

[35] B. R. Choe, Y. J. Lee. Commuting Toeplitz operators on the harmonic Bergman space. *Michigan Math. J.*, 46(1):163–174, 1999.

[36] B. R. Choe, Y. J. Lee. Commutants of analytic Toeplitz operators on the harmonic Bergman space. *Integral Equations Operator Theory*, 50(4):559–564, 2004.

[37] B. R. Choe, Y. J. Lee, and K. Na. Toeplitz operators on harmonic Bergman spaces. *Nagoya Math. J.*, 174:165–186, 2004.

[38] B. R. Choe, Y. J. Lee, K. Nam, and D. Zheng. Products of Bergman space Toeplitz operators on the polydisk. *Math. Ann.*, 337(2):295–316, 2007.

[39] B. R. Choe, K. S. Nam. Note on commuting Toeplitz operators on the pluriharmonic Bergman space. *J. Korean Math. Soc.*, 43(2):259–269, 2006.

[40] M.-D. Choi. A simple C^*-algebra generated by two finite–order unitaries. *Can. J. Math.*, 31(4):867–880, 1979.

[41] J. A. Cima, Ž. Čučković. Compact Toeplitz operators with unbounded symbols. *J. Operator Theory*, 53(2):431–440, 2005.

[42] L. A. Coburn. Singular integral operators and Toeplitz operators on odd spheres. *Indiana Univ. Math. J.*, 23(5):433–439, 1973.

[43] L. A. Coburn. Toeplitz operators, quantum mechanics, and mean oscillation in the Bergman metric. In *Operator theory: operator algebras and applications, Part 1 (Durham, NH, 1988)*, volume 51 of *Proc. Sympos. Pure Math.*, pages 97–104. Amer. Math. Soc., Providence, RI, 1990.

[44] L. A. Coburn. On the Berezin-Toeplitz calculus. *Proc. Amer. Math. Soc.*, 129(11):3331–3338, 2001.

[45] R. Courant. *Dirichlet's principle, conformal mapping, and minimal surfaces.* Springer-Verlag, New York-Heidelberg.

[46] Ž. Čučković. Commutants of Toeplitz operators on the Bergman space. *Pacific J. Math.*, 162(2):277–285, 1994.

[47] Ž. Čučković. Commuting Toeplitz operators on the Bergman space of an annulus. *Michigan Math. J.*, 43(2):355–365, 1996.

[48] Ž. Čučković. Berezin versus Mellin. *J. Math. Anal. Appl.*, 287(1):234–243, 2003.

[49] Ž. Čučković, D. S. Fan. Commutants of Toeplitz operators on the ball and annulus. *Glasgow Math. J.*, 37(3):303–309, 1995.

[50] Ž. Čučković, N. V. Rao. Mellin transform, monomial symbols, and commuting Toeplitz operators. *J. Funct. Anal.*, 154(1):195–214, 1998.

[51] J. Cuntz. Simple C^*-algebras generated by isometries. *Commun. Math. Phys.*, 57:173–185, 1977.

[52] R. E. Curto, P. S. Muhly. C^*-algebras of multiplication operators on Bergman spaces. *J. Funct. Anal.*, 64:315–329, 1985.

[53] N. Das. Norm of Toeplitz operators on the Bergman space. *Indian J. Pure Appl. Math.*, 33(2):255–267, 2002.

[54] J. Dauns, K. H. Hofmann. *Representation of Rings by Sections*, volume 83 of *Memoirs Amer. Math. Soc.* AMS, 1968.

[55] Ch. Davis. Generators of the ring of bounded operators. *Proc. Amer. Math. Soc.*, 6:970–972, 1955.

[56] J. Dixmier. *Les C^*-algebres et Leurs Représentations.* Gauthier-Villars, Paris, 1964.

[57] R. G. Douglas. Local Toeplitz operators. *Proc. London Math. Soc.*, 36:243–272, 1976.

[58] R. G. Douglas. *Banach Algebra Techniques in Operator Theory.* Academic Press, New York, 1972.

[59] Ronald G. Douglas. *Banach Algebra Techniques in the Theory of Toeplitz Operators.* AMS, Providence, Rhode Island, 1973.

[60] M.J. Dupré. The classification and structure of C^*-algebra bundles. *Memoirs AMS*, (222):1–77, 1979.

[61] V. Dybin, S. Grudsky. *Inrtoduction to the theory of Toeplitz operators with infinite index.* Birkhäuser Verlag, Basel, 2002.

[62] A. Dzhuraev. On the theory of systems of singular integral equations over a bounded plane domain. *Soviet Math. Dokl.*, 20(6):1178–1182, 1979.

[63] A. Dzhuraev. Multikernel functions of a domain, kernel operators, singular integral operators. *Soviet Math. Dokl.*, 32(1):251–253, 1985.

[64] A. Dzhuraev. *Methods of Singular Integral Equations.* Longman Scientific & Technical, 1992.

[65] N. Elias. Toeplitz operators on weighted Bergman spaces. *Integral Equations Operator Theory*, 11(3):310–331, 1988.

[66] M. Engliš. A note on Toeplitz operators on Bergman spaces. *Comment. Math. Univ. Carolin.*, 29(2), 1988.

[67] M. Engliš. Some density theorems for Toeplitz operators on Bergman spaces. *Czechoslovak Math. J.*, 40(115)(3):491–502, 1990.

[68] M. Engliš. Density of algebras generated by Toeplitz operator on Bergman spaces. *Ark. Mat.*, 30(2):227–243, 1992.

[69] M. Engliš. Functions invariant under the Berezin transform. *J. Funct. Anal.*, 121(1):233–254, 1994.

[70] M. Engliš. Berezin quantization and reproducing kernels on complex domains. *Trans. Amer. Math. Soc.*, 348(2):411–479, 1996.

[71] M. Engliš. Compact Toeplitz operators via the Berezin transform on bounded symmetric domains. *Integral Equations Operator Theory*, 33(4):426–455, 1999.

[72] M. English. Toeplitz operators and the Berezin transform on H^2. *Linear Algebra Appl.*, 223–224:171–204, 1995.

[73] N. Faour, S. Yousef. Toeplitz operators associated with Bergman and Hardy spaces. *Riv. Mat. Univ. Parma (4)*, 15:51–62, 1989.

[74] N. S. Faour. Toeplitz operators on Bergman and Hardy spaces. *Houston J. Math.*, 8(2):185–191, 1982.

[75] N. S. Faour. Toeplitz operators on Bergman spaces. *Rend. Circ. Mat. Palermo (2)*, 35(2):221–232, 1986.

[76] M .V. Fedoriuk. *Saddle-Point Mathod.* Nauka, Moskow, 1977.

[77] J.M.G. Fell. The structure of algebras of operator fields. *Acta Math.*, 106:233–280, 1961.

[78] T. Finck, S. Roch, and B. Silbermann. Two projection theorems and symbol calculus for operators with massive local spectra. *Math. Nachr.*, 162:167–185, 1993.

[79] G. R. Ford. *Automorphic Functions.* Chelsea Publishing Company, New York, N. Y., 1951.

[80] I. Gohberg, On an application of the theory of normed rings to singular integral equations. *Uspehi Matem. Nauk*, 7:149–156, 1952.

[81] I. Gohberg, S. Goldberg, and M. A. Kaashoek. *Basic classes of linear operators*. Birkhäuser Verlag, Basel, 2003.

[82] I. Gohberg, N. Krupnik. On the algebra generated by Toeplitz matrices. *Funct. Anal. Appl*, 3:119–127, 1969.

[83] I. Gohberg, N. Krupnik. On the algebra generated by one–dimensional singular integral operators with piece–wise continuous coefficients. *Funct. Analysis Appl.*, 4(3):193–201, 1970.

[84] I. Gohberg, N. Krupnik. On the algebra generated by one-dimensional singular integral operators with piecewise continuous coefficients. *Funct. Anal. Appl*, 4:193–201, 1970.

[85] I. Gohberg, N. Krupnik. *One–dimensional linear singular integral equations. Vol. 2. General theory and applications*. Birkhäuser Verlag, Basel, 1992; Russian original: Shtiintsa, Kishynev 1973.

[86] I. S. Gradshteyn, I. M. Ryzhik. *Tables of Integrals, Series, and Products*. Academic Press, New York, 1980.

[87] S. Grudsky, A. Karapetyants, and N. Vasilevski. Toeplitz operators on the unit ball in $\mathbf{C}^n$ with radial symbols. *J. Operator Theory*, 49:325–346, 2003.

[88] S. Grudsky, A. Karapetyants, and N. Vasilevski. Dynamics of properties of Toeplitz operators on the upper half-plane: Hyperbolic case. *Bol. Soc. Mat. Mexicana*, 10:119–138, 2004.

[89] S. Grudsky, A. Karapetyants, and N. Vasilevski. Dynamics of properties of Toeplitz operators on the upper half-plane: Parabolic case. *J. Operator Theory*, 52(1):185–204, 2004.

[90] S. Grudsky, A. Karapetyants, and N. Vasilevski. Dynamics of properties of Toeplitz operators with radial symbols. *Integr. Equat. Oper. Th.*, 20(2):217–253, 2004.

[91] S. Grudsky, R. Quiroga-Barranco, and N. Vasilevski. Commutative C^*-algebras of Toeplitz operators and quantization on the unit disk. *J. Funct. Anal.*, 234(1):1–44, 2006.

[92] S. Grudsky, N. Vasilevski. Bergman-Toeplitz operators: Radial component influence. *Integr. Equat. Oper. Th.*, 40(1):16–33, 2001.

[93] S. Grudsky, N. Vasilevski. Anatomy of the C^*-algebra generated by Toeplitz operators with piece-wise continuous symbols. Reporte Interno #384, Departamento de Matemáticas, CINVESTAV del I.P.N., Mexico City, 27 p, 2008.

[94] S. Grudsky, N. Vasilevski. On the structure of the C^*-algebra generated by Toeplitz operators with piece-wise continuous symbols. *Complex Analysis Operator Theory* (To appear), 2008.

[95] S. M. Grudsky, N. L. Vasilevski. Toeplitz operators on the Fock space: radial component effects. *Integral Equations Operator Theory*, 44(1):10–37, 2002.

[96] K. Guo, D. Zheng. Toeplitz algebra and Hankel algebra on the harmonic Bergman space. *J. Math. Anal. Appl.*, 276(1):213–230, 2002.

[97] R. Hagen, S. Roch, and B. Silbermann. *C^*-Algebras and Numerical Analysis*. Marcel Dekker, Inc., New York, Basel, 2001.

[98] P. Halmos. Two subspaces. *Trans. Amer. Math. Soc.*, 144:381–389, 1969.

[99] P. R. Halmos. *A Hilbert Space Problem Book*. Springer–Verlag, Berlin Heidelberg New York Tokyo, 1982.

[100] W. K. Hayman, P. B. Kennedy. *Subharmonic functions*. Academic Press, London - New York, 1976.

[101] J. He, L. Peng. Hankel and Toeplitz type operators on the unit disk. *Sci. China Ser. A*, 45(9):1154–1162, 2002.

[102] H. Hedenmalm, B. Korenblum, and K. Zhu. *Theory of Bergman Spaces*. Springer Verlag, New York Berlin Heidelberg, 2000.

[103] K. H. Hofmann. Representations of algebras by continuous sections. *Bull. Amer. Math. Soc.*, 78(3):291–373, 1972.

[104] J. Hu, S. Sun, X. Xu, and D. Yu. Reducing subspace of analytic Toeplitz operators on the Bergman space. *Integral Equations Operator Theory*, 49(3):387–395, 2004.

[105] I. S. Hwang. Hyponormal Toeplitz operators on the Bergman space. *J. Korean Math. Soc.*, 42(2):387–403, 2005.

[106] I. S. Hwang, J. Lee. Hyponormal Toeplitz operators on the Bergman space. II. *Bull. Korean Math. Soc.*, 44(3):517–522, 2007.

[107] C. Jiang, Y. Li. The commutant and similarity invariant of analytic Toeplitz operators on Bergman space. *Sci. China Ser. A*, 50(5):651–664, 2007.

[108] Q. T. Jiang, L. Z. Peng. Toeplitz and Hankel type operators on the upper half-plane. *Integral Equations Operator Theory*, 15(5):744–767, 1992.

[109] S. H. Kang. Berezin transforms and Toeplitz operators on the weighted Bergman spaces of the half-plane. *Bull. Korean Math. Soc.*, 44(2):281–290, 2007.

[110] S. H. Kang, J. Y. Kim. Toeplitz operators on weighted analytic Bergman spaces of the half-plane. *Bull. Korean Math. Soc.*, 37(3):437–450, 2000.

[111] A. N. Karapetyants. The space $\mathrm{BMO}^p_\lambda(\mathbb{D})$, compact Toeplitz operators with $\mathrm{BMO}^1_\lambda(\mathbb{D})$ symbols on weighted Bergman spaces, and the Berezin transform. *Izv. Vyssh. Uchebn. Zaved. Mat.*, (8):76–79, 2006.

[112] A. N. Karapetyants, A. V. Golikov. The Berezin transform and radial operators in weighted Bergman spaces in the unit disk. *Vladikavkaz. Mat. Zh.*, 7(2):55–63, 2005.

[113] Yu. I. Karlovich. Pseudodifferential operators with compound slowly oscillating symbols. In: *"The extended field of operator theory"*. Operator Theory: Advances and Applications, Vol. 171, Birkhäuser, Basel-Boston-Berlin, pages 189–224, 2007.

[114] Yu. I. Karlovich, L. Pessoa. Algebras generated by the Bergman and anti-Bergman projections and by multiplications by piecewise continuous functions. *Integral Equations Operator Theory*, 52(2):219–270, 2005.

[115] Yu. I. Karlovich, L. Pessoa. C^*-algebras of Bergman type operators with piecewise continuous coefficients. *Integral Equations Operator Theory*, 57(4):521–565, 2007.

[116] K. Kasuga. Examples of compact Toeplitz operators on the Bergman space. *Hokkaido Math. J.*, 30(3):697–703, 2001.

[117] J. R. Klauder, B.-S. Skagerstam, editors. *Coherent states*. World Scientific Publishing Co., Singapore, 1985. Applications in physics and mathematical physics.

[118] W. Klingenberg. *A Course in Differential Geometry*. Springer, New York, 1978.

[119] I. Kolář, P. Michor, and J. Slovák. *Natural Operations in Differential Geometry*. Springer-Verlag, 1993.

[120] B. Korenblum, K. Zhu. An application of Tauberian theorems to Toeplitz operators. *J. Operator Theory*, 339:353–361, 1995.

[121] B. Korenblum, M. Stessin. On Toeplitz-invariant subspaces of the Bergman space. *J. Funct. Anal.*, 111(1):76–96, 1993.

[122] J. W. Lark, III. Spectral theorems for a class of Toeplitz operators on the Bergman space. *Houston J. Math.*, 12(3):397–404, 1986.

[123] Y. J. Lee. Pluriharmonic symbols of commuting Toeplitz type operators on the weighted Bergman spaces. *Canad. Math. Bull.*, 41(2):129–136, 1998.

[124] Y. J. Lee. Essentially commuting Toeplitz operators on the weighted Bergman spaces. *Far East J. Math. Sci. (FJMS)*, 1(1):145–166, 1999.

[125] Y. J. Lee. Compact radial operators on the harmonic Bergman space. *J. Math. Kyoto Univ.*, 44(4):769–777, 2004.

[126] J. Lehner. *Discontinuous Groups and Automorphic Functions*. American Mathematical Society, Providence, R.I., 1964.

[127] P. Lin, R. Rochberg. Trace ideal criteria for Toeplitz and Hankel operators on the weighted Bergman spaces with exponential type weights. *Pacific J. Math.*, 173(1):127–146, 1996.

[128] C. Liu. A "deformation estimate" for the Toeplitz operators on harmonic Bergman spaces. *Proc. Amer. Math. Soc.*, 135(9):2867–2876, 2007.

[129] Y. M. Liu. A new characterization of compactness for Toeplitz and Hankel operators on weighted Bergman spaces. *J. Math. Res. Exposition*, 19(suppl.):238–240, 1999.

[130] Y. M. Liu. Schatten class of Toeplitz operators on weighted Bergman spaces. *Chinese Ann. Math. Ser. A*, 20(1):53–56, 1999.

[131] M. Loaiza. Algebras generated by the Bergman projection and operators of multiplication by piecewise continuous functions. *Integral Equations Operator Theory*, 46(2):215–234, 2003.

[132] M. Loaiza. On an algebra of Toeplitz operators with piecewise continuous symbols. *Integr. Equat. Oper. Th.*, 51(1):141–153, 2005.

[133] I. Louhichi, E. Strouse, and L. Zakariasy. Products of Toeplitz operators on the Bergman space. *Integral Equations Operator Theory*, 54(4):525–539, 2006.

[134] Y. F. Lu. A class of Toeplitz operators on the Bergman space. *Dongbei Shida Xuebao*, (3):8–12, 1996.

[135] Y. F. Lu. Localization of Toeplitz operators on a Bergman space. *Dongbei Shida Xuebao*, 32(2):7–10, 2000.

[136] Y. F. Lu, S. L. Sun. Locality and the Fredholm property for Toeplitz operators on the Bergman space. *Acta Math. Sinica (Chin. Ser.)*, 40(5):695–700, 1997.

[137] Y. F. Lu, S. L. Sun. Toeplitz operators on the Bergman space of an annulus. *Chinese Ann. Math. Ser. A*, 19(5):559–566, 1998.

[138] Y. Lu. Localization of Toeplitz operators on Bergman spaces. *Northeast. Math. J.*, 17(4):461–468, 2001.

[139] Y. Lu. Commuting of Toeplitz operators on the Bergman spaces of the bidisc. *Bull. Austral. Math. Soc.*, 66(2):345–351, 2002.

[140] W. Lusky. Toeplitz operators on generalized Bergman-Hardy spaces. *Math. Scand.*, 88(1):96–110, 2001.

[141] G. McDonald. Toeplitz operators on the ball with piecewise continuous symbol. *Illinois J. Math.*, 23(2):286–294, 1979.

[142] G. McDonald, C. Sundberg. Toeplitz operators on the disk. *Indiana Math. J.*, 28:595–611, 1979.

[143] J. Miao. Toeplitz operators on harmonic Bergman spaces. *Integral Equations Operator Theory*, 27(4):426–438, 1997.

[144] J. Miao. Toeplitz operators with bounded radial symbols on the harmonic Bergman space of the unit ball. *Acta Sci. Math. (Szeged)*, 63:639–645, 1997.

[145] A. Michalak. Essential norms of Toeplitz operators on Bergman-Hardy spaces on the unit disk. *Funct. Approx. Comment. Math.*, 28:211–220, 2000. Dedicated to Włodzimierz Staś on the occasion of his 75th birthday.

[146] S. G. Mikhlin, S. Proessdorf. *Singular Integral Operators*. Springer Verlag, Berlin, Heidelberg, 1986.

[147] Yu. P. Moskaleva, Yu. S. Samoĭlenko. Systems of n subspaces and representations of ∗-algebras generated by projections. *Methods Funct. Anal. Topology*, 12(1):57–73, 2006.

[148] G. J. Murphy. C^*-*algebras and Operator Theory*. Academic Press, Inc., Boston San Diego New York London, 1990.

[149] T. Nakazi, R. Yoneda. Compact Toeplitz operators with continuous symbols on weighted Bergman spaces. *Glasg. Math. J.*, 42(1):31–35, 2000.

[150] M. Nishio, N. Suzuki, and M. Yamada. Toeplitz operators and Carleson measures on parabolic Bergman spaces. *Hokkaido Math. J.*, 36(3):563–583, 2007.

[151] B. O'Nell. *Semi-Riemannian Geometry*. Academic Press, 1983.

[152] J.-D. Park. Bounded Toeplitz products on the Bergman space of the unit ball in $\mathbb{C}^n$. *Integral Equations Operator Theory*, 54(4):571–584, 2006.

[153] G. K. Pedersen. Measure theory for C*-algebras, II. *Mat. Scand.*, 22:63–74, 1968.

[154] L. Peng, R. Rochberg, and Z. Wu. Orthogonal polynomials and middle Hankel operators on Bergman spaces. *Studia Math.*, 102(1):57–75, 1992.

[155] L. Z. Peng, C. X. Xu. Jacobi polynomials and Toeplitz-Hankel type operators. *Complex Variables Theory Appl.*, 23(1-2):47–71, 1993.

[156] L. Z. Peng, G. K. Zhang. Middle Hankel operators on Bergman space. In *Function spaces (Edwardsville, IL, 1990)*, volume 136 of *Lecture Notes in Pure and Appl. Math.*, pages 325–336. Dekker, New York, 1992.

[157] L. Peng, M. W. Wong. Toeplitz operators on the upper half plane. *Panamer. Math. J.*, 7(3):1–33, 1997.

[158] A. M. Perelomov. *Generalized Coherent States and Their Applications*. Springer-Verlag, Berlin, New York, 1986.

[159] V. S. Pilidi. A local method in the theory of linear operator equations of the type of bisingular integral equations. (Russian). *Mathematical analysis and its applications*, III:81–105, 1971.

[160] S. Pott, E. Strouse. Products of Toeplitz operators on the Bergman spaces A_α^2. *Algebra i Analiz*, 18(1):144–161, 2006.

[161] R. Quiroga-Barranco, N. Vasilevski. Commutative algebras of Toeplitz operators on the Reinhardt domains. *Integral Equations Operator Theory*, 59(1):67–98, 2007.

[162] R. Quiroga-Barranco, N. Vasilevski. Commutative C^*-algebras of Toeplitz operators on the unit ball, I. Bargmann-type transforms and spectral representations of toeplitz operators. *Integral Equations Operator Theory*, 59(3):379–419, 2007.

[163] R. Quiroga-Barranco, N. Vasilevski. Commutative C^*-algebras of Toeplitz operators on the unit ball, II. Geometry of the level sets of symbols. *Integral Equations Operator Theory*, 59(1):89–132, 2008.

[164] V. Rabinovich, N. Vasilevski. Bergman-Toeplitz and pseudodifferential operators. In *Complex Analysis and Related Topics. Operator Theory, Advances and Applications*, Vol. 114, Birkhäuser, Basel-Boston-Berlin, pages 207–234, 2000.

[165] I. Raeburn, A. M. Sinclair. The C^*-algebra generated by two projections. *Math. Scand.*, 65(2):278–290, 1989.

[166] R. Raimondo. Toeplitz operators on the Bergman space of the unit ball. *Bull. Austral. Math. Soc.*, 62(2):273–285, 2000.

[167] J. Ramírez, I. M. Spitkovsky. On the algebra generated by a poly-Bergman projection and a composition operator. In *Factorization, singular operators and related problems (Funchal, 2002)*, pages 273–289. Kluwer Acad. Publ., Dordrecht, 2003.

[168] M. Rieffel. C^*-algebras associated with irrational rotation. *Pacific J. Math.*, 90(2):415–429, 1981.

[169] S. Roch, B. Silbermann. Representations of noncommutative Banach algebras by continuous functions. *Algebra and Analysis*, 3(4):171–185, 1991.

[170] W. Rudin. *Function Theory in the Unit Ball of $\mathbb{C}^n$*. Springer-Verlag, New York, Heidelberg, Berlin, 1980.

[171] Sangadji, K. Stroethoff. Compact Toeplitz operators on generalized Fock spaces. *Acta Sci. Math. (Szeged)*, 64:657–669, 1998.

[172] D. Sarason. Functions of vanishing mean oscillation. *Trans. Amer. Math. Soc.*, 207:391–405, 1975.

[173] D. Sarason. Toeplitz operators with piecewise quasicontinuous symbols. *Indiana Univ. Math. J.*, 26(5):817–838, 1977.

[174] D. J. Saunders. *The Geometry of Jet Bundles.* London Mathematical Society Lecture Note Series, no 142. Cambridge University Press, Cambridge, 1989.

[175] K. Seddighi. Toeplitz operators on Hardy-Bergman spaces. *Bull. Iranian Math. Soc.*, 14(1):22–30, 1987.

[176] B. V. Shabat. *Introduction to Complex Analysis,* Part I, and Part II. "Nauka", Moscow, 1976.

[177] I. B. Simonenko. A new general method for studying linear operator equations of singular integral equation type, I. *Izv. Akad. Nauk SSSR, Ser. Math.*, 29:567–586, 1965.

[178] I. B. Simonenko. A new general method for studing linear operator equations of singular integral equation type, II. *Izv. Akad. Nauk SSSR, Ser. Math.*, 29:757–782, 1965.

[179] J. Spielberg. Free–product groups, Cuntz–Krieger algebras, and covariant maps. *Int. J. Math.*, 2(4):457–476, 1991.

[180] K. Stroethoff. Essentially commuting Toeplitz operators with harmonic symbols. *Canad. J. Math.*, 45(5):1080–1093, 1993.

[181] K. Stroethoff. Compact Toeplitz operators on Bergman spaces. *Math. Poc. Camb. Phil. Soc.*, 124:151–160, 1998.

[182] K. Stroethoff. Compact Toeplitz operators on weighted harmonic Bergman spaces. *J. Austal. Math. Soc. (Series A)*, 64:136–148, 1998.

[183] K. Stroethoff, D. Zheng. Toeplitz and Hankel operators on Bergman spaces. *Trans. Amer. Math. Soc.*, 329(2):773–794, 1992.

[184] K. Stroethoff, D. Zheng. Products of Hankel and Toeplitz operators on the Bergman space. *J. Funct. Anal.*, 169(1):289–313, 1999.

[185] K. Stroethoff, D. Zheng. Invertible Toeplitz products. *J. Funct. Anal.*, 195(1):48–70, 2002.

[186] K. Stroethoff, D. Zheng. Bounded Toeplitz products on the Bergman space of the polydisk. *J. Math. Anal. Appl.*, 278(1):125–135, 2003.

[187] K. Stroethoff, D. Zheng. Bounded Toeplitz products on Bergman spaces of the unit ball. *J. Math. Anal. Appl.*, 325, 2007.

[188] D. Suárez. Approximation and symbolic calculus for Toeplitz algebras on the Bergman space. *Rev. Mat. Iberoamericana*, 20(2):563–610, 2004.

[189] D. Suárez. The Toeplitz algebra on the Bergman space coincides with its commutator ideal. *J. Operator Theory*, 51(1):105–114, 2004.

[190] D. Suárez. Approximation and the n-Berezin transform of operators on the Bergman space. *J. Reine Angew. Math.*, 581:175–192, 2005.

[191] S. L. Sun, Y. J. Wang. The commutants of a class of analytic Toeplitz operators on Bergman spaces. *Acta Sci. Natur. Univ. Jilin.*, (2):4–8, 1997.

[192] S.. Sun, Y. Lu. Localization of Toeplitz operators on Bergman space. *Northeast. Math. J.*, 12(3):373–380, 1996.

[193] S. Sun, Y. Wang. Reducing subspaces of certain analytic Toeplitz operators on the Bergman space. *Northeast. Math. J.*, 14(2):147–158, 1998.

[194] V. S. Sunder. N Subspaces. *Can. J. Math.*, XL(1):38–54, 1988.

[195] M. Takesaki, J. Tomiyama. Application of fiber bundles to certain class of C^*-algebras. *Tohoku Math. J., ser 2*, 13:498–522, 1961.

[196] I. Vaĭs. Algebras that are generated by two idempotents. In *Seminar Analysis (Berlin, 1987/1988)*, pages 139–145. Akademie-Verlag, Berlin, 1988.

[197] J. Varela. *Duality of C^*-algebras, Memories Amer. Math. Soc.*, volume 148, pages 97–108. AMS, Providence, Rhode Island, 1974.

[198] N. Vasilevski. On a general local principle for C^*-algebras. *Izv. VUZ North-Caucasian Region, Natural Sciences, Special Issue, "Pseudodifferential operators and some problems of mathematical physics"*, 234:34–42, 2005 (Russian).

[199] N. Vasilevski. On Toeplitz operators with piecewise continuous symbols on the Bergman space. In: *"Modern Operator Theory and Applications"*. *Operator Theory, Advances and Applications*, Vol. 170, Birkhäuser, Basel-Boston-Berlin, pages 229–248, 2007.

[200] N. L. Vasilevski. Banach algebras that are generated by certain two-dimensional integral operators. I. (Russian). *Math. Nachr.*, 96:245–255, 1980.

[201] N. L. Vasilevski. On the algebra generated by two-dimensional integral operators with the Bergman kernel and piecewise continuous coefficients. *Soviet Math.Dokl*, 28(1):191–194, 1983.

[202] N. L. Vasilevski. Banach algebras generated by two-dimensional integral operators with a Bergman kernel and piecewise continuous coefficients. I. *Soviet Math. (Izv. VUZ)*, 30(3):14–24, 1986.

[203] N. L. Vasilevski. Banach algebras generated by two-dimensional integral operators with a Bergman kernel and piecewise continuous coefficients. II. *Soviet Math. (Izv. VUZ)*, 30(3):44–50, 1986.

[204] N. L. Vasilevski. Local principle in operator theory. In *Linear Operators in Functional Spaces*, Abstracts of lectures at North-Caucasus regional Conference, pages 32–33, Grozny, Russia, 1989.

[205] N. L. Vasilevski. Convolution operators on standard CR–manifold. II. Algebras of convolution operators on the Heisenberg group. *Integr. Equat. Oper. Th.*, 19:327–348, 1994.

[206] N. L. Vasilevski. C^*-algebras generated by orthogonal projections and their applications. *Integr. Equat. Oper. Th.*, 31:113–132, 1998.

[207] N. L. Vasilevski. On Bergman-Toeplitz operators with commutative symbol algebras. *Integr. Equat. Oper. Th.*, 34:107–126, 1999.

[208] N. L. Vasilevski. On the structure of Bergman and poly–Bergman spaces. *Integr. Equat. Oper. Th.*, 33:471–488, 1999.

[209] N. L. Vasilevski. The Bergman space in tube domains, and commuting Toeplitz operators. *Doklady, Mathematics*, 61(3):9–12, 2000.

[210] N. L. Vasilevski. Bergman space on tube domains and commuting Toeplitz operators. In *Proceedings of the Second ISAAC Congress, Vol. 2 (Fukuoka, 1999)*, volume 8 of *Int. Soc. Anal. Appl. Comput.*, pages 1523–1537, Dordrecht, 2000. Kluwer Acad. Publ.

[211] N. L. Vasilevski. Bergman spaces on the unit disk. In: *Clifford Analysis and Its Applications*, F. Brackx et al.(eds.), pages 399–409, 2001.

[212] N. L. Vasilevski. Toeplitz operators on the Bergman spaces: Inside-the-domain effects. *Contemp. Math.*, 289:79–146, 2001.

[213] N. Vasilevski. Toeplitz operators on the Bergman space. In *Factorization, singular operators and related problems (Funchal, 2002)*, pages 315–333. Kluwer Acad. Publ., Dordrecht, 2003.

[214] N. L. Vasilevski. Bergman space structure, commutative algebras of Toeplitz operators and hyperbolic geometry. *Integr. Equat. Oper. Th.*, 46:235–251, 2003.

[215] N. Vasilevski. Poly-Bergman spaces and two-dimensional singular integral operators. In *The extended field of operator theory*, volume 171 of *Oper. Theory Adv. Appl.*, pages 349–359. Birkhäuser, Basel, 2007.

[216] N. L. Vasilevski, I. M. Spitkovsky. On the algebra generated by two projectors. *Dokl. Akad. Nauk. UkSSR*, A(8):10–13, 1981.

[217] N. L. Vasilevskiĭ, S. M. Grudskiĭ, and A. N. Karapetyants. Dynamics of properties of Toeplitz operators on weighted Bergman spaces. *Sib. Èlektron. Mat. Izv.*, 3:362–383, 2006.

[218] N. B. Vasiliev. C^*-algebras with finite-dimensional irreducible representations. *Soviet Math. Surv.*, 21(1):135–154, 1966.

[219] I. N. Vekua. *Generalized analytic functions*. Pergamon Press, London, Paris, Frankfurt, 1966.

[220] Y. Weiss. On Banach algebras generated by two idempotents. In *Algebraic methods in operator theory*, pages 90–97. Birkhäuser Boston, Boston, MA, 1994.

[221] J. Xiao. Compactness of both Toeplitz and Hankel operators on Bergman space A^2. *Chinese J. Contemp. Math.*, 15(2):135–143, 1994.

[222] F. Xu. Toeplitz operators on the Bergman space. *Dongbei Shida Xuebao*, (4):14–17, 1996.

[223] F. Xu, Y. F. Lu. Analytic Toeplitz operators on the Bergman space $L_a^2(D)$. *Dongbei Shida Xuebao*, (4):21–24, 1993.

[224] Q. X. Xu. Toeplitz and Hankel operators whose symbols are radial functions. *J. Fudan Univ. Natur. Sci.*, 32(2):155–160, 1993.

[225] S. Z. Xu. The main properties of Toeplitz and Hankel operators on a Bergman space. *J. Fudan Univ. Natur. Sci.*, 31(4):413–424, 1992.

[226] J. X. Yan, D. H. Yu. Compactness of Toeplitz operators on a weighted Bergman space and their Berezin transforms. *Sichuan Daxue Xuebao*, 36(2):211–219, 1999.

[227] R. Yoneda. Compact Toeplitz operators on Bergman spaces. *Hokkaido Math. J.*, 28(3):563–576, 1999.

[228] R. Younis, D. Zheng. Algebras on the unit disk and Toeplitz operators on the Bergman space. *Integral Equations Operator Theory*, 37(1):106–123, 2000.

[229] D. H.i Yu, S. H. Sun, and Z. G. Dai. The Fredholm theory of Toeplitz operators on Bergman space. *Northeast. Math. J.*, 4(4):405–411, 1988.

[230] T. Yu. Carleson measures and Toeplitz operators on the weighted Bergman spaces. *Acta Anal. Funct. Appl.*, 8(4):369–376, 2006.

[231] T. Yu. Essential spectra of Toeplitz operators on weighted Bergman spaces. *Acta Math. Sinica (Chin. Ser.)*, 49(2):357–362, 2006.

[232] T. Yu, G. Cao. Toeplitz operators on the Bergman space of multi-connected domains. *Northeast. Math. J.*, 17(4):487–493, 2001.

[233] T. Yu, S. L. Sun. Compact Toeplitz operators on the weighted Bergman spaces. *Acta Math. Sinica (Chin. Ser.)*, 44(2):233–240, 2001.

[234] X. Zeng. Toeplitz operators on Bergman spaces. *Houston J. Math.*, 18(3):387–407, 1992.

[235] D. Zheng. Hankel operators and Toeplitz operators on the Bergman space. *J. Funct. Anal.*, 83(1):98–120, 1989.

[236] D. Zheng. Semi-commutators of Toeplitz operators on the Bergman space. *Integral Equations Operator Theory*, 25(3):347–372, 1996.

[237] K. Zhu. VMO, ESV, and Toeplitz operators on the Bergman space. *Trans. Amer. Math. Soc.*, 302:617–646, 1987.

[238] K. Zhu. Positive Toeplitz operators on weighted Bergman space of bounded symmetric domains. *J. Oper. Theory*, 2:329–357, 1988.

[239] K. Zhu. Multipliers of BMO in the Bergman metric with applications to Toeplitz operators. *J. Funct. Anal.*, 87(1):31–50, 1989.

[240] K. Zhu. *Operator Theory in Function Spaces*. Marcel Dekker, Inc., 1990.

[241] K. Zhu. Schatten class Toeplitz operators on weighted Bergman spaces of the unit ball. *New York J. Math.*, 13, 2007.

[242] N. Zorboska. The Berezin transform and radial operators. *Proc. Amer. Math. Soc.*, 131(3):793–800, 2003.

[243] N. Zorboska. Toeplitz operators with BMO symbols and the Berezin transform. *Int. J. Math. Math. Sci.*, (46):2929–2945, 2003.

List of Figures

Index

Operator Theory: Advances and Applications (OT)

Edited by
Israel Gohberg, Tel Aviv University, Israel

This series is devoted to the publication of current research in operator theory, with particular emphasis on applications to classical analysis and the theory of integral equations, as well as to numerical analysis, mathematical physics and mathematical methods in electrical engineering.